ESO ASTROPHYSICS SYMPOSIA
European Southern Observatory

Series Editor: Jacqueline Bergeron

Physics and Astronomy ONLINE LIBRARY

http://www.springer.de/phys/

Springer-Verlag Berlin Heidelberg GmbH

ESO ASTROPHYSICS SYMPOSIA
European Southern Observatory

Series Editor: Jacqueline Bergeron

G. Meylan (Ed.), **QSO Absorption Lines**
Proceedings, 1994. XXIII, 471 pages. 1995.

D. Minniti, H.-W. Rix (Eds.), **Spiral Galaxies in the Near-IR**
Proceedings, 1995. X, 350 pages. 1996.

H. U. Käufl, R. Siebenmorgen (Eds.), **The Role of Dust in the Formation of Stars**
Proceedings, 1995. XXII, 461 pages. 1996.

P. A. Shaver (Ed.), **Science with Large Millimetre Arrays**
Proceedings, 1995. XVII, 408 pages. 1996.

J. Bergeron (Ed.), **The Early Universe with the VLT**
Proceedings, 1996. XXII, 438 pages. 1997.

F. Paresce (Ed.), **Science with the VLT Interferometer**
Proceedings, 1996. XXII, 406 pages. 1997.

D. L. Clements, I. Pérez-Fournon (Eds.), **Quasar Hosts**
Proceedings, 1996. XVII, 336 pages. 1997.

L. N. da Costa, A. Renzini (Eds.), **Galaxy Scaling Relations: Origins, Evolution and Applications**
Proceedings, 1996. XX, 404 pages. 1997.

L. Kaper, A. W. Fullerton (Eds.), **Cyclical Variability in Stellar Winds**
Proceedings, 1997. XXII, 415 pages. 1998.

R. Morganti, W. J. Couch (Eds.), **Looking Deep in the Southern Sky**
Proceedings, 1997. XXIII, 336 pages. 1999.

J. R. Walsh, M. R. Rosa (Eds.), **Chemical Evolution from Zero to High Redshift**
Proceedings, 1998. XVIII, 312 pages. 1999.

J. Bergeron, A. Renzini (Eds.), **From Extrasolar Planets to Cosmology: The VLT Opening Symposium**
Proceedings, 1999. XXVIII, 575 pages. 2000.

A. Weiss, T. G. Abel, V. Hill (Eds.), **The First Stars**
Proceedings, 1999. XIII, 355 pages. 2000.

Series homepage – http://www.springer.de/phys/books/eso/

Jacqueline Bergeron Alvio Renzini (Eds.)

From Extrasolar Planets to Cosmology: The VLT Opening Symposium

Proceedings of the ESO Symposium
Held at Antofagasta, Chile, 1–4 March 1999

Plenary Sessions:
Science in the VLT Era and Beyond
edited by Jacqueline Bergeron and Alvio Renzini

Parallel Workshop 1:
Clusters of Galaxies at High Redshift
edited by Bernard Fort and Piero Rosati

Parallel Workshop 2:
Star-Way to the Universe
edited by Rolf Kudritzki and Luca Pasquini

Parallel Workshop 3:
From Extrasolar Planets to Brown Dwarfs
edited by Michel Mayor and Francesco Paresce

Springer

Volume Editors

Jacqueline Bergeron
Alvio Renzini

European Southern Observatory
Karl-Schwarzschild-Strasse 2
85748 Garching, Germany

Series Editor

Jacqueline Bergeron

European Southern Observatory
Karl-Schwarzschild-Strasse 2
85748 Garching, Germany

Library of Congress Cataloging-in-Publication Data applied for.

Die Deutsche Bibliothek - CIP-Einheitsaufnahme

From extrasolar planets to cosmology : the VLT Opening Symposium ; proceedings of the ESO symposium, held at Antofagasta, Chile, 1 - 4 March 1999 / Jacqueline Bergeron ; Alvio Renzini (ed.). - Berlin ; Heidelberg ; New York ; Barcelona ; Hong Kong ; London ; Milan ; Paris ; Singapore ; Tokyo : Springer, 2000
(ESO astrophysics symposia)

ISBN 978-3-662-30945-2 ISBN 978-3-540-46517-1 (eBook)
DOI 10.1007/978-3-540-46517-1

Originally published by Springer-Verlag Berlin Heidelberg New York in 2000.
Softcover reprint of the hardcover 1st edition 2000

Typesetting: Camera-ready by the authors/editors
Cover design: Erich Kirchner, Heidelberg

Printed on acid-free paper SPIN: 10720961 55/3144/du - 5 4 3 2 1 0

Preface

The VLT Opening Symposium marked the beginning of the VLT era. It took place just one month before the start of regular observations with the first VLT unit telescope, ANTU, while first light was achieved on the second VLT unit telescope, KUEYEN, ahead of schedule on the night of 4 March. Visits to the Paranal Observatory were organized each evening during the first three days of the Symposium to give the participants the opportunity to see the VLT in operation.

The Symposium was dedicated to science opportunities with the VLT and has provided a forum to the 201 participants for discussing current and new projects in several rapidly evolving fields. To ensure both scientific breadth and depth, the Symposium consisted of Plenary Sessions on "Science in the VLT Era and Beyond" accompanied by three parallel workshops: "Clusters of Galaxies at High Redshift", "Star-Way to the Universe" and "From Extrasolar Planets to Brown Dwarfs". This volume is a collection of the papers presented at the Symposium.

All major areas of astronomical research were discussed in the Plenary Sessions, ranging from where we stand in cosmology to the new frontiers in the solar system. The last Plenary Session was devoted to science at millimeter and submillimeter wavelengths with the new giant telescope project ALMA, and the synergy between the VLT and the NGST. Workshop 1, "Clusters of Galaxies at High Redshift", focussed on the different ways and new prospects of finding clusters of galaxies at $z \gtrsim 1$, the cosmological constraints derived from the existence of high-redshift clusters, the population of galaxy cluster members and the gravitational lensing by distant compact clusters which magnifies our view of the distant universe. Workshop 2, "Star-Way to the Universe", concentrated on the use of different stellar populations as distance, age, and abundance indicators, and thus on the resulting constraints on the age of the universe, the star-formation history of various types of galaxies, and the role of different stellar populations in the chemical evolution of galaxies. Workshop 3, "From Extrasolar Planets to Brown Dwarfs", presented the extraordinary progress made in the discovery of sub-stellar mass companions orbiting solar-type stars. Discussions focussed on the prospects for extending the search for planets to stars of all types, and to planets with small masses and a wide range of separations from their parent stars.

We are very grateful to the Universidad Católica del Norte and its rector, Professor J.A. Music Tomicic, for hosting the Symposium. Thanks are due in particular to Luis Barrera, Director of the Institute of Astronomy, and Osvaldo Muñiz, Director of Research and Technical Division. We are also greatly indebted to Danielle Alloin and Daniel Hofstadt of the Local Organizing Committee, as well as to Georgette Hubert, Andrea Lagarini and Christina Stoffer for their efficient help with the organization and logistics of the Symposium, and to Pamela Bristow for her assistance in preparing these proceedings.

Garching, December 1999

Jacqueline Bergeron
Alvio Renzini

Contents

PLENARY SESSIONS: SCIENCE IN THE VLT ERA AND BEYOND

Edited by Jacqueline Bergeron and Alvio Renzini

Science with FORS
I. Appenzeller, R. Bender, H. Böhnhardt, S. Cristiani, M. Dietrich, K. Fricke, W. Fürtig, W. Gässler, R. Gilmozzi, R. Häfner, R. Harke, J. Heidt, H.-J. Hess, U. Hopp, W. Hummel, K. Jäger, P. Jürgens, R.-P. Kudritzki, M. Kümmel, K.-H. Mantel, D. Mehlert, W. Meisl, C. Moellenhoff, B. Muschielok, H. Nicklas, A. Renzini, P. Rosati, G. Rupprecht, R. Saglia, W. Seifert, S. Seitz, J. Spyromilio, O. Stahl, T. Szeifert, K. Tarantik 3

Science with ISAAC
A. Moorwood 12

The Most Distant Galaxies
M. Pettini, C.C. Steidel, K.L. Adelberger, A.E. Shapley, M. Dickinson, M. Giavalisco 20

Nuclear Activity and Galaxy Evolution
G. Miley 32

Supernovae at High Redshift
B. Leibundgut 43

Cosmic Star Formation History and the Brightness of the Night Sky
P. Madau 52

Galaxy Centers
R. Bender 61

New Frontiers in the Solar System
M.A. Barucci 70

Science with the Atacama Large Millimeter Array
K.M. Menten 78

Synergy Between the VLT and the NGST
H.S. Stockman 94

Clusters of Galaxies at High Redshift
B. Fort 102

PARALLEL WORKSHOP 1: CLUSTERS OF GALAXIES AT HIGH REDSHIFT

Edited by Bernard Fort and Piero Rosati

OPTICAL/IR STUDIES OF GALAXY CLUSTERS

Galaxy Populations in High Redshift Clusters
S.A. Stanford 115

One of the Highest Redshift X–Ray Selected Clusters of Galaxies
I. Lehmann, G. Hasinger, R. Giacconi, J.E. Gunn, M. Schmidt, D.P. Schneider, T. Stanke, J. Trümper, J. Wambsganss, D. Woods, G. Zamorani 121

Intracluster Planetary Nebulae
M. Arnaboldi, M. Capaccioli, H. Ford, K.C. Freeman, R. Kudritzki, O. Gerhard, R. Méndez 126

GRAVITATIONAL LENSING AND CLUSTER DYNAMICS

Probing Distant Galaxies with Lensing Clusters
R. Pelló 131

Gravitational Redshift and Cluster Masses
T. Broadhurst, E. Scannapieco 138

Velocity Dispersion Analysis of CNOC Clusters
S. Borgani 143

X-RAY, RADIO, SZ STUDIES OF CLUSTERS

A Sunyaev-Zel'dovich Effect Survey for High Redshift Clusters
J.J. Mohr, J.E. Carlstrom, G.P. Holder, W.L. Holzapfel, M.K. Joy, E.M. Leitch, E.D. Reese 150

Probing the Large-Scale Structure with the REFLEX Cluster Survey
H. Böhringer, L. Guzzo, C.A. Collins, P. Schuecker, S. Schindler, W. Voges, D.M. Neumann, R. Cruddace, S. DeGrandi, G. Chincarini, T.H. Reiprich, A.C. Edge, H.T. MacGillivray, P. Shaver 157

The Cluster Population in a Complete Sample of Optically Identified RASS X-Ray Sources
I. Appenzeller, R. Kneer, F.-J. Zickgraf, J. Krautter, I. Thiering 164

ICM PHYSICS

The Metal Enrichment History of Clusters of Galaxies
A. Renzini 168

The Connection of X-Ray Clusters of Galaxies with the Early Star Formation Histories
N. Menci, A. Cavaliere 176

The Connection Between X-Ray Clusters and Star Formation
P. Tozzi, C. Norman 180

Clustering at High Redshift
S. Cristiani 184

LSS, CLUSTERING AND WIDE FIELD SURVEYS

EIS – An Imaging Survey for VLT Science
L. da Costa 192

The Space Missions WAXS/WFXT and SWIFT: New Targets for the VLT
G. Chincarini 198

The VST – VLT Survey Telescope
M. Arnaboldi, M. Capaccioli, D. Mancini, R. Scaramella, G. Sedmak, R. Kurz 204

PARALLEL WORKSHOP 2: STAR-WAY TO THE UNIVERSE

Edited by Rolf Kudritzki and Luca Pasquini

KUEYEN: A Unique Machine for Stellar Studies
L. Pasquini 211

DISTANCE INDICATORS

Improvement of Cepheids as Distance Indicators
W.P. Gieren, P. Fouqué, J. Storm 217

Eclipsing Binaries as Precise Distance Indicators
J.V. Clausen 225

Magellanic Cloud Eclipsing Binaries: Distances and Dimensions
J. Pritchard, V. Niemela 232

Blue Supergiants – Starway to Virgo and Fornax
R.P. Kudritzki 236

Extragalactic Planetary Nebulae and What to Do with Them
R.H. Méndez 242

Towards Understanding the Planetary Nebula Luminosity Function
P. Marigo, A. Weiss, M.A.T. Groenewegen, L. Girardi ... 248

HII Galaxies as Cosmological Probes
J. Melnick, R. Terlevich, E. Terlevich, F. Selman 252

The Hubble Constant from the Fornax Cluster Distance
T. Richtler, G. Drenkhahn, M. Gómez, W. Seggewiss 259

Lunar Occultations, Setting the Stage for VLTI: The Case Study of CW-Leo (aka IRC+10216)
H.U. Käufl, B. Stecklum, S. Richter, A. Richichi 264

AGE INDICATORS

Absolute Ages of Globular Clusters and the VLT
F. Pont 269

The Galactic Bulge: Prospects for the VLT
R.M. Rich 275

Globular Clusters in the Galactic Bulge
S. Ortolani 281

Studying the Stellar Populations of the Local Group with VLT
E. Tolstoy 287

Fine Structure of the Red Clump in Local Group Galaxies
L. Girardi 294

Old Stars in Dwarf Irregular Galaxies: WLM and NGC3109
D. Minniti, M. Rejkuba, A.A. Zijlstra, M.V. Alonso, M. Gregg 298

High Mass Class I Sources in M 17
R. Chini, M. Nielbock, M. Jütte 303

WR Stars in Star Forming Regions
W. Schmutz, W.D. Vacca 307

Supernova Rates
E. Cappellaro, F. Patat 313

ABUNDANCE INDICATORS

Massive Stars and Their Role in Spectra and Abundances of High Redshift Galaxies
A. Maeder 317

**Low and Intermediate Mass Stars:
Learning Theory from Observations**
C. Chiosi 323

**Abundance Determinations in Massive Blue Stars
Within the Local Group**
A. Herrero, R.-P. Kudritzki, S.J. Smartt, K. Venn,
N. Przybilla, D.J. Lennon, J.K. McCarthy,
M.I. Monteverde 331

**Abundances in Sagittarius:
Present State and Perspectives for the Use of VLT**
P. Bonifacio 338

The Old Population of the Magellanic Clouds
M. Spite, V. Hill 345

The Deep Thick Disk as Seen by UVES
K. Fuhrmann 351

**Abundances of Intermediate/Low Mass Stars
in the Galactic Bulge**
B. Barbuy 357

Abundances and Kinematics of Halo Stars
P.E. Nissen 363

Tidal Tails Around Galactic Globular Clusters
G. Meylan, S. Leon, F. Combes 369

**PARALLEL WORKSHOP 3:
FROM EXTRASOLAR PLANETS
TO BROWN DWARFS**

Edited by Michel Mayor and Francesco Paresce

FORMATION AND EVOLUTION: THEORY

**Diversity of Planetary Systems:
Formation Scenarios and Unsolved Problems**
J.J. Lissauer, D.N.C. Lin 377

Solar Nebulae and Replenished Dust Disks with Planets
P. Artymowicz 391

Evolution of Low-Mass Stars and Substellar Objects. Contribution to the Galactic Mass Budget
G. Chabrier, I. Baraffe, F. Allard, P.H. Hauschildt 399

Extrasolar Giant Planets: Masses and Luminosities from In-Situ Formation Theories
G. Wuchterl 408

FORMATION AND EVOLUTION: OBSERVATIONS

Imaging Exosolar Dust Disks
P. Kalas 415

The Stellar Mass Function
F. Paresce 422

Aspects of the Substellar Mass Function
G. Basri 429

Planetary Microlensing: Present Status and Long-Term Goals
P.D. Sackett 436

Approaching the Deuterium Burning Limit in the σ Orionis Young Cluster
M.R. Zapatero Osorio, V. Béjar, E.L. Martín, R. Rebolo, G. Basri 446

TECHNIQUES: INDIRECT DETECTIONS; ASTROMETRY AND TRANSITS

From PTI to the Keck Interferometer
G.T. van Belle 450

Astrometry with the VLT Interferometer
A. Quirrenbach 462

PRIMA – The VLTI Dual Feed System
A. *Glindemann*, S. Lévêque 468

High Precision Interferometry with the NPOI
D. Mozurkewich 474

Extra-Solar Planets with GAIA
M.G. Lattanzi, A. Sozzetti, A. Spagna 479

Searching for Planets with the Space Interferometry Mission
S. Unwin 492

Extrasolar Planets Transits: Detection and Follow-Up
J. Schneider 499

TECHNIQUES: DIRECT DETECTIONS

2MASS Brown Dwarfs and a First Estimate of the Substellar Mass Function
J. Liebert, I.N. Reid, J.D. Kirkpatrick, A. Burrows, J.E. Gizis, A. Burgasser, C.C. Dahn, D.G. Monet, R. Cutri, C. Beichman, M. Skrutskie 505

Brown Dwarfs in Open Clusters
M.R. Zapatero Osorio 511

Adaptive Optics Imaging of Faint Companions: Test Case of MWC480
L. Close, D. Potter, A. Tokunaga 518

Observations of Disks and Faint Companions with NAOS
A.M. Lagrange, D. Mouillet 521

VISIR and the Detection of Exo-Planets
P.-O. Lagage 528

The Darwin Space Interferometer
A. Penny, A. Leger, R. Laurance, M. Fridlund 535

TECHNIQUES: RADIAL VELOCITY SEARCHES AND INTERFEROMETRIC DETECTIONS

Radial Velocity Searches for Extra Solar Planets from Keck and McDonald Observatories
A.P. Hatzes, W.D. Cochran 539

Extrasolar Planets in the Southern Hemisphere: The CORALIE Survey
D. Queloz, M. Mayor, D. Naef, N. Santos, S. Udry, M. Burnet, B. Confino 548

Use of the 3.6 m Telescope at La Silla for Planetary Detection
M. Kürster 556

Direct Observation of Hot Jupiters and Other Faint Companions with the VLTI
V. Coudé du Foresto 560

Direct Detection of Hot Extrasolar Planets Using Differential Interferometry
B. Lopez, R.G. Petrov 565

The CORALIE Planet-Search Sample
S. Udry, M. Mayor, D. Queloz, D. Naef, N. Santos 571

The HARPS Project
F. Pepe, M. Mayor, W. Benz, J.-L. Bertaux, J.-P. Sivan, D. Queloz, S. Udry 572

Author Index 573

List of Participants

Name	Institution
AFONSO, Cristina	CEA Saclay, DAPNIA/SPP cafonso@cea.fr
ALCAINO, Gonzalo	Instituto Isaac Newton, Santiago inewton@reuna.cl
ALLARD, France	CRAL - ENS, Lyon fallard@cral.ens-lyon.fr
ALLOIN, Danielle	ESO, Santiago dalloin@eso.org
ALVARADO, Franklin	Instituto Isaac Newton, Santiago inewton@reuna.cl
ALVES, João	ESO, Garching jalves@eso.org
APPENZELLER, Immo	Landessternwarte Heidelberg & MPIA iappenze@lsw.uni-heidelberg.de
ARAGÓN-SALAMANCA, Alfonso	Inst. of Astronomy, Univ. of Cambridge aas@ast.cam.ac.uk
ARDILA, David	Univ. of California, Berkeley ardila@garavito.berkeley.edu
ARNABOLDI, Magda	Osserv. Astron. di Capodimonte magda@na.astro.it
ARTYMOWICZ, Pawel	Stockholm Observatory pawel@astro.su.se
BARBUY, Beatriz	IAG-USP, Sao Paulo barbuy@orion.iagusp.usp.br
BARUCCI, Maria Antonietta	Observatoire de Paris, DESPA, Meudon barucci@obspm.fr
BASRI, Gibor	Univ. of California, Berkeley basri@soleil.berkeley.edu

BENDER, Ralf	Universitäts-Sternwarte München bender@usm.uni-muenchen.de
BERGERON, Jacqueline	ESO, Garching jbergero@eso.org
BERRUYER, Nicole	Observatoire de la Côte d'Azur, Nice nicole@obs-nice.fr
BOFFIN, Henri M.J.	Royal Observatory of Belgium, Brussels Henri.Boffin@oma.be
BÖHNHARDT, Hermann	ESO, Paranal hboehnha@eso.org
BÖHRINGER, Hans	MPI für extraterrestr. Physik, Garching hxb@mpe.mpg.de
BONIFACIO, Piercarlo	Osservatorio Astronomico di Trieste bonifaci@oat.ts.astro.it
BOOTH, Roy	Onsala Space Observatory roy@oso.chalmers.se
BORGANI, Stefano	INFN - Trieste borganis@ictp.trieste.it
BREWER, James	ESO, Chile jbrewer@eso.org
BREYSACHER, Jacques	ESO, Garching jbreysac@eso.org
BROADHURST, Tom	Univ. of California, Berkeley tjb@wibble.berkeley.edu
BUSS, Jeremy	CTIO, La Serena buss@ctios4.ctio.noao.edu
CAMPUSANO, Luis	Universidad de Chile, Santiago lcampusa@calan.das.uchile.cl
CAPPELLARO, Enrico	Osservatorio Astronomico di Padova cappellaro@pd.astro.it
CATALANO, Francesco A.	Università di Catania, Ist. di Astronomia fcatalano@alpha4.ct.astro.it
CAVALIERE, Alfonso	II Università di Roma, Dip. di Fisica cavaliere@roma2.infn.it
CESARSKY, Catherine	CEA / Direction des Sciences de la Matière, Saclay catherine.cesarsky@cea.fr
CHABRIER, Gilles	CRAL - ENS, Lyon chabrier@ens-lyon.fr

CHINCARINI, Guido	Univ. di Milano & Osservatorio Astronomico di Brera guido@merate.mi.astro.it
CHINI, Rolf	Ruhr-Univ. Bochum, Astron. Inst. chini@astro.ruhr-uni-bochum.de
CHIOSI, Cesare	Università di Padova, Dip. di Astronomia chiosi@pd.astro.it
CLAUSEN, Jens	Copenhagen Univ. Obs., Niels Bohr Inst. jvc@astro.ku.dk
CLOSE, Laird	ESO, Garching lclose@eso.org
CONNERS, Andrew	AMAT conners-andrew@amat.com
CORSINI, Enrico Maria	Università di Padova, Dip. di Astronomia corsini@pd.astro.it
COUDÉ DU FORESTO, Vincent	Observatoire de Paris, DESPA, Meudon foresto@obspm.fr
COUTURES, Christian	C.E.A., DAPNIA/SPP / ESO, Santiago ccouture@eso.org
CRISTIANI, Stefano	ESO, Garching scristia@eso.org
CRUZALÈBES, Pierre	Observatoire de la Côte d'Azur, Grasse Pierre.Cruzalebes@obs-azur.fr
CURÉ, Michel	Universidad de Valparaiso (Chile) mcure@uv.cl
DA COSTA, Luiz	ESO, Garching ldacosta@eso.org
DAIGNE, Gérard	Observatoire de Bordeaux daigne@observ.u-bordeaux.fr
DE BERGH, Cathérine	Observatoire de Paris, DESPA, Meudon debergh@mesiob.obspm.fr
DELAHODDE, Catherine	ESO, Santiago cdelahod@eso.org
DEMARCO, Ricardo	CTIO, La Serena demarco@ctios4.ctio.noao.edu
DENNEFELD, Michel	Institut d'Astrophysique, Paris dennefel@iap.fr

DIERICH, Patrick	Observatoire de Paris, Meudon dierich@despa.obspm.fr
DOUBLIER, Vanessa	ESO, La Silla vdoublie@eso.org
ELS, Sebastian	ESO, Santiago sels@eso.org
ELST, Eric W.	Royal Observatory of Belgium, Brussels ericelst@helios.oma.be
ENDL, Michael	ESO, Santiago mendl@eso.org
FORT, Bernard	Institut d'Astrophysique, Paris fort@iap.fr
FORVEILLE, Thierry	Obs. de Grenoble, Lab. d'Astrophysique Thierry.Forveille@obs.ujf-grenoble.fr
FRANX, Marijn	Leiden Observatory franx@strw.LeidenUniv.nl
FUHRMANN, Klaus	Universitäts-Sternwarte München fuhrmann@usm.uni-muenchen.de
GALLART, Carme	Yale University / Universidad de Chile carme@ociw.edu
GIACCONI, Riccardo	ESO, Garching dg@eso.org
GIAVALISCO, Mauro	STScI, Baltimore mauro@ociw.edu
GIEREN, Wolfgang	Univ. de Concepcion (Chile) wgieren@coma.cfm.udec.cl
GIOIA, Isabella	IRA-CNR, Bologna and IfA, Hawaii gioia@galileo.ifa.hawaii.edu
GIRARDI, Leo	Università di Padova, Dip. di Astronomia lgirardi@pd.astro.it
GLINDEMANN, Andreas	ESO, Garching aglindem@eso.org
GOLDMAN, Bertrand	CEA Saclay, DAPNIA/SPP bgoldman@cea.fr
GREENE, Jenny	CTIO, La Serena greene@ctios4.ctio.noao.edu
GREGGIO, Laura	Osservatorio Astronomico di Bologna greggio@astbo4.bo.astro.it

GUILLOT, Tristan	Observatoire de la Côte d'Azur, Nice guillot@obs-nice.fr
GUROVICH, Sebastian	Graduate School of Astronomy & Astrophysics, Buenos Aires / Mount Stromlo & Siding Spring Obs. sgurovic@hotmail.com
HAINAUT, Olivier	ESO, La Silla ohainau@eso.org
HARDY, Eduardo	NRAO, Santiago ehardy@nrao.edu
HATZES, Artie	Univ. of Texas at Austin artie@zorba.as.utexas.edu
HERRERO, Artemio	Instituto de Astrofísica de Canarias (IAC) ahd@ll.iac.es
HOARD, Donald W.	CTIO, La Serena dhoard@noao.edu
HUBRIG, Swetlana	Univ. of Potsdam shubrig@astro.physik.uni-potsdam.de
JABLONKA, Pascale	Observatoire de Paris, DAEC, Meudon jablonka@obspm.fr
JOGUET, Benoît	ESO, Santiago bjoguet@eso.org
JØRGENSEN, Henning	Astronomical Observatory, Copenhagen henning@astro.ku.dk
KALAS, Paul	STScI, Baltimore kalas@stsci.edu
KÄUFL, Hans Ulrich	ESO, Garching hukaufl@eso.org
KERBER, Florian	ST-ECF, Garching fkerber@eso.org
KIM-QUIJANO, Jessica	CTIO, La Serena kim@ctios4.ctio.noao.edu
KLOSE, Sylvio	Thüringer Landessternwarte, Tautenburg klose@tls-tautenburg.de
KRAUTTER, Joachim	Landessternwarte Heidelberg jkrautte@lsw.uni-heidelberg.de
KUDRITZKI, Rolf-Peter	Universitäts-Sternwarte München kudritzki@usm.uni-muenchen.de

KUNTH, Daniel	Institut d'Astrophysique, Paris kunth@iap.fr
KÜRSTER, Martin	ESO, Chile mkurster@eso.org
LAGAGE, Pierre-Olivier	CEA Saclay, DAPNIA/SPP lagage@cea.fr
LAGRANGE, Anne Marie	Laboratoire Astrophysique de Grenoble lagrange@obs.ujf-grenoble.fr
LAMBAS, Diego G.	IATE - Córdoba Astronomical Obs. dgl@oac.uncor.edu
LE POOLE, Rudolf	Sterrewacht Leiden lepoole@strw.leidenuniv.nl
LÉGER, Alain	Institut d'Astrophysique Spatiale, Université Paris-Sud leger@ias.fr
LEHMANN, Ingo	Astrophysikalisches Institut Potsdam ilehmann@aip.de
LEIBUNDGUT, Bruno	ESO, Garching bleibund@eso.org
LÉNA, Pierre	Université & Observatoire de Paris lena@obspm.fr
LENZEN, Rainer	MPI für Astronomie, Heidelberg lenzen@mpia-hd.mpg.de
LIDMAN, Chris	ESO, Chile clidman@eso.org
LIEBERT, James	STScI, Baltimore liebert@stsci.edu
LIN, Douglas	Univ. of California, Santa Cruz lin@ucolick.org
LINDBLAD, Per Olof	Stockholm Observatory po@astro.su.se
LISSAUER, Jack	NASA/Ames Research Center lissauer@ringside.arc.nasa.gov
LOISEAU, Sacha	Jet Propulsion Lab., Pasadena loiseau@huey.jpl.nasa.gov
LOPEZ, Bruno	Observatoire de la Côte d'Azur, Nice lopez@obs-nice.fr

LUB, Jan	Sterrewacht Leiden lub@strw.leidenuniv.nl
LUCCHIN, Francesco	Università di Padova, Dip. di Astronomia lucchin@pd.astro.it
MADAU, Piero	STScI, Baltimore madau@stsci.edu
MAEDER, André	Observatoire de Genève andre.maeder@obs.unige.ch
MARANO, Bruno	Università di Bologna marano@astbo3.bo.astro.it
MARCHIS, Franck	ESO, Chile fmarchis@eso.org
MARDONES, Diego	Universidad de Chile, Santiago dmardones@das.uchile.cl
MARIGO, Paola	Università di Padova, Dip. di Astronomia marigo@pd.astro.it
MAYOR, Michel	Observatoire de Genève michel.mayor@obs.unige.ch
MAZA, José	Universidad de Chile, Santiago jmaza@das.uchile.cl
McCAUGHREAN, Mark	Astrophysikalisches Institut Potsdam mjm@aip.de
McGRUDER, Charles	Western Kentucky Univ., Dept. of Physics & Astronomy McGruder@wku.edu
MELLIER, Yannick	Institut d'Astrophysique, Paris mellier@iap.fr
MELNICK, Jorge	ESO, La Silla jmelnick@eso.org
MENCI, Nicola	Osserv. Astron. di Roma, Monteporzio menci@coma.mporzio.astro.it
MENDEZ, Roberto H.	Universitäts-Sternwarte München mendez@usm.uni-muenchen.de
MENTEN, Karl M.	MPI für Radioastronomie, Bonn kmenten@mpifr-bonn.mpg.de
MEYLAN, Georges	ESO, Garching gmeylan@eso.org
MILEY, George	Sterrewacht Leiden miley@strw.leidenuniv.nl

MINNITI, Dante	P. Universidad Catolica de Chile, Santiago dante@astro.puc.cl, dminniti@llnl.gov
MIRABEL, Félix	CEA Saclay, DAPNIA/SPP mirabel@discovery.saclay.cea.fr
MOHR, Joseph	Univ. of Chicago, Dept. of Astronomy & Astrophysics mohr@rockie.uchicago.edu
MOORWOOD, Alan	ESO, Garching amoor@eso.org
MORALES OLIVARES, Robert	Univ. of Pierre and Marie Curie, Paris VI rmorales@caramail.com
MOUTOU, Claire	ESO, La Silla cmoutou@eso.org
MOZURKEWICH, David	Naval Research Laboratory, Washington mozurk@rsd.nrl.navy.mil
MURPHY, Miguel	Univ. Catolica del Norte, Antofagasta mmurphy@socompa.ucn.cl
MUSELLA, Ilaria	Osserv. Astron. di Capodimonte ilaria@na.astro.it
NAEF, Dominique	Observatoire de Genève dominique.naef@obs.unige.ch
NISSEN, Poul Erik	Univ. of Aarhus, Inst. of Physics & Astronomy pen@obs.aau.dk
NORDSTRÖM, Birgitta	Niels Bohr Inst. for Astronomy, Copenhagen birgitta@astro.ku.dk
NYMAN, Lars-Åke	ESO/SEST, La Silla lnyman@eso.org
ORTOLANI, Sergio	Univ. di Padova, Dip. di Astronomia ortolani@pd.astro.it
PARESCE, Francesco	ESO, Garching fparesce@eso.org
PASQUINI, Luca	ESO, Garching lpasquin@eso.org
PATAT, Ferdinando	ESO, La Silla fpatat@eso.org

PEDREROS, Mario	Universidad de Tarapaca, Physics Dept. mpedrero@vitor.faci.uta.cl
PELLO, Roser	Observatoire Midi-Pyrennées roser@obs-mip.fr
PENNY, Alan	Rutherford & Appleton Laboratories alan.penny@rl.ac.uk
PETR, Monika	ESO, Paranal mpetr@eso.org
PETTINI, Max	Inst. of Astronomy, Univ. of Cambridge pettini@ast.cam.ac.uk
PIOTTO, Giampaolo	Università di Padova, Dip. di Astronomia piotto@pd.astro.it
PIZZELLA, Alessandro	ESO, La Silla apizzell@eso.org
PONT, Frédéric	Observatoire de Genève frederic.pont@obs.unige.ch
PRITCHARD, John	Univ. of Canterbury, Christchurch, New Zealand j.pritchard@phys.canterbury.ac.nz
QUELOZ, Didier	Jet Propulsion Lab., Pasadena didier@huey.jpl.nasa.gov
QUINN, Peter	ESO, Garching pjq@eso.org
QUINTANA, Hernan	P. Universidad Catolica de Chile, Santiago hquintana@astro.puc.cl
QUIRRENBACH, Andreas	CASS, Univ. of California, San Diego aquirrenbach@ucsd.edu
RAEDSCHELDERS, Guy	BELGACOM-BELCOMLAB, Kessel-Lo, Belgium guy.raedschelders@is.belgacom.be
RAUCH, Thomas	Universität Tübingen, Institut für Astronomie & Astrophysik rauch@astro.uni-tuebingen.de
REIMERS, Dieter	Hamburger Sternwarte dreimers@hs.uni-hamburg.de
REISENEGGER, Andreas	P. Universidad Catolica de Chile, Santiago areisene@astro.puc.cl
RENZINI, Alvio	ESO, Garching arenzini@eso.org

RICH, R. Michael	Univ. of California, Los Angeles rmr@ciel.astro.ucla.edu
RODRIGUEZ, Marcela	Instituto Isaac Newton, Santiago inewton@reuna.cl
ROSATI, Piero	ESO, Garching prosati@eso.org
ROTH, Miguel	Las Campanas Observatory, Carnegie Inst., La Serena miguel@lco.cl
RÖTTGERING, Huub	Sterrewacht Leiden rottgeri@strw.leidenuniv.nl
RUBIO, Monica	Universidad de Chile, Santiago mrubio@das.uchile.cl
RUIZ, Maria Teresa	Universidad de Chile, Santiago mtruiz@das.uchile.cl
SACKETT, Penny	Kapteyn Astronomical Inst., Groningen psackett@astro.rug.nl
SCHMUTZ, Werner	ETH Zürich, Inst. of Astronomy schmutz@astro.phys.ethz.ch
SCHNEIDER, Jean	Observatoire de Paris, Meudon schneider@obspm.fr
SCHNEIDER, Peter	MPI für Astrophysik, Garching peter@mpa-garching.mpg.de
SCHOELLER, Markus	ESO, Garching mschoell@eso.org
SCHREIER, Ethan J.	STScI, Baltimore schreier@stsci.edu
SEGGEWISS, Wilhelm	Astron. Inst. Univ. Bonn, Observatorium Hoher List seggewis@astro.uni-bonn.de
SEKIGUCHI, Tomohiko	ESO, Santiago tsekiguc@eso.org
SELMAN, Fernando	ESO, Santiago fselman@eso.org
SPITE, Monique	Observatoire de Paris, Meudon Monique.Spite@obspm.fr
STANFORD, Adam	UC-Davis, IGPP, Livermore adam@igpp.llnl.gov

STERZIK, Michael	ESO, Chile msterzik@eso.org
STOCKMAN, Hervey (Peter)	STScI, Baltimore stockman@stsci.edu
SUNYAEV, Rashid	MPI für Astrophysik, Garching sunyaev@mpa-garching.mpg.de
TARENGHI, Massimo	ESO, Garching mtaren gh@eso.org
TOLSTOY, Eline	ESO, Garching etolstoy@eso.org
TOZZI, Paolo	Johns Hopkins Univ., Dept. of Physics & Astronomy paolo@pha.jhu.edu
UDRY, Stéphane	Observatoire de Genève stephane.udry@obs.unige.ch
UNWIN, Steve	Jet Propulsion Lab., Pasadena unwin@huey.jpl.nasa.gov
VAN BELLE, Gerard	Jet Propulsion Lab., Pasadena gerard@huey.jpl.nasa.gov
VAN DER KRUIT, Pieter	Kapteyn Astronomical Inst., Groningen vdkruit@astro.rug.nl
VERA MEGE, Rafael	Univ. de Concepcion (Chile) rvera@udec.cl
VIARD, Elise	ESO, Garching eviard@eso.org
VON DER LÜHE, Oskar	Kiepenheuer-Institut für Sonnenphysik, Freiburg i. Breisgau ovdluhe@kis.uni-freiburg.de
WACHTER, Stefanie	CTIO, La Serena swachter@noao.edu
WAELKENS, Christoffel	Katholieke Universiteit Leuven, Instituut voor Sterrenkunde christoffel@ster.kuleuven.ac.be
WALKER, Alistair	CTIO, La Serena awalker@noao.edu
WEST, Richard M.	ESO, Garching rwest@eso.org
WHITE, Simon D.M.	MPI für Astrophysik, Garching swhite@mpa-garching.mpg.de

WUCHTERL, Günther	Inst. für Astronomie der Universität Wien wuchterl@astro.univie.ac.at
ZAPATERO-OSORIO, María R.	Instituto de Astrofísica de Canarias (IAC) mosorio@ll.iac.es
ZINNECKER, Hans	Astrophysikalisches Institut Potsdam hzinnecker@aip.de

PLENARY SESSIONS:

SCIENCE IN THE VLT ERA AND BEYOND

Edited by

Jacqueline Bergeron and Alvio Renzini

Science with FORS

I. Appenzeller[1,6], R. Bender[3], H. Böhnhardt[5], S. Cristiani[4], M. Dietrich[1], K. Fricke[2], W. Fürtig[1], W. Gässler[3], R. Gilmozzi[4], R. Häfner[3], R. Harke[2], J. Heidt[1], H.-J. Hess[3], U. Hopp[3], W. Hummel[3], K. Jäger[2], P. Jürgens[2], R.-P. Kudritzki[3], M. Kümmel[1,6], K.-H. Mantel[3], D. Mehlert[1], W. Meisl[3], C. Moellenhoff[1], B. Muschielok[3], H. Nicklas[2], A. Renzini[4], P. Rosati[4], G. Rupprecht[4], R. Saglia[3], W. Seifert[1], S. Seitz[3], J. Spyromilio[5], O. Stahl[1], T. Szeifert[1], and K. Tarantik[3]

[1] Landessternwarte, Heidelberg, Germany
[2] Universitäts-Sternwarte Göttingen, Germany
[3] Universitäts-Sternwarte München, Germany
[4] ESO Garching, Germany
[5] ESO Paranal, Chile
[6] Max-Planck-Institut für Astronomie, Heidelberg, Germany

Abstract. In September 1998 the first one of the two FORS multimode focal plane instruments for the ESO VLT has been installed at the unit telescope Antu. Test observations of faint objects, such as extragalactic stars, galaxy clusters, gravitational arcs, and high-redshift galaxies and quasars obtained during two commissioning runs and a short science verification period confirm the high optical quality and efficiency of FORS and demonstrate the great potential of FORS and the VLT for imaging, photometry, quantitative spectroscopy, polarimetry and spectropolarimetry.

1 The FORS Instruments

The two FORS (**Fo**cal **R**educer **S**pectrograph) instruments for the ESO VLT are focal reducer cameras which can be used for imaging, low-resolution longslit spectroscopy and multiple object spectroscopy (MOS) in the wavelength range 330 to 1100 nm. Both instruments have a field of view of alternatively 6.8×6.8 arcmin^2 with a 0.2 arcsec per pixel image scale (standard resolution mode) or 3.4×3.4 arcmin^2 with 0.1 arcsec per detector pixel (high resolution mode). FORS I, which has been installed at the VLT unit telescope Antu in September 1998, can also be used to carry out high-accuracy linear and circular polarimetry and spectropolarimetry over the whole FOV in the standard and high resolution modes. FORS II, which will be installed at the VLT unit telescope Kueyen in October 1999, will allow us to do, in addition to normal low-resolution spectral observations, also echelle spectroscopy . Detailed technical descriptions of the FORS instruments can be found in (e.g.) [1], [2], [3], [4], [5], [6] [7].

The FORS instruments were built by a consortium consisting of the Landessternwarte (State Observatory) Heidelberg and the University Observatories in Göttingen and München, Germany, in close cooperation with ESO.

They were designed to do photometry, quantitative spectroscopy, and polarimetry of the faintest objects observable with the VLT ([8]). Hence much effort was devoted to develop an optical system where, even under optimal seeing conditions, the PSF of real images is not significantly affected by optical aberrations or the effects of mechanical flexure. Moreover, in order to permit an optimal and accurate sky subtraction for quantitative MOS spectroscopy and spectropolarimetry of faint targets we developed a correspondingly designed assembly of individually movable slit jaws. Apart from providing a high systematic accuracy due to the precisely machined slit jaws this assembly allows us to adjust instantaneously the slit width to the seeing conditions to achieve an optimal S/N. For less precise survey-type spectroscopy of a larger numbers of objects FORS II will have (in addition to the individually movable slits) a mask exchange facility permitting conventional MOS spectroscopy with laser cut focal plane slit masks.

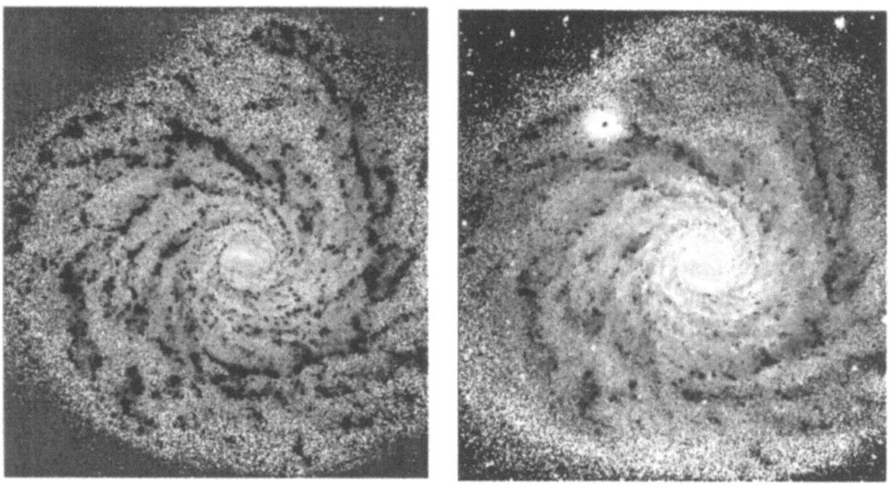

Fig. 1. U-B (left image) and V-I (right image) color index images of NGC 1232. The dark patches in the U-B image represent the clearly defined star forming regions, while the V-I image shows the very narrow dust lanes of this giant spiral galaxy. These images are based on FORS U, B, V and I frames with (exposure times between 3 and 10 minutes) obtained during the FORS commissioning period

2 Examples of First Results Obtained with FORS

Regular operation of the VLT and FORS I will start on April 1st, 1999. Therefore, so far there exist only few observations with the new instrument which were obtained during two commissioning runs in September and December

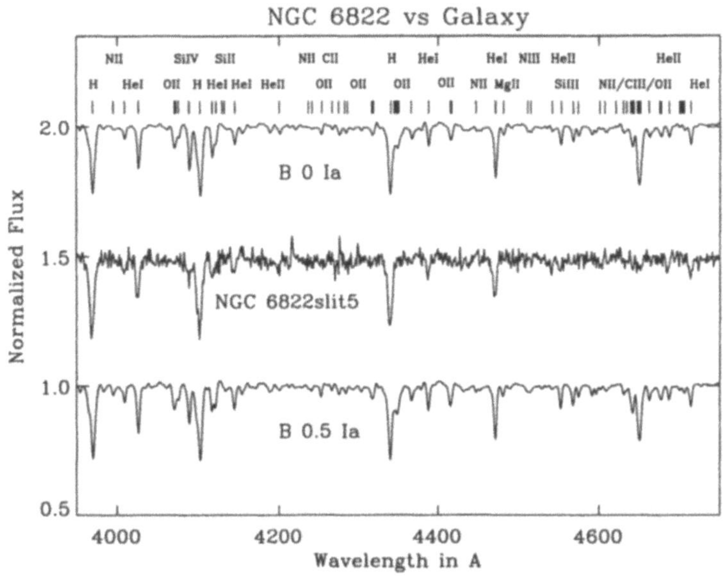

Fig. 2. Spectrum of a V = 18.4 individual star in the irregular galaxy NGC 6822 (central tracing) observed with FORS in the MOS mode. The exposure time was 30 minutes. For comparison the spectra of two galactic stars of similar effective temperature are also included (upper and lower tracing). Note the striking difference in the strength of the metallic lines, while the H and He lines show similar strengths in all three spectra

1998 (devoted mainly to technical tests, for details see [9]) and a brief science verification run in January 1999. However, these observations already demonstrate that the design goals of FORS 1 have been reached and that images, spectra and polarization measurements of very high quality can be obtained with this device. Star images observed with FORS 1 often had better FWHM values than predicted by the Paranal seeing monitor and stellar image FWHM values as low as 0.25 arcsec have been recorded.

The publicly best known results of the FORS commissioning runs are various spectacular true-color images of spiral galaxies (such as NGC1232 and NGC 1365), of galaxy clusters (such as Abell 370) and of planetary nebulae (such as M27) published as ESO press releases and through the ESO Web page (http://www.eso.org). As these images are readily available through the Web, no example will be reproduced here. Instead, in this paper we will concentrate on some of the quantitative scientific information contained in the commissioning and science verification data.

Judging from its appearance in newspapers, on posters, on T-shirts, etc. the most popular FORS image obtained so far seems to be the 5-color image of the giant spiral galaxy NGC 1232, obtained during a period of good seeing on September 21, 1998. The color image (available through the Web) shows

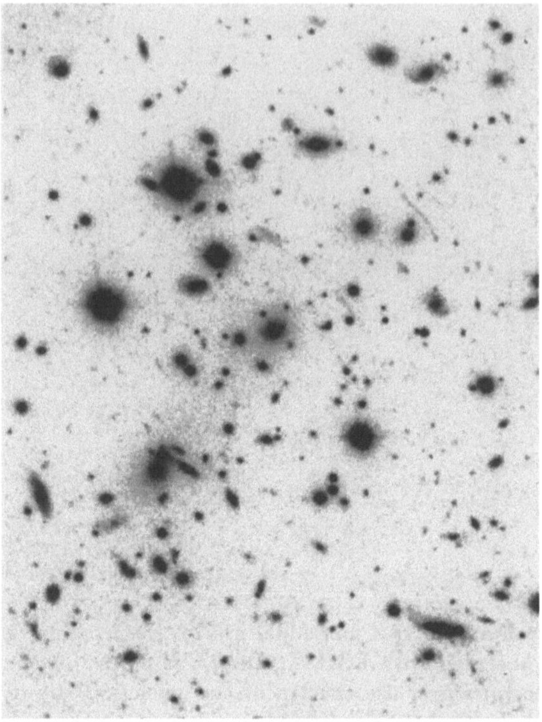

Fig. 3. Gunn-g image of the z = 0.29 galaxy cluster 1ES0657-558. The exposure time was 10 minutes. Note the gravitational arcs in the upper right part of the image

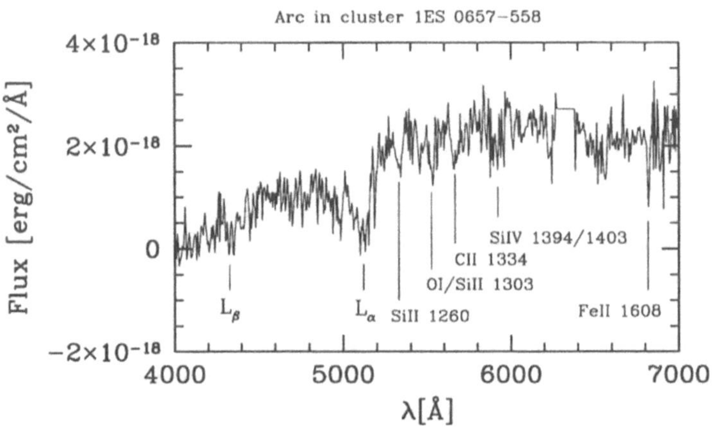

Fig. 4. Spectrum of the prominent narrow gravitational arc in Figure 3

an unprecedented wealth of details of this system. In order to illustrate that such images also have a great potential for quantitative work we reproduce in Fig. 1 two color-index images calculated from the same frames. The U-B image outlines (in black) very well the clearly defined star forming regions of this galaxy, while the V-I image demonstrates the presence of numerous very narrow dust lanes (presumably shock fronts) in this system.

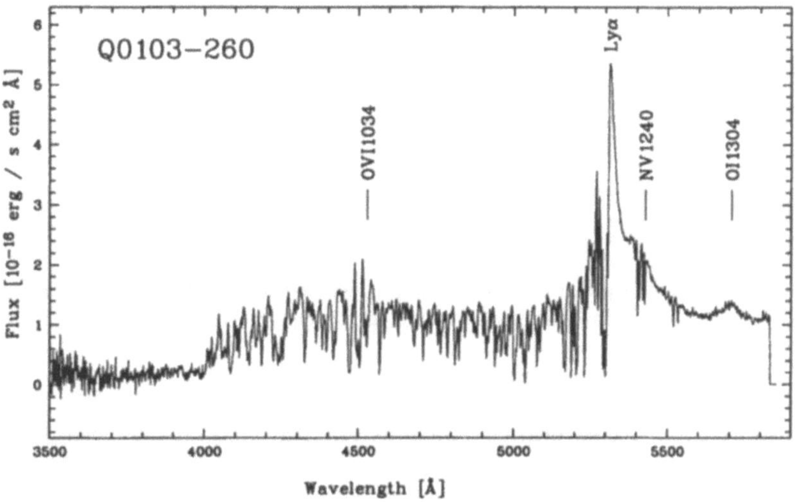

Fig. 5. Medium resolution (R = 2500) FORS spectrum of the Lyman forest of the Quasar Q0103-260

During a period of even better seeing we obtained well resolved broad-band and Hα images of the Local Group galaxy NGC 6822. (The Hα images are again available through the ESO Web site). While these images were taken mainly to test the performance of FORS in crowded fields, we also used the opportunity to take medium resolution MOS spectra of a selection of bright individual stars in this extragalactic system. An example is reproduced in Fig. 2, where spectra of two galactic stars of similar effective temperature are included for comparison. As shown by this figure, the expected differences in the strength of the metallic lines (due to the different chemical composition of NGC 6822) are clearly visible and measurable on FORS spectra. Extrapolating from these results we expect that quantitative spectroscopy of individual bright blue stars will (at somewhat lower, but still adequate, spectral resolution) be possible with FORS to at least the distance of the Virgo cluster.

Among the more distant objects observed so far with FORS are several interesting galaxy clusters. Fig. 3 shows as an example a (1.7$\times$2.4 arcmin^2) section of a Gunn-g frame with the X-ray bright cluster 1ES0675-558. Even

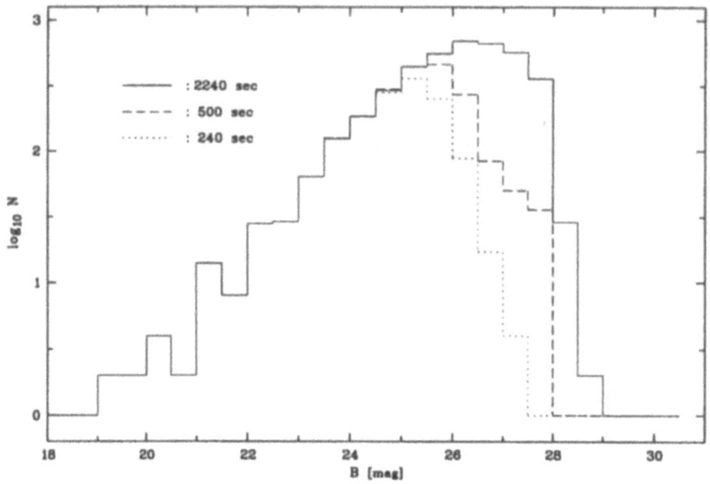

Fig. 6. The number of galaxies per 0.5 mag. detected in the (Bessell) B-band with a S/N > 10 as a function of the magnitude for three different exposure times

on this short exposure several conspicuous gravitational arcs are visible. Test observations during the commissioning runs showed that FORS has a great potential for the spectroscopy of such faint arcs. This is demonstrated by Fig. 4, where we reproduce the FORS spectrum (based on three 30 minutes exposures) of the prominent narrow, almost straight arc in the upper right (NW) part of Fig. 3. As shown by the figure, the arc is, in fact, the lensed image of a z = 3.23 galaxy.

The most distant objects observed so far with FORS are several high redshift quasars, including the z = 5.0 object Q0338+0021 [10] . Fig. 5 shows as an example a spectrum of the z = 3.36 quasar Q0103-260. This spectrum, obtained using a 0.4 arcsec slit width (giving a resolution of R = 2500, corresponding to $\Delta v \approx 120$ kms^{-1} or 28 kms^{-1} in the object's rest frame) shows very well the rich Ly forest of this quasar. In the red FORS I can reach with this slit width a spectral resolution of about 4400. In FORS II the limiting resolution is expected to be (at least) about 7000.

In order to test the capability of FORS for the photometry of faint galaxies we obtained during the first commissioning run a number of frames of a field free of bright foreground stars or galaxy clusters near the South Galactic Pole (the "FORS Deep Field" or FDF). Fig. 6 shows the number of galaxies per 0.5 mag. brightness interval detected with the extraction program FOCAS for a photometric S/N of 10 to 15 (depending on the surface brightness distribution) and a seeing of 0.9 arcsec. As shown by the figure at this (modest) seeing after 2240 sec of exposure time most galaxies are detected with a S/N adequate for quantitative photometry down to a (Bessell) B magnitude

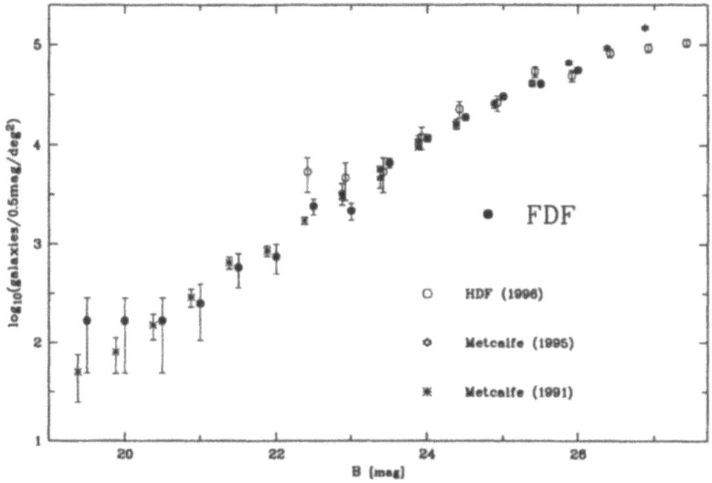

Fig. 7. Comparison of galaxy densities observed with FORS in the "FORS Deep Field" (FDF) with results taken from the literature

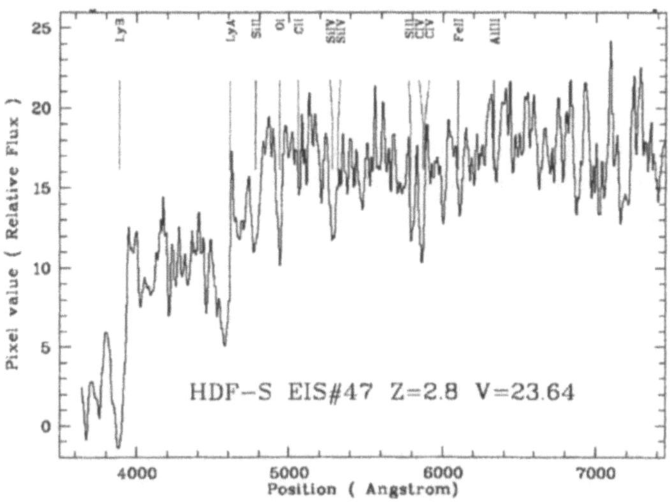

Fig. 8. FORS MOS spectrum of the high-redshift galaxy HDF-S EIS # 47

of about 27.0. As illustrated in Fig. 7, the galaxy densities from the FORS counts agree well with those derived in other ground-based or HST counts.

Finally, in Fig. 8 and 9 we present two examples of FORS MOS spectra of faint high-redshift galaxies identified photometrically in the ESO Imaging Survey (EIS) HDF-S field (Da Costa et al. [11]) and the AXAF Deep Field (Rengelink et al. [12]). The total exposure times were 3 hours for EIS 47 (observed during the second commissioning run in December 1998 at an air mass of about 2) and 5.5 hours for AXAF 107 (observed during the Science

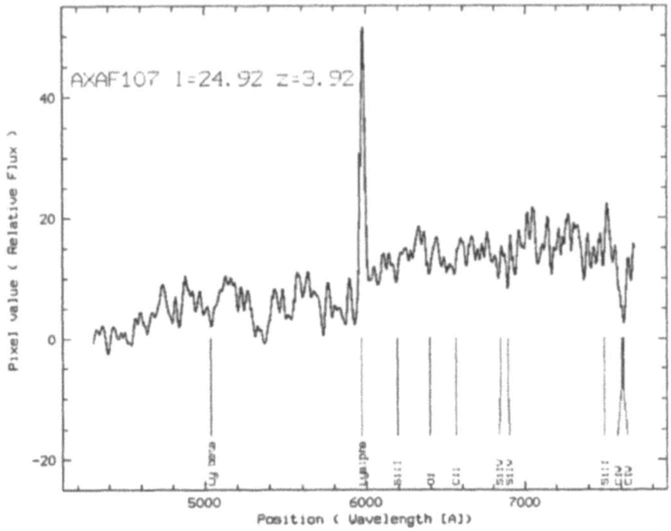

Fig. 9. FORS MOS spectrum of the galaxy # 107 in the AXAF Deep Field

Verification period in January 1999). Both spectra demonstrate the potential of FORS for the spectroscopy of galaxies to at least 25th mag.

Space limitations do not allow us to include here results of the polarimetric and spectropolarimetric observations of various types of objects which we carried out during the commissioning runs. An example of these results (circular spectropolarimetry of the magnetic White Dwarf GD 229) can be found in [9].

3 The Scientific Potential of FORS

Among the various powerful faint-object focal plane instruments built (or under construction) for the new large telescopes FORS has - at present- several unique or particularly well developed capabilities. Among these are (1) a MOS with high quality and instantly adjustable slits, (2) a uniform high image quality from 330 to 1100 nm, (3) very low straylight and ghosts, (4) IF filter imaging without field effects, and (5) the capability to do accurate polarimetry and spectropolarimetry in all Stokes parameters over the whole FOV. It is obvious that the greatest potential of FORS is in those types of observations which make use of these particular opportunities. Therefore, we expect that during the next years FORS will be used most extensively for the quantitative spectroscopy of distant galaxies and other faint objects, in order to derive chemical compositions, stellar populations and dynamical data. Also important will the photometry of *very* faint targets of all types, from solar system objects to arcs and very distant galaxies or protogalaxies. The "wide field" interference filter capability of FORS will permit the detection of new faint emission line sources. Finally, we hope that the high-performance polarimetric capability of FORS I will contribute significantly to the progress in our understanding of the non-thermal and magnetic processes in the distant universe.

We would like to thank the staff of the Paranal observatory for the efficient and kind help during the commissioning and science verification runs. This work was supported by the German Federal Ministry for Education and Research (grants nos. 053HD50A, 053GO10A, and 053MU104).

References

1. Appenzeller, I., Rupprecht, G. (1992) The Messenger **67**, 18
2. Appenzeller, I., Duensing, K.H., Gong, S., Hess, H.-J., Kiesewetter, S., Möhler, S., Muschielok, B., Nicklas, H., Östreicher, R., Roth, M., Rupprecht, G., Seifert, W., Stahl, O.(1992) in: 'Conference on Progress in Telescope and Instrumentation Technologies', ed. M.H. Ulrich, ESO Conference Proceedings **42**, 577
3. Seifert, W., Mitsch, W., Nicklas, H., Rupprecht, G. (1994) in: 'Instrumentation in Astronomy VIII', eds. C. Crawford and E.R. Craine, SPIE Proceedings **Vol. 2198**, 213
4. Böhnhardt, H., Möhler, S., Hess, H.-J., Kiesewetter, S., Nicklas, H.(1995) in: 'Scientific and Engineering Frontiers of 8-10m Telescopes', eds. M. Iye and T. Nishimura, Universal Academic Press Inc. Tokyo, 199
5. Möhler S., Seifert, W., Appenzeller, I., Muschielok, B. (1995) in: Calibrating and Understanding HST and ESO Instruments', ed. P. Benvenuti, ESO Conference and Workshop Proceedings **53**, 149
6. Nicklas, H., Seifert, W., Böhnhardt, H., Kiesewetter-Köbinger, S., Rupprecht, G. (1996) in: Optical Telescopes of Today and Tomorrow', ed. A. Ardeberg, SPIE Proceedings **2871**, 1222
7. Szeifert, T., Appenzeller, I., Fürtig, W., Seifert, W., Stahl, O., Böhnhardt, H., Gässler, W., Häfner R., Hess, H.-J., Mantel K.-H., Meisl, W., Muschielok, B., Tarantik, K., Harke, R., Jürgens, P., Nicklas, H., Rupprecht, G. (1998) in: 'Optical Astronomical Instrumentation', SPIE Proc. **3355**, 20
8. Appenzeller, I., Stahl, O., Kiesewetter, S., Kudritzki, R.-P., Nicklas, H., Rupprecht, G. (1997) in: 'The Early Universe with the VLT', ed. J. Bergeron, ESO Astrophysics Symposia Proceedings, Springer-Verlag, p. 35
9. Appenzeller, I., Fricke, K., Fürtig, Gässler, W., Häfner R., Harke, R., Hess, H.-J., Hummel, W., Jürgens, P., Kudritzki, R.-P., Mantel K.-H., Meisl, W., Muschielok, B., Nicklas, H., Rupprecht, W., Seifert, W., Stahl, O., G. Szeifert, T., Tarantik, K., (1998) The Messenger **94**, 1
10. Fan, X., Strauss, M.A., Schneider, D.P., et al. (1999) ApJ, in press (July 1999).
11. Da Costa, L., Nonino, M., Rengelink, R. et al. (1998) Astron. Astrophys. in press (AstroPh/98/12105)
12. Rengelink, R. et al. (1998), preprint (AstroPh/98/12190)

Science with ISAAC

Alan Moorwood

European Southern Observatory, Karl-Schwarzschild-Str. 2,
D-85748 Garching bei München, Germany

Abstract. ISAAC, the Infrared Spectrometer and Array Camera, offers imaging, polarimetric and spectroscopic capabilities between 0.9 and 5μm and is the first VLT instrument fully developed by ESO. It was installed at the Nasmyth B focus of UT1 (Antu) in November 1998 and started routine observations in Service and Visitor mode for the community in April 1999. Programmes awarded time in this first operational period range from studies of solar system bodies to the most distant galaxies and confirm both the interest and diversity of the science expected be produced by ISAAC. Some illustrations of this have already appeared in the form of the images and spectra obtained during commissioning which have been displayed in The Messenger and at ESO's Web site.

1 Introduction

ISAAC (Infrared Spectrometer and Array Camera) offers imaging, polarimetry and spectroscopic capabilities between 0.9 and 5μm. It comprises two cameras optimized for the 1-2.5μm (SW) and 2.5-5μm (LW) spectral ranges and equipped with Rockwell 1024x1024 Hg:Cd:Te and SBRC 256x256 (to be upgraded to 1024x1024) InSb arrays respectively. These cameras are used to either re-image the telescope focal plane or an intermediate spectrum produced by the grating spectrometer. A visible slit viewer or the infrared imaging mode itself are used for spectroscopic target acquisition.

ISAAC is the first VLT instrument fully developed by ESO. It was installed at the Nasmyth B focus of UT1(Antu) in November 1998 and, following commissioning runs in November/December 1998 and February/March 1999, started routine Service and Visitor Mode operations in April 1999. Figure 1 shows ISAAC installed at the telescope. More information on the instrument, its commissioning activities and examples of the images and spectra obtained can be found in Moorwood et al.(1998, 1999) and various press releases posted at ESO's Web site (http://www.eso.org/outreach/press-rel).

2 Observing Capabilities

ISAAC offers the following observational modes:

- Imaging through broad and narrow band filters
  - $\lambda\lambda$ 0.9-2.5μm with a 2.5x2.5' f.o.v and 0.147" pixels
  - $\lambda\lambda$ 2.5-5μm with a 40x40" max. f.o.v and 0.08 and 0.16" pixels

Fig. 1. ISAAC installed at the Nasmyth B focus of Antu. The instrument itself is attached to the adapter/rotator to the left while the unit attached to the platform on the right includes the control electronics cabinets, closed cycle cooler compressor and the cable co-rotator system used to feed cables and cooling hoses to and from the rotating instrument.

- Imaging polarimetry
  - $\lambda\lambda$ 0.9-2.5μm using a Wollaston prism with broad/narrow band filters
- Spectroscopy at Rs$\simeq$ 500 and 3000 and slit widths s of 0.3, 0.6, 1 and 2" and max. slit length of 2'.

3 Performance

Results of the commissioning tests have confirmed that ISAAC essentially meets or exceeds its specified scientific performance. In particular, it has been demonstrated that the overall Paranal/telescope/instrument combination can yield direct images with around 0.25-0.3" FWHM quality in both the SW and LW arms. In the case of the LW observations, both images and spectra were obtained using chopping together with field stabilisation at the telescope M2 unit and nodding of the telescope. The performance in this particular mode, the most challenging in terms of instrument/telescope interaction, was particularly pleasing. Also very satisfying are the extremely low read noise ($\simeq$ 5e) and dark current (30 e/hr) values achieved with the SW detector in operation at the telescope and which allow background limited medium resolution spectroscopy even between the OH sky lines shortward of 2μm where the sky continuum corresponds to only $\simeq$ 0.1 e/s/pix at the detector! How this all potentially translates into scientific results is discussed further below with reference to the commissioning data.

4 Science Potential of ISAAC

ISAAC was conceived at the beginning of the 1990's as a 'workhorse' instrument with scientific applications foreseen in most areas of astronomy from the solar system to distant galaxies. Near infrared imaging and spectroscopy of high redshift galaxies was one of the specific areas highlighted as potentially gaining most from such an instrument on a large telescope. More specifically it was proposed that spectroscopy of redshifted 'visible' emission lines; deep searches for extremely red primeval galaxies; studies of cluster evolution etc could be very important. As this was proposed at a time which pre-dated the discovery of 'normal' high z galaxies the test results obtained and the relatively large amount of observing time allocated to such programmes in the first operational period is rather gratifying. In particular, the obvious interest and need for broad and narrow band infrared imaging surveys now justifies the building of such a physically large instrument to achieve a 2.5x2.5' field! This was much less obvious at the time of the instrument definition when infrared arrays were still in their infancy (both size and performance) and infrared astronomy was still largely directed to the study of single objects. It is also important to note that this 'wide field' imaging capability both allows ISAAC to find its own faint spectroscopic targets and to center selected objects in the slit. This is of interest not only for studies of the high z Universe but also e.g in searching for low mass stars and brown dwarfs in the galaxy and faint solar system objects. The original claim that ISAAC could contribute in most areas of astronomy appears to remain true and the following is my own personal and updated list of some of the most important topics which can be addressed with it. Observations in most of these areas and some in addition have already been scheduled for the first 6 month period of operation (P63).

Examples of ISAAC Science Topics

- Solar system - outer bodies, Centaurs
- Brown dwarfs - detection, spectroscopy(molecular bands)
- Star formation - low mass IMF; embedded outflow sources
- High z supernovae - light curves, spectroscopy
- Active galaxies
  - 'coronal' line spectroscopy to probe EUV continuum
  - Nuclear starbursts
  - Gaseous/stellar bars as BH fuelling mechanism
- Ultraluminous Infrared Galaxies
  - Starburst vs AGN
  - Mass distribution in mergers
- High redshift galaxies/clusters
  - Gravitational lensing
  - Photometric redshifts (with FORS)+ spectroscopic confirmation
  - High z emission line galaxies - NB filter surveys; spectroscopy for extinction; star formation rate; mass (linewidths, rotation curves).

There is no space here, nor am I personally qualified to enter into detailed prophecies as to what might be discovered in all these areas. This will largely depend anyway on the skill and imagination of our community of users. That data of increased depth and quality will become available through the ISAAC/VLT/Paranal site combination, however, is already evident from some of the results obtained during commissioning. This, despite the fact that there was relatively little time available for astronomical observations as such (commissioning is mostly devoted to more mundane tests of image quality, sensitivity, backgrounds, flexure, software interfacing, observation blocks etc.) and integration times so far have rarely exceeded 1hr as a result. I will limit myself, therefore, to comments in just a few of these areas where commissioning observations have been made and displayed on the Web and/or the Messenger. The relevant raw data are in the process of being made publicly accessible.

4.1 Star Forming Regions

A colour composite image of a visually obscured ($A_v \simeq 60$ mag.) region of star formation near to the optical HII region RCW38 has appeared not only on the Web and in the Messenger but on the covers of many astronomical journals around the world. This is largely, I believe, because the image is visually appealing - on a par with 'real' visible images - thus illustrating the maturity of infrared imaging now reached with the latest generation of large format infrared arrays. The science is, nevertheless, different. Despite the relatively short integration times of a few minutes/band, this image goes several magnitudes deeper than the hydrogen burning limit for young stars at this distance of $\simeq$ 2kpc and should allow studies of the total stellar content plus possible detection of sub-stellar objects in this field. A spectroscopic example is provided by HH212 whose K band spectrum is reproduced in Moorwood et al (1999). Obtained under good seeing conditions with a 2"x 2' slit this shows 'images' of the remarkably symmetrical bi-polar ejection of 'bullets' from this embedded protostar simultaneously in several of the ro-vibrational lines of molecular hydrogen.

4.2 Nearby Galaxies

Relatively few observations of nearby galaxies were obtained during commissioning - FORS does generally yield prettier images in this field! The colour composite image of the interacting galaxy system ESO202-G23 obtained by combining an ISAAC H(1.6μm) band image with B and R images obtained with the VLT Test Camera, however, indicates that such combinations could be of increasing interest (Moorwood et al (1999) and ESO Press Release 19/98, http://www.eso.org/outreach/press-rel/pr-1998). By increasing the spectral coverage in such composites one exposes a wider range of the phenomena present - in this case greater visibility of the dust obscured central region plus the ionization cone, a star forming region and even several background galaxies seen through the system. The image of NGC1365 which appeared in Moorwood et al (1998) and on the Web together with a selection of visible images (ESO Press Photos 08a-e/99)is also interesting. The actual ISAAC image shown combines LW Ks(2.16μm) and L(3.8μm) band images - the latter being in the heavily thermal infrared where few images of galaxies have been obtained to date due to the combined problems of sensitivity and detector saturation. New information in this image includes the detection of visually obscured circumnuclear star clusters and two supernova remnants only seen previously at radio wavelengths.

4.3 Gravitational Lensing

Two examples of this were illustrated amongst the ISAAC 1st light images (ESO Press Release 19/98, http://www.eso.org/outreach/press-rel/pr-1998). One is a J+K colour composite of the gravitationally lensed quasar system MG0414+0534 showing, within a region only 2" across, the z=0.96 galaxy plus the four (two not quite resolved) images of the z=2.64 quasar and a faint arc. One of the main scientific interests in follow-up imaging and spectroscopy of this and other such systems is the possibility for determining the Hubble constant. The second example is the image of the z = 0.3 galaxy cluster CL2244-02 obtained by combining an ISAAC Ks band image with visible images obtained with the VLT Test Camera. This reveals not only the well known prominent, blue gravitational arc near to the cluster centre but also an extremely red arc further out which was not, in fact, detected in the visible and has only been seen faintly in the I band with HST. Follow-up spectroscopy of the red arc is now needed to confirm if it is at much higher redshift.

4.4 The High Redshift Universe

One of the few images published so far in black and white is reproduced here in Fig. 2. It is a 45 min Ks band image of the 2.5x2.5' field centred on the z=3.14 radio galaxy MRC0316-257 which also contains several other galaxies at the same redshift. Technically, it is of interest that the image quality after combining 45 1min exposures is around 0.5" and that galaxies can be seen down to Ks $\simeq$ 22. The optical quality and uniformity (the image is not flat fielded) is also evident. One purpose of taking this image was for a follow-up ISAAC spectrum of the radio galaxy and one of the other galaxies suspected to be at the same redshift. The interest in ISAAC spectroscopy of high z radio and other galaxies is, however, better illustrated by the spectra of the z = 2.4 radio galaxy MRC0406-244. Fig. 3 is a low resolution spectrum covering most of the 1-2.5μm spectral range showing several of the 'visible' emission lines from [OII] to [SII] seen in optical spectra of low z objects. Fig. 4 is a medium resolution spectrum covering red-shifted H_β and [OIII] which

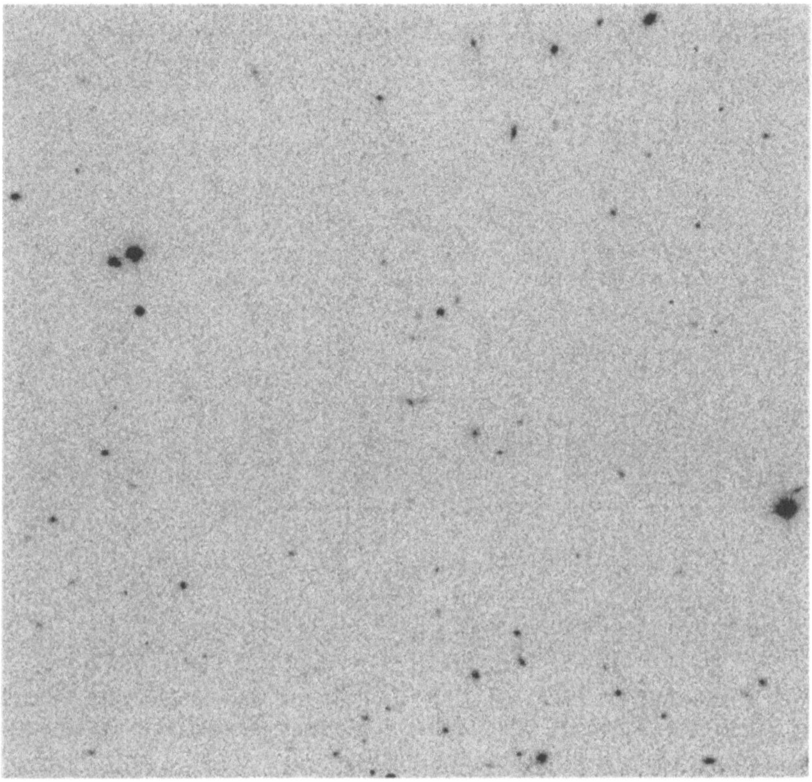

Fig. 2. Ks (2.2μm) image of the 2.5x2.5 arcmin field centered on the z = 3.14 radio galaxy MRC0316-257. This image is a composite of 45 jittered exposures of 1min and exhibits a FWHM image quality $\simeq$ 0.5" and reaches around Ks $\simeq$ 22 mag.

shows better the velocity structure and also illustrates the possibility at this resolving power of exploiting the extremely dark sky background between the OH lines. These spectra, although of a relatively bright object, illustrate also the interest in using ISAAC to study galaxies in general at z $\simeq$ 1-3 where the most prominent 'visible' emission lines have to be observed in the infrared. One technical application is simply redshift determination and related checks on the reliability of photometric redshifts. Astrophysically, however, infrared spectroscopy allows an extension of the use of the Balmer lines to measure star formation rates from z=0 to beyond 3. This means that a single tracer, which suffers lower extinction than the generally used UV continuum and which provides a better measure of the instantaneous star formation rate, can be used to cover the redshift range where the strongest evolution appears to have taken place. Combining NB filter imaging, to find candidate emission line galaxies, with spectroscopic follow-up offers a very promising use of ISAAC to distinguish between starburst and AGN activity;

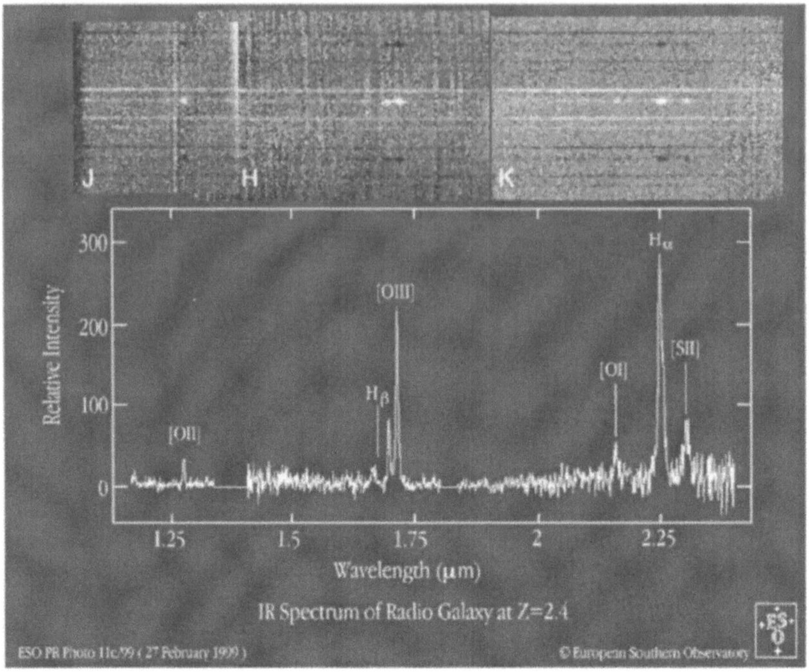

Fig. 3. J,H,K band low resolution (R $\simeq$ 500) 2D and extracted spectra of the z=2.4 radio galaxy MRC0406-244. Also visible below the radio galaxy on the 2D spectra are the continua of two extremely red galaxies.

measure the star formation density at high z; study clustering and determine the mass and metal abundance of individual galaxies. ISAAC spectra of z $\simeq$ 0.6 galaxies from the CFRS survey have been shown in Moorwood et al (1999) and H_α in a z = 2.18 candidate star forming galaxy discovered in a preparatory deep, large area, NB imaging survey conducted with SOFI at the ESO NTT appears to have been detected during the commissioning thus confirming that such programmes are essentially viable.

5 Conclusions

ISAAC has been installed and commissioned at the ESO VLT with results which suggest that it will fulfill its promise of delivering exciting new discoveries in many areas of astronomy in the coming years. In particular, its combination of 'wide' field infrared imaging and spectroscopy offers a potentially powerful new capability for studying star formation at high redshift.

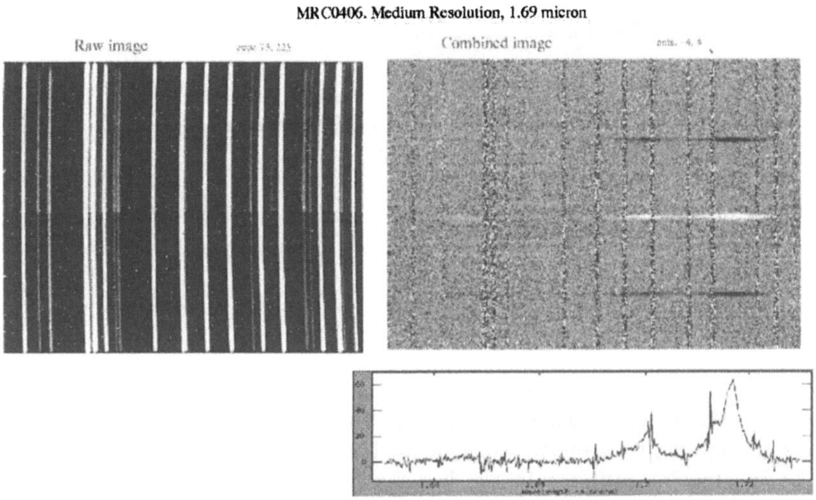

Fig. 4. Medium resolution (R $\simeq$ 3000) spectrum around 1.7μm of the z=2.4 radio galaxy MRC0406-244. The upper left panel shows a raw 2D spectrum which is dominated by OH sky lines; the upper right panel shows the sky subtracted 2D spectrum and the lower panel the extracted 1D spectrum. The vertical bands remaining in the sky subtracted spectra are due to the increased photon noise at the position of the OH sky lines.

6 Acknowledgements

As this contribution has emphasised science and astronomical commissioning data I wish to particularly thank the ISAAC Instrument Scientist, Jean-Gabriel Cuby who was primarily responsible for the commissioning and took and reduced much of the data referred to here. Thanks are also due to Chris Lidman who was also heavily involved, to Jason Spyromilio, Monika Petra and to Roberto Gilmozzi who, amongst other things, produced the beautiful IR/visible colour composite images referred to and available on the Web. I would also like to take this opportunity to thank again the main technical members of the team who built ISAAC and without whom none of these results would have been possible - P. Biereichel, J. Brynnel, R. Conzelmann, B. Delabre, A. van Dijsseldonk, G. Finger, H. Gemperlein, T. Herlin, G. Huster, J. Knudstrup, J.-L. Lizon, H. Mehrgan, M. Meyer, G. Nicolini, A. Silber and J. Stegmeier.

References

1. Moorwood, A.F.M., et al. 1998, The Messenger, 94, 7
2. Moorwood, A.F.M., et al. 1999, The Messenger, 95, 1

The Most Distant Galaxies

Max Pettini[1], Charles C. Steidel[2], Kurt L. Adelberger[2], Alice E. Shapley[2], Mark Dickinson[3], and Mauro Giavalisco[3]

[1] Institute of Astronomy, Madingley Road, Cambridge CB3 0HA, England
[2] Palomar Observatory, California Institute of Technology, MS 105–24, Pasadena, CA 91125, USA
[3] Space Telescope Science Institute, 3700 San Martin Drive, Baltimore, MD 21218, USA

Abstract. With the versatility and power of the VLT telescopes European astronomers are now set to take the study of the high redshift universe to new frontiers. Despite spectacular progress in the last few years in identifying normal galaxies at large look-back times, we have not yet isolated the epoch when the first stars formed and the universe was re-ionised. Even a $z = 3$, where galaxies have now been catalogued in their hundreds, detailed studies of their stellar populations await the availability of suitable instrumentation, such as ISAAC. In this article we review recent progress in the study of galaxies at high redshifts, focusing in particular on results which point the way for future work with the VLT.

1 Introduction

The discovery of galaxies at progressively higher redshifts has advanced in the last few years at a pace which must surely have surprised even the optimists among us. 'Normal' galaxies, with spectra which have close counterparts in the local universe, are now being identified at redshifts well in excess of $z = 5$ by a variety of techniques which include Lyα emission, broad-band surveys in the optical and infrared, searches at sub-mm wavelengths, follow-up of γ-ray bursts, finding 'old' galaxies at $z \sim 2$, as well as serendipitous discoveries of sources gravitationally lensed by foreground clusters of galaxies. Galaxies have now overtaken QSOs as the holders of the record for the most distant objects in the universe, and there may well be more to this development than simply chance due to small-number statistics, as we shall see in a moment. Claims have been made for galaxies at redshifts as high as $z \sim 7$ and possibly even $z > 10$.

While the search for the most distant galaxies satisfies a basic human curiosity, and knowing that such galaxies exist at all has far-reaching consequences for many aspects of observational cosmology, it is only through the study of large samples of objects selected according to well-defined criteria that we can throw light on the early stages of galaxy formation and growth of structure. Of the many avenues pursued in this search, it is the Lyman break technique—a particularly simple form of photometric selection tuned to the passage of the Lyman limit through a chosen filter—which has proved most effective in isolating large numbers of galaxies whose properties could then be studied in a wholesale manner. Given the space limitations, this review deals primarily with recent developments in the application of the Lyman break technique to ground-based surveys at redshifts $z \simeq 3$ and 4, where spectroscopically confirmed galaxies now number several hundreds (Steidel et al. 1999).

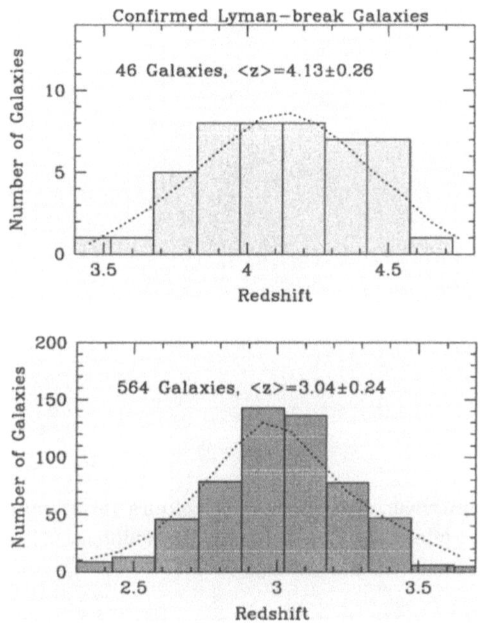

Fig. 1. Redshift histograms for spectroscopically confirmed Lyman break galaxies as of February 1999. The dotted curves are the predicted redshift distributions for our model selection function; this model is designed to reproduce at the same time the observed colour distribution as a function of redshift.

2 Large-Scale Properties of Lyman Break Galaxies

2.1 The Colour of Lyman Break Galaxies

The sample of Lyman break galaxies has now grown to a size sufficiently large to determine the selection functions of our surveys and address quantitatively the issue of incompleteness. Using Monte-Carlo simulations we have obtained estimates not only of the usual "detection incompleteness" due to the declining probability of detecting a galaxy at the faintest magnitudes (because of photometric errors and blending with foreground objects), but also for "colour incompleteness" (if photometric errors place galaxies outside our colour selection window), and even "template incompleteness" which would arise if some fraction of high redshift galaxies had true colours that lay outside our selection window.

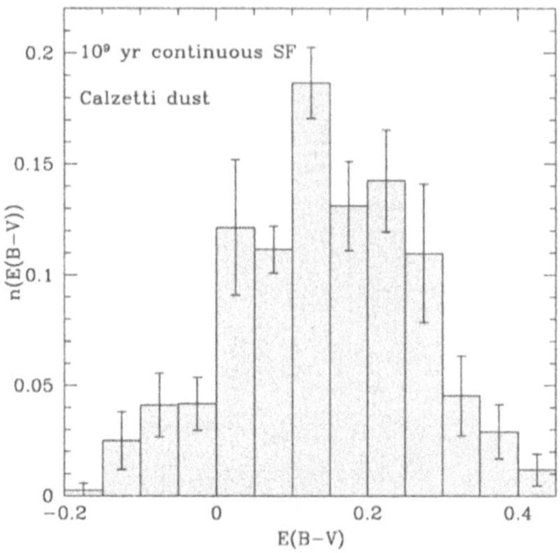

Fig. 2. The inferred intrinsic distribution of colours for Lyman Break galaxies at $z = 3$. The error bars reflect the field-to-field dispersion.

This last point is particularly important and merits further discussion. Figure 2 shows the distribution of intrinsic galaxy spectral shapes for the $z = 3$ sample of LBGs. In this treatment we have chosen to describe the distribution of UV spectral slopes in terms of the dust reddening parameter $E(B-V)$. This is clearly an oversimplification, in that some variation in the spectral slope may also result from different ages and fluctuations in interstellar and intergalactic absorption, but the exact physical causes of the observed spread of galaxy colours is irrelevant for the present purposes. The important point to get across is that we can estimate the incompleteness of our Lyman break selection technique as a function of $E(B-V)$ because we are sensitive to different amounts of reddening at different redshifts. Even after correcting for the bias against reddened objects, we find from Figure 2 that the comoving number density of galaxies with $E(B-V) \sim 0.4$ is less than one tenth of that of galaxies with $E(B-V) \sim 0.1$. Overall, the distribution of UV spectral slopes of LBGs at $z = 3$ is very similar to that of nearby star forming galaxies. The main conclusion from Figure 2 is that there seem to be few galaxies with such extreme red (or blue) colours to be excluded altogether by our $U_n, G, \mathcal{R}$ photometric selection criteria.

2.2 Luminosity Functions at $z = 3$ and 4

Armed with our incompleteness corrections we can now estimate the effective volume of our surveys as a function of apparent magnitude,

$$V_{\mathrm{eff}}(m) \equiv \int dz\, p(m,z) dV/dz, \qquad (1)$$

where $p(m, z)$ is the probability that a galaxy of apparent magnitude m at redshift z will be detected in our images and recognised as a Lyman break object, and $dz\, dV/dz$ is the comoving volume per square arcminute at redshift z. From these effective volumes it is then straightforward to compute luminosity functions from the observed surface densities of U_n and G dropouts.

The luminosity functions are reproduced in Figure 3 for an Einstein-de Sitter cosmology, together with data from the Hubble Deep Field North (Madau, Pozzetti, & Dickinson 1998). The agreement in both the shape and the normalisation of the bright end of the luminosity functions at $z = 3$ and $z = 4$ is evident from Figure 3—note that this result depends only mildly on the cosmology adopted. For all cosmologies considered, the observed UV-luminosity density integrated to the luminosity equivalent of $I_{AB} = 25.0$ at $z = 4.13$ (the completeness limit of our G-dropout sample) shows no significant change between $\langle z \rangle = 4.13$ and $\langle z \rangle = 3.04$ (the median redshifts of our surveys).

This conclusion is in apparent conflict with several analyses based upon the HDF-N alone, which have claimed a steep decline of the luminosity density with increasing redshift between these two epochs (by a factor of 4 in the original Madau et al. 1996 analysis, later refined to a factor of ~ 2.5 by Madau et al. 1998). However, as can be seen from Figure 3, there is really no conflict at the bright end of the luminosity function, where the ground-based and *HST* results overlap. The HDF still provides the only data on fainter LBGs at these high redshifts. So, one possibility is that the decline in the star formation rate density deduced by Madau and colleagues is due to a change in the faint end slope of the luminosity function between $z = 3$ and 4. Another explanation, which we consider more plausible, is that the HDF analyses may have underestimated the comoving density of star forming galaxies at $z = 4$ simply due cosmic variance which can be significant when probing volumes as small as that of the HDF. We already know from our ground-based surveys, which cover an area of sky 160 times larger than that of the HDF-N (although obviously to a shallower depth), that Lyman break galaxies are highly clustered at $z = 3$ (Adelberger et al. 1998). If we assume that the HDF B_{450}-dropouts have the same clustering strength as brighter LBGs at $z = 3$, we would expect an HDF-sized field ($\simeq 5$ arcmin^2) to contain an average of (23 ± 8) $z = 4$ candidates. The observed number of B_{450}-dropouts, 13, is therefore inconsistent with the expected number at only the 1.3σ level. We conclude that, even at the faint magnitudes probed by the HDF, the null hypothesis that the luminosity functions of galaxies at $z = 3$ and 4 are *the same* cannot be ruled out with much confidence.

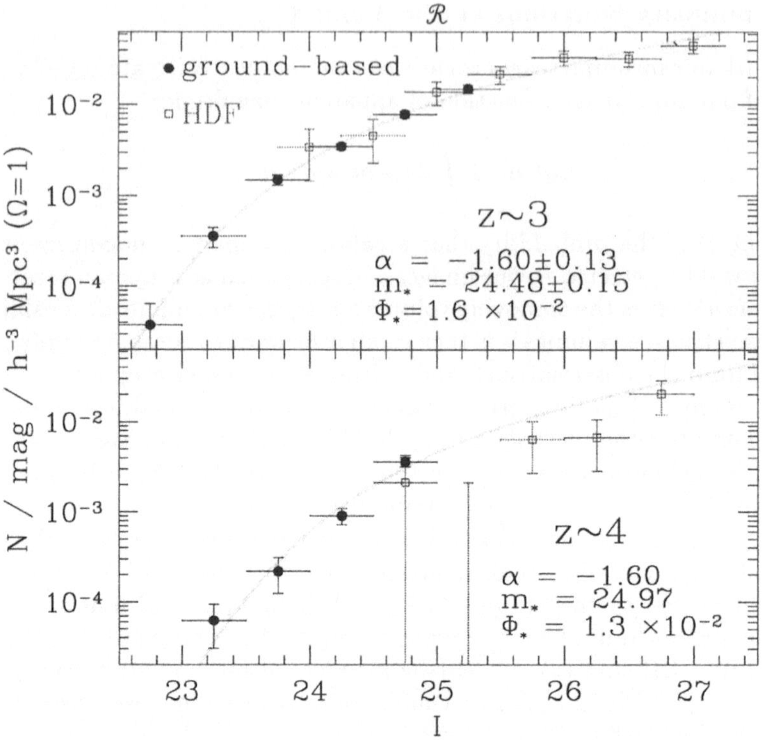

Fig. 3. Composite ground-based and HDF-N luminosity distributions for $z = 3$ and $z = 4$ LBGs. The results of a Schechter (1976) function fit to the $z = 3$ luminosity distribution is shown in the top panel. In the bottom panel, the curve shown is the same as the $z = 3$ LF shifted by 0.49 magnitudes (the relative distance modulus between $\langle z \rangle = 3.04$ and $\langle z \rangle = 4.14$ for an Einstein-de Sitter cosmology) with the normalisation adjusted to 80% of the $z = 3$ value. The error bars for the HDF-N points implicitly assume no clustering of the LBGs (whereas the ground-based points do include this effect) so that the "true" errors bars are probably larger.

3 Redshift Evolution of the Star Formation Rate

Figure 4 shows the results of integrating luminosity functions from surveys at different redshifts (as referenced in the caption) uniformly down to $0.1L^*$. This plot of the 'cosmic star formation history', first constructed by Madau et al. (1996), has changed considerably over the last couple of years and will undoubtedly undergo further revisions as future surveys, including those to be carried out with the VLT, improve our knowledge of distant galaxies. The details of the luminosity function are not the only uncertainty and the appropriate correction for dust obscuration is at least as important. The points in the lower panel of Figure 4 were obtained by assuming that Calzetti's (1997)

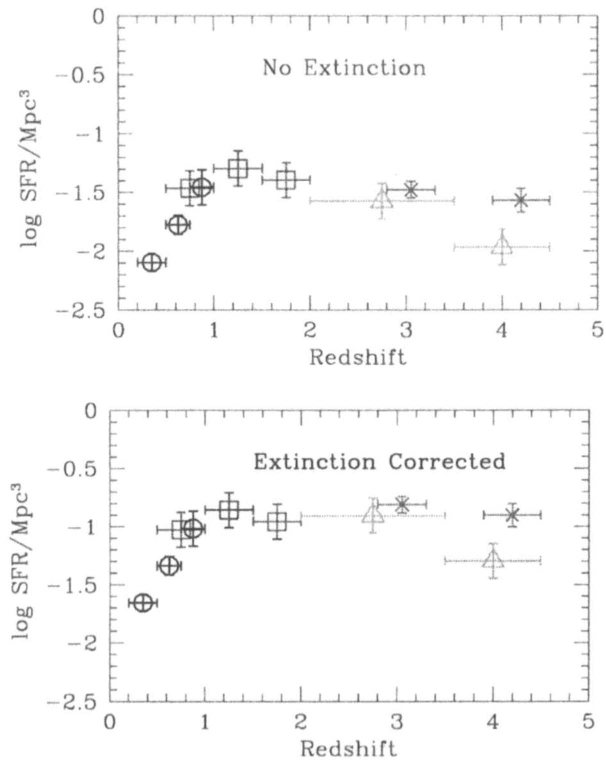

Fig. 4. The UV luminosity density as a function of redshift, following Madau et al. (1996) and using, like those authors, $H_0 = 50$ km s^{-1} Mpc^{-1}; $q_0 = 0.5$. The different data points are from the work by Lilly et al. (1996) [circles]; Connolly et al. (1997) [squares]; and Madau et al. (1998) [triangles]. The new measurements for LBGs at $z = 3$ and 4 are shown as crosses.

attenuation curve and the typical $E(B-V)= 0.15$ of LBGs at $z = 3$ (Figure 2) apply throughout.[1] This 'blanket' assumption actually turns out to be at least plausible; it reproduces the observed redshift distribution of the G-dropouts and is consistent with the degree of dust extinction deduced at $z \leq 1$ from the comparison of Hα and near-UV luminosities (e.g. Tresse & Maddox 1998; Glazebrook et al. 1999).

If, even with the above caveats, we assume that the lower panel in Figure 4 is approximately correct, several interesting consequences follow. First, the new data show little evidence for a peak in the universal star formation rate at $z = 1-2$ (as claimed by several authors, most recently Yan et al.

[1] The corresponding upward corrections to the star formation rate density are by factors of 4.7 at $z > 2$ and of 2.7 at $z < 2$, reflecting the relative extinction at 1500 Å and 2800 Å respectively.

1999). Rather, the emissivity for comoving volume due to young stars remains essentially flat for all redshifts $z > 1$. Thus it is not clear that 'epoch of galaxy formation' has yet been identified—the onset of substantial star formation in galaxies may well have occurred at $z > 4.5$. A flat behaviour in Figure 4 is actually not unexpected in hierarchical models of galaxy formation, and one can indeed find early predictions to this effect (e.g. White & Frenk 1991), but now apparently sets star-forming galaxies apart from AGN for which a peak in the comoving number density at $z = 2 - 3$ seems well established (e.g. Dunlop 1997; Shaver et al. 1998). One consequence of this redshift evolution in the relative numbers of star-forming galaxies and QSOs is that it now seems more likely that stars, rather than AGN, were responsible for re-ionising the universe at $z > 5$, although the escape fraction of Lyman continuum photons is still undetermined (Madau 1999).

As a consistency check, we may ask what is the total mass in stars produced over the Hubble time by integrating the star formation rate in Figure 4. For an Einstein-de Sitter cosmology we find:

$$\int_0^{13\ Gyr} \dot{\rho}_* \, dt \simeq 3.3 \times 10^8 \ M_\odot \ \mathrm{Mpc}^{-3} = 0.0043 \ \rho_{\mathrm{crit}} \approx \Omega_{\mathrm{stars}} \rho_{\mathrm{crit}} \qquad (2)$$

where $\rho_{\mathrm{crit}} = 7.7 \times 10^{10} \ h_{50}^2 \ M_\odot \ \mathrm{Mpc}^{-3}$ and Ω_{stars} is the fraction of the closure density contributed by stars at $z = 0$ (Fukugita, Hogan, & Peebles 1998). Thus, within the rough accuracy with which this accounting can be done, the star formation activity we see is apparently sufficient to produce the entire stellar content, in disks and spheroids, of the present day universe. Note also that $\approx 1/4$ of today's stars were made before $z = 2.5$ (the uncertain extrapolation of $\dot{\rho}_*$ beyond $z = 4$ makes little difference). As pointed out by Pettini (1999), the metal production associated with the star formation activity before $z = 2.5$ is sufficient to enrich the entire baryonic content of the universe to an average metallicity $Z \approx 1/25 Z_\odot$. Only a fraction of these metals has been detected directly, however. Damped Lyα systems, the Lyα forest, and the Lyman break galaxies themselves perhaps account for $\approx 20\%$ of the comoving density of metals produced by $z = 2.5$. Since metal-enriched outflows appear to be a common feature of starburst galaxies at high as well as redshifts (Heckman et al. 1998; Pettini et al. 1998), the 'missing' metals may well be found far from their production sites, possibly in hot gas in galactic halos and proto-clusters.

4 Narrow-Band Imaging

We have recently explored the possibility of extending searches for high-z galaxies to fainter magnitudes than those reached with the Lyman break technique, by targeting Lyα emission. Although surveys based on Lyα emission were generally unsuccessful in the 1980s, we now realise that detection of this spectral feature was often just below the thresholds which could be reached with the instrumentation available at that time. More recent efforts have been rewarded with the discovery of the most distant galaxies known at present (e.g. Hu, McMahon, & Cowie 1999). Our pilot observations have been

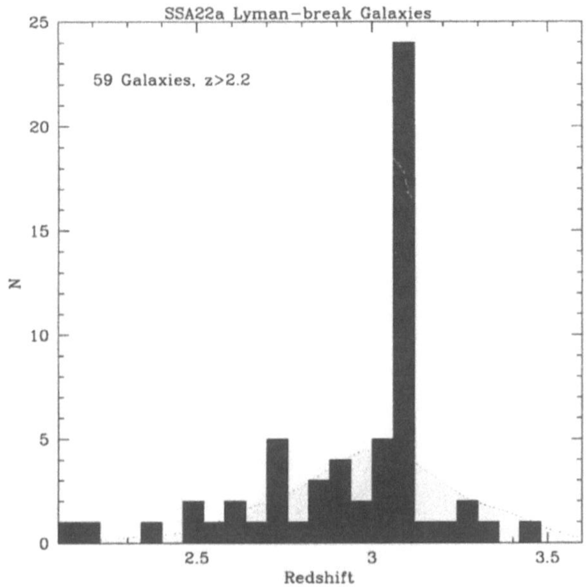

Fig. 5. Redshift histogram of spectroscopically confirmed Lyman break galaxies in the 9×9 arcmin field also imaged in the narrow-band 4970/80 filter. Redshifts have been binned so that the prominent "spike" is centred at $\langle z \rangle = 3.09$, the median redshift of the galaxies in the structure, with a bin width corresponding to the redshift range probed by the 80 Å wide filter at $z \sim 3$. The density enhancement relative to the overall survey selection function (shown by the light, smooth histogram) is a factor of 6 ± 1.2 (Steidel et al. 1998).

directed towards the SSA22 field where we had previously discovered a high over-density of galaxies at $\langle z \rangle = 3.09$ (see Figure 5) . Using a specially designed interference filter with a bandpass centred at 4970 Å (corresponding to Lyα at $z = 3.09$) and bandwidth FWHM = 80 Å, we imaged a $\sim 9 \times 9$ arcmin region of the sky (sampled with 2048 x 2048 0.283 arcsec pixels) for a total integration time of 15.8 hours using the prime focus camera of the Palomar 200-inch telescope; the seeing in the stacked image was 1.2 arcsec FWHM. This image is sufficiently deep that galaxies belonging to the structure at $\langle z \rangle = 3.09$ can be identified not only by an excess of flux (this being the conventional strategy of narrow-band searches), but also by a *deficit* of signal in the narrow-band image relative to the continuum, depending on whether Lyα is in emission or in absorption. As a result, we have now identified 162 ob-

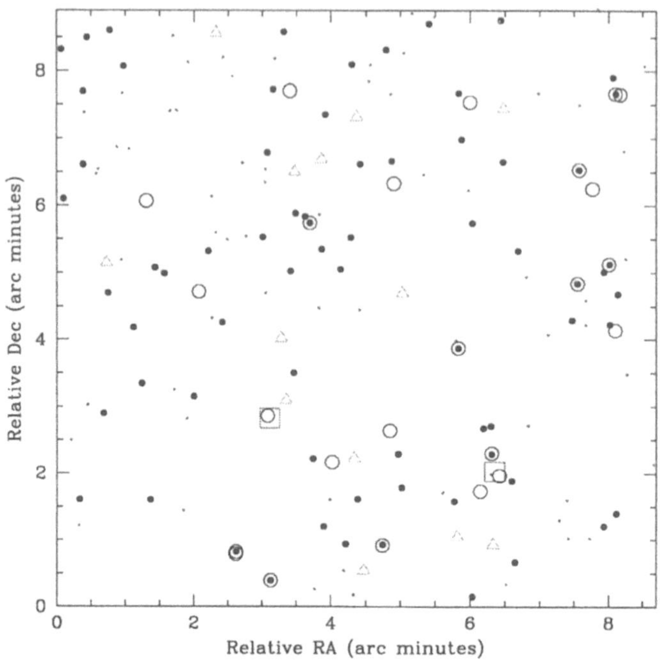

Fig. 6. A map (North is up, East to the left) of the confirmed and suspected "spike" objects in the SSA22 field. Open circles represent LBGs with spectroscopic redshifts in the range $3.06 \leq z \leq 3.12$; larger and smaller dots represent respectively brighter and fainter narrow-band excess objects; open triangles are candidates selected on the basis of flux *deficit* in the narrow-band image (Lyα is in absorption); and the large open squares are the two giant Lyα line emitting "blobs". In the volume probed there is a total of 162 objects which are either known or likely to be members of the over-density at $\langle z \rangle = 3.09$.

jects which are either confirmed or highly likely members of this proto-cluster; their distribution in the 9 × 9 arcmin field is shown in Figure 6. The density of emission line candidates is approximately six times higher than that of published 'blind' narrow-band searches to comparable depth and at similar redshifts (Cowie & Hu 1998), but this is the same degree of density enhancement which we had found to apply to Lyman break galaxies (that is objects selected in the UV continuum) in this region (Steidel et al. 1998). Thus it appears that the overdensity in the "spike" persists to galaxies at least two magnitudes fainter in the continuum than the faint limit of the Lyman break technique. Although capable of reaching further down the luminosity function, Lyα imaging is clearly *not* the best way to assemble complete samples of high redshift galaxies. As is now well know (e.g. Charlot & Fall 1993; Chen & Neufeld 1994; Kunth et al. 1998), Lyα is subject to complicated selection effects, so that its strength is not well correlated with more physically relevant quantities, such as the UV continuum luminosity. From our data it appears that only about 20 – 25 % of all galaxies at a given UV continuum luminosity would be flagged as narrow-band excess objects based on the typical rest frame equivalent width limit of $W_0(\text{Ly}\alpha) \simeq 80$ Å; in the remainder Lyα emission is too weak to make narrow-band selection effective. This fraction does not appear to depend significantly on continuum luminosity—we find no evidence for a greater fraction of large Lyα line equivalent widths at faint continuum luminosities, despite a widespread perception that this may be the case.

On the other hand, narrow-band searches can be extremely useful for mapping out, in an albeit incomplete way, known or suspected structures at high redshift, thanks to their sensitivity to fainter sources. The surface density of emission line objects within over-densities can be extremely high, as can be appreciated from Figure 6, allowing detailed topological analysis of the galaxy distribution at high redshifts.

4.1 Giant Lyα Nebulae

And of course there is always the thrill of the unexpected discovery, whenever a new technique is tested out in earnest. Two out of the 162 objects in Figure 6 are not point sources, but are diffuse patches of Lyα emission which have sizes of $\sim 100\,h^{-1}$ kpc ($\approx$ 15 arcsec angular size) and Lyα luminosities of $\sim 10^{44}$ erg s^{-1}—that is they are 20 – 40 times larger and more luminous than the typical Lyα emitter in the field. These emission 'blobs' are shown on an expanded scale in Figure 7. The only previously known examples of such Lyα nebulae are associated with high redshift radio galaxies (McCarthy 1993; van Ojik et al. 1997), but the 'blobs' in SSA22 are radio-quiet to a detection level three orders of magnitude below that typical of powerful high redshift radio galaxies! Even optically, the discrete objects positionally coincident with, or near, the blobs (see Figure 7) have continuum luminosities with are two to three magnitudes fainter than powerful radio galaxies at $z = 3$. The nature of these nebulosities remains unclear. Possibly they are excited by continuum sources that are heavily obscured along our line of sight—K_s-band imaging with NIRC does show a very red source spatially coincident with the central knot of Blob 1 (Steidel et al. 2000). Alternatively, they may be high redshift analogues of the massive cooling flows observed in X-ray clusters at low redshift, although the mass deposition rate implied by the Lyα luminosities may be unreasonably high unless each baryon recombines many times (Heckman et al. 1989). Whatever their nature, we may have stumbled on a phenomenon which, for all we know, may be more common than realised hitherto (given the lack of suitable observations) and which, like QSOs and other AGN, may mark out regions of large matter over-density.

Inevitably, this brief review has been very selective and there has been little opportunity to mention many other aspects of the study of the high

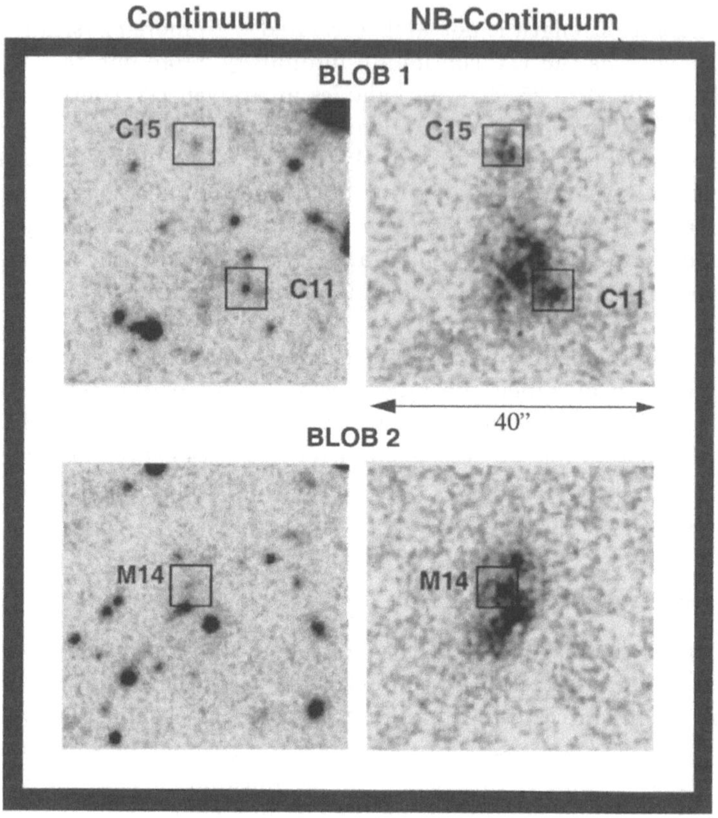

Fig. 7. Portions of the continuum and continuum-subtracted narrow-band images showing the two prominent Lyα emitting nebulae. The boxes indicate Lyman break galaxies which are known members of the $z = 3.09$ overdensity.

redshift universe—such as QSO absorption line spectroscopy, for instance—which have been highly successful in the last few years and produced results just as exciting as those described here. What is clear, however, is that we stand at the brink of a new era in observational cosmology when, with the VLT and other large telescopes, we will at last be able to unravel the properties of the first stars and galaxies which signalled the end of the 'dark age' of our universe.

Max Pettini would like to express his sincere thanks to the organisers of this meeting for a most enjoyable week in Antofagasta, and his congratulations to everyone associated with the VLT project which truly heralds the beginning of a new era for European astronomy.

References

1. Adelberger, K.L., Steidel, C.C., Giavalisco, M., Dickinson, M., Pettini, M., & Kellogg, M. 1998, ApJ, 505, 18
2. Calzetti, D. 1997, AJ, 113, 162
3. Charlot, S., & Fall, S.M. 1993, ApJ, 415, 580
4. Chen, W.L., & Neufeld, D.A. 1994, ApJ, 432, 567
5. Connolly, A.J., Szalay, A.S., Dickinson, M., SubbaRao, M.U., & Brunner, R.J. 1997, ApJ, 486, L11
6. Cowie, L.L., & Hu, E.M. 1998, AJ, 115, 1319
7. Dunlop, J. 1997, in Cosmology with the New Radio Surveys, ed. M. Bremer et al. (Dordrecht: Kluwer) (astro-ph/9704294)
8. Fukugita, M., Hogan, C.J., & Peebles, P.J.E. 1998, ApJ, 503, 518
9. Glazebrook, K., Blake, C., Economou, F., Lilly, S., & Colless, M. 1999, MNRAS, 306, 843
10. Heckman, T.M., Baum, S.A., van Breugel, W.J.M., & McCarthy, P.J. 1989, ApJ, 338, 48
11. Heckman, T.M., Robert, C., Leitherer, C., Garnett, D.R., & van de Rydt, F. 1998, ApJ, 503, 646
12. Hu, E.M., McMahon, R.G., & Cowie, L.L. 1999, ApJ, 522, L9
13. Kunth, D., Mas-Hesse, J.M., Terlevich, E., Terlevich, R., Lequeux, J., & Fall, S.M. 1998, A& A, 334, 11
14. Lilly, S.J., LeFèvre, O., Hammer, F., & Crampton, D. 1996, ApJ, 460, L1
15. Madau, P. 1999, this volume (astro-ph/9907268)
16. Madau, P., Ferguson, H.C., Dickinson, M. Giavalisco, M., Steidel, C.C. & Fruchter, A. 1996, MNRAS, 283, 1388
17. Madau, P., Pozzetti, L., & Dickinson, M. 1998, ApJ, 498, 106
18. McCarthy, P.J. 1993, ARAA, 31, 639
19. Pettini, M. 1999, in Chemical Evolution from Zero to High Redshift, ed. J. Walsh & M. Rosa (Berlin:Springer), in press (astro-ph/9902173)
20. Pettini, M., Kellogg, M., Steidel, C.C., Dickinson, M., Adelberger, K.L., & Giavalisco, M. 1998, ApJ, 508, 539
21. Schechter, P. 1976, ApJ, 203, 297
22. Shaver, P., Hook, I.M., Jackson, C.A., Wall, J.V., & Kellerman, K.I. 1998, in Highly Redshifted Radio Lines, ed. C. Carilli et al. (astro-ph/9801211)
23. Steidel, C.C., Adelberger, K.L., Dickinson, M., Giavalisco, M., Pettini, M., & Kellogg, M. 1998, ApJ, 492, 428
24. Steidel, C.C., Adelberger, K.L., Giavalisco, M., Dickinson, M., & Pettini, M. 1999, ApJ, 519, 1
25. Steidel, C.C., Adelberger, K.L., Shapley, A.E., Pettini, M., Dickinson, M., & Giavalisco, M. 2000, ApJ, in press (astro-ph/9910144)
26. Tresse, L., & Maddox, S.J. 1998, ApJ, 495, 691
27. van Ojik, R., Roettgering, H.J., Miley, G.K., & Hunstead, R.W. 1997, A&A, 317, 358
28. White, S.D.M., & Frenk, C.S. 1991, ApJ, 379, 25
29. Yan, L., McCarthy, P.J., Freudling, W., Teplitz, H.I., Malamuth, E.M., Weymann, R.J., & Malkan, M.A. 1999, ApJ, 519, 47

Nuclear Activity and Galaxy Evolution

George Miley

Leiden Observatory, Postbox 9513, 2300 RA Leiden, The Netherlands

Abstract. This talk discusses the connection between nuclear activity and galaxy evolution. First, the evidence for a fundamental connection is reviewed. Then, the properties of distant radio galaxies are summarized. It is argued that these objects play a particularly important role in galaxy evolution. The last part of the talk addresses some fundamental questions that will be tackled during the next decade with the VLT, VLTI and other available instruments.

Preamble

This is no ordinary meeting. We are here to celebrate the VLT. As a member of the astronomical community who has had the privilege of visiting Paranal several times over the last 5 years, I know that the VLT would not have become the marvelous machine that we see today without enormous sacrifices, by a large number of people, way beyond the call of duty.

So before going to the subject of my talk, I pay tribute to the staff of ESO at every level, DGs, engineers, scientists, managers and administrators and to the hundreds of workers from Chile and elsewhere who joined together to obtain funding, inspire, design and build the VLT.

Thank you all very much.

1 Introduction

The subject of my talk is the connection between nuclear activity and galaxy evolution. I shall first discuss the evidence that a fundamental connection exists. Then I shall proceed to describe some properties of distant radio galaxies and argue that these objects play a particularly important role in galaxy evolution. The last part of my talk will address some relevant fundamental questions that will be tackled during the next decade with the VLT, VLTI and other instruments available to us.

We know for more than half a century that violent activity can exist in the nuclei of galaxies. In 1943 Carl Seyfert detected broad emission lines from the nuclei of a handful of galaxies, with line widths corresponding to velocities of several thousand km/s. A few years later radio galaxies were discovered. Their symmetric double-lobed appearance was attributed to explosive events in the nuclei of their host galaxies. Since then studies of the diverse species of active galaxies have become an important branch of astrophysics.

Since the discovery of active nuclei, the prevalent view amongst astronomers has been that these objects are important laboratories for studying high-energy astrophysical phenomena and useful tracers of the high-redshift Universe. AGN, as relatively rare and exotic objects, were not regarded as very important in the evolution of the general population of galaxies. Recently this view has begun to change.

2 Importance of Nuclear Activity in Galaxy Evolution

There are now several pieces of evidence that nuclear activity plays a fundamental role in the evolution of galaxies.

(i) Nuclear activity is widespread. It is seen not only in quasars, radio galaxies and other species of **active** galaxies, but signatures of nuclear activity (e.g. weak radio cores, faint emission line regions/liners) are seen frequently in apparently "normal" galaxies. In addition, dynamic evidence of present or previous nuclear activity (massive black holes) has been found in several otherwise "passive" galaxies (e.g. M31, M33, Milky Way).

(ii) Black hole masses and spheroids. A correlation has been reported between the deduced masses of the black holes in galaxies and the luminosities/masses of the associated ellipticals and bulges (e.g. Richstone et al. 1998). Assuming that the black holes are the remnants of previous quasar activity, this indicates a connection between the formation of the spheroidal component and the birth of quasars.

(iii) The quasar population explosion. During the first 20% of the lifetime of the Universe, between 1.5 and 2.5 Gyr after the Big Bang, there was a dramatic increase and decrease in the population of bright quasars and radio galaxies – almost a delta function on a linear-linear plot of their space density versus time. Their peak space density was several hundred times greater than the present value. It is probable that every large galaxy underwent at least one period of quasar activity during this epoch, the "quasar era" of the Universe, $3 > z > 1.5$.

(iv) Relation of global star formation to AGN evolution. Recent studies of the global change in the star formation rate indicate that most galaxies formed their first stars between 1 and 2 Gyr, i.e. close to or slightly before the peak of the quasar era (see talk by Piero Madau at this symposium). Just as with the behaviour of the AGN space density, the global star formation rate declined by at least an order of magnitude between the quasar epoch and the present. Again it is particularly revealing to view this behaviour on a linear-linear plot in contrast to the log-log plot on which it is usually displayed.

(v) Evidence for jet-induced star formation. About a decade ago, it was found (surprisingly) that the rest-frame optical/UV continuum light of distant radio galaxies ($z \gtrsim 0.6$) is roughly aligned with the axes of their associated radio sources (Chambers et al. 1987; McCarthy et al. 1987). The first explanation of this alignment effect was that it was due to shocks associated with the outward propagation of the radio jets that trigger or enhance star formation (e.g. Rees 1989). Other mechanisms that were proposed subsequently include (i) scattering of light from a hidden quasar by electrons or dust and (ii) nebular continuum emission associated with strong emission line regions associated with the radio jet. No single mechanism for the alignment effect can account for all the observed properties. For example, the frequent occurrence of optical polarisation suggests that dust scattering can produce an aligned component, while such a model cannot explain similar bending behaviour sometimes observed between the optical and radio structures or several cases where the infrared continuum emission redwards of the rest frame 4000 A break is observed to be aligned with the radio source. Moreover, Keck spectra of a few high-redshift radio galaxies (e.g. 4C 41.17 at $z = 3.8$) show absorption features similar to those of nearby starburst galaxies indicating that distant radio galaxies are sometimes undergoing substantial star formation. Taken together these data suggest that jet-induced star formation is significant in distant radio galaxies.

Studies of several low-redshift radio galaxies provide further strong evidence that jet-induced star formation can occur. These include NGC 541/Minkowski's Object (3C40) (van Breugel et al. 1985), 3C 285 (van Breugel and Dey 1993) and the closest radio galaxy Centaurus A (Graham 1998). Such effects can be expected to be much more prevalent during the quasar era of the Universe, when every galaxy may well have undergone quasar activity.

Taken together, the results summarized above provide overwhelming evidence that *nuclear activity has been an important factor in the evolution of the general galaxy population.* The corollary is that any successful model of galaxy evolution must provide an explanation of the effects and role of nuclear activity. A more speculative conclusion is that nuclear activity (jet-induced star formation) may well have played a significant role in triggering or enhancing star formation in galaxies during the first 3Gyr after the Big Bang.

3 Distant Radio Galaxies as Probes of Galaxy Evolution

In the second part of my talk I shall concentrate on distant radio galaxies and attempt to convince you that they represent a particularly important case of nuclear activity in galaxy evolution. High redshift radio galaxies (HzRGs, $2 < z < 5$) are important laboratories for studying the early Universe (for reviews see McCarthy 1993; Röttgering and Miley 1996). A comprehensive survey of the field can be found in the proceedings of the recent Amsterdam Colloquium on "The Most Distant Radio Galaxies" (eds. Röttgering et al. 1999). Amongst the reasons for the cosmological interest of HzRGs are:

(i) as probes of massive galaxy formation. From their K-band luminosities it can be inferred that HzRGs are amongst the most massive distant galaxies. There is mounting evidence that they are undergoing vigorous star formation and that they evolve into massive ellipticals in the local Universe. In several HzRGs the UV continuum is dominated by young stars, with estimated star-formation (SF) rates of up to 1000 $M_\odot$ per year (Dey et al. 1997). Sub-millimetre continuum dust emission has been solidly detected in 4 of the highest redshift galaxies (Röttgering et al. 1999) which also imply similar large SF rates. On dynamical time scales $\sim 10^8$ years) the stars would settle and evolve into a fully developed ellipticals at $z \sim 1$. Indeed, Best et al. (1997a, 1997b) concluded that the magnitudes, colours and profiles of 3CR radio galaxies at $z \sim 1$ are consistent with the host of the radio galaxy being a massive $M_\star \sim 5 \times 10^{11}$ $M_\odot$ elliptical galaxy that formed at redshifts of 3–5.

(ii) as probes of proto-clusters. The evidence that HzRGs are often located in dense (proto-) cluster environments is strong and includes (i) an excess of companion galaxies (Röttgering et al. 1996; Pascarelle et al. 1996; Le Fevre et al. 1996); (ii) the measurement of large radio rotation measures (RM) for the HzRGs, indicating that their host galaxies are surrounded by hot magnetized cluster gas (Carilli et al. 1997; see also Athreya et al. 1998); (iii) the X-ray detection of the HzRG with the highest RM, 1138-262 at $z = 2.2$ (Carilli et al. 1998), indicating that this object is located in a hot cluster gas. The luminosity is comparable to that of a nearby rich cluster.

There is additional strong evidence that distant radio galaxies are the massive central elliptical galaxies at the centres of proto-clusters of galaxies. The HST images of several HzRGs (Pentericci et al. 1997; van Breugel et al. 1998) show tens of continuum clumps embedded in giant Ly-alpha halos, reminiscent of simulations of forming cD galaxies, according to hierarchical models. Following the well-known adage and popular version of Occam's razor that something that "waddles like a duck and quacks like a duck is a duck", I shall henceforth assume that ***distant radio galaxies are forming massive galaxies in proto clusters.*** Because radio galaxies have been observed out to redshifts > 5, they are potential tracers of the earliest proto-clusters.

4 Diagnostic Power of Distant Radio Galaxies

Radio galaxies have several distinct emitting components which provide diagnostics about different physical constituents of the early Universe. A non-exhaustive list of these is given in Table 1, together with the main facilities that will be accessible by us during the next few years for studying them. Besides serving as luminous beacons, the components are spatially resolvable from the ground and therefore provide geometrical information about the internal processes within the objects. The fact that these various diagnostics are *simultaneously* present and that the *interrelationships and interactions between them* can be studied make distant radio galaxies unique laboratories for probing the early Universe.

Table 1. Diagnostics from distant radio galaxies

OBSERVABLE COMPONENT	CONSTITUENT	PROPERTIES	MAIN OBSERVATIONAL FACILITY
UV/Optical Continuum	Stars	Stellar Populations and their Evolution	VLT: Optical (FORS) VLT: Near IR (ISAAC) HST: (Advanced Camera /ACS)
	Dust	Scattering Properties	
Optical Emission Lines	Ionized Gas ($10^4 - 10^5$ K)	Kinematics, Mass, Densities, Temperatures, Ionization, Ages	VLT: Optical (FORS, UVRES) VLT: Near IR (ISAAC)
Optical Absorption Lines	Neutral Hydrogen (Lyman α)	Kinematics, Mass, Column Density	VLT: Optical (FORS + UVRES) VLT: Near IR (ISAAC)
IR - Millimetre Continuum	Dust	Temperature, Mass	JCMT: mm (SCUBA), mm arrays VLT: Mid-IR (VISIR)
IR- Millimetre Lines	Molecular Gas	Temperatures, Densities, Masses, Stellar Evolution	JCMT, mm arrays VLT (VISIR)
Radio Continuum	Relativistic Plasma	Magnetic Fields, Ages, Energetics, Pressures, Particle acceleration, Jet collimation and propagation	VLA: Imaging + Polarization VLBI: Imaging (EVN, JIVE)
	Thermal Plasma	Densities, magnetic fields, (Faraday Depolarization)	
Radio Absorption Lines	Neutral Hydrogen	Kinematics, Spin temperatures, Column Densities	Westerbork, VLA, VLBI (EVN, JIVE)

An efficient technique for finding the high redshift galaxies is to concentrate on objects with ultra-steep radio spectra (USS). Now more than 150 radio galaxies with redshifts > 2 are known and most of them were found using the USS technique, many with the aid of an ESO Key Programme. They include a recently discovered object with a probable redshift of 5.2, which would be the most distant AGN known at this time (van Breugel et al. 1999).

5 Some Outstanding Questions

There are many open questions about nuclear activity and its place in galaxy evolution. Several of these will be addressed by the VLT, by VLTI and by the battery of other new instruments that will be accessible to astronomers during the next decade, including the Advanced Camera for Surveys (ACS) due to be fitted to the HST in late 2000, the LSA/MMA and the NGST. The VLT will be particularly powerful in studying the star formation history of radio galaxies and for investigating the kinematics and physical conditions in their halos of ionized and neutral gas (Röttgering and Miley, 1996).

Slit spectroscopy will address the nature of the stellar populations, the kinematics and physical conditions within the gas halos and multi-object spectroscopy at the highest redshifts with FORS will be used to search for clustering around HzRGs. Multicolour imaging and narrow-band imaging with e.g. FORS and ISAAC will also investigate the environment of HzRGs and measure the statistics of Lyα and Hα galaxies. VLTI will probe the inner nuclear regions of AGN, delineating the dust, synchrotron jets and emission line regions closer to the massive black hole powerhouse.

Some of the relevant fundamental questions include:

(i) Are HzRGs located in dense (proto-) clusters and if so what are the properties of the associated large-scale structures? Searches for cluster galaxies at similar redshift to the HzRG can be carried out via the Lyman break technique with deep FORS ISAAC and ACS/HST imaging in several colours of the surrounding regions. In addition, the VLT will provide two complementary tools for detecting and studying high-redshift clusters. Spectroscopy in Lyα and Hα will obtain redshifts for faint Lyman-break and 4000 Å - break objects and narrow-band imaging in Lyα and Hα will detect faint emission-line galaxies at suspected cluster redshifts. The forthcoming X-ray satellites AXAF and XMM will allow the hot gas in the associated clusters to be detected and spatially resolved out to redshifts of ~ 4. Follow-up spectroscopy and imaging of cluster candidates would be an exciting new diagnostic of the state of the early Universe. For example, measured velocity dispersions > 1000 km/s for a $z > 2$ cluster would imply a mass as large as $M \geq 10^{14}\ M_\odot$ and values of $\Omega \ll 1$ and/or $\Lambda > 0$.

(ii) Are HzRGs massive central-cluster galaxies formed via the hierarchical assembly of Lyman break galaxies? The clumpy morphologies seen in the HST images of HzRGs are ***strikingly similar to simulations of forming brightest cluster galaxies***, made on the basis of the hierarchical models. The sizes, profiles and luminosities of the clumps are consistent with those of Lyman break galaxies.

By providing information on the ***spatial distribution of stellar populations and their kinematics***, the VLT should constrain the history of HzRG formation. Keck spectra on a few objects have already revealed evidence of young stars. It is particularly important to make deeper optical and infrared spectra across several distant radio galaxies with the VLT to define their properties and compare these with those of the dropout objects and HST UV spectra of nearby starburst galaxies. Analysis of the UV spectra of starburst galaxies show that while low-ionization lines are emitted by the ISM, strong high ionization resonance lines such as SiIV $\lambda1400$ and CIV $\lambda1550$ are probably due to stellar winds (e.g. Heckman 1998) and may well have P-Cygni profiles.

The HST and VLT will map the profiles, colours and colour distributions of the clumps in detail. Specifically, the range of colour distributions for the aligned component and the central galaxy will be measured. If the clumps in HzRGs are redder than the Lyman break galaxies this would be evidence that they are older and that galaxies in a cluster environment form earlier than those in the field.

(iii) What are the origin and fate of giant ($\sim$ 100 kpc) gas halos associated with HzRGs and what role do they play in the formation of massive galaxies? One of the most remarkable features of distant radio galaxies is the 100 kpc-sized luminous halos of ionized gas that surround them (e.g. van Ojik et al. 1997). It is still unclear whether the gas halo is the product of accretion or cooling flow processes and to what extent and how the gas is eventually converted to stars. A gas halo has a characteristic mass of 10^9 $M_\odot$ and is typically composed of 10^{12} clouds, each having a size of about 40 light days, i.e. comparable with that of the solar system. Individual clouds may well delineate an early formation stage of stars.

Deep Absorption features in the Lyα profiles of more than half the cases studied and 15% have $N_{HI} > 10^{19}$ cm^{-2}. The Lyα absorption provides a tool for studying and spatially resolving neutral gas, which we have shown to also be associated with the large ionized gas halos.

These large gas halos are an important diagnostic for studying the kinematic properties of gas in the early Universe. For example, systematic velocity gradients found in the faint outer regions of 1243+036 with $z = 3.6$ (van Ojik et al. 1996) indicate the presence of a disk-like structure with a spatial scale of 150 kpc, rotating around a dynamical mass of 10^{12} $M_\odot$.

Lyman α studies by the VLT can (i) measure rough kinematics in the fainter quiescent regions of the halos to constrain the dynamical masses and (ii) separate and measure kinematics for the ionized (emission) and neutral (absorption) components of the gas in the inner regions. Studying the relationship of the ionized and neutral gas will constrain geometrical and kinematics models for the gas halos.

In addition, the VLT can make slit spectroscopic measurements of fainter lines than hitherto was possible. In the optical FORS will measure the spatial distributions of profiles for lines such as CIV λ1549, CIII] λ1909, with resolutions of 300 and above. A comparison of intensities, profiles and spatial distributions for the various species will constrain the properties of the emitting and absorbing gas. In the infrared ISAAC will provide comparison data for Hα and [OIII].

The detailed ionization properties of the gas clumps can be compared with the distributed stellar population data. Comparison between VLT measurements of spatially-extended emission and absorption lines will constrain the geometry and ionization structure of the gas (van Ojik et al. 1997). By establishing how much of the ionization is caused by nuclear photoionization and how much is due to localized "in situ" phenomena, the stellar populations and properties of an eventual obscured AGN can be constrained. One of the most important questions to investigate is whether and to what extent the gas in the halos is being converted into stars.

(iv) Is nuclear activity a source of star formation in the early Universe? Following my discussion in Section 2, it is important to investigate and quantify the link between nuclear activity and star formation. During the next few years the VLT and HST can be used to search for differences in properties between the clumpy material along the radio jet and elsewhere in the host galaxy. The excellent polarimetric characteristic and high sensitivity of FORS and ISAAC are well suited to determining the wavelength dependence of polarization in a sample of distant radio galaxies, thereby disentangling the relative contributions of scattering and direct starlight to the alignment effect in HzRGs. By demonstrating the presence of a young stellar population associated with synchrotron jets, the VLT could provide unambiguous evidence that jet-induced star formation is occurring at high redshifts and a definite link between AGN and star formation. Additional information on this question could be provided by stellar population studies of regions in nearby radio galaxies in which star formation is associated with radio jets. For example ACS/HST studies of Centaurus A should constrain the IMF produced in jet-induced star formation processes.

(v) Why are powerful radio sources associated with AGN in ellipticals but not with AGN in spiral galaxies? Although we see several of the signatures of AGN in spiral galaxies (e.g. broad and narrow emission line regions massive central regions), no powerful radio sources are observed in spiral hosts. This longstanding but unsolved question may imply differences in the characteristics of black holes in disk galaxies or in the properties of the circumnuclear medium. Possible explanations are that in spiral galaxies (i) the black hole rotation axes do not remain fixed in direction for long enough for radio jets to be formed (ii) insufficient material is accreted or (iii) propagation of the jet is impeded. To investigate such matters optical/IR studies of the nuclear regions with high spatial resolution (e.g. VLTI) will be extremely important (e.g. Miley, 1996).

VLTI baselines (to $\sim$ 130 m) will give structural information on scales down to $\sim$ 15 mas at N-band, 3.5 mas at K-band and 0.9 mas at V-band. The corresponding maximum linear resolution is of course a function of redshift and varies from 100 pc for high-redshift objects at N-band to better than 0.1 pc for a nearby AGN at $z \sim 0.001$. Taking their characteristic linear sizes into account, three of the nuclear "building blocks" of AGN are particularly relevant for study with VLTI. These are (i) the mid-IR warm dust component, (ii) synchrotron jets and (iii) nearby broad-line regions. Longer baselines will be needed eventually, particularly for probing the morphologies and kinematics of the broad-line regions.

In the case of the radio AGN, kinematic information and its relation with the radio morphologies would provide information about the details (energetics etc) of how superluminal synchrotron jets interact with the nuclear gas. The most fundamental driver for studying the kinematics of the BLR is the information it might provide about the enclosed mass on the scale of a few tenths of parsec, M and the resultant constraints on the presumed black hole at the centre. Taking the simplified formula, the mass, $M \sim Rv^2/G$, where v is the velocity measured at radius R. Scaling R by the resolution difference between VLTI and the HST, would imply that VLTI can measure masses a hundred times more effectively.

The usefulness of VLTI for studying AGN will be determined by the sensitivity that can be attained. Because of the ability to combine phase referencing with adaptive optics (e.g. with PRIMA), AGN within a few tens of arcseconds from bright stars should be observable with sensitivities comparable with those of the individual 8m telescopes and promise to be fruitful targets for such studies.

(vi) At what epoch did the first AGN and their parent massive black holes form? There are several observational tools for exploring this intriguing question. First, on the global scale, the evolution of AGN space density and star formation must be more accurately determined and compared. Much larger samples of high-redshift AGN and other distant galaxies are needed and will be obtained. These will allow the VLT and other large optical/IR telescopes to push to larger redshifts.

In this context I mention the technique that we have been using to pinpoint the most distant radio galaxies by concentrating on radio sources with extremely steep radio synchrotron spectra. We have recently begun a study of a "hyper-steep" spectra sample of radio sources ($\alpha < -1.3$; $S_\nu \sim \nu^\alpha$), about 1% of radio sources in low-frequency surveys. Confirmation of the power of this technique is provided by Keck observations carried out in December 1998 (van Breugel et al. 1999). A radio source with a spectral index of -1.6 was identified with a galaxy at a redshift of 5.2, making it the most distant AGN known.

Secondly, VISIR on the VLT, the LSA/MMA and the NGST should help establish the epoch at which heavy elements, dust and molecular gas are first produced and place this in the context of AGN evolution. Are these fundamental building blocks of galaxies only produced in stars, according to the tenets of conventional wisdom, or do active nuclei also play a direct role in their manufacture?

Thirdly, as discussed above, spatially studying the star formation history of HzRGs should constrain the various time scales involved.

Fourthly, optical/IR interferometry will delineate the central regions of galaxies on the scale of a parsec or smaller, possibly providing clues to the process by which matter is accreted on to the black holes and how this can be placed into the general context of galaxy evolution.

6 Conclusions

In this talk I have given you a personal perspective on the connection between nuclear activity and galaxy formation. There are two main points that I wish to leave you with, namely, (i) nuclear activity is more important in galaxy evolution than is usually supposed and (ii) there are several intriguing questions about the subject which we shall be able to tackle with the new astronomical facilities that will become available during the next decade.

While preparing for this talk I read some of the old papers in the field and discovered that the importance of nuclear activity to the general scheme of galaxy evolution was long ago recognized by a few "activists". I therefore conclude by citing from an important and prophetic paper which was published in 1963, more than 35 years ago, by Geoffrey and Margaret Burbidge and Alan Sandage, but which has been somewhat forgotten (Burbidge et al. 1963). Referring to ideas that Ambartsumian aired a few years earlier, they wrote: "... the many arguments which now suggest strongly that *violent events occur in the nuclei of galaxies*, that they are fairly frequent and may *have many effects on the structure and evolution of galaxies.*" "If violent events occur frequently in some types of galaxies and a large amount of material is ejected from the nuclear regions, it is possible that some of this *material may condense into stars* in the halo regions."

References

1. Athreya, R.M., Kapahi, V.K., McCarthy, P.J., van Breugel, W., 1998, A&A, 329, 809
2. Best, P.N., Longair, M.S., Röttgering, H.J.A., 1997a, MNRAS, 286, 785
3. Best, P.N., Longair, M.S., Röttgering, H.J.A., 1997b, MNRAS, 292, 758
4. Burbidge, G., Burbidge, M., Sandage, A., 1963, Rev. Mod. Physics, 35, 947

5. Carilli et al., 1997, ApJ Suppl., 109, 1
6. Carilli et al., 1998, ApJ Lett., 496, L57
7. Chambers, K.C., Miley, G.K., van Breugel, W., 1987, Nature 329, 604
8. Dey, A., van Breugel, W., Vacca, W., Antonucci, R., 1997, ApJ, 490, 698
9. Graham, J.A., 1998, ApJ, 502, 245
10. Heckman, T., 1998, Proc. Amsterdam Colloquium on "The Most Distant Radio Galaxies" p. 227, eds. H.J.A. Röttgering, P.N. Best and M.D. Lehnert, Royal Netherlands Academy of Arts and Sciences, ISBN 90-6984-238-6
11. Le Fevre, O., et al., 1996, ApJ Lett., 471, L11
12. McCarthy, P.J., Van Breugel, W., Spinrad, H., Djorgovski, S. 1987, ApJ (Lett.), 321, L29
13. McCarthy, 1993, Ann. Reviews A&A, 31, 639
14. Miley, G.K., 1997, in Proc. ESO June 1996 Workshop on "Science with the VLT Interferometer", ed. F. Paresce, p. 91–99, Springer-Verlag, ISBN: 3540632646
15. Pascarelle, S.M., et al., 1996, Nature, 383, 45
16. Pentericci et al., 1997, HST Images and properties of the most distant radio galaxies, AA in press, astro-ph/9809056
17. Rees, M.J., 1989, MNRAS, 239, 1p
18. Richstone, D., Ajhar, E.A., Bender, R., Bower, G., Dressler, A., Faber, S.M., Fillipenko, A.V., Gebhardt, K., Green, R., Ho, L.C., Kormendy, J., Lauer, T.R., Magorrian, J., Tremaine S., 1998, Nature, 395, A14–A19
19. Röttgering H.J.A., and Miley G.K., 1996, in The Early Universe with the VLT, ed. J. Bergeron, pp. 285–299, Springer-Verlag
20. Röttgering, H.J.A., West, M., Miley, G., Chambers, K., 1996, A&A, 307, 376
21. Röttgering, H.J.A., Best, P.N., Lehnert, M.D. (eds.), 1999, Proc. Colloquium on "The Most Distant Radio Galaxies", Royal Netherlands Academy of Arts and Sciences, ISBN 90-6984-238-6
22. van Breugel, W., Filippenko, A.V., Heckman, T., Miley, G., 1985, ApJ, 293, 83
23. van Breugel, W.J.M., Dey, A., 1993, ApJ, 414, 563
24. van Breugel, W., Stanford, A., Dey, A., Miley, G., Stern, D., Spinrad, H., Graham, J., and McCarthy, P., 1998, Proc. Amsterdam Colloquium on "The Most Distant Radio Galaxies" p. 49, eds. H.J.A. Röttgering, P.N. Best and M.D. Lehnert, Royal Netherlands Academy of Arts and Sciences, ISBN 90-6984-238-6
25. van Breugel, W., De Breuck, C., Stanford, S.A., Stern, D., Röttgering, H., Miley, G., A Radio Galaxy at $z = 5.19$, submitted to ApJ Lett.
26. van Ojik, R., Röttgering, H., Carilli, C., Miley, G., Bremer, M., 1996, A&A, 313, 25
27. van Ojik, R., Röttgering, H.J.A., Miley, G.K., Hunstead, R., 1997, A&A, 317, 358

Supernovae at High Redshift*

Bruno Leibundgut

European Southern Observatory, Karl-Schwarzschild-Strasse 2,
D-85748 Garching, Germany

Abstract. Distant supernovae provide a convenient tool for cosmology. As bright beacons with small variation in their peak luminosity Type Ia Supernovae can be used to determine distances and derive cosmological parameters. Beyond the Hubble constant, which still requires the knowledge of the absolute luminosity, the cosmic deceleration can be measured from relative distances. Two groups have undertaken this experiment with the conclusion that the distances are too large to be accommodated within matter-dominated world models. The distances can be explained by additional acceleration due to a cosmological constant.

Supernovae also provide direct evidence of the stellar death rate at significant look-back times. The evolution of the supernova rate as a function of redshift can give direct evidence of the star formation history throughout the observable universe. These studies are emerging as a main challenge for large telescopes.

1 Introduction

The observations of distant supernovae has become common place with the advent of large-aperture telescopes and large-field imaging cameras. It is now possible to find supernovae, classify them by a suitable spectrum, and follow their light curves in several filters out to redshifts of about 1. The near future will see further advances in redshifts and look-back time to probe even earlier epochs of star formation. Since supernovae are stellar end-products they can trace star formation more directly than any other method.

Future observational programs will employ all known supernova types and use their specific characteristics for cosmological purposes. It is of course the extreme luminosity of Type Ia Supernovae (hereafter SNe Ia) which opened the distant universe to supernovae first.

To employ any astronomical object for the derivation of cosmological parameters requires sufficient physical understanding of its properties in the

* I report results of the High-z Supernova Search Team. The team includes B. Schmidt (MSSSO), J. Spyromilio and P. Woudt (ESO), M. Phillips (LCO), N. Suntzeff, R. Schommer and C. Smith (CTIO), A. Clocchiatti (Universidad Catolica, Santiago), M. Hamuy (Steward Obs.), R. Kirshner, P. Garnavich, S. Jha and P. Challis (Harvard Univ.), C. Hogan, C. Stubbs, A. Diercks and D. Reiss (Univ. Washington), A. Filippenko and A. Riess (Univ. California, Berkeley), R. Gilliland (STScI), and J. Tonry (Hawaii). More information is available at `http://cfa-www.harvard.edu/cfa/oir/Research/supernova/HighZ.html`

nearby universe. Evolutionary effects must then be considered and, if necessary, corrected. In particular, the use of SNe Ia as standard candles demands a sufficient understanding of the explosions and their physical background to reliably apply them to cosmological problems.

The understanding of the local ($z < 0.1$) supernovae forms the basis for the cosmological results. There are basically four independent indications that SNe Ia form a rather homogeneous group of objects. These are the Hubble diagram ($\log z$ *vs.* m), the absolute luminosity at maximum as determined from direct distance measurements (mostly Cepheid stars), the uniform light curves and spectral evolution, and the fact that the peak luminosity of SNe Ia can be 'corrected' with a distance independent quantity, namely the decline rate after maximum light. We will discuss each of these features in the following.

Several Hubble diagrams of nearby SNe Ia have been published in the past few years. They all demonstrate the linear expansion to a high degree (e.g. Tammann & Leibundgut 1990, Hamuy et al. 1996b, Riess et al. 1996, Saha et al. 1999), although deviations from the linear expansion within a recession velocity of about 7000 to 10000 km/s have been claimed (Zehavi et al. 1998, Tammann 1999). The Hubble diagrams show a remarkably small scatter of the supernovae around the expansion line, which provides ample evidence that they all seem to reach a very similar absolute luminosity.

Another, more direct, route is the measurement of the absolute luminosities of SNe Ia. This direct comparison provides a further test of the uniformity of the standard candle assumption for SNe Ia. Such a program has been undertaken with the determination of Cepheid distances to now eight nearby SNe Ia (Saha et al. 1999). It is remarkable that almost all objects reached the same luminosity to within 20%.

The very uniform appearance of SNe Ia is further evidence for their great similarity. Although a strong explosion with the complete disruption of a star gives rise to the supernovae, the temporal evolution of the light emission follows a rather narrow parameter space. Deviations are observed, but it is amazing that the chaotic explosion mechanism and the complicated energy release yield the uniform events we observe.

Several observable parameters of SNe Ia seem to be correlated. The first, and foremost, such relation was the decline vs. peak luminosity dependence (Phillips 1993, Hamuy et al. 1996a, Riess et al. 1996, Riess et al. 1998, Phillips et al. 1999). In the meantime it has been found that the color (Hamuy et al. 1996b), the expansion velocity of the ejecta (Mazzali et al. 1998), the line strengths (Nugent et al. 1995), and the galaxy type (Hamuy et al. 1996a, Riess et al. 1996) also correlate with the peak luminosity.

After a correction with these distance independent observables, the scatter around the Hubble line becomes very small and is of the order of the photometric uncertainties (Hamuy et al. 1996b, Riess et al. 1996, Phillips et al. 1999, Suntzeff et al. 1999, Jha et al. 1999). This remarkable fact, has boosted the use of SNe Ia as distance indicators and seems by now accepted by most people.

Yet, there are a number of unexplained features of SNe Ia exists and an accepted theory for the explosion and the radiation transport in the ejecta is still missing. The very relation between the peak luminosity and the decline rate is not understood (e.g. Höflich et al. 1996). The secondary peak in the *RIJHK* light curves (Ford et al. 1993, Suntzeff 1996, Vacca & Leibundgut 1996) is likely due to the complicated release of photons in the expanding fireball. The bolometric light curves show this shoulder as well (Contardo et al. 1999). The spectrum is not formed in a regular photosphere, but is the result of complicated scattering of photons out of the ejecta. The input energy is coming from the radioactive decays of ^{56}Ni and its daughter product ^{56}Co, which is converted into a pseudo-continuum through down scattering in many lines (e.g. Pinto & Eastman 1999).

The bolometric luminosity of SNe Ia can differ among individual objects up to a factor of two (Contardo et al. 1999). Extreme cases, like SN 1991bg (Filippenko et al. 1992, Leibundgut et al. 1993, Turatto et al. 1996) and SN 1997n (Turatto et al. 1998), can be several times fainter.

Several independent determinations of Ni masses of SN Ia explosions exist by now (Spyromilio et al. 1992, Bowers et al. 1997, Cappellaro et al. 1997, Mazzali et al. 1998, Contardo et al. 1999). A fairly large range of Ni masses is inferred for the individual explosions. The theoretical understanding of this range in Ni masses is not very complete, although a number of explosion mechanisms has been proposed and could be responsible for what we observe (e.g. Höflich & Khokhlov 1996).

We have thus a fairly mottled view of SNe Ia as distance indicators. On the one hand a strong empirical basis for the use of SNe Ia as 'modified standard candles' has emerged, while the detailed physics of the events are largely undetermined.

2 Distant Type Ia Supernovae

Supernovae are rare. Although they can outshine their parent galaxies, they maintain this luminosity only for a brief time, which makes their detection at significant redshifts ($z > 0.1$) very difficult. Only dedicated wide-field searches which reach a limiting magnitude of at least $R \approx 23$ can hope to find a sufficient number of SNe Ia out to redshifts of about 0.5 (Perlmutter et al. 1995, 1997, 1999, Schmidt et al. 1998). More distant objects become increasingly difficult to find. Searching several square degrees of sky during a single night and comparing them with observations obtained several weeks before can provide about a dozen candidates which then are classified through spectroscopy (Riess et al. 1998, Perlmutter et al. 1999).

The two teams working in this field are now routinely discovering about 20 objects in a search run (Schmidt et al. 1997, Perlmutter et al. 1998b). A total of 53 supernovae with $z > 0.15$ have by now been published (Nørgaard-Nielsen et al. 1989, Riess et al. 1998, Perlmutter et al. 1999).

The High-z Supernova Search Team (Schmidt et al. 1998) has followed a strategy of trying to characterize the distant supernovae in as many aspects as possible. We have obtained light curves in at least two filters to detect any dust absorption in the host galaxy (Garnavich et al. 1998a, Schmidt et al. 1998, Riess et al. 1998). All objects have been positively identified as SNe Ia from spectra, and doubtful cases could be investigated in detail (Riess et al. 1998). The Supernova Cosmology Project (Perlmutter et al. 1995, 1997, 1998a, 1999) has amassed a sample of 42 distant supernovae. Many of these supernovae have a classification from spectroscopy and also two filter observations, but some are lacking this information.

Both teams developed the search and analysis software independently. The distant supernova samples are completely independent and there exists only a small overlap in the nearby supernova set, which is drawn partially from the Calan-Tololo supernova search (Hamuy et al. 1996a). The High-z team currently enjoys a larger nearby sample (29 supernovae).

It has to be noted that the cosmology depends on the local sample just as much as it does on the distant one. Since the measurement of the deceleration is relative, the nearby sample provides the essential comparison set.

The main result is that the distant supernovae (mean redshift for both groups is around 0.5) are too dim when compared to the local sample (Figure 1). It is important to note that this result is based entirely on the observational data and does not depend on any world model. The brightness difference to a cosmology with $\Omega_M = 0.2, \Omega_\Lambda = 0$ is plotted in the lower panel. It is clear that the local sample is insensitive to the exact cosmology and provides the zero-point for the various world model lines. The distant supernovae are about 0.25 mag fainter (Riess et al. 1998). The exact same result is found by the Supernova Cosmology Project (Perlmutter et al. 1999).

Various corrections have been applied in the analysis to arrive at Figure 1. They include the exact photometry of the supernova on the irregular galaxy background (achieved through subtraction of a galaxy template image), the K-correction to convert the magnitudes from the observer system into rest-frame wavelengths, the $(1+z)$ correction for time dilation (Leibundgut et al. 1996, Goldhaber et al. 1997), the light curve fitting, and the correction to the luminosity for light curve shape. All the details are given in Schmidt et al. (1998), Riess et al. (1998), Woudt et al. (2000) and Perlmutter et al. (1999).

The faintness of the distant supernovae can be explained in several different ways. Proposed explanations include evolution of the supernovae, dust with absorption properties different from Galactic dust, and the effect of a cosmological constant. More exotic explanations have been proposed as well, but will not be discussed here.

Evolution has been the downfall of every proposed standard candle in the past. SNe Ia are not fully understood and it is quite conceivable that they have

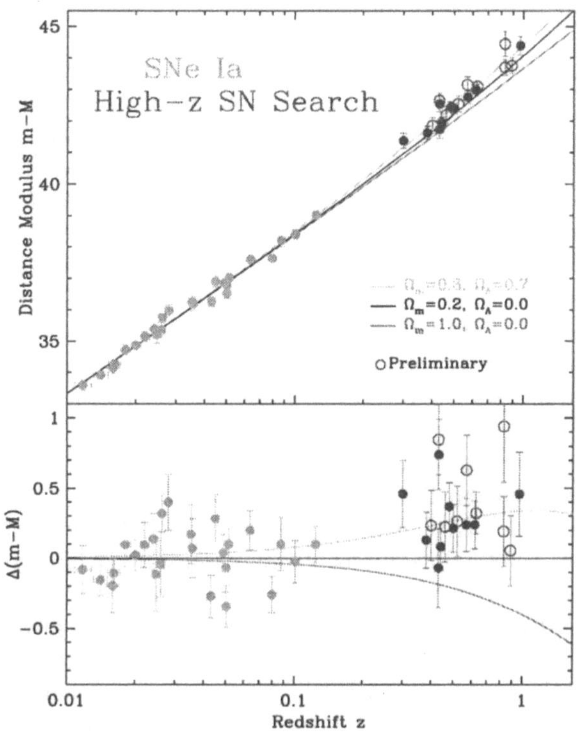

Fig. 1. Hubble diagram of the High-z SN Search Team

evolved over half the current age of the universe. However, the basic energy source of SNe Ia is the thermo-nuclear explosion and explanations have been sought in abundance differences which lead to slightly different explosion energetics (Höflich et al. 1998, Dominguez et al. 1999, Umeda et al. 1999). None of these abundance differences are very compelling at the moment, but can not be excluded. However, the effect is expected to be strongest in the local supernova sample, which presumably contains progenitors of the largest age range. The fact, that the luminosity correction works so well in this sample (Schmidt et al. 1998) gives us confidence, that it also would do so for the distant supernovae. A systematic sampling difference is largely diminished by the correction, but is also not seen in the data (Riess et al. 1998, Perlmutter et al. 1999). No other clear signatures of differences in the light curves or the spectral evolution have been detected so far, although the quality of the spectroscopy is not always sufficient to draw stringent conclusions. The High-z SN Team has now shifted its emphasis to observe higher quality spectra to explore any evolutionary signatures. A rest-frame *I* light curve of SN 1999Q (Riess et al. 2000) also displays the familiar secondary maximum of SNe Ia. The first sign of a possible systematic difference between local and distant supernovae has been reported in their rise times (Riess et al. 1999).

Dust has messed up many astrophysical measurements. We still do not have a very good handle on dust outside our own Milky Way although indications have been presented that at least in nearby galaxies the optical properties are not too different from the solar neighborhood. Dust has been invoked to explain the faintness of the distant supernovae (Aguirre 1999a, b). The dust properties have to be such that no or very little reddening can be observed. The precaution taken by the observers to obtain a reddening estimate has payed off. By extending the wavelength baseline (to include the rest-frame I-band) it will be possible to test this hypothesis soon (Riess et al. 2000). It has to be noted that different dust characteristics would have wide-ranging implications for other cosmological results, such as the inferred star formation history (e.g. Madau et al. 1996, Madau, this volume) or quasar luminosities (e.g. Shaver et al. 1997).

A further explanation for the faintness of the distant supernovae is the existence of a cosmological constant (Riess et al. 1998, Garnavich et al. 1998b, Perlmutter et al. 1999). Baring any strong evolutionary effects or grey dust, the distant supernovae do not allow a solution with positive matter density for any model with $\Omega_\Lambda = 0$ (Fig. 1). Hence, a cosmological constant (or another source of energy, e.g. Caldwell et al. 1998) is unavoidable (Caroll et al. 1991, Turner & Tyson 1999). With the currently measured distances the expansion must have accelerated during the last ~5 Gyr.

3 Cosmological Implications

The supernova result has some surprising implications for our understanding of the universe. One very interesting consequence of the accelerated expansion concerns the estimate of the dynamic age of the universe. Since the expansion was slower in the past, the inferred age is larger than in an Einstein-de Sitter or massless model. With a Hubble constant of $H_0 = 65$ km s^{-1} Mpc^{-1} an age of about 14 Gyr is inferred. This for the first time comfortably contains the age of the oldest known stellar components of the universe (Chaboyer et al. 1998), thus ending a debate which had started when the expansion of the universe was discovered by Edwin Hubble (1929).

Recent measurements of the cosmic microwave background have provided a first hint of the first Doppler peak in the CMB power spectrum (Hancock et al. 1998, Lineweaver 1998). Since the CMB constraint is nearly orthogonal to the supernova result (White 1998), the two combined give fairly strict constraints of the values of Ω_M and Ω_Λ (Lineweaver 1998, Garnavich et al. 1998b, Perlmutter et al. 1999). Interestingly enough the value found for Ω_M corresponds very well with the matter density inferred from clusters of galaxies (Turner & Tyson 1999). We thus have a consistent picture of cosmology at the moment.

Since the CMB result favors a flat geometry of the universe, inflationary models are en vogue again. On the other hand, severe problems arise from a cosmological constant: the value of the vacuum energy has to be tuned after the inflationary phase and lacks any explanation at the moment.

For a flat universe, the equation of state can be determined (Lineweaver 1998, Garnavich et al. 1998b, Perlmutter et al. 1999). The currently favored values would exclude cosmic strings or textures as dominating energy sources and find that the cosmological constant is the best explanation.

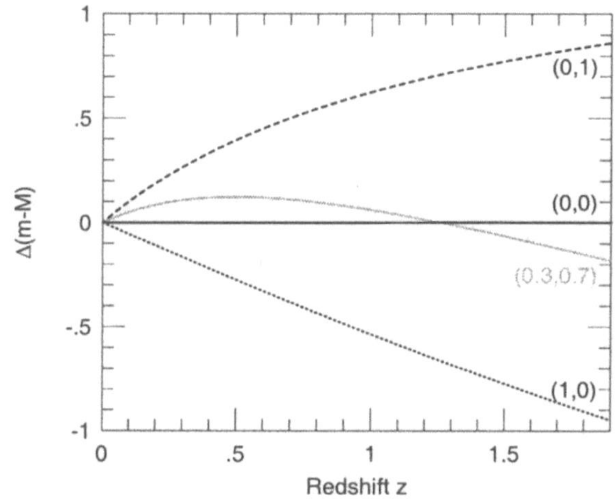

Fig. 2. The relative distance modulus for different cosmological models $(\Omega_M, \Omega_\Lambda)$

The existence of the cosmological constant can be fairly easily checked with further observations of SNe Ia. At early times the universe was dominated by matter and only later has been turned around by the cosmological constant. Hence, at sufficiently large lookback times, i.e. redshifts, we should find a decelerated universe (Fig. 2). With the current best fit parameters we find that supernovae near $z \approx 0.8$ should be the relatively faintest, while at redshifts >1.3 they should appear brighter. This non-monotonic behavior can be observed and is very unlike any of the other possible explanations. It has to be expected that evolutionary effects grow as a function of redshift and also the dust contents of the universe would have to be fine-tuned to mimic the predicted luminosity behavior for a cosmological constant.

4 Distant Core-Collapse Supernovae as Cosmological Probes

After the successful observations of distant SNe Ia it can be expected that other supernova types may provide important cosmological information as well. Supernova rates will tell us something about the progenitor systems.

In particular, the short fuse for core-collapse supernovae provides a simple way to measure the star formation rate at significant redshifts (Madau et al. 1998, Dahlén & Fransson 1999).

There are several ways how to observe distant core-collapse supernovae. An interesting proposal is to observe their X-ray/UV shock breakout (Chugai et al. 1999), which essentially depends on the size of the progenitor star. These are the most luminous moments of a SN II. Time dilation stretches the approximately day-long flash to several days at significant redshifts. The flashes have the advantage that they could be observed in the observer's optical window.

The advent of large telescopes opens up the infrared domain for faint objects. This is necessary to reach the rest-frame optical wavelengths of these supernovae as their UV flux is strongly suppressed (Cappellaro & Turatto, 1995, Kirshner et al. 1993) due to very strong Fe line blanketing. Especially the range from $0.5 < z < 4$ is interesting as the most intensive phase of star formation (Madau et al. 1998, Dahlén & Fransson 1999).

As the brightest signatures of individual stars supernovae are interesting *stellar* cosmological probes. Supernovae have an important role to play for our understanding of the constituents of the universe and their history.

References

1. Aguirre, A. N. 1999a, ApJ, 512, L19'
2. Aguirre, A. N. 1999b, ApJ, in press (astro-ph/9904319)
3. Bowers, E. J. C., et al. 1997, MNRAS, 290, 663
4. Caldwell, R. Dave, R., & Steinhardt, P. J. 1998, Phys. Rev. Lett., 80, 1582
5. Cappellaro, E., Turatto, M., & Fernley, J. 1995, ESA-SP 1189
6. Cappellaro, E., et al. 1997, A&A, 328, 203
7. Carroll, S. M., Press, W. H., & Turner, E. L. 1992, ARA&A, 30, 499
8. Chaboyer, B., Demarque, P., Kernan, P. J., & Krauss, L. M. 1998, ApJ, 494, 96
9. Chugai, N. N., Blinnikov, S. I., & Lundqvist, P. 1999, in Future Directions of Supernova Research: Progenitors to Remnants, eds. S. Cassisi and P. Mazzali, Mem. Soc. Astr. It, in press (astro-ph/9901298)
10. Contardo, G., Leibundgut, B., & Vacca, W. D. 1999, A&A, submitted
11. Dahlén, T., & Fransson, C., 1999, A&A, in press (astro-ph/9905201)
12. Dominguez, I., Höflich, P., Straniero, O., & Wheeler, J. C. 1999, in Future Directions of Supernova Research: Progenitors to Remnants, eds. S. Cassisi and P. Mazzali, Mem. Soc. Astr. It, in press (astro-ph/9905047)
13. Filippenko, A. V., et al. 1992, AJ, 104, 1543
14. Ford, C. H., et al. 1993, AJ, 106, 1101
15. Garnavich, P. M., et al. 1999a, ApJ, 493, L53
16. Garnavich, P. M., et al. 1999b, ApJ, 509, 74
17. Goldhaber, G., et al. 1997, in Thermonuclear Supernovae, ed. P. Ruiz-Lapuente et al., (Dordrecht: Kluwer), 777
18. Höflich, P., Wheeler, J. C., & Thielemann, F.-K., 1998, ApJ, 495, 617

19. Höflich, P., & Khokhlov, A. 1996, ApJ, 457, 500
20. Höflich, P., et al. 1996, ApJ, 472, L81
21. Hamuy, M., et al. 1996a, AJ, 112, 2391
22. Hamuy, M., et al. 1996b, AJ, 112, 2398
23. Hancock, S., Rocha, G., Lasenby, A. N., & Gutierrez, C. M. 1998, MNRAS, 294, L1
24. Hubble, E. 1929, Proc. N. A. S., 15, 168
25. Jha, S., et al. 1999, ApJS, in press (astro-ph/9906220)
26. Kirshner, R. P., et al. 1993, ApJ, 415, 589
27. Leibundgut, B., et al. 1993, AJ, 105, 301
28. Leibundgut, B., et al. 1996, ApJ, 466, L21
29. Lineweaver, C. H. 1998, ApJ, 505, L69
30. Madau, P., et al. 1996, MNRAS, 283, 1388
31. Madau, P., Della Valle, M., & Panagia, N., 1998, MNRAS, 297, L17
32. Mazzali, P., et al. 1998, ApJ, 499, L49
33. Nørgaard-Nielsen H. U., et al. 1989, Nature, 339, 523
34. Perlmutter, S., et al. 1995, ApJ, 440, L41
35. Perlmutter, S., et al. 1997, ApJ, 483, 565
36. Perlmutter, S., et al. 1998a, Nature, 391, 51
37. Perlmutter, S., et al. 1998b, IAU Circ. 6881
38. Perlmutter, S., et al. 1999, ApJ, 517, 565
39. Phillips, M. M. 1993, ApJ, 413, L105
40. Phillips, M. M., Lira, P., Suntzeff, N. B., Schommer, R. A., Hamuy, M., & Maza, J. 1999, AJ, in press (astro-ph/9907052)
41. Pinto, P. A., & Eastman, R. G. 1999, ApJ, submitted (astro-ph/9611195)
42. Riess, A. G., Press, W. H., Kirshner, R. P. 1996, 473, 88
43. Riess, A. G., et al. 1998, AJ, 116, 1009
44. Riess, A. G., Filippenko, A. V., Li, W., & Schmidt, B. P. 1999, AJ, submitted (astro-ph/9907038)
45. Riess, A. G., et al. 2000, in preparation
46. Saha, A., et al. 1999, ApJ, in press (astro-ph/9904389)
47. Schmidt, B. P., et al. 1997, IAU Circ. 6602
48. Schmidt, B. P., et al. 1998, ApJ, 507, 46
49. Shaver, P. A., Wall, J. V., Kellermann, K. I., Jackson, C. A., & Hawkins, M, R. S. 1996, Nature, 384, 439
50. Spyromilio, J., et al. 1992, MNRAS, 258, 53
51. Suntzeff, N. B. 1996, in IAU Colloquium 145: Supernovae and Supernova Remnants, eds. R. McCray and Z. Wang, (Cambridge: Cambridge University Press), 41
52. Suntzeff, N. B., et al. 1999, AJ, 117, 1175
53. Tammann, A. G. 1999, in General Relativity, ed. T. Piran (Singapore: World Scientific), in press (astro-ph/9805013)
54. Tammann, G. A., & Leibundgut, B. 1990, A&A, 236, 9
55. Turatto, M., et al. 1996, MNRAS, 283, 1
56. Turatto, M., et al. 1998, AJ, 116, 2431
57. Turner, M. S., & Tyson, J. A. 1999, Rev. Modern Phys., 71, 145
58. Umeda, H., Nomoto, K., Kobayashi, C., Hachisu, I., & Kato, M. 1999, ApJ, in press (astro-ph/9906192)
59. Vacca, W. D., & Leibundgut, B. 1996, ApJ, 471, L37
60. White, M. 1998, ApJ, 506, 495
61. Woudt, P., et al. 2000, in preparation
62. Zehavi, I., Riess, A., G., Kirshner, R. P., & Dekel, A. 1998, ApJ, 503, 483

Cosmic Star Formation History and the Brightness of the Night Sky

Piero Madau

Institute of Astronomy, University of Cambridge, Cambridge CB3 0HA, UK

Abstract. I review the constraints imposed by the observed extragalactic background light (EBL) on the history of the stellar birthrate in galaxies. At faint magnitudes, the logarithmic slope of the galaxy counts is flatter than 0.4 in all seven *UBVIJHK* optical bandpasses of the *Hubble Deep Field-South* imaging survey. The integration of the number counts provides a lower limit to the surface brightness of the optical extragalactic sky of 15 $\mathrm{nW\,m^{-2}\,sr^{-1}}$, comparable to the intensity of the far-IR background from *COBE* data. If the initial mass function has a Salpeter slope with a lower mass cutoff consistent with observations of M subdwarf disk stars, a lower limit of $\Omega_* > 0.005\,I_{50}$ (at Hubble constant 50 $\mathrm{km\,s^{-1}\,Mpc^{-1}}$) is derived for the visible (processed gas + stars) mass density needed to generate an extragalactic background light (EBL) at a level of $50\,I_{50}$ $\mathrm{nW\,m^{-2}\,sr^{-1}}$. The current 'best-guess' estimate to Ω_* is $0.012\,I_{50}$, about 16% of the nucleosynthetic baryon density. The contribution of quasar activity to the observed EBL is unlikely to exceed 20%.

1 Introduction

Recent progress in our understanding of faint galaxy data made possible by the combination of *Hubble Space Telescope* (*HST*) deep imaging and ground-based spectroscopy has dramatically increased our knowledge of the evolution of the stellar birthrate in optically-selected galaxies from the present-epoch up to $z \approx 4$ [12], [31], [21]. The explosion in the quantity of information available on the high-redshift universe at optical wavelengths has been complemented by the measurement of the far-IR/sub-mm background by DIRBE and FIRAS on board the *COBE* satellite [19], [15], [27], by the detection of distant ultraluminous sub-mm sources with the SCUBA camera [28], [9], and by theoretical progress made in understanding how intergalactic gas follows the dynamics dictated by dark matter halos until radiative, hydrodynamic, and star formation processes take over [2], [30], [24]. The IR data have revealed the 'optically-hidden' side of galaxy formation, and shown that a significant fraction of the energy released by stellar nucleosynthesis is re-emitted as thermal radiation by dust [8], [18]. The underlying goal of all these efforts is to understand the growth of cosmic structures, the internal properties of galaxies and their evolution, and ultimately to map the star formation history of the universe from the end of the cosmic 'dark age' to the present epoch.

In this talk I will focus on the galaxy number-apparent magnitude relation and its first moment, the integrated galaxy contribution to the extragalactic background light (EBL). The logarithmic slope of the differential galaxy counts is a remarkably simple cosmological probe of the history of stellar birth in galaxies, as it must drop below 0.4 to yield a finite value for the EBL. The recently released *Hubble Deep Field-South* (HDF-S) images, together with other existing *HST* and ground-based observations, provide a unique dataset to estimate the spectrum and amplitude of the optical EBL from discrete sources. Together with the far-IR/sub-mm background, the optical EBL is an indicator of the total luminosity of the universe, as the cumulative emission from young and evolved galactic systems, as well as from active galactic nuclei (AGNs), is recorded in this background. As such it provides, for a given initial mass function, a quantitative estimate of the baryonic mass that has been processed by stars throughout cosmic history.

Unless otherwise stated, an Einstein-de Sitter (EdS) cosmology ($\Omega_M = 1$, $\Omega_\Lambda = 0$) with $H_0 = 100\,h\ \mathrm{km\,s^{-1}\,Mpc^{-1}}$ will be adopted in this talk. All magnitudes will be given in the AB system. The work presented here has been done in collaboration with L. Pozzetti.

2 Galaxy Counts

The HDF-S dataset includes deep near-IR NICMOS images and the deepest observation ever made with the STIS 50CCD filterless imaging mode. The galaxy sample used here was extracted from version 1 of the HDF-S catalog on ftp://archive. stsci.edu/pub/hdf_south/version1/. At near-IR wavelengths (in the F110W, F160W, and F222M bandpasses, corresponding to the J, H, and K filters), it consists of 425 objects detected in the $J+H$ image, over a field of $50'' \times 50''$. The 50CCD (corresponding roughly to a $V+I$ filter) STIS catalog consists of 674 objects detected again over a field of the same size.

Figure 1 shows the HDF-N and -S galaxy counts compiled directly from the catalogs, as a function of AB isophotal magnitudes in the $UBVIJHK$ bandpasses for all galaxies with signal-to-noise ratio $S/N > 3$ within the band. No correction for detection completeness have been made. A compilation of existing *HST* and ground-based data is also shown [22], [25]. All magnitudes have been corrected to the AB system, while the second order colour corrections for the differences in the filter effective wavelengths have not been applied to the ground-based data (for the typical colours of galaxies in the HDF these corrections are less than 0.1 mag). The HDF optical counts agree well with previous surveys, to within 20% in the magnitude range $22 < m_{\rm AB} < 26$. One should note, however, that different algorithms used for 'growing' the photometry beyond the outer isophotes of galaxies can significantly change the magnitude of faint galaxies. According to [3], roughly 50% of the flux from resolved galaxies with $V > 23$ mag lie outside the standard-sized apertures used by photometric packages. An extragalactic sky pedestal created by the overlapping wings of resolved galaxies may also contribute significantly to the sky level, and would be undetectable except

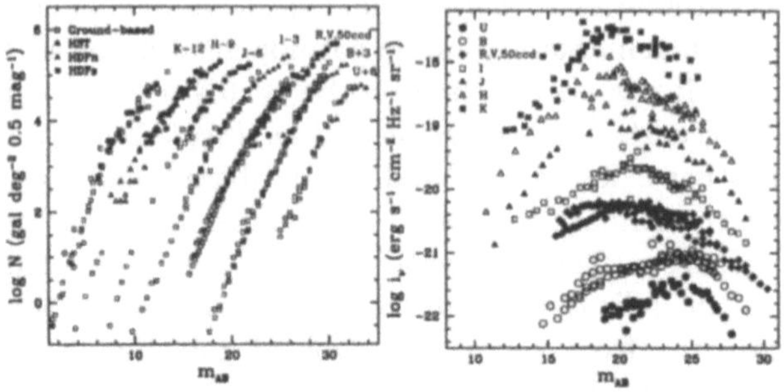

Fig. 1. *Left:* Differential $UBVIJHK$ galaxy counts as a function of AB magnitudes. The sources of the data points are given in the text. Note the decrease of the logarithmic slope $d\log N/dm$ at faint magnitudes. *Right:* Extragalactic background light per magnitude bin, $i_\nu = 10^{-0.4(m_{\rm AB}+48.6)} N(m)$, as a function of U (*filled circles*), B (*open circles*), V (*filled pentagons*), I (*open squares*), J (*filled triangles*), H (*open triangles*), and K (*filled squares*) magnitudes. For clarity, the $BVIJHK$ measurements have been multiplied by a factor of 2, 6, 15, 50, 150, and 600, respectively [22].

by absolute surface photometry [3]. Also, at faint magnitude levels, distant objects which are brighter than the nominal depth of the catalog may be missed due to the $(1+z)^4$ dimming factor. All these systematic errors are inherent in *HST* faint-galaxy photometry; as a result, our estimate of the integrated flux from resolved galaxies will typically be too low, and must be strictly considered as a *lower limit*.

3 The Extragalactic Background Light

The contribution of known galaxies to the optical EBL can be calculated directly by integrating the emitted flux times the differential number counts down to the detection threshold. I have used the compilation of ground-based, *HST*, and HDF data shown in Figure 1 to compute the integrated flux at $0.35 \lesssim \lambda \lesssim 2.2\,\mu$m.

In all seven bands, the slope of the differential number-magnitude relation is flatter than 0.4 above $m_{\rm AB} \sim 20$ (25) at near-IR (optical) wavelengths, and this flattening appears to be more pronounced at the shorter wavelengths. The leveling off of the counts is clearly seen in Figure 1, where the function $i_\nu = 10^{-0.4(m_{\rm AB}+48.6)} N(m)$ is plotted against apparent magnitude in all bands. While counts having a logarithmic slope $d\log N/dm_{\rm AB} = \alpha \geq 0.40$ continue to add to the EBL at the faintest magnitudes, it appears that the HDF survey has achieved the sensitivity to capture the bulk of the near-ultraviolet, optical, and near-IR extragalactic light from discrete sources. The flattening at faint apparent magnitudes cannot be due to the reddening of distant sources as their Lyman break gets redshifted into the blue passband, since the fraction of Lyman-break galaxies at (say) $B \approx 25$ is only of order 10% [32]. Moreover, an absorption-induced loss of sources cannot explain the similar change of slope in the $V, I, J, H,$ and K bands. While this suggests that the surface density of optically luminous galaxies is leveling off beyond $z \sim 1.5$, one should worry about the possibility of a significant amount of light being missed at faint magnitudes.

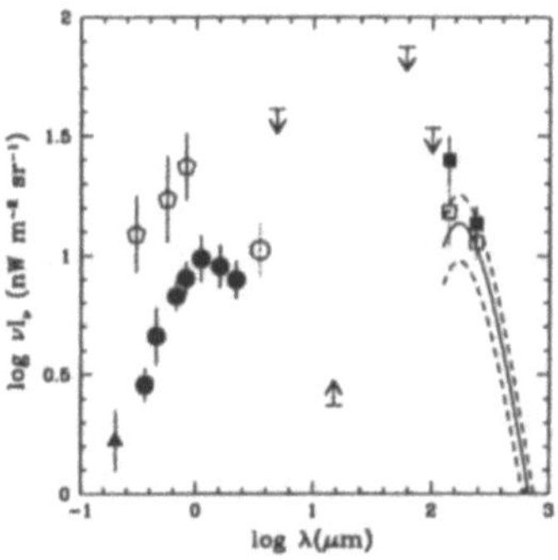

Fig. 2. Spectrum of the optical extragalactic background light from resolved sources as derived from a compilation of ground-based and space-based galaxy counts in the *UBVIJHK* bands (*filled dots*), together with the FIRAS 125–5000 μm (*solid and dashed lines*) and DIRBE 140 and 240 μm (*filled squares*) detections [19], [15]. The *empty squares* show the DIRBE points after correction for WIM dust emission [20]. Also plotted (*filled triangle*) is a FOCA-UV point at 2000 Å [1], and a tentative detection at 3.5 μm (*empty dot*) from *COBE*/DIRBE observations [7]. The empty pentagons at 3000, 5500, and 8000 Å are recent detections from absolute photometry [3]. Upper limits are from [19], the lower limit from [10]. Note that the values obtained by integrating the brightness of resolved galaxies are strict lower limits to the EBL intensity.

The spectrum of the optical EBL is shown in Figure 2, together with the recent results from *COBE*. The value derived by integrating the galaxy counts down to very faint magnitude levels (because of the flattening of the number-magnitude relation most of the contribution to the optical EBL comes from relatively bright galaxies) implies a lower limit to the EBL intensity in the 0.2–2.2 μm interval of $I_{\rm opt} \approx 15\,{\rm nW\,m^{-2}\,sr^{-1}}$. Including the tentative detection at 3.5 μm by [7] would boost $I_{\rm opt}$ to $\approx 19\,{\rm nW\,m^{-2}\,sr^{-1}}$. Recent estimates of the optical EBL at 3000, 5500, and 8000 Å from absolute surface photometry by [3] lie between a factor of two to three higher than the integrated light from galaxy counts. Applying this correction factor to the range 3000–8000 Å gives an optical EBL intensity in excess of $25\,{\rm nW\,m^{-2}\,sr^{-1}}$ in the interval *0.2–3.5 μm*. The *COBE*/FIRAS [15] measurements yield $I_{\rm FIR} \approx$

14 $\mathrm{nW\,m^{-2}\,sr^{-1}}$ in the 125–2000 μm range. When combined with the DIRBE [19], [27], [20] points at 140 and 240 μm, one gets a far-IR background intensity of $I_{\rm FIR}(140-2000\,\mu{\rm m}) \approx 20\ \mathrm{nW\,m^{-2}\,sr^{-1}}$. The residual emission in the 3.5 to 140 μm region is poorly known, but it is likely to exceed 10 $\mathrm{nW\,m^{-2}\,sr^{-1}}$ [8]. A 'best-guess' estimate to the total EBL intensity observed today is then

$$I_{\rm EBL} = 55 \pm 20\ \mathrm{nW\,m^{-2}\,sr^{-1}}. \tag{1}$$

In the rest of my talk, I will adopt a reference value for the background light associated with star formation activity over the entire history of the universe of $50\,I_{50}\,\mathrm{nW\,m^{-2}\,sr^{-1}}$.

4 EBL from Quasar Activity

A direct estimate of the contribution of quasars to the EBL depends on the poorly known bolometric correction and the possible existence of a distant population of dusty AGNs with strong intrinsic absorption, as invoked in many models for the X-ray background. These Type II QSOs, while undetected at optical wavelengths, could contribute significantly to the far-IR background. It is in principle possible to bypass some of the above uncertainties by weighing the local mass density of black holes remnants [29]. Recent dynamical evidence indicates that supermassive black holes reside at the center of most nearby galaxies. The available data (see Fig. 3) show a correlation (but with a large scatter) between bulge and black hole mass, with $M_{\rm BH} \approx 0.006\,M_{\rm sph}$ as a best-fit [23]. The mass density in old spheroidal populations today is estimated to be $\Omega_{\rm sph} h = 0.0018^{+0.0012}_{-0.00085}$ [16], implying a mean mass density of quasar remnants today

$$\rho_{\rm BH} = 3 \pm 2 \times 10^6\,h\ \mathrm{M_\odot\,Mpc^{-3}}. \tag{2}$$

Since the observed energy density from all quasars is equal to the emitted energy divided by the average quasar redshift, the total contribution to the EBL from accretion onto black holes is

$$I_{\rm BH} = \frac{c^3}{4\pi}\frac{\eta\rho_{\rm BH}}{\langle 1+z\rangle} \approx 4 \pm 2.5\ \mathrm{nW\,m^{-2}\,sr^{-1}}\ \eta_{0.05}\frac{2.5}{\langle 1+z\rangle} \tag{3}$$

($h = 0.5$), where $\eta_{0.05}$ is the efficiency of accreted mass to radiation conversion (in units of 5%). Therefore, unless dust-obscured accretion onto supermassive black holes is a very efficient process ($\eta \gg 0.05$), a population of quasars peaking at $z \sim 1.5 - 2$ is expected to make a contribution to the brightness of the night sky not exceeding 10–20% [14], [21].

5 The Stellar Mass Density Today

With the help of some simple stellar population synthesis tools we can now set a lower limit to the total stellar mass density that produced the observed

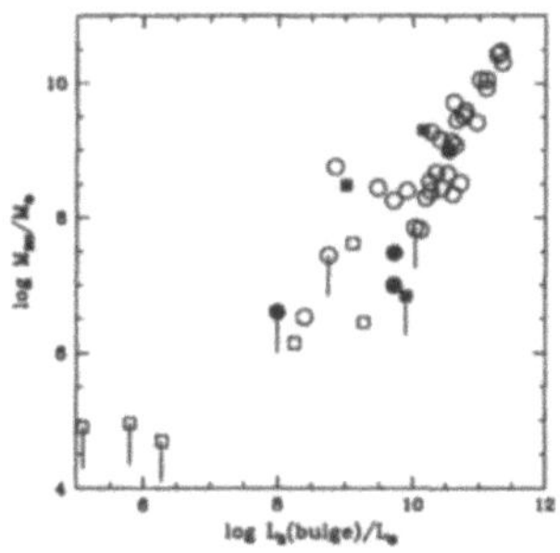

Fig. 3. Black hole mass distribution against the bulge luminosity of their host galaxies [23]. Arrows indicate upper limits on $M_{\rm BH}$. The symbols denote different galaxy types: *empty circles* (E), *filled squares* (S0), *filled circles* (Sab), and *empty squares* (Sbc-Scd).

sky brightness and constrain the cosmic history of star birth in galaxies. One of the most serious uncertainties in this calculation has always been the lower cutoff (usually treated as a free parameter) of the initial mass function (IMF). Observations of M subdwarfs stars with the *HST* have recently shed some light on this issue, showing that the IMF in the Galactic disk can be represented analytically over the mass range $0.1 < m < 1.6$ (here m is in solar units) by $\log \phi(m) = \text{const} - 2.33 \log m - 1.82 (\log m)^2$ ([17], hereafter GBF). For $m > 1$ this mass distribution agrees well with a Salpeter function. Observations of normal Galactic star-forming regions also show some convergence in the basic form of the IMF at intermediate and high masses, a power-law slope that is consistent with the Salpeter value [13]. In the following I will use a 'universal' IMF with the GBF form for $m < 1$, matched to a Salpeter slope for $m \geq 1$; the mass integral of this function is 0.6 times that obtained extrapolating a Salpeter function down to $0.1\,M_\odot$.[1]

As shown in Figure 4, the *bolometric* luminosity as a function of age τ of a simple stellar population (a single generation of coeval, chemically homogeneous stars having total mass M, solar metallicity, and the above IMF) can be well approximated by

$$L(\tau) = \begin{cases} 1200\,L_\odot \frac{M}{M_\odot} & \tau \leq 2.6 \text{ Myr;} \\ 0.7\,L_\odot \frac{M}{M_\odot} \left(\frac{\tau}{1\,\rm Gyr}\right)^{-1.25} & 2.6 \leq \tau \leq 100 \text{ Myr;} \\ 2.0\,L_\odot \frac{M}{M_\odot} \left(\frac{\tau}{1\,\rm Gyr}\right)^{-0.8} & \tau > 100 \text{ Myr} \end{cases} \tag{4}$$

[1] The bolometric light contributed by stars less massive than $1\,M_\odot$ is quite small for a 'typical' IMF. The use of the GBF mass function at low masses instead of Salpeter leaves then the total radiated luminosity of the stellar population virtually unaffected.

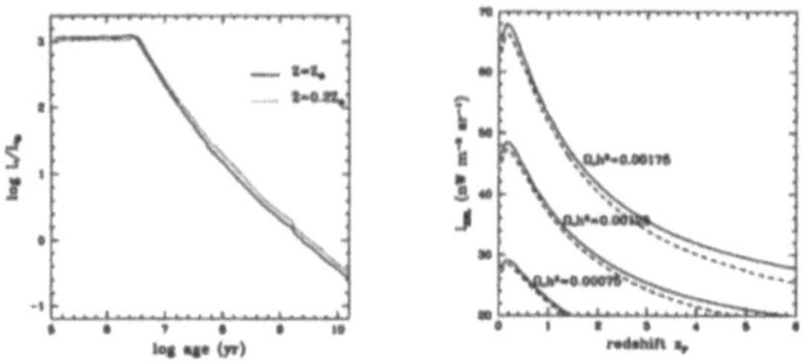

Fig. 4. *Left:* Synthetic [4] bolometric luminosity versus age of a simple stellar population having total mass $M = 1\,M_\odot$, metallicity $Z = Z_\odot$ (*solid line*) and $Z = 0.2\,Z_\odot$ (*dotted line*), and a GBF+Salpeter IMF (see text for details). ***Right:*** EBL observed at Earth from the instantaneous formation at redshift z_F of a stellar population having the same IMF, solar metallicity, and mass density $\Omega_* h^2 = 0.00175, 0.00125$, and 0.00075, as a function of z_F. *Solid curves:* EdS universe with $h = 0.5$. *Dashed curves:* Λ-dominated universe with $\Omega_M = 0.3$, $\Omega_\Lambda = 0.7$, and $h = 0.65$.

(cf [6]). Over a timescale of 13 Gyr (the age of the universe for an EdS cosmology with $h = 0.5$), about 1.3 MeV per stellar baryon are radiated away. This number depends only weakly on the assumed metallicity of stars. In a stellar system with arbitrary star formation rate per comoving cosmological volume, $\dot{\rho}_*$, and formation epoch t_F, the comoving bolometric emissivity at time t is given by the convolution integral

$$\rho_{\rm bol}(t) = \int_0^t L(\tau)\dot{\rho}_*(t-\tau)d\tau. \tag{5}$$

The total background light observed at Earth ($t = t_H$) is

$$I_{\rm EBL} = \frac{c}{4\pi}\int_{t_F}^{t_H} \frac{\rho_{\rm bol}(t)}{1+z}dt, \tag{6}$$

where the factor $(1+z)$ at the denominator is lost to cosmic expansion when converting from observed to radiated (comoving) luminosity density. To set a lower limit to the present-day mass density, Ω_*, of processed gas + stars (in units of the critical density $\rho_{\rm crit} = 2.77 \times 10^{11}\,h^2\,{\rm M_\odot\,Mpc^{-3}}$), consider now a scenario where all stars are formed instantaneously at redshift z_F. The background light that would be observed at Earth from such an event is shown in Figure 4 as a function of z_F for $\Omega_* h^2 = 0.00075, 0.00125, 0.00175$ (corresponding to 9, 6.5, and 4 percent of the nucleosynthetic baryon density, $\Omega_b h^2 = 0.0193 \pm 0.0014$ [5]), and two different cosmologies. A couple of points are worth noting here: (1) the time evolution of the luminosity radiated by a simple stellar population makes the dependence of the observed EBL from z_F much shallower than the $(1+z_F)^{-1}$ lost to cosmic expansion, as the energy output from stars is spread over their respective lifetimes; and

(2) in order to generate an EBL at a level of $50\,I_{50}\,\mathrm{nW\,m^{-2}\,sr^{-1}}$, one requires $\Omega_* h^2 > 0.00125\,I_{50}$ (for an EdS universe with $h = 0.5$), hence a mean mass-to-blue light ratio today of $\langle M/L_B\rangle_* > 3.5\,I_{50}$ for a present-day blue luminosity density of $\rho_B = 2.0 \times 10^8\,h\,\mathrm{L_\odot\,Mpc^{-3}}$ [11]. As shown in Figure 4, the dependence of these estimates on the cosmological model is rather weak. With the adopted IMF, about 30% of this mass will be returned to the interstellar medium in 10^8 yr after intermediate-mass stars eject their envelopes and massive stars explode as supernovae. This 'return fraction', R, becomes 50% after about 10 Gyr.

A visible mass density at the level of the above lower limit, $0.00125h^{-2}\,I_{50}$, while able to explain the measured sky brightness, requires all the stars that give origin to the observed light to have formed at $z_F \approx 0.2$, and is, as such, rather implausible. A more realistic scenario appears to be one where the star formation density evolves as

$$\dot{\rho}_*(z) = 0.23\,\frac{e^{3.4z}}{e^{3.8z} + 44.7}\ \mathrm{M_\odot\,yr^{-1}\,Mpc^{-3}}. \tag{7}$$

This model fits reasonably well all measurements of the UV-continuum and Hα luminosity densities from the present-epoch to $z = 4$ after an extinction correction of $A_{1500} = 1.2$ mag ($A_{2800} = 0.55$ mag) is applied to the data [21], and produce a total EBL of the right magnitude ($I_{50} = 1$). Since about half of the present-day stars are formed at $z > 1.3$ (hence their contribution to the EBL is redshifted away), the resulting visible mass density is $\Omega_* h^2 = 0.0031\,I_{50}$ ($\langle M/L_B\rangle_* = 8.6\,I_{50}$). Note that this estimate ignores the recycling of returned gas into new stars.

The observed EBL therefore requires that between 7% and 16% of the nucleosynthetic baryons are today in the forms of stars, processed gas, and their remnants. According to the most recent census of cosmic baryons, the mass density in stars and their remnants observed today is $\Omega_s h = 0.00245^{+0.00125}_{-0.00088}$ [16], corresponding to a mean visible mass-to-blue light ratio of $\langle M/L_B\rangle_s = 3.4^{+1.7}_{-1.3}$ ($h = 0.5$) (about 70% of this mass is found in old spheroidal populations). While this is about a factor of 2.5 smaller than the visible mass density predicted by equation (7), efficient recycling of ejected material into new star formation would tend to reduce the apparent discrepancy in the budget. Alternatively, the gas returned by stars may be ejected into the intergalactic medium. With an IMF-averaged yield of returned metals of $y \approx 1.5\,Z_\odot$,[2] the predicted mean metallicity at the present epoch is $y\Omega_*/\Omega_b = 0.25\,Z_\odot$, in good agreement with the values inferred from cluster abundances [26].

References

1. Armand, C., Milliard, B., & Deharveng, J.-M. 1994, A&A, 284, 12

[2] Here we have taken $y \equiv \int m p_{zm} \phi(m) dm \times [\int m\phi(m) dm]^{-1}$, the stellar yields p_{zm} of [33], and a GBF+Salpeter IMF.

2. Baugh, C. M., Cole, S., Frenk, C. S., & Lacey, C. G. 1998, ApJ, 498, 504
3. Bernstein, R. A., Freedman, W. L., & Madore, B. F. 1999, in preparation
4. Bruzual, A. C., & Charlot, S. 1993, ApJ, 405, 538
5. Burles, S., & Tytler, D. 1998, ApJ, 499, 699
6. Buzzoni, A. 1995, ApJS, 98, 69
7. Dwek, E., & Arendt, R. G. 1999, ApJ, in press
8. Dwek, E., et al. 1998, ApJ, 508, 106
9. Eales, S., et al. 1999, ApJ, 515, 518
10. Elbaz, D., et al. 1998, in The Universe as seen by ISO, ed. P. Cox & M. F. Kessler (Noordwijk: ESA Pub.), in press
11. Ellis, R. S., et al. 1996, MNRAS, 280, 235
12. Ellis, R. S. 1997, ARA&A, 35, 389
13. Elmegreen, B. G. 1998, in Unsolved Problems in Stellar Evolution, ed. M. Livio (Cambridge: Cambridge University Press), in press
14. Fabian, A. C., & Iwasawa, K. 1999, MNRAS, in press
15. Fixsen, D. J., et al. 1998, ApJ, 508, 123
16. Fukugita, M., Hogan, C. J., & Peebles, P. J. E. 1998, ApJ, 503, 518
17. Gould, A., Bahcall, J. N., & Flynn, C. 1996, ApJ, 465, 759 (GBF)
18. Guiderdoni, B., et al. 1997, Nature, 390, 257
19. Hauser, M. G., et al. 1998, ApJ, 508, 25
20. Lagache, G., et al. 1999, A&A, 344, 322
21. Madau, P. 1999, in Proceedings of the Nobel Symposium on Particle Physics and the Universe, Physica Scripta, in press
22. Madau, P., & Pozzetti, L. 1999, in preparation
23. Magorrian, G., et al. 1998, AJ, 115, 2285
24. Nagamine, K., Cen, R., & Ostriker, J. P. 1999, ApJ, submitted
25. Pozzetti, L., et al. 1998, MNRAS, 298, 1133
26. Renzini, A. 1997, ApJ, 488, 35
27. Schlegel, D. J., Finkbeiner, D. P., & Davis, M. 1998, ApJ, 500, 525
28. Smail, I., Ivison, R. J., & Blain, A. W. 1997, ApJ, 490, L5
29. Soltan, A. 1982, MNRAS, 200, 115
30. Somerville, R. S., Primack, J. R., & Faber, S. M. 1999, MNRAS, in press
31. Steidel, C. C., et al. 1998, in The Birth of Galaxies, ed. B. Guiderdoni, F. R. Bouchet, Trinh X. Thuan, & Tran Thanh Van (Gif-sur-Yvette: Edition Frontieres), in press
32. Steidel, C. C., et al. 1996, ApJ, 462, L17
33. Tsujimoto, T., et al. 1995, MNRAS, 277, 945

Galaxy Centers

Ralf Bender

Universitäts-Sternwarte der Ludwig-Maximilians-Universität,
Scheinerstr. 1, D-81679 München, Germany

Abstract. Recent progress in the analysis of the central regions of nearby bulges and ellipticals is reviewed. Topics covered are: central density profiles, black holes, shapes of galaxy centers, relations between central and global properties of spheroids, influence of black holes and mergers on core properties. A formation scenario which links centers and their host galaxies is discussed.

1 Introduction

In the past, most studies of galaxy centers have focused on the phenomenon of nuclear activity. Only more recently, the centers of nearby 'normal' galaxies found increasing attention because improved spatial resolution both on the ground and from the Hubble Space Telescope finally allowed to address key questions concerning their structure and dynamics: What is the structure of galaxy centers and how large is the variety? How is it linked to the presence of massive black holes? Do all galaxy centers contain massive black holes? Do centers evolve dynamically? How are centers linked to their host galaxies? Do centers influence the orbital structure of the galaxy as a whole and how? What can galaxy centers tell us about galaxy formation?

The VLT with its large suite of excellent instruments and the VLT interferometer provide powerful new tools for the analysis of galactic centers. Instruments like CONICA, which will allow to take adaptively corrected images of very high resolution using a laser guide star, or SINFONI, an integral field spectrograph with low order adaptive optics and medium spectral resolution, will be capable to produce many new exciting results already in the next few years. Especially, integral field spectroscopy with excellent spatial resolution will be an extremely efficient tool to analyse the structure of galaxy centers, *e.g.* [11,38].

This review summarizes some key findings concerning the centers of nearby bulges and ellipticals. The reason for this restriction is that I believe that our knowledge about these systems changed most dramatically in recent years. In addition, a complete review covering also pure disk systems and active galactic nuclei would be beyond the scope of these proceedings.

I refer the reader to the following recent reviews and papers which give more details on the various topics addressed here: Faber et al. [12], Kormendy & Richstone [24], Bender [8], Kormendy [25], Richstone et al. [35], Merritt [31]. These papers also contain a more complete set of references than could be given here.

2 The Radial Density Distribution of Stars in Galactic Centers

Before observations of sufficient spatial resolution existed, the stellar density distribution in galaxy centers was generally assumed to be spherically symmetric and to follow an isothermal or King-Michie model [18]. Such properties were believed to be the outcome of violent relaxation during the formation of a galaxy.

In their pioneering observations, Kormendy [20] and Lauer [26] could resolve the cores of many nearby bulges and ellipticals. They found that the surface brightness profiles of bright ellipticals do indeed often flatten out toward the center as expected from King-type cores. The sizes of the cores as measured by the core radius (operationally defined as the radius where the surface brightness has dropped to half its central value) and their central surface brightnesses showed good correlations with the luminosities and velocity dispersions of the galaxies. Interestingly, ellipticals and spheroids followed a correlation distinctly different from the one of dwarfs and spiral disks indicating a connection between the properties of galaxy centers and the formation of their host galaxies [21].

Hubble Space Telescope imaging broadly supports the ground-based results and correlations but modifies them in important ways [12]. The sequence of spheroids separates into two categories. Either the surface brightness profile shows a clear break where the slope changes significantly and produces a cuspy core (similar to an isothermal flat core but with non-zero central slope), or, the surface brightness profile is a steep power-law without any break. Cuspy-core galaxies follow the ground-based correlations while power-law galaxies remain unresolved. At a radius of about 10pc the density of power-law galaxies are typically a factor 1000 denser than cuspy-core galaxies [12].

Most notably, the two categories defined by the core properties are matched by the two classes of ellipticals and bulges which had already been identified previously on the basis of their isophotal shapes, rotational properties and radio or x-ray emission [2,4,33,12].

3 Two Categories of Spheroids, Two Categories of Galaxy Centers

Summarizing all the information available one obtains the following categorization of ellipticals and bulges [2,4,7,22,12,25], see also Figure 1:

(1) Bulges in early-type spirals and S0s (see Kormendy for a definition of proper bulges) and disky ellipticals are rotationally supported and generally show no prominent nuclear activity or radio emission. Also, their X-ray emission seems to be dominated by discrete sources. They have few globular clusters per unit luminosity [19]. *All bulges and disky Es seem to have high*

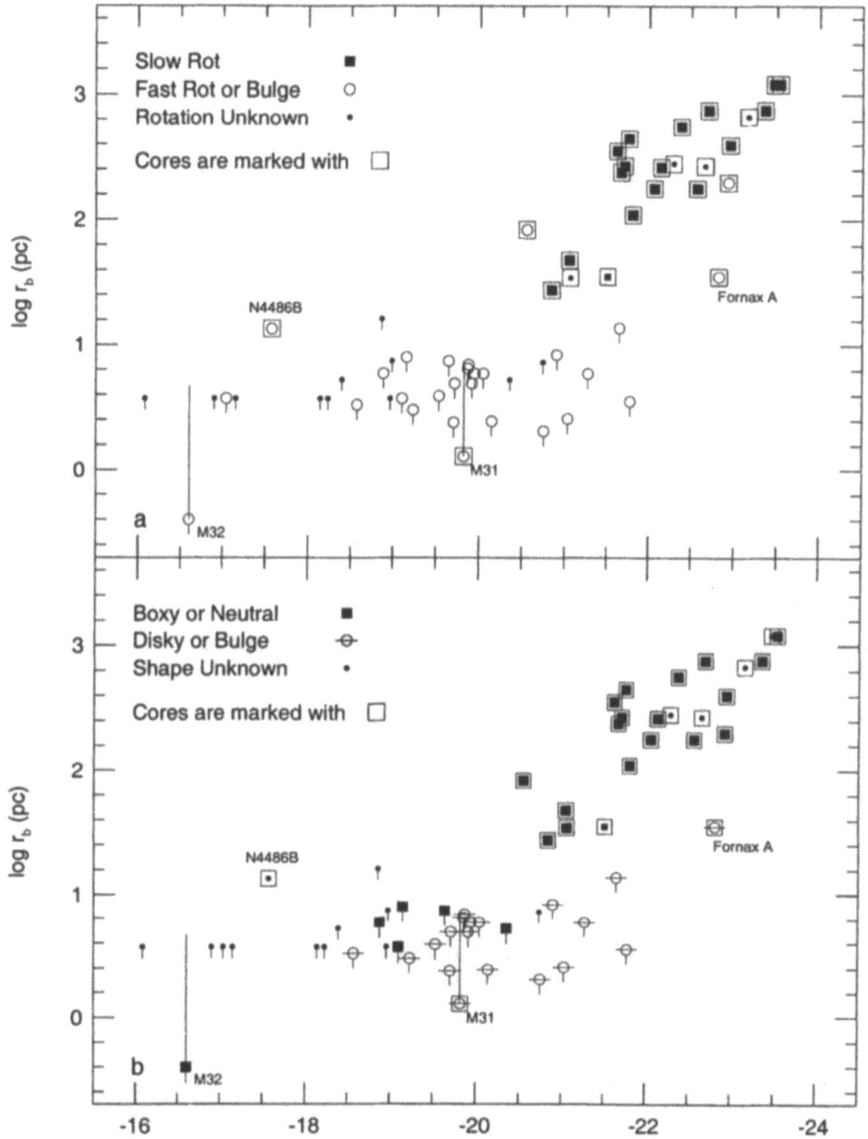

Fig. 1. Break radii r_b of bulges and ellipticals against their luminosities. Objects with different rotational properties and isophotal shapes are marked by different symbols. Hubble type S0-Sb are classed as disky. Diagram kindly provided by S.M. Faber [12].

density centers with power-law profiles. These data indicate that gas and dissipation must have been essential for the formation of bulges and disky Es. Dissipation is needed to form disks and also to create power-law centers of high densities.

(2) Boxy Es, on the other hand, are mostly supported by anisotropic velocity dispersions and frequently exhibit kinematic peculiarities, like minor axis rotation and kinematically decoupled central regions. Such peculiarities are observed in 1/3 to 1/2 of boxy ellipticals. Boxy ellipticals are on average more massive than disky ellipticals and bulges. They often also show nuclear radio emission and are surrounded by gaseous luminous X-ray halos. The globular cluster frequency of boxy ellipticals is usually high, the globular cluster systems can have bimodal color distributions [19]. ***Boxy, slowly rotating Es generally have cuspy, low density cores.*** The properties of boxy Es are a natural by-product of merging of star-dominated systems [15,1,7].

These two categories of spheroids suggest key differences in the formation of boxy cuspy-core galaxies and disky power-law galaxies. A possible formation scenario will be discussed in Section 7.

4 Massive Black Holes in Galaxy Centers

Most if not all spheroids appear to contain massive black holes. The recent Hubble Space Telescope measurements confirm ground-based results which appear to be quite robust. Though the limits on the central mass concentrations are usually not tight enough to demonstrate undoubtedly that the nuclear massive dark objects are indeed black holes (with the exception of a few cases, especially the Galactic center [13]), this is at least the most plausible conclusion. Black hole masses seem to be roughly proportional to the spheroid mass, not to the total galaxy mass *i.e.* $M_{BH} \approx 0.005 M_{sph}$. The black hole mass density inferred from dynamical measurements is in agreement with quasar energetics. Results on massive central black holes are summarized in several recent reviews, *e.g.* [24,16,35].

Black holes as well as high density power-law centers may have significant influence on the orbit distribution of stars [14,10,40]. Any sufficiently concentrated central mass will depopulate box orbits and make galaxies more spherical/axisymmetric. Box orbits are the backbone of triaxial objects (see e.g. [29,30]). Scattering of box orbits would preserve nuclear disks and kinematically decoupled central components but would isotropize the velocity distribution and reduce the triaxiality of the galaxy. Adiabatic growth of black holes via accretion can cause a steepening of central density profile [41].

Little is known so far about the influence of massive black holes and the consequence of high central stellar densities on the evolution of the stellar populations in galaxy centers and near black holes.

5 The Shape of Galaxy Centers

The importance of orbit scattering can be tested by comparison of the photometric structures of the innermost regions of spheroids and the spheroids themselves. Orbit scattering should make the inner isophotes rounder and, if boxiness is caused at least partially by box orbits, also more elliptical or disky [40]. To check this prediction, [9] analysed WFPC2 images of elliptical and S0 galaxies from the HST archive. It turns out that around the break radius in cuspy-core galaxies or the innermost available radius in power-law galaxies none of the objects is boxy anymore, the isophote shapes are either elliptical or disky, see Figure 2. This result supports preliminary findings by Nieto et al. [34]. On the other hand, ***isophotes do not become less flattened towards smaller radii*** which would be expected if orbit scattering is in action. Actually, there is only a weak correlation between the flattening of the centers and main bodies. Nevertheless, these findings are mostly compatible with orbit scattering. More data and simulations are needed to draw more reliable conclusions, see also [36].

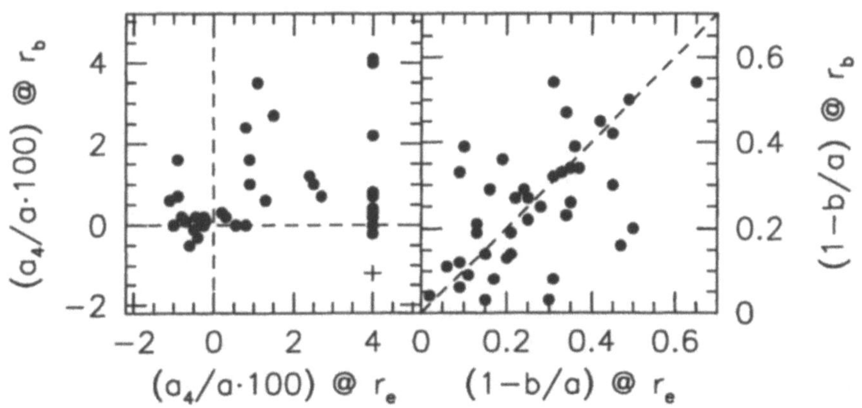

Fig. 2. Comparison between the a_4/a parameter around the effective radius r_e and the central break radius r_b (if the center had a power-law profile we used the innermost radius where a_4/a was measurable). Objects with $a_4/a > 0.04$ were plotted at $a_4/a = 0.04$. (b) Comparison between the ellipticity $1 - b/a$ around r_e and r_b. Isophotes around the break radius are elliptical or disky and have similar mean flattening as in the outer regions.

The key question arising from this analysis is how many galaxy centers are actually disk-like (thin or thick). It is striking that the nearest spiral galaxy, M 31, has a non-symmetric double nucleus which can be well modeled by an *elliptic disk around a massive black hole* [28,39,23]. Other examples of such objects exist [27] and indeed most galaxy centers are not yet resolved enough to answer this question.

6 The Origin of Cores

Why do different categories of galaxy centers exist in spheroids? A possible scenario for the formation of galaxy centers has been developed in Faber et al. [12], see also [25]. The basic ideas are the following. Power-law centers are formed in gas-rich mergers, which is consistent with the properties of their host galaxies [6]. Cuspy-core galaxies would then be formed in gas-free mergers. However, this proposition is not as compelling because of the following reasons. Both hierarchical galaxy formation models as well as observed and simulated mergers between large and small galaxies do not produce cores. It is especially unclear why the inner dense parts of compact satellite galaxies, which should be accreted at least once in the life time of a giant elliptical, should not survive the merging and fill-in any pre-existing core [17]. Faber et al. [12] proposed that massive black holes may provide an explanation for the production and survival of cuspy cores. The coalescence of binary black holes during galaxy merging will eject stars efficiently from the central region yielding a flat or cuspy core. Massive black holes could also heat and disperse accreted compact satellites explaining why companions cannot survive in the fluffy cores of boxy Es.

The problem of this model that it does not explain why some galaxies have cuspy cores and others power-law profiles. Obviously, another process besides ejection of stars through black holes must be at work. Different amounts of dissipation are a possibility or the energy output from an active nucleus which prevents the accumulation of gas in the center [25]. In fact, dissipation must have been present in the formation of cuspy core objects, too, because some of them show kinematically decoupled disk-like components in their centers. These components have masses up to several $10^9 M_\odot$ and can only have formed dissipatively (e.g. NGC 4365, NGC 4406, see [5,7]).

7 A Possible Formation Scenario for Spheroids and their Centers

A scenario for the formation of the different categories of spheroids and their corresponding centers within a hierarchical model could have two free parameters mainly: central black hole mass and gas content of the merger progenitors. Both parameters roughly scale with galaxy mass but may also depend on environment.

The critical merging events in the live of a galaxy are the equal mass mergers. They will mostly determine the galaxy's structure. Unequal mass or minor mergers only cause small alterations in the structure of the more massive partner. Unequal mass mergers will also largely preserve the central structure. If a core existed in the more massive partner the black hole will prevent the core from being filled in by a more compact companion. If both partners have power-law cores, then this feature may survive, too, as long as one black hole is significantly smaller than the other one (which likely is the case if the two objects have very different masses). Diskiness and rotational flattening of the main body is also preserved for mergers with mass ratios below 1:3 [32].

Now, assume a similar mass merger with large amounts of gas, probably small spheroids in the progenitors and correspondingly low central black hole masses. This will produce a disky elliptical or a bulge plus disk system. At first, the merging black hole binary would eject stars from the center during the ongoing merger and tend to produce a core. However, gas will be continuously dumped on the center causing a nuclear starburst and the formation of a cuspy core. The black hole(s) will accrete gas and nuclear activity may set in but it will not strong enough to completely ionize the nuclear gas and stop the starburst. After the merger is complete and relaxed, The high central concentration of stars (and gas) plus the black hole will still allow efficient orbit scattering and make the galaxy more isotropic and axisymmetric (this could be one reason for the rotational flattening of bulges and disky Es but likely not the only one). Especially near the center, nonaxisymmetry is rapidly reduced finally suppressing the accretion of mass onto the black hole. This could be an explanation why nuclear activity is lower in disky Es and bulges than in massive Es. The steep central gradients in density and potential are also not favorable for the focusing of jets which may be one reason why they are not frequent in bulges and disky Es. Last but not least, the large amounts of gas present before the merging will allow to form a large scale disk again once the merger is complete.

Let's consider the other option, a merger between two massive spheroids or ellipticals of similar mass. The nuclear black hole masses will be large. Furthermore, because such objects have experienced many mergers and associated starbursts before, the gas content of the progenitors is small (i.e. the progenitors will be similar to present-day E or Sa galaxies). The merging black holes will again eject stars from the center. But contrary to gas-rich mergers and because of the high black hole masses which cause high nuclear energy output, no power-law center can form, a (cuspy) core is left over. A starburst is still possible, but it will take place in the region surrounding the core, not in the core itself. This may also be the origin of decoupled kinematical components, typically having a radius of a few hundred parsecs. The only small amount of gas also results in a predominantly dissipationless merger of the main bodies of the galaxies, probably causing boxy isophotes, slow rotation and modest triaxiality [37,32]. The latter may help to continuously feed the black holes with material leading to more nuclear activity. The *flat central density distribution* allows easier jet formation. Only after many orbital time scales, orbit scattering by the black hole will reduce triaxiality and the possibility for accretion. The steepness of the cusp in the core may depend on how much the mass of the black hole grew adiabatically after the merger [41].

It is evident that this scenario is highly speculative and can only serve as a working hypothesis. Hopefully, the VLT and the other modern telescopes on the ground and in space will soon provide exciting new results on the centers of galaxies.

Acknowledgements

I thank the Nuker team for fruitful and exciting discussions and collaborations on galaxy centers and black holes. Sandy Faber kindly supplied Figure 1. Thanks also go to ESO for organizing a stimulating VLT Opening Conference. My research is supported by the Deutsche Forschungsgemeinschaft (SFB 375), by the Bundesministerium für Forschung und Technologie and by the Max-Planck-Gesellschaft.

References

1. Barnes, J.E., Hernquist, L. 1996, ApJ 471, 115
2. Bender, R. 1988a, A&A 193, L7
3. Bender, R., Surma, P., Döbereiner, S., Möllenhoff, C., Madejsky, R. 1989, A&A 217, 35
4. Bender, R. 1990, in *Dynamics and Interactions of Galaxies*, ed. R. Wielen, Springer Verlag, Heidelberg, p.232
5. Bender, R., Surma, P. 1992, A&A 258, 250
6. Bender, R., Burstein, D., Faber, S. 1992, ApJ 399, 462
7. Bender, R., 1997, in *The nature of elliptical galaxies*, Eds. M. Arnaboldi, G.S. Da Costa, P. Saha, ASP Conf. Ser. 116, 11
8. Bender, R. 1999 in *Galaxy Dynamics*, Eds. D.R. Merritt, M. Valluri, J.A. Sellwood ASP Conf. Series, in press
9. Bender, R., Tymann, R. 2000, in preparation
10. Dubinski, J. 1994, ApJ 431, 617
11. Emsellem, E., Bacon, R., Monnet, G., Poulain, P. 1996, A&A 312, 777
12. Faber, S.M., Tremaine, S., Ajhar, E., Byun, Y.-I., Dressler, A., Gebhardt, K., Grillmair, C., Kormendy, J., Lauer, T., Richstone, D. 1997, AJ, 114, 1771
13. Genzel, R., Eckart, A., Ott, T., Eisenhauer, F. 1997, MNRAS 291, 219
14. Gerhard, O., Binney, J. 1985, MNRAS, 216, 467
15. Hernquist, L., Barnes, J.E. 1991, Nature 354, 210
16. Ho, L. 1998, in *Observational Evidence for Black Holes in the Universe*, ed. S. Charkrabarti, Dordrecht, Kluwer
17. Holley-Bockelmann, K., Richstone, D. 1999, preprint, astro-ph 9811231
18. King, I.R. 1966, AJ, 71, 64
19. Kissler-Patig, M., Forbes, D.A., Minniti, D. 1998, MNRAS 298, 1123
20. Kormendy, J. 1985a, ApJ 292, L9

21. Kormendy, J. 1985b, ApJ 295, 73
22. Kormendy, J., Bender, R. 1996, ApJ 464, L119
23. Kormendy, J., Bender, R. 1999, ApJ, in press
24. Kormendy, J., Richstone, D. 1995, ARAA 33, 581
25. Kormendy, J. 1999, in *Galaxy Dynamics*, Eds. D.R. Merritt, M. Valluri, J.A. Sellwood ASP Conf. Series, in press
26. Lauer, T. 1985, ApJ 292, 104L
27. Lauer, T. Tremaine, S., Ajhar, E.A., Bender, R., Dressler, A., Faber, S.M., Gebhardt, K., Grillmair, C.J., Kormendy, J., Richstone, D. 1998, ApJ 471, L79
28. Lauer, T., Faber, S.M., Ajhar, E.A., Grillmair, C.J., Scowen, P.A. 1998, ApJ 506, 33
29. Merritt, D. 1997, in *The nature of elliptical galaxies*, Eds. M. Arnaboldi, G.S. Da Costa, P. Saha, ASP Conf. Ser. 116, 32
30. Merritt, D., Quinlan, G.D. 1998, ApJ, 498, 625
31. Merritt, D. 1999 in *Galaxy Dynamics*, Eds. D.R. Merritt, M. Valluri, J.A. Sellwood ASP Conf. Series, in press
32. Naab, T., Burkert, A., Richstone, D. 1999, ApJ, in press
33. Nieto, J.-L., Bender, R., Surma, P. 1991a, A&A 244, L37
34. Nieto, J.-L., Bender, Arnaud, J., Surma, P. 1991b, A&A 244, L25
35. Richstone, D. *et al.* 1998, 395, A14
36. Ryden, B. 1999, in *Galaxy Dynamics*, Eds. D.R. Merritt, M. Valluri, J.A. Sellwood ASP Conf. Series, in press
37. Steinmetz, M. 1995, in *Galaxies in the Young Universe*, ed. H. Hippelein, Springer Verlag, Heidelberg
38. Thatte, N. *et al.* 1998, SPIE 3353, 704
39. Tremaine, S. 1995, AJ, 110, 628
40. Valluri, M., Merritt, D. 1998, ApJ 506, 686
41. van der Marel, R. 1999, ApJ 117, 744

New Frontiers in the Solar System

Maria Antonietta Barucci

Observatoire de Paris, DESPA, 92195 Meudon Principal Cedex, France

Abstract. Nearly 130 Trans–Neptunian Objects (TNOs) have been discovered since the first one was found in 1992. On the basis of the discovery statistics, more than 100,000 objects with diameter larger than 100 km, are estimated to orbit in a radial zone extending outward from the orbit of Neptune. These objects, located so far from the Sun, are expected to be well–preserved fossils of the protoplanetary disk and they contain the least thermal perturbed material accessible to direct investigation. Their faintness represents a formidable challenge for the astronomers and only the use of 8–10 m class telescope will allow us to obtain a more precise picture of the solar system frontiers. The state of the art on the knowledge of these objects will be presented and discussed.

1 Introduction

The picture of the solar system with Pluto as last planet at ~ 40 AU and the Oort cloud far away at $\sim 40{,}000$ AU is not anymore valid. Pluto is not anymore the last object visible in our planetary system. After the first discovery of 1992 QB1 (1), 124 similar bodies have been discovered so far. Jewitt (2) estimate that more than 100,000 objects (with diameter larger than 100 km) orbit between 30 and 50 AU. We have now the observational evidence than trans–Neptunian solar system contains a vast population of icy bodies. These objects, classified as asteroids due to the lack of activity evidence, do not have yet a real name. Some called them Trans–Neptunian Objects (TNOs), some call them just Kuiper objects, some others Edgeworth–Kuiper objects. Edgeworth (3) was the first to form the hypothesis that a large number of small bodies could exist beyond the orbit of Pluto; successively, Edgeworth (4) and Kuiper (5) speculated that an alternative source of short period comets can exist just beyond the outer planets. I will adopt in this paper the term Trans–Neptunian Objects (TNOs) which is the least controversial of the possibilities. Oort (6) proved that long–period comets come from a spherical cloud (later called Oort cloud) and assumed that short–period comets were derived from the long–period objects captured by giant planets. This assumption was cast in doubt by Wipple (7) and many others. Fernandez (8) reproposed a trans–neptunian source of the short–period comets, and Duncan et al. (9) added motivation to search for this source. Luu and Jewitt (10) started an observational program to find this type of objects followed by Kowal (11), and many others. After five years of searching the first object was detected (1) and the number of new discoveries increased rapidly. In this paper we will present the orbital, physical and compositional characters of these objects.

2 TNO Orbital Characteristics

The known TNOs are divided in three dynamical classes: classical objects, resonances objects and scattered objects. In Fig. 1 the distribution of the TNO semi–major axis vs. eccentricity and in Fig. 2 the distribution of the TNO semi-major axis vs. inclination are reported. The locations of mean–motion resonances are represented in both plots by vertical dashed lines.

The Classical Objects have circular orbits with small eccentricities (mean value 0.07) and with an inclination which do not exceed 32°. Their orbits have semi–major axes 42 AU $\leq a \leq$ 47 AU and they do not seem associated with resonances.

The Resonance Objects are more than 35 % of the all known objects which are located mainly in the 3:2 mean motion resonance with Neptune at $a = 39$ AU. These objects are called Plutinos for their orbital similarity with Pluto. The apparent abundance can be affected by observational bias. Some of the Plutinos have perihelia inside Neptune's orbit as well as Pluto. The eccentricities range from 0.1 to 0.34 and the inclinations not larger than 20°. The other resonances (as the 4:3, 5:3, and 2:1) seem also to be populated even if at lower level. Numerical experiments (12) show that more than 50% of initially non-resonant objects are captured into mean-motion resonances.

The Scattered Objects are represented for the moment only by 1996 TL66. Its orbit has $a = 85$ AU, $e = 0.58$ and $i = 24^o$. Its perihelion is at 35 AU while its aphelion is at 135 AU. Luu et al. (13) suggested that more than 6000 objects can have orbits similar to that of 1996 TL66. New discoveries of this type of objects will provide constrains on the mass radial distribution in the solar nebula.

The inner edge of TNOs is well defined by Neptune, while the outer edge remains unknown. We do not know if these objects are radially distributed from the region of planets up to the Oort cloud. Due to the observational limits it is very difficult to detect the external edge and to determine the belt substructure.

3 Population Statistics

The objects discovered up to now have a limit magnitude of about 26.6. The absolute magnitude range between 4.5 and 9.8. Since they are not resolved and their albedoes are unknown, the diameter estimation is strongly affected. If an albedo value of 0.05 is assumed, the diameter of these objects range between 740 km and 60 km for the fainter. On the basis of observation statistics it is possible to compute the Cumulative Luminosity Function (CLF) of this population inside the 30–50 AU region. The CLF is the number of TNOs per unit area of sky brighter than a given magnitude, measured as a function of the magnitude. The CLF reflects both the size distribution and radial distance distribution of these objects. The CLF is described by the

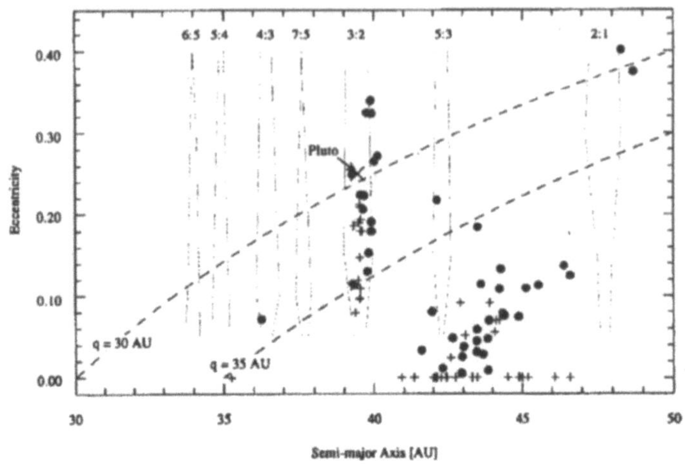

Fig. 1. Semi–major axis vs. orbital eccentricity for the first 85 discovered TNOs. Solid circles represent multi–opposition objects, pluses are for single opposition objects while cross is for Pluto. 1996 TL66 falls outside and is omitted from the plot. Plot is from Jewitt (14).

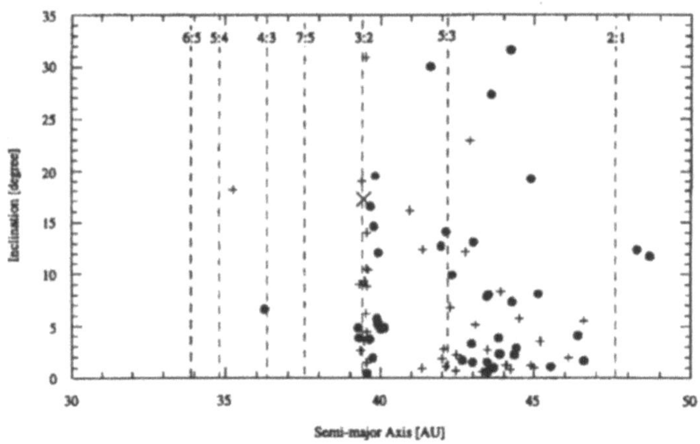

Fig. 2. Distribution of semi–major axis vs. orbital inclination for the known TNOs. The reported symbol are like in Fig. 1. Plot is from Jewitt (14).

power law:

$$\log\left[S(m_R)\right] = \alpha(m_R - m_o)$$

where $S(m_R)$ is the number of objects per square degree brighter than red magnitude m_R, and α and m_o are constants. Least square fits of the CLF (Fig. 3) give $\alpha = 0.54 \pm 0.04$ and $m_o = 23.20 \pm 0.10$ ($20 \leq m_R \leq 26.6$) (15). Gladman et al. (16) using a different subset of data, including some

HST controversial observations which lead to an estimation of larger number of TNOs, found $\alpha = 0.76 \pm 0.11$. Given the size distribution, it is possible to calculate the total mass. Following Luu and Jewitt (15), the total mass between 30 and 50 AU is estimated at ~ 0.1 M_{Earth}. This 0.1 M_{Earth} inside the 50 AU includes $\sim$ 100,000 objects larger than 100 km. If extrapolated to small size, the total number of TNOs larger than 1 km diameter is larger than 10^{11} while in the other directions the distribution seems to extend smoothly to Pluto (diameter = 2300 km) and allows the possibility that other Pluto–size objects await discovery.

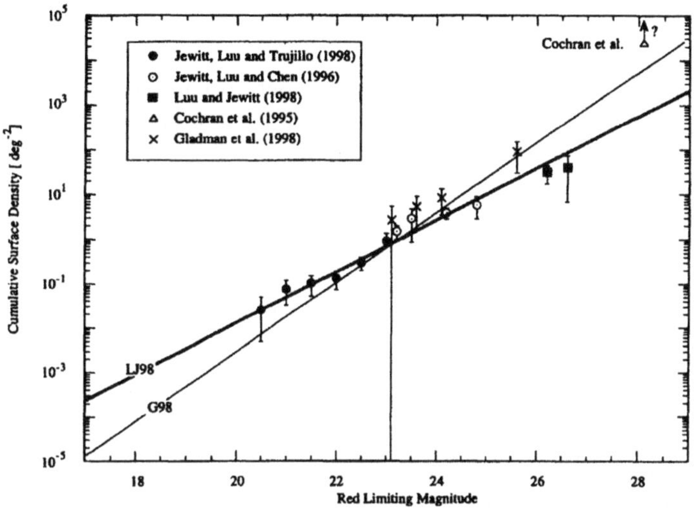

Fig. 3. The plotted line is the fitted cumulative function from Jewitt (14).

We know the surface mass density up to Neptune and if we extrapolate with a power law the surface density over 30–50 AU, the surface mass density becomes 25 M_{Earth} which is much larger than the estimated one. The present estimated mass of TNOs in the region 30–50 AU is too small, in fact more than 10 M_{Earth} of material are needed to grow Pluto–size objects. More than 99% of the initial mass has been lost and lost mechanisms remain unidentified.

4 Short-Period Comets and Centaurs

Two populations of objects are related to the Trans-Neptunian Objects: the Centaurs and short–period comets. The short–period comets are most likely originated in the TNO disk; it is not clear which part of the disk is the main source. Morbidelli (17) suggests that chaotic zones at the edges of resonances represent one possible source.

The Centaurs, which orbits have semi–major axes falling between those of Jupiter ($a = 5$ AU) and Neptune ($a = 30$ AU), are the best candidates of TNOs escaped. Long–term orbital integrations of the TNOs suggest that perturbation by giant planets or collisions (18,19,20) provide source of Centaurs. Their orbits are unstable with a dynamical lifetime measured in millions of years. To date only nine objects have been discovered with these particular orbits: Jedicke and Herron (21) estimate that more than 2000 Centaurs with absolute magnitude range H = -4 to 10.5 exist, having intermediate size, between the typical comet diameter (1–20 km) and Pluto (2300 km) one. 2060 Chiron , the first object of this group discovered, was classified asteroid even if Kowal (22) labeled it *Centaurian object* for its particular orbit. Chiron is the only object of this group which shows cometary activity (23). It is possible that Centaurs are ice–rich bodies that can sublimate under right conditions and therefore evolve into short–period comets. The Centaurs represent the link between TNOs and the short–period comets; being closer, and so brighter than the typical TNOs, they are much easy to be studied.

5 Physical Observations

The TNO's physical and chemical properties are still poorly known because of the intrinsic faintness of these distant objects. Observations with good signal to noise ratio are extremely difficult and requires considerable telescope resources as well as very careful data reduction and calibration. CCD imaging is the easier way to observe these faint objects especially when the precision of the orbital position is not very well known. An indication of rotational period is available only for few objects ranging between 6 and 12 hours (24).

Optical (25,26,27) and near–infrared colors (28) of these objects have been obtained and exhibit a wide diversity, spanning from neutral color up to strongly reddened. This diversity implies a different surface composition. Even if it is difficult to infer the surface composition on the only basis of broad–band colors, it seems realistic to interpret the neutral color as the one of a dirty water ice surface, while the reddening is a consequence of the presence of carbon–rich compounds like organics and tholins. In the visible region Centaurs and TNOs have very similar spectral characteristics (29) and this similarity supports the hypothesis that Centaurs are Trans–Neptunian objects injected into giant planet crossing orbits. Tegler and Romanishin (30), on the basis of B, V, and R colors obtained for nine TNOs and four Centaurs, found two groups: one of neutral color and the other very red, while Barucci et al. (27) analysing the colors (B,V,R,I) available to date for a large sample (22 TNOs) found a more complex and non homogeneous population composition (Fig. 4).

Near–infrared spectroscopy (up to 2.5 μm) has been attempted so far only for two of TNOs (1993 SC and 1996 TL66) and for three Centaurs (2060 Chiron, 5145 Pholus and 1997 CU29). Chiron and Pholus are the brightest, so their spectra are of good quality, while the others, even observed with the Keck I are very noisy. The spectra of 2060 Chiron (31) and of 1996 TL66 (32) show a flat featureless behaviour similar to that of dirty water ice. The spectrum of 1993 SC shows features very similar to those present in the spectra of Triton and Pluto, particularly the feature at 2.2 μm interpreted as absorption by hydrocarbon ices. The Centaur 5146 Pholus (33) seems to contain organic compounds like methanol ice, water ice and some silicates. The other Centaur 1997 CU29 (34,35) shows a spectrum with absorption due to water ice probably in amorphous state. Summarizing, even if the quality of all these spectra is very poor, their behaviour shows that these objects are very different from each other. The differences can be due to different degrees of collisional evolution. In fact as shown by numerical simulation (36) the collision evolution seems to have played a substantial role especially in the inner part of the TNOs. Collisions could have reworked the object surface and could have revealed fresh interior material. Different degrees of space weathering alteration due to cosmic ray bombardment on surfaces of different ages (from freshly excavated material to old crust) can be at the origin of the observed differences.

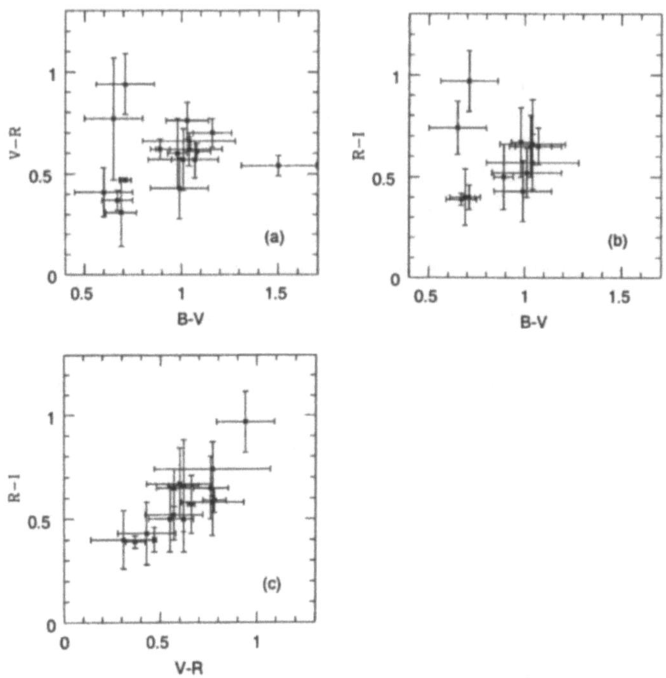

Fig. 4. a) B–V vs. V–R colors, **b)** B–V vs. R–I colors and **c)** V–R vs. R–I colors are plotted, from Barucci et al. (27).

6 Conclusion

TNOs can be considered fossil remnants of the solar nebula survived over the age of the solar system. In fact the material formed in the outer solar system would not have truncated sharply right at Neptune's orbit.

The discovery of these objects in our solar system leads to a natural hypothesis that similar objects may exist around other stars. The study of these bodies, as well as of dust star disks (e.g. HR 4796A) gives important constraints on planetary system formation.

TNOs represent one of the most interesting and primitive population of solar system objects. Because of their faintness and the instrumental limitations, the properties of these objects are very poorly known and much more observations are necessary. The 8–10 m class telescopes will allow us to carry out systematic observational studies of the physical and chemical nature of this population. With new discoveries, a more and more precise picture of the solar system frontiers will be available in the near future.

References

1. Jewitt D.C. and Luu J.X. (1993) Discovery of the candidate Kuiper belt object 1992 QB1. Nature **362**, 730–732
2. Jewitt D.C. (1999) The outer Solar System. ESO BOSS98 proceeding, in press
3. Edgeworth K.E. (1943) The evolution of our planetary system. Journal of British Astron. Soc. **53**, 181–188
4. Edgeworth K.E. (1949) The origin and the evolution of the solar system. MNRAS **109**, 600–609
5. Kuiper G.P. (1950) On the origin of the solar system, In Astrophysics, ed J.A. Hynek, New York, McGraw-Hill, pp 375–424
6. Oort J.H. (1950) The structure of the cloud of comets surrounding the solar system and a hypothesis concerning its origin. Bull. Astr. Inst. Netherlands **11**, 91–110
7. Whipple F.L. (1964) Evidence for a comet belt beyond Neptune. Proc. Nat. Acad. Sci. **51**, 711–718
8. Fernandez J. (1980) On the existence of a comet belt beyond Neptune. MNRAS **192**, 481–491
9. Duncan L., Quinn T., Tremaine S. (1988) The origin of short period comets. Ap.J. **328**, 69–73
10. Luu J.X., Jewitt D.C. (1988) A two–part search for slow–moving objects. Astron.J. **95**, 1256–1262
11. Kowal C. (1989) A solar system survey. Icarus **77**, 118–123
12. Malhotra R. (1996) The phase space structure near Neptune resonances in the Kuiper belt. Astron.J. **111**, 504–516.
13. Luu J., Marsden B.G., Jewitt D., Trujillo C.A., Hargenrother C.W., Chen J., Offutt W.B. (1997) A new dynamical class of object in the outer solar system. Nature **387**, 573–575
14. Jewitt D.C. (1999) Kuiper Blet Objects. Ann. Rev. of Earth and Planetary Science, in press

15. Luu J.X., Jewitt D.C. (1998) Deep Imaging of the Kuiper belt with the Keck 10 m telescope. Ap.J. **502**, L91–L94
16. Gladman B., Kavelaars I.J., Nicholson P.D., Loredo T.J., Burns J.A. (1998) Pencil–beam surveys for faint Trans–Neptunian Objects. Astron.J. **116**, 2042–2054
17. Morbidelli A. (1997) Chaotic diffusion,and the origin of comets from the 3/2 resonance in the Kuiper belt. Icarus **127**, 1–12
18. Stern S.A.(1995) Collisional timescales in the Kuiper disk and their implications. Astron.J. **110**, 856–868
19. Duncan M., Levison H., Budd S. (1995) The dynamical structure of the Kuiper belt. Astron.J. **110**, 3073–3081
20. Stern S.A., Campins (1996) Chiron and the Centaurs: escapees from the Kuiper Belt. Nature **382**, 507–510
21. Jedicke R., Herron J.D. (1997) Observational Constraints on the Centaur Population. Icarus **127**, 494–507
22. Kowal, C.T. (1996) Asteroids: their nature and utilization, New York: Wiley
23. Meech K.J., Belton M.J. (1990) The Atmosphere of 2060 Chiron. Astron.J. **100**, 1323–1338
24. Hainaut O., Delahodde C., Boehnhart H., Dotto E., Barucci A., Meech K., Bauer J., West R. and Doressoundiram (1999) Physical Properties of TNO 1996 TO66. Astron.Astroph., submitted
25. Luu J.X. and Jewitt D.C. (1996) Color diversity among the Centaurs and Kuiper belt objects. Astron.J. **112**, 2310–2318
26. Green S.F., McBride N, O'Ceallaigh D., Fitzsimmons A., Williams I., Irwin M. (1997) Surface reflectance properties of distant Solar system bodies. MNRAS **290**, 186–192
27. Barucci M.A. Doressoundiram A., Tholen D., Fulchignoni M., Lazzarin M. (1999) Spectrophotometric observations of Edgeworth-Kuiper Belt objects. Icarus in press
28. Jewitt D.C, Luu J.X. (1998) Optical-Infrared spectral diversity in the Kuiper Belt. Astron.J. **115**, 1667–1670
29. Barucci M.A., Lazzarin M. (1999) Optical observations of Trans-neptunian and Centaur objects. ESO MBOSS98 proceedings, in press
30. Tegler S.C., Romanishin W. (1998) Two distinct populations of Kuiper–belt objects. Nature **392**, 49–50
31. Luu J. X., Jewitt D.C., Cloutis E. (1994) Near Infrared Spectroscopy of Primitive Solar System Objects. Icarus **109**, 133–144
32. Luu J.X., Jewitt D.C (1998) Optical and Infrared reflection spectrum of Kuiper belt object 1996 TL66. AP.J. **494**, 117–120
33. Cruikshank D., Roush T., Bartholomew M., Geballe T., Pendleton Y., White S., Bell J., Davies J., Owen T., de Bergh C., Tholen D., Bernstein M., Brown R., Tryka K., and Dalle Ore C. (1998) The composition of Centaur 5145 Pholus. Icarus **135**, 389–408
34. Brown R.H., Cruikshank D.P., Pendleton Y., Veeder G.J. (1998) Identification of water ice on the Centaur 1997 CU26. Science **280**, 1430–1431
35. Brown M.E., Koresko C.D. (1998) Detection of water ice on the centaur 1997 CU26. Ap.J. **505**, L65–L67
36. Davis D.R. and Farinella P. (1998) Collision evolution of Edgeworth-Kuiper Belt Objects. Icarus **125**, 50–60

Science with the Atacama Large Millimeter Array

Karl M. Menten

Max-Planck-Institut für Radioastronomie,
Auf dem Hügel 69, D-53121 Bonn, Germany

Abstract. The Atacama Large Millimeter Array (ALMA) will allow millimeter- and submillimeter-wavelength observations with unprecedented sensitivity and a spatial resolution similar to the diffraction limit of large optical/infrared telescopes. In this paper the planned instrument is described and its scientific potential is illustrated with a number of examples from various fields of astronomy.

1 Introduction – A Bit of History

For more than a decade, plans have been developed by the astronomical communities of various countries for a next generation interferometric array working at millimeter and submillimeter wavelengths. In the USA, planning for a large Millimeter Array (MMA) started in the mid-1980s, motivated by the scientific success of the two millimeter-wavelength arrays operated by the University of California at Berkeley and Caltech. In Europe, discussions on a Large Southern Array (LSA) began in the early 1990s, at about the same time when the Japanese community started to work on a Large Millimeter and Submillimeter Array (LMSA).

The Chilean high Atacama desert was considered as a site for all three of these projects. Because of this and the fact that the number of baselines of an aperture synthesis array grows approximately proportionally to the number of its antenna elements squared (plus many other overwhelming synergy reasons), in 1999 the European and US projects decided on a joint design and development phase with the goal of constructing and operating a joint instrument, the Atacama Large Millimeter Array (ALMA). Japan is expected to become an equal partner in ALMA in the near future.

In the following, we briefly review, in §2, some millimeter- and submillimeter-wavelength astronomy and interferometry fundamentals. In §3 we describe the basic instrumental parameters and performance specifications of ALMA. In §4 we demonstrate ALMA's potential by briefly describing its impact on several scientific questions from various fields of astronomy.

Up-to-date information on the ALMA project can be found on World Wide Web sites at the European Southern Observatory (ESO)[1] and the US National Radio Astronomy Observatory (NRAO).[2]

[1] http://www.eso.org/projects/alma/
[2] http://www.alma.nrao.edu/

2 (Sub)millimeter Astronomy and Interferometry

2.1 Thermal Emission from Dust and Molecules

The appearance of the sky at meter and centimeter wavelengths is largely dominated by continuum emission produced by *non-thermal* processes, such as synchrotron radiation. In contrast, at millimeter and submillimeter wavelengths, *thermal* emission rapidly gains importance with decreasing wavelength (or increasing frequency, ν). For example, black body emission from optically thick compact sources such as the photospheres of a wide variety of stars becomes detectable for sensitive instruments. For these the flux density, S_ν, varies as ν^2.

(Sub)mm continuum emission from the interstellar medium is dominated by grey body radiation from cold and warm dust grains. In the Rayleigh-Jeans regime, the flux density, S_ν, of a dust source at frequency ν is proportional to $\nu^{2+\beta}$, where β has a value between 1 and 2. Given, first, this steep spectral index and, second, the increasing absorption by the Earth's atmosphere toward higher frequencies, one finds that from ground-based observatories dust emission is optimally detected between frequencies of 200 and 360 GHz, corresponding to wavelengths of 1.5 and 0.83 mm.

The total dust emission from a galaxy has a spectral energy distribution that peaks in the far-infrared (around 3 THz or 100 μm). For "normal" spirals, far-infrared dust emission accounts for a significant part of the total luminosity (de Jong et al. 1984; Braine & Hughes 1999) and completely dominates the energy output of ultraluminous infrared galaxies (ULIRGS, see Sanders & Mirabel 1996). Most interestingly, for high redshifts, z, the dust emission peak shifts from the far-infrared to submillimeter and millimeter wavelengths. The resulting enhancement of the (sub)millimeter emission (over)compensates the dimming due to the source's distance (see, e.g., Franceschini et al. 1991; Blain & Longair 1993). This "negative K correction" allows detection of dust in ultraluminous galaxies at $z > 4$ (see §4.5).

Apart from dust, the main astronomical interest in the (sub)millimeter regime derives from the fact that it contains the rotational spectra of many astrophysically interesting molecules. (Sub)mm molecular lines are used as probes for an extremely wide range of astrophysical environments: from cold molecular clouds at a temperature, T, of 10 K and with a molecular hydrogen (H_2) density, n_{H_2}, of 10^2 cm^{-3} to the innermost parts of the circumstellar envelopes around red giant stars, where $T \approx 1500$ K and $n_{H_2} \approx 10^{10}$ cm^{-3}. In particular, the carbon monoxide (CO) molecule, whose rotational lines are observed *in all of these environments*, is widely used to map the global molecular content of the Milky Way and external galaxies as a tracer of the 10^4 times more abundant, but difficult to observe, H_2. High redshift CO emission has been detected toward $\approx$ 10 sources with redshifts between 2.3 and 4.7 (see Barvainis 1999).

2.2 (Sub)millimeter Interferometry Today

Submillimeter single dish telescopes with diffraction-limited resolutions better than $\approx 10''$ become prohibitively expensive to build. Therefore, interferometric methods provide the only practical means for obtaining (sub)arcsecond resolution. The first millimeter wavelength interferometer arrays were brought into operation in the 1980s. Today, a total of four such instruments are working highly successfully, namely the Hat Creek Array operated by the Berkeley-Illinois-Maryland (BIMA) consortium, the millimeter array of the Owens Valley Radio Observatory (OVRO) operated by Caltech, the Nobeyama Millimeter Array (NMA) of the National Astronomical Observatory of Japan, and the interferometer operated by the Institute for Radio Astronomy at Millimeter Wavelengths (IRAM) on the Plateau de Bure in the French Alps. These arrays operate at frequencies between 80 and 270 GHz and contain 9×6 m (BIMA), 5×10 m (OVRO), 6×10 m (NMA), and 5×15 m (IRAM). The IRAM instrument is the one with the highest sensitivity for point-like sources. For an overview of millimeter interferometry, see the proceedings edited by Ishiguro & Welch (1994).

At higher frequencies, pioneering work was done with an ad hoc two element interferometer consisting of the Caltech Submillimeter Observatory (CSO) 10.4 m telescope and the 15 m James Clerk Maxwell Telescope (JCMT) on Mauna Kea, Hawaii (Carlstrom et al. 1994). The first multi-element submillimeter interferometer, the Submillimeter Array (SMA, see Moran 1998), currently under construction on Mauna Kea, will consist of eight 6 m antennas.

3 ALMA – The Next Generation (Sub)millimeter Interferometer

ALMA will revolutionize the field of (sub)millimeter astronomy by opening new vistas in wavelength coverage, spatial resolution, sensitivity, and mapping fidelity. Very diverse demands for high surface brightness sensitivity on the one hand and very high spatial resolution on the other will be met by relocating the array antennas in various configurations by means of transport vehicles. An artist's impression of one relatively compact configuration is shown in Fig. 1.

The excellent site on the Llano de Chajnantor near San Pedro de Atacama (Fig. 1) allows submillimeter operation most of the time and permits placement of antennas for optimized *uv*-coverage over an area of size several kilometers. Longer baselines can be realized by moving antennas to adjoining areas.

In Table 1, ALMA's basic parameters are compared to that of the IRAM interferometer, an instrument that in many respects represents today's state-of-the-art. ALMA's much larger collecting area, wider bandwidth, and the

Fig. 1. The *top* panel shows an artist's impression of several of ALMA's antennas in a compact configuration. The *bottom* panel presents a panoramic view of the 5000 meter high Llano de Chajnantor in the high Atacama desert of northern Chile

combination of advanced receiver technology and superior site result in a continuum sensitivity that is ≈ 100 times higher than that of existing instruments.

Table 1. ALMA in Comparison to a State-of-the-Art Instrument

	ALMA	IRAM Interferometer
Antennas	64×12 m	5×15 m
Collecting Area	7000 m^2	1000 m^2
Maximum baseline	10 km	500 m
Highest resolution	$0\overset{''}{.}01$	$0\overset{''}{.}5$
Frequency coverage	30 – 900 GHz	80 – 115 & 210 – 240 GHz
Bandwidth	2×8 GHz	0.4 GHz
Altitude	5000 m	2500 m

In the following, we address some of the most important aspects of (sub)-millimeter interferometry by discussing the limitations of today's instruments and describing the new possibilities opened by ALMA.

- **Submillimeter operation** is precluded for existing arrays by the relatively low altitudes at which they reside, resulting in significant atmospheric absorption and poor phase coherence, and/or the limited surface accuracy of the telescope dishes. On the site selected for ALMA, long-range monitoring over a five year period has shown that *excellent* atmospheric conditions, i.e. less than 1 mm of precipitable water vapor, allowing operations in all of the submillimeter windows, prevail for $\approx 50\%$ of the time. ALMA's antenna specifications call for an rms surface accuracy of 25μm, allowing efficient submillimeter operation in all the atmospheric windows below 1 THz.
- The highest **spatial resolution** obtainable with today's mm-interferometers is of order $0\farcs5$, more than an order of magnitude worse than the values achievable with the Very Large Array (VLA), the world's premier radio interferometer, at its shortest operating wavelength (0.7 cm) and with large optical telescopes using adaptive optics or speckle methods. Achieving better resolving power by moving the telescopes further apart is limited by the phase coherence that can be achieved on poor sites. Methods to overcome this limitation, such as fast switching between the target source and a calibrator source on a few second time cycle, are limited by the mechanics of today's telescopes. With ALMA, spatial resolutions better than $0\farcs02$ will be achieved, comparable to the diffraction limit of an 8 meter telescope at optical wavelengths. ALMA's design considers the necessary requirements in signal transmission technology and allows for fast switching, which may be needed to achieve the highest angular resolutions.
- Given their total collection areas, existing arrays have a limited **brightness sensitivity**, which would preclude imaging of all but the hottest sources with resolutions significantly better than those attainable today even if the telescopes could be moved further apart. As demonstrated in Fig. 2, ALMA's combination of collecting area, wider receiving bandwidth, better receivers and site will allow continuum imaging with sub-Kelvin sensitivity at resolutions of $0\farcs02$ or better.
- With the small numbers of telescopes comprising existing interferometers, **mapping** complex emission distributions at adequate dynamic range requires observations in multiple configurations, i.e. baseline combinations, which are obtained by moving telescopes to different locations, making this a long duration process. In contrast, much of the power of the VLA derives from the excellent "snapshot" performance made possible by its 27 antennas yielding 351 baselines. The 2016 baselines formed by ALMA's 64 antennas will provide very high dynamic range and permit rapid imaging.

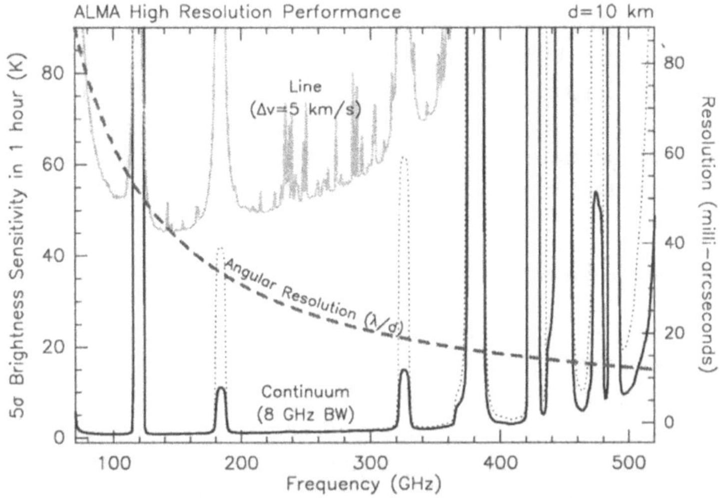

Fig. 2. ALMA's brightness temperature sensitivity at high spatial resolution. The *bottom curves* represent 5 times the 1σ rms noise level (see *left hand ordinate*) that can be reached with ALMA for continuum observations (8 GHz bandwidth, 2 polarizations) within 1 hour of integration time, assuming precipitable water vapor (PWV) contents of 0.8 mm (*solid black line*) and 1.3 mm (*dotted line*). PWV values equal to or lower than 0.8 and 1.3 mm are found at the ALMA site for 25 and 50% of the time, respectively. The *grey curve* shows the 5σ brightness temperature sensitivity reached in 1 hour for a spectral line observation with a resolution of 5 km s^{-1} and 0.8 mm of PWV. All sensitivity calculations assume a source elevation of 45° and a receiver temperature equal to $\frac{5h\nu}{k}$. The *dashed line* indicates the angular resolution (λ/d) corresponding to a maximum baseline length, d, of 10 km (see *right hand ordinate*)

4 Science with ALMA – Examples

With its sensitivity and resolving power, ALMA will allow ground-breaking observations *relevant to all fields of astrophysics.* (Sub)millimeter observations with ALMA will become important cornerstones of multi-wavelength studies, which for optimal scientific return require comparable resolution in the different wavelength regimes. In particular, the synergy between ALMA and the ESO Very Large Telescope (VLT) in many areas of astronomy has been summarized by Lequeux (1996).

A comprehensive overview of ALMA's scientific potential can be found in the proceedings of the "Science with Large Millimetre Arrays" Workshop edited by Shaver (1996). Downes (1999) reviews recent results and future directions of (sub)millimeter interferometry and also discusses the fundamental technological specifications for ALMA and the main science drivers.

Rather than presenting another listing of ALMA's capabilities, in the following we *sketch out* a few interesting problems from various fields of astronomy, from solar system research to cosmology, that could quite easily be addressed by ALMA, but would be practically impossible to tackle with current-day instruments. A preference has been given to examples that demonstrate ALMA's excellent point source sensitivity. Observing time estimates are approximate and assume 1 mm of precipitable water vapor receiver noise temperatures equal to $5\frac{h\nu}{k}$.

4.1 The Outer Solar System – Weather on Pluto and Charon?

Recently, a series of ground-breaking new observations have revolutionized our knowledge of the Pluto–Charon system (for up-to-date reviews and references, see the book by Stern & Tholen 1997). Highlights include the definitive discovery of an atmosphere containing methane (CH_4) and possibly other, heavier constituents, such as nitrogen (see, e.g., Owen et al. 1993), as well as spatially resolved albedo maps (Fig. 3). These were derived from observations gathered during a period of mutual occultations which occurred between 1985 and 1990 when the Pluto–Charon system's orbital plane passed through the line of sight (Buie et al. 1992; Young et al. 1999). Furthermore, imaging with the Faint Object Camera (FOC) of the Hubble Space Telescope (HST) by Albrecht et al. (1994) resulted in definitive determinations of diameters and direct maps of the albedo distribution.

Fig. 3. *B* band albedo distributions of Pluto and its moon Charon reconstructed from multiple mutual occultation events using a maximum entropy method. (Image courtesy Marc W. Buie/Lowell Observatory)

At millimeter wavelengths, the Pluto–Charon system was first detected by Altenhoff et al. (1988) using the IRAM 30 m telescope. These observations implied an astonishingly low brightness temperature, T_B, of only 39 ± 4 K, which is somewhat below the equilibrium temperature, T_{eq}, maintained by absorbed solar heating and thermal emission, and significantly below the values inferred from IRAS observations, which gave a temperature between

55 and 60 K (Sykes et al. 1987). The actual value of the temperature is of great importance for all models of Pluto's atmosphere, since all of its plausible constituents (CH_4, N_2, and CO) have sublimation temperatures in the range in question. The difference between the temperatures inferred from the IRAS and the IRAM results may be explained by a bimodal surface temperature distribution (Jewitt 1994). In this scenario most of the surface is covered by nitrogen ice at a low temperature of 35 K, while 20% is at a higher temperature between 55 and 72 K. In the latter regions the nitrogen ice has evaporated and it seems natural to identify them with the optically dark patches inferred from available albedo maps.

Using ALMA it will be possible to *image* the surface temperature distribution on Pluto. At 345 GHz, a 1σ rms noise level of 6 μJy will be achievable in 1^{h} of integration At the spatial resolution of $0\farcs02$ (10 km baseline) necessary to resolve the disks of Pluto and Charon, which have diameters of $0\farcs14$ and $0\farcs08$, respectively, the corresponding 1σ rms level in brightness temperature is $\approx$ 0.2 K! As discussed, these temperature determinations using ALMA will provide input of fundamental importance for all models of Pluto's and Charon's surfaces and atmospheres.

Using ALMA it will also be possible to detect (sub)millimeter emission from numerous Kuiper–Belt objects in a matter of minutes.

4.2 Imaging of Jupiter-Like Planets Around Nearby Stars

As extensively discussed elsewhere in these proceedings, searches for extra-solar planets around nearby stars are of the greatest current and future interest. High precision astrometry with ALMA will allow detection of Jupiter-mass planets by the reflex motion of their host stars. However, using ALMA, it will also be possible to *directly* image Jupiter-like planets and their host stars out to distances of 5 pc. Brown dwarfs will be detectable to significantly greater distances.

For example, Jupiter at the distance of α Centauri ($D = 1.34$ pc) would have a 345 GHz flux density of 6 μJy, while α Cen A (G2 V) itself is expected to have a flux density of 19 mJy at this frequency. A Jupiter-like planet moving on the same orbit around α Cen A as the real Jupiter around the Sun would, at maximum elongation, appear at a projected distance of $3\farcs9$ from α Cen A. Using ALMA, this planet could be detected at the 5σ level in $\approx$ 20 hours. The dynamic range of $\approx$ 10000 (i.e. maximum signal over rms noise) necessary for this observation should be easily reachable, given that adaptive calibration (referred to as "self calibration" by radio astronomers) can be employed using α Cen A as a reference.

Self calibration will be possible as long as the reference star is detected on each baseline at a few σ within a coherence time. Even for coherence times as short as a few seconds, we find that self calibration with solar-type stars will be possible out to distances of 5 pc. Once coherence is achieved with self calibration, long integrations are possible; e.g. 250 hours of observing time would be required for a 5σ detection of the "Jupiter" at $D = 5$ pc. Other possible target stars observable by ALMA within or around that distance include τ Cet (G8 VI, $D = 3.6$ pc), σ Pav (G6 V, $D = 5.7$ pc), Sirius (A1 V, $D = 2.7$ pc), Procyon (F5 V, $D = 3.5$ pc), and Altair (A7 V, $D = 5.1$ pc).

4.3 Planetary System Formation in Protostellar Accretion Disks

Recent HST observations have provided spectacular direct evidence for accretion disks around newly formed stars (McCaughrean & O'Dell 1996; Padgett et al. 1999). However, high resolution submillimeter observations are necessary to understand the evolution of dust grains into planetesimals during the first million years of a proto-solar system disk.

Fig. 4. Hydrodynamic model of a protostellar accretion disk in which a giant protoplanet is forming (Bryden et al. 1999). The newly formed protostar resides (invisibly) at the center of the accretion disk and a Jupiter-mass protoplanet orbits around it at the Jupiter-Sun distance. A gap in the disk is cleared out by the protoplanet. Using ALMA it will be easy to image such gaps in protostellar disks in nearby star-forming regions. In the plot disk surface density is coded as "height"

Millimeter-wavelength emission from both dust and molecules in circumstellar accretion disks is convincingly detected with today's interferometers (Dutrey et al. 1994; Wilner et al. 1996; Rodríguez et al. 1998; Guilloteau et al. 1999). In particular, state-of-the-art observations using the VLA now afford resolutions that, in the nearest star-forming regions such as the Taurus molecular cloud complex ($D = 140$ pc), correspond to spatial scales or ≈ 5 AU, which is ≈ 10 times better than what can be achieved with today's millimeter interferometers. However, using the VLA, dust can only be detected at 0.7 cm, this instrument's shortest operating wavelength, whereas, given its steep spectral index, dust emission is much stronger and optimally studied in the (sub)millimeter regime. Here, ALMA, which will allow spatially resolved analyses of the temperature and density structure of the planet-forming regions of such disks.

Most excitingly, using ALMA it may be possible to directly observe the interaction of newly formed planets with their accretion disks. Fig. 4 shows the results of a numerical simulation of a protostellar accretion disk in which a Jupiter-mass planet has formed at a distance from the central protostar that is equal to Jupiter's distance from the Sun in our solar system (Bryden et al. 1999). The planet clears out a gap $\approx$ 1.5 AU wide, in which the surface density drops sharply. In a protoplanetary disk in the Taurus complex the cleared-out ring would have an angular diameter of $0\farcs07$ and a width of $0\farcs01$, which will be resolvable by ALMA (see Fig. 2).

Multi-wavelength observations with ALMA will spatially resolve circumstellar accretion disks, yielding density and temperature profiles. This and its potential to detect rings and gaps caused by giant planet formation, will make ALMA a unique tool for studying many of the most important aspects of planetary system formation.

4.4 Elemental Depletion and Dust Formation in Red Giant Stars

During their time on the asymptotic giant branch (AGB), stars lose mass at rates between a few times 10^{-8} and 10^{-4} $M_{\odot}$ yr^{-1}. Dust produced in these circumstellar outflows is an important source of material return to the interstellar medium.

Existing mm-wavelength interferometers provide detailed maps of the emission distributions of numerous molecules in the outer envelopes of these stars, i.e., at distances of hundreds to thousands of stellar radii from the central star (see, e.g., Lucas & Guélin 1999). The latest of these maps are of such high quality that it will be of great interest to compare them with state-of-the-art models of the chemistry in these regions, which is driven by the interstellar radiation field with initial conditions set by the equilibrium chemistry in the stellar atmosphere (see, e.g., Millar & Herbst 1994).

ALMA's high resolution and superb brightness sensitivity will allow detailed studies of the innermost regions of the circumstellar envelopes around mass-losing red giant stars (Fig. 5). With ALMA it will be possible to resolve the (sub)millimeter photospheres (see Reid & Menten 1997) of many nearby stars. Using high excitation spectral lines as probes it will be possible to measure molecular abundance profiles throughout the hot ($T > 1000$ K) inner envelope, which today is only probed by infrared absorption and radio and submillimeter maser lines. ALMA will provide direct determinations of the spatial scales on which refractory molecules condense into dust grains and thus provide fundamental new information on elemental depletion and dust formation processes.

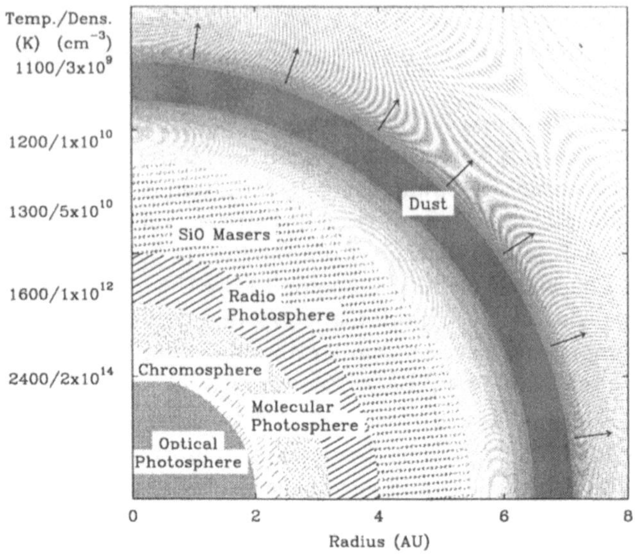

Fig. 5. The innermost envelope of a long-period variable star (from Reid & Menten 1997). Indicated are, starting from the origin and moving outward, the optical photosphere, a possible chromosphere, the molecular photosphere, the radio photosphere, a molecular region from which, e.g., SiO maser emission arises, and the dust formation region. Once dust forms, it is accelerated by radiation pressure away from the star. Distance from the center of the star is shown on the horizontal axis, while the vertical axis gives temperature and density profiles for a static model atmosphere

4.5 Star Formation in the Early Universe

The recent detection of a submillimeter background (Puget et al. 1996) and its resolution into discrete sources of dust emission (Hughes et al. 1998; Barger et al. 1998) indicates that determinations of the star-forming activity in the early universe derived from optical studies alone, in particular of the Hubble Deep field (HDF, Madau et al. 1996), may considerably underestimate star formation rates at $z > 1$. Currently, mapping with multi-element bolometer cameras, such as the Submillimeter Common User Bolometer Array (SCUBA) on the JCMT, is the most efficient way to find candidate sources and has led to the detection of more than two dozen sources that are almost certainly highly redshifted ($z > 1$) ultraluminous infrared galaxies with infrared luminosities that are several times higher than that of the archetypical ULIRG Arp 220 (Barger et al. 1999).

Rapid technological developments will lead to increasingly larger bolometer arrays. While current systems typically have several dozens of elements, arrays with thousands of bolometers are currently being planned. One might therefore think that this multiplexing advantage together with the highly competitive sensitivity of a bolometer compared to a coherent receiver would

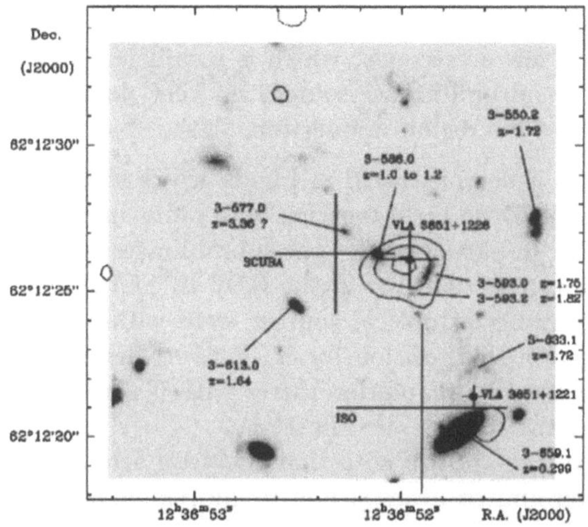

Fig. 6. Multi-wavelength view of part of the Hubble Deep Field (from Downes et al. 1999). The *greyscale* display shows the area of the BVI image of the Hubble Deep Field around the submillimeter source HDF 850.1 detected with SCUBA by Hughes et al. 1998. The *medium size* and *large crosses* denote the positions and uncertainties of the SCUBA source and a 15μm source detected by ISO, while the *small crosses with dots* mark radio sources detected with the VLA. The *contour lines* represent the 2$''$ resolution IRAM interferometer map of the 1.3 mm dust emission from HDF 850.1. Several optical objects are identified with their photometric redshift, except for 3–577.0, which has a tentative spectroscopic redshift of 3.36. See Downes et al. 1999 for references

greatly favor bolometer arrays over ALMA. While, indeed, single dish observations with bolometer arrays will produce large area maps in which many luminous sources will be detected, the most interesting questions in this field will only be solved by ALMA for the following reasons:

- Bolometer arrays will reach the **confusion** limit once sensitivity levels reached are only a factor of a few below those already achieved by deep SCUBA observations ($\approx$ 1 mJy at 850μm). While confusion is a *fundamental* limit for single dish bolometer observations (see Blain et al. 1998), it will virtually not affect ALMA observations. Even in low resolution configurations, ALMA's synthesized beam will have an area that is hundreds of time smaller than the diffraction-limited single dish beam, thus spatially isolating confusing sources.
- Many of the **stability** problems limiting single-dish integrations do not affect interferometer observations. With ALMA, hundreds of hours long integrations will be possible, allowing detection of the faint end of the luminosity function.
- ALMA will be able to spatially **resolve sources** and thus probe their morphology.

- Using a single dish, it is very difficult to achieve **position accuracy** better than a few arcseconds, which is insufficient to make secure identifications of (sub)millimeter sources on very deep optical, infrared, or radio images of the region in question.

The last point is demonstrated in Fig. 6, which shows part of the optical image of the Hubble Deep Field together with radio, infrared, and submillimeter identifications. In particular, for the submillimeter source HDF 850.1, the strongest of the sources detected in the HDF by SCUBA at 850μm (Hughes et al. 1998), a number of optical sources were within the single dish error box. However, given the position for HDF 850.1 determined by Downes et al. (1999) using the IRAM interferometer with $0\farcs3$ precision, none of these turned out to be the correct identification.

As a practical remark, we note that ALMA's field of view is determined by the diffraction-limited beam size of its individual antennas, i.e. 20″ for a 12 m diameter dish at 850μm wavelength. Thus, covering the area of the HDF mapped with SCUBA would require $\approx$ 500 half beamwidth-spaced pointings. The $\approx$ 1 mJy noise level reached by SCUBA over a 50 hour observation would be reached in less than a minute, so that an observation of the full mosaic would require a total observing time of several hours only. Given the long integration potential of interferometers described above, very deep ALMA observations will provide crucial information on the star formation history of the early universe.

4.6 Zooming in on Nearby Starburst Cores

The starburst cores of ultra-luminous infrared galaxies (ULIRGs) are hidden from direct observation at all wavelengths shortward of the far-infrared range. In the archetypical ULIRG Arp 220, the dust is optically thick even at far-infrared wavelengths. Analysis of ISO observations of fine structure lines from various atoms and ions, in particular the absence of high-ionization lines, leads to the conclusion that Arp 220's prodigious luminosity of $1.5 \times 10^{12}\ L_\odot$ is indeed produced by a massive starburst (star formation rate 100 $M_\odot$ yr^{-1}) and not by nuclear activity (Genzel et al. 1998).

Most interestingly, this conclusion is reinforced by recent milliarcsecond resolution radio imaging with an intercontinental Very Long Baseline Interferometry (VLBI) network (Smith et al. 1998). No high brightness continuum core, which could be identified as an AGN, is found. These observations "resolve out" most of the radio continuum emission, but result in the detection of more than a dozen unresolved sources, which are likely to be luminous radio supernovae (Fig. 7), providing evidence for Arp 220's violent star formation activity.

Given the high dust opacities and the fact that sub-arcsecond resolution observations at far-infrared wavelengths have to await future space interferometry missions (see Leisawitz et al. 1999), radio and (sub)millimeter interferometry provides the only means for detailed studies of the excitation and

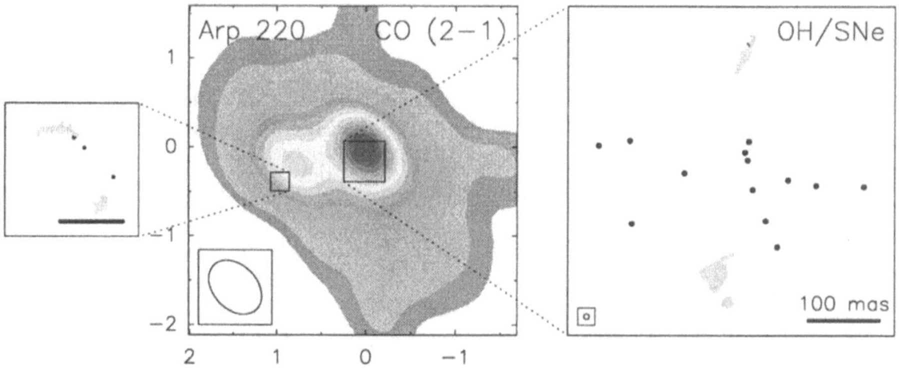

Fig. 7. The *center* panel shows a map of CO $J = 2 - 1$ emission toward the central region of the ultra-luminous infrared galaxy Arp 220 made by Downes & Solomon 1999 using the IRAM interferometer. The FWHM size of their restoring beam is shown in the ***lower left hand*** corner. Position offsets are in arcseconds relative to $(\alpha, \delta)_{J2000} = 15^h34^m57\overset{s}{.}2246, +23°30'11\overset{''}{.}564$. The *left* and ***right hand*** panels show, as indicated, blow-ups of the regions in which Lonsdale et al. 1998 and Smith et al. 1998 have imaged OH megamaser emission and continuum emission from numerous radio supernovae with the VLBA at a resolution of $\approx$ 8 milliarcseconds (see ***left hand*** corner of the ***right hand*** map. Horizontal bars represent 100 milliarcseconds. The unresolved supernovae are marked as ***black dots***, while the outlines of the regions showing megamaser emission are indicated in ***grey***. Note that most of the megamaser emission is "resolved out" by the VLBA

dynamics of the interstellar medium in the cores of ULIRGs in the local universe. Recent imaging of CO emission toward a sample of ULIRGs has shown that these galaxies have nuclear disks or rings (Downes & Solomon 1999). In the case of Arp 220, enhanced CO emission is observed from two regions, indicating that these are the dominantly active starburst sites, which is consistent with the VLBI observations (Fig. 7). However, the $\approx 0\overset{''}{.}5$ resolution attainable today with the IRAM interferometer precludes detailed studies of the molecular clouds in Arp 220 and the interactions of newly formed and exploding stars with their environment. As shown in Fig. 2 ALMA will allow studies of warm molecular material with a resolution of $\sim 0\overset{''}{.}02$, which corresponds to $\approx$ 7 pc at Arp 220's distance of 75 Mpc, allowing resolution of single molecular clouds. Suitable tracers will include the higher rotational lines of CO, HCN, and the CI fine structure lines.

I would like to thank Malcolm Walmsley for useful suggestions and Mark Reid for a careful reading of the manuscript. I am grateful to Geoffrey Bryden and Dennis Downes for sending me material used in Figs. 4 and 7 in electronic form.

References

1. Albrecht, R. et al. (1994) High resolution imaging of the Pluto–Charon system with the Faint Object Camera of the Hubble Space Telescope. ApJ **435**, L75–L78
2. Altenhoff, W. J. et al. (1988) First radio astronomical estimate of the temperature of Pluto. A&A **190**, L15–L17
3. Barger, A. J. et al. (1998) Submillimetre-wavelength detection of dusty star-forming galaxies at high redshift. Nature **394**, 248–251
4. Barger, A. J., Cowie, L. L., Sanders, D. B. (1999) Resolving the submillimeter background: the 850 micron galaxy counts. ApJ **518**, L5–L8
5. Barvainis, R. (1999) Molecular emission lines from high redshift AGNs. in *Highly Redshifted Radio Lines*, eds. C. L. Carilli, S. J. E. Radford, K. M. Menten, & G. I. Langston (ASP: San Francisico), 39–44
6. Blain, A. W., Ivison, R. J., Smail, I. (1998) Observational limits to source confusion in the millimetre/submillimetre waveband. MNRAS **296**, L29–L33
7. Blain, A. W., Longair, M. S. (1993) Submillimetre cosmology. MNRAS **264**, 509–521
8. Braine, J., Hughes, D. H. (1999) The 10 GHz – 10 THz spectrum of a normal spiral galaxy. A&A **344**, 779–786
9. Bryden, G., Chen, X., Lin, D. N. C., Nelson, R. P., Papaloizou, J. C. B. (1999) Tidally induced gap formation in protostellar disks: gap clearing and suppression of protoplanetary growth. ApJ **514**, 344–367
10. Buie, M. W., Tholen, D. J., Horne, K. (1992) Albedo maps of Pluto and Charon: initial event results. Icarus **97**, 211–227
11. Carlstrom, J. E., Hills, R. E., Lay, O. P., Force, B., Hall, C. G., Phillips, T. G., Schinckel, A. E. (1994) The CSO – JCMT submillimeter interferometer. in Ishiguro & Welch (1994), 35–40
12. De Jong, T. et al. (1984) IRAS observations of Shapley-Ames galaxies. ApJ **278** L67–L70
13. Downes, D. (1999) High resolution millimeter and submm astronomy: recent results and future directions. Rev. Mod. Astr. **12**, 69–96
14. Downes, D. et al. (1999) Proposed identification of Hubble Deep Field submillimeter source HDF 850.1. A&A **347**, 809–820
15. Downes, D., Solomon, P. M. (1998) Rotating nuclear rings and extreme starbursts in ultraluminous galaxies. ApJ **507**, 615–654
16. Dutrey, A., Guilloteau, S., Simon, M. (1994) Images of the GG Tauri rotating ring. A&A **286**, 149–159
17. Franceschini, A., Toffolatti, L., Mazzei, P., Danese, L., De Zotti, G. (1991) Galaxy counts and contributions to the background radiation from 1μm to 1000μm. A&A **89**, 285–310
18. Genzel, R. et al. (1998) What powers ultraluminous IRAS galaxies? ApJ **498**, 579–605
19. Guilloteau, S., Dutrey, A., Simon, M. (1999) GG Tauri: the ring world. A&A **348**, 570–578
20. Hughes, D. H. et al. (1998) High-redshift star formation in the Hubble Deep Field revealed by a submillimetre-wavelength survey. Nature **394**, 241–247
21. Ishiguro, M., Welch, W. J. (eds.) (1994) IAU Coll. 140, *Astronomy with Millimeter and Submillimeter Wave Interferometry*, (ASP: San Francisco)

22. Jewitt, P. (1994) Heat from Pluto. AJ **107**, 372–378
23. Leisawitz, D., et al. (1999) The submillimeter probe of the evolution of cosmic structure (SPECS). in *The Physics and Chemistry of the Interstellar Medium*, eds. V. Ossenkopf, J. Stutzki, & G. Winnewisser (GCA-Verlag: Herdecke), 454–457
24. Lequeux, J. (1996) The synergy between a Large Southern Array and the Very Large Telescope of ESO. in Shaver (1996), 337–339
25. Lonsdale, C. J., Lonsdale, C. J., Diamond, P. J., Smith, H. E. (1998) Symmetric parsec-scale OH megamaser structures in Arp 220. ApJ **493**, L13–L16
26. Lucas, R., Guélin, M. (1999) Millimeter-wave interferometry of circumstellar envelopes. in IAU Symp. 191, *Asymptotic Giant Branch Stars*, eds. T. Le Bertre, A. Lèbre, & C. Waelkens (San Francisco: ASP), 305–314
27. Madau, P., Ferguson, H. C., Dickinson, M. E., Giavalisco, M., Steidel, C. C., Fruchter, A. (1996) High-redshift galaxies in the Hubble Deep Field: colour selection and star formation history to $z \sim 4$. MNRAS **283**, 1388–1404
28. McCaughrean, M. J., O'Dell, C. R. (1996) Direct imaging of circumstellar disks in the Orion nebula. AJ **111**, 1977–1986
29. Millar, T. J., Herbst, E. (1994) A new chemical model of the circumstellar envelope surrounding IRC+10216. A&A **288**, 561–571
30. Moran, J. M. (1998) Submillimeter array. in *Advanced Technology MMW, Radio, and Terahertz Telescopes*, ed. T. G. Phillips, Proc. SPIE **3357**, 208–219
31. Owen, T. C. et al. (1993) Surface ices and the atmospheric composition of Pluto. Science **261**, 745–748
32. Padgett, D. L., Brandner, W., Stapelfeldt, K. R., Strom, S. E., Terebey, S., Koerner, D. (1999) Hubble Space Telescope/NICMOS imaging of disks and envelopes around very young stars. AJ **117**, 1490–1504
33. Puget, J.-L., Abergel, A., Bernard, J.-P., Boulanger, F., Burton, W. B., Desert, F.-X., Hartmann, D. (1996) Tentative detection of a cosmic far-infrared background with COBE. A&A **308**, L5–L8
34. Reid, M. J., Menten, K. M. (1997) Radio photospheres of long-period variable stars. ApJ **476**, 327–346
35. Rodríguez, L. F., et al. (1998) Compact protoplanetary disks around the stars of a young binary system. Nature **395**, 355–357
36. Sanders, D. B., Mirabel, I. F. (1996) Luminous infrared galaxies. ARA&A **34**, 749–792
37. Shaver, P. A. (ed.) (1996) *Science with Large Millimetre Arrays – Proceedings of the ESO-IRAM-NFRA-Onsala Workshop* (Springer: Berlin, Heidelberg)
38. Smith, H. E., Lonsdale, C. J., Lonsdale, C. J., Diamond, P. J. (1998) A starburst revealed – Luminous radio supernovae in the nuclei of Arp 220. ApJ **493** L17–L21
39. Stern, S. A., Tholen, D. J. (eds.) (1997) *Pluto and Charon* (University of Arizona Press: Tucson)
40. Sykes, M. V., Cutri, R. M., Lebofsky, L. A., Binzel, R. P. (1987) IRAS serendipitous survey observations of Pluto and Charon. Science **237**, 1336–1340
41. Wilner, D. J., Ho, P. T. P., Rodríguez, L. F. (1996) Subarcsecond VLA observations of HL Tauri: imaging the circumstellar disk. ApJ **470**, L117–L121
42. Young, E. F., Galdamez, K., Buie, M. W., Binzel, R. P., Tholen, D. J. (1999) Mapping the variegated surface of Pluto. ApJ **117**, 1063–1076

Synergy Between the VLT and the NGST

H.S. Stockman

Space Telescope Science Institute, 3700 San Martin Drive,
Baltimore MD 21218, USA

Abstract. Three major astronomical facilities will address the distant universe and processes of star formation in the next decade: the Next Generation Space Telescope (NGST), the Very Large Telescope as a complex (VLT) and an interferometer (VLTI), and, at longer wavelengths, the Large Southern Array/Millimeter Array (LSA/MMA). Each facility exploits natural windows and low backgrounds to achieve a unique view of the universe. The initial complement of VLT instruments are defined and are under development, but those for NGST are still under study by NASA/ESA/CSA teams. The NGST Ad Hoc Science Working Group (ASWG) has compiled a representative list of scientific programs to guide the design and development of the observatory. We compare these programs to the goals of the VLT/VLTI to illustrate complementary areas of strength and to suggest paths for scientific progress and instrumental improvement.

1 Introduction

The "core" science goals for NGST were enunciated in the report of the HST & Beyond Committee, chaired by Alan Dressler [1]. In their executive summary, the committee emphasized that a large (> 4 m dia. (diameter)), passively cooled telescope would enable the study of the origins and evolution of galaxies. Working in the near infrared (NIR) and limited only by the scattering and emission of the zodiacal dust cloud, the soon-to-be-called NGST would be able to detect the earliest star bursts at very high redshift (z >10) as well as the assembly of galaxies in the epoch z ~1–5. The compelling images (0.4–0.9 μm) of the Hubble Deep Field North (HDFN) [2] have been complemented with NIR bands using NICMOS and the recent survey in the south (HDFS) [3]. These images and the spectroscopic followups with the Keck telescopes [4]and the VLT verification program [5] have only strengthened the case for a large, space telescope optimized in the NIR. To provide additional guidance in the formulation phase of NGST (previously referred to as Phase A/B), the ASWG has compiled an extensive set of scientific studies, the Design Reference Mission, which represent the key science enabled by NGST (http://www.ngst.stsci.edu/).

After the initial concept study in 1996 [6] and Pre-Phase A studies by the TRW Corp [7] and Ball Aerospace [8], the architecture for NGST is qualitatively defined in preparation for the selection of the formulation phase studies prior to construction and launch in 2007–2008. All viable concepts involve a large, deployable primary mirror (~ 8 m dia.) and an even larger sunshade to permit the entire telescope structure and instrument complement to cool passively to approximately $\sim 30\,\mathrm{K}$. NGST will be launched into a halo orbit about the second Lagrange point, L2. The key elements of the NGST are shown in Fig. 1.

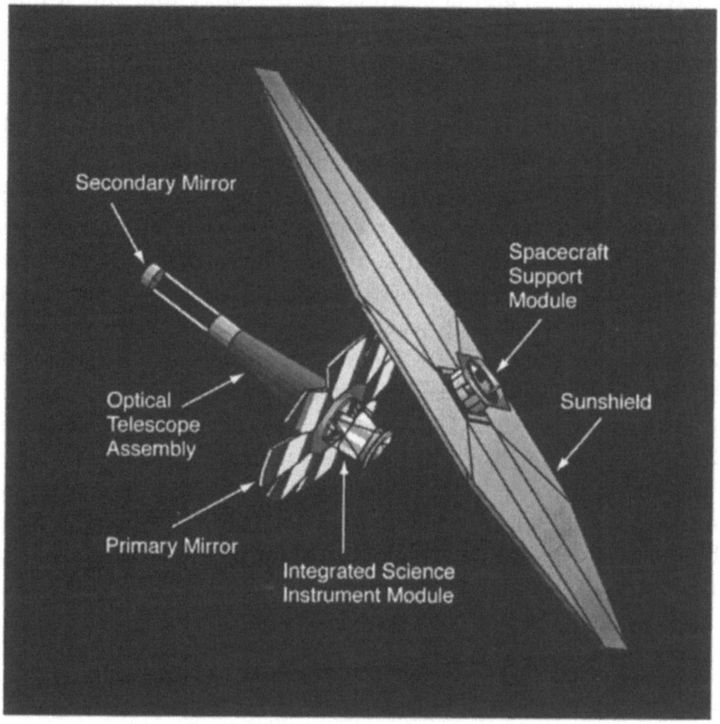

Fig. 1. The Yardstick NGST concept (GSFC/NASA)

2 NGST Science

The DRM programs have five major themes: the structure and geometry of the Universe; the origins and evolution of galaxies; the Milky Way and its neighbors; the physics of star formation; and the origins and evolution of planetary systems. These themes address the higher goal of understand the origin of the Milky Way, our Solar System, and the astronomical origins of life. We expect that the VLT will pursue many of the same goals and, with its powerful complement of instruments, will be especially well-suited for in-depth studies of sources initially discovered by HST, the VLT Survey Telescope (VST [9].) and, in the latter portions of the next decade, by NGST. In this section, we discuss two of these themes in more detail and indicate the areas of common interest.

2.1 The Origins and Evolution of Galaxies

Since the HST & Beyond report, HST and large telescopes such as the Keck and the VLT have made substantial progress in detecting early galactic structures and what they may imply. Perhaps the best evidence is that the highest confirmed redshift, $z = 5.6$, is ascribed to a galaxy and not a quasar [10]. The UV luminosity density of high redshift galaxies can be interpreted in terms of star formation rates and, when it is adjusted for dust obscuration, is remarkably flat in the epoch $z \sim$ 1-4. ([11], [12]). Indeed, the amount of star formation may be sufficient to maintain the ionization of the Universe at epochs prior to the ultraviolet contributions by AGN [13].

The goals for HST and ground-based telescopes will be to detect galaxies at ever high redshifts and better characterize the number densities and morphologies of galaxies at redshifts, $z \sim 1$. The former goal will involve specialized surveys to find unusually red objects, H-I > 4–5. Such objects will be important to understanding when the assembly of spheroid structures is complete. The latter goal will involve moderate resolution NIR spectroscopy ($R \sim 5000$) to work between the OH lines and to measure the dynamics/mass of Milky Way galaxies when they were 5–6 Gyr old. These galaxies will be $H_{\mathrm{AB}} \sim$ 21–23, with a half-light radius of *sim* 0.5 arcsec [14]. NIR spectrometers such as the VLT ISAAC will be perfectly suited to such studies, using the superb natural seeing at Paranal. Because of the OH "forest", such studies will highlight galaxies and clusters with redshifts where the chief diagnostic lines may be cleanly separated from sky emission.

The NGST goals for this theme are:

- Detect the formation of the earliest stars, the earliest stellar structures.
- Survey the numbers and morphologies of progenitors of galaxies such as the Milky Way between $1 < z < 5$. Obtain star formation rates, metallicities, and masses of the major components (~1–10% the mass of the Milky Way).
- Study the evolution of galaxies in the dense environments of groups and clusters ($1 < z < 5$).
- Determine the number and morphologies of dust-enshrouded, star formation regions in Milky Way galaxies and their progenitors to $z \sim 2$. These regions emit much of their luminosity in the MIR from warm dust in the photo- dissociation regions (PDR) surrounding young star clusters.

In Fig. 2, we display the potential sensitivity of NGST for a NGST deep field exposure (10^6 s in a broad band). Also shown are predicted spectra of young starbursts (10^6 solar masses) and older, established progenitors, 0.01 L_*, 10^8 solar masses and 10^8 yr. $\Omega_{\mathrm{M}} = 0.2, \Omega_\Lambda = 0.0$). With these sensitivities, NGST should detect thousands of very high redshift starbursts and Milky Way progenitors in every $4' \times 4'$ field. In the HDFN and HDFS NICMOS observations, we have already detected the population of galaxies whose spectra can be studied with NGST. These sources are faint, $H_{\mathrm{AB}} \sim$ 29, are small with a half-light-radius of 0.1 arcsec. To provide discrimination between low redshift dwarf galaxies and high redshift progenitors, the wavelength range of NGST has been extended to 0.6 μm. Galaxies with $z > 5$ will have their spectra heavily absorbed by the Lyman- alpha forest and will "drop-out" of the NGST R band images. Galaxies at intermediate redshifts will be readily identified by their spectral energy distributions [15].

We demonstrate the difference between potential VLT objects and NGST targets in Fig. 3. In this HDFS NICMOS image (50″ × 50″), we have placed 1.5 arcsec circles upon all targets with integrated magnitude $H_{\mathrm{AB}} < 23$. Although the brighter galaxies (presumably with $z \sim 1$) spill over such a large aperture, the surface brightness outside the aperture is too faint for spectral observations, except perhaps with NGST. The smaller targets have high central surface brightnesses and fit within the 0.4 arcsec apertures drawn on the image. There are approximately 100 targets, $H_{\mathrm{AB}} = 27$–29 within the 1′ × 1′ field.

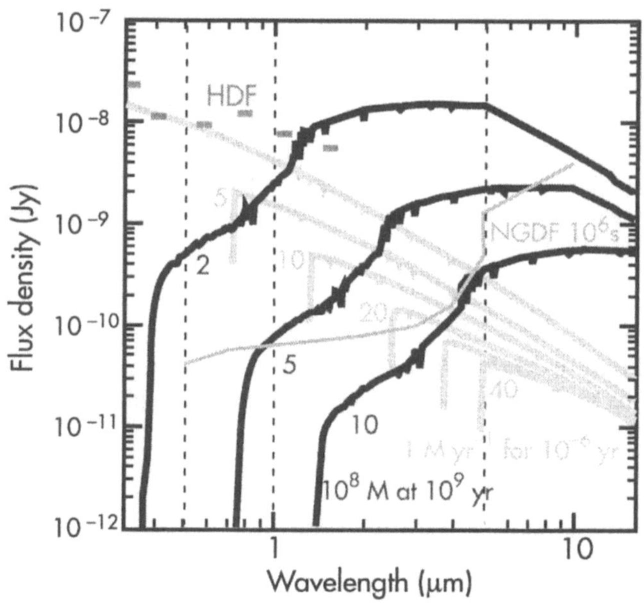

Fig. 2. The sensitivity of a NGST Deep Field image. R = 3, 10^6 s. Spectra of young starbursts and older, established galactic progenitors are also shown. The cosmology adopted is pessimistic: $\Omega_{\mathrm{M}} = 0.2$, $\Omega_{\Lambda} = 0.0$ (Lilly, U. Toronto)

Recent discoveries of "extremely red objects" have revealed a population of relatively bright NIR sources that show very little emission in visible

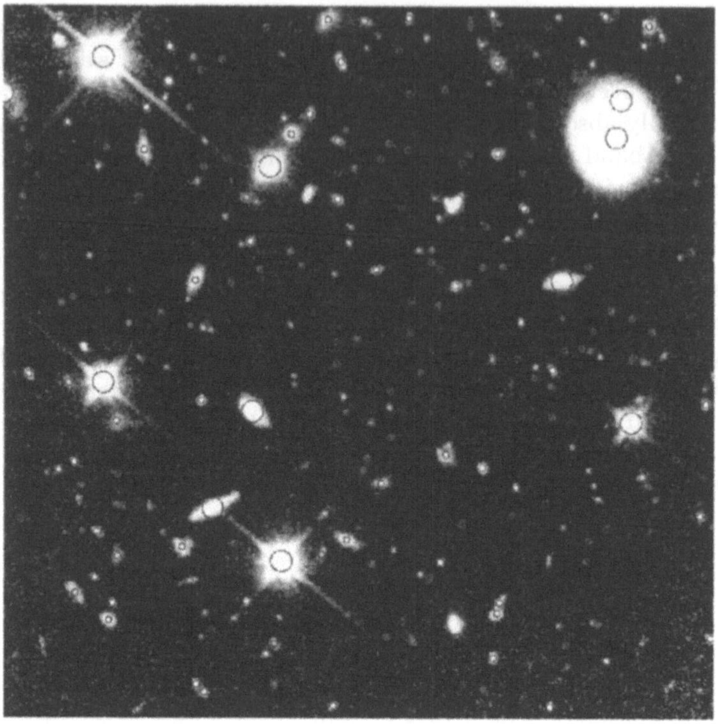

Fig. 3. The NICMOS image of the HDFS field[15]. Large, 1.5″ diameter apertures are placed over all sources brighter than $H_{\mathrm{AB}} \sim 23$. Small apertures, 0.5″ aperture, are placed over sources with $H_{\mathrm{AB}}= 24$–29. The HAB field is complete to $H_{\mathrm{AB}} = 29.5$. (Stiavelli, STScI)

bands. These sources, which are also very compact, may be quiescent spheroids at moderate redshift,$z = 2$–3, or luminous star forming regions at high redshift,$z > 9$–10. We expect that Keck and the VLT will distinguish between these two scenarios.

2.2 Star and Planet Forming Regions

Both NGST and VLTI can utilize special capabilities in the NIR and MIR to study the physics of star and planet formation. The goal of these studies is to understand the process of star formation and the origin of the initial mass function. The DRM contains programs that measure the inflow and outflow rates of protostars, class 0–I, by observing MIR emission line strengths and shape and by actually resolving the inner regions of the accretion disk. The NGST will be able to establish the period during which gas is cleared from the proto- planetary nebula, the end of giant planet formation. The LSA/MMA will probe the cooler gas, farther from the central star. At 10 μm wavelengths, the VLTI will resolve the most central regions and perhaps the origins of the powerful jets.

As discovered by IRAS, many young stars display evidence of remnant protostellar disks. In the MIR, NGST will be capable of detecting such disks about all stars in the Solar neighborhood. These ancient remnants of the solar nebulae may well provide evidence – gaps – of planet formation. Using a graded Lyot stop, NGST would be capable of detecting the MIR emission from a Jupiter mass planet around stars out to 8 parsec (a list of 50 candidates to be winnowed by the Space Interferometer mission (SIM)).

3 Scientific Instrumentation

The scientific instruments on NGST should take full advantage of the diffraction-limited angular resolution, the relatively wide field of view, the low background, and the perfect transparency of space. One dozen science teams are studying a wide variety of instruments and NGST-specific technologies. The results of these studies will be written reports and papers presented at the NGST Technology and Science Challenge conference at the Woods Hole conference center, September 16–16, 1999. NASA, ESA, and CSA will decide the international contributions to the mission based upon the recommendations of the ASWG, cost, and the technological readiness of the designs. In 1998, the ASWG concluded that the two highest priorities for instrument capabilities are wide-field NIR imaging – probably extended to 0.6 μm – and some form of wide-field NIR spectroscopy. However there are several viable solutions to those instruments, including concepts such as an Imaging Fourier Transform Spectrometer (IFTS) that combine the two functions. There is also growing interest in the astronomy community in MIR astronomy. NGST would be the first MIR mission after SIRTF and has the potential to revolutionize astronomy in this wavelength region by providing several orders of magnitude increases in sensitivity over other planned facilities.

However, the NGST instruments will be significantly more focused than their ground-based cousins. Mechanisms, detector arrays, dispersing elements, and other features will severely constrained by the overall budgets for development and operations. To meet the cost mark, the NGST science and management teams must find ways to achieve the major scientific goals of NGST while striving for simplicity and economy. It is not possible to achieve the "no-moving" parts simplicity of the SIRTF mission. But it is clear that the NGST instruments must be simpler than those for HST. In Table 1 [16], we indicate which of the planned VLT instruments are similar to those considered most important for NGST. The three areas where NGST is likely not to compete with the VLT are very wide field imaging and spectroscopy with natural seeing and fields larger than 10/arcmin, visible light spectroscopy, and high resolution NIR spectroscopy (R >5000, 1–2.5 μm). Except for the regions of strong atmospheric absorption, ground-based telescopes should have performance comparable (or superior) to that of NGST in these areas.

Table 1. First Generation VLT Instruments

Instrument	Range (μm	Description	Comments
CONICA	1–5	AO Camera/Spec.	NGST-like
NAOS	1–5	AO System	NGST-like
ISAAC	1–5	WF Camera/Spec.	Complementary
FLAMES	0.3–2.2	Fibre MOS(s)	Complementary
SINFONI	1–2.2	AO Integral Fld.	NGST-like
UVES	0.3–1.1	Hi-Res Echelle	Complementary
VISIR	8.0–24.0	Cam/Spectro.	NGST-like
CRIRES	1–5	Hi-Res Echelle	Complementary
FORS	0.33–1.1	WF Camera/Spec.	Complementary
VIMOS	0.36–1.0	WF Slit MOS	Complementary
NIRMOS	1–1.8	WF Slit MOS	NGST-like

4 Synergy Between NGST and the VLT

NGST has already benefited from the engineering and technologies used on the VLT. The VLT and NNT are arguably the best example of active optics to control the overall wavefront of a large, inherently flexible telescope. The experience with the segmented optical design of the Keck telescopes has also shown the advantages of replacing the mechanical strength of the reflecting surface with stronger support structures. However, the philosophy and sophistication of end- to-end engineering models are a hallmark of the VLT project. Such models are indispensable for coherent, objective design and management of modern space- and ground-based telescopes.

The scientific discoveries from VLT will have profound influence on the selection and design of the NGST science instruments in 2001–2002 and setting the scientific context of NGST and the LSA/MMA in 2005 prior their operation. The large and deep redshift surveys will establish the nature of the populations that NGST will study and will provide selected fields for NGST surveys of deep fields and high redshift clusters. Finally, the search for high redshift supernovae with the VST will establish whether ground-based surveys may be useful for the discovery of UV precursors of supernovae at high redshifts. Because high-redshift supernovae may be relatively rare in an

NGST $4' \times 4'$ field, the VST/VLT may be essential for extending the study of cosmology and high- redshift metal production with NGST.

What can the NGST do for the VLT and other ground-based observatories? The most obvious benefits are the technologies being researched for the NGST and other NASA Origins mission: ultra-lightweight mirrors, cryogenic actuators, micro- mirror and micro-aperture arrays, and low-noise, large format NIR and MIR detectors. The latter will become increasingly important for the VLT for high- resolution spectroscopy of faint objects. Low dark currents and readnoise performance of NIR detectors will become comparable to that of CCDs and will have the high quantum efficiency of materials such as InSb. This performance will enable the resolution and reduction of the atmospheric OH lines with almost no loss in overall efficiency. All NIR spectroscopy with the VLT and AO will benefit. We also can imagine Lyman-alpha forest observations and IGM studies of rare, high redshift quasars at $z \sim 8$–10, if the Universe is ionized at these epochs!

5 Acknowledgements

NASA, ESA, and CSA support the NGST program. We have made liberal use of the scientific goals for NGST developed by the Ad Hoc Science Working Group and encourage the reader to learn more about the NGST project at the NASA and STScI websites: http://www.ngst.nasa.gov/ and http://www.ngst.stsci.edu/. We are also grateful for the invitation to join the ESO staff and the astronomers of Europe in celebrating the opening of the VLT, which soon will be the world's premier ground-based observatory. Bon voyage.

References

1. Dressler, A. (1996) Exploration and the Search for Origins: A Vision for Ultraviolet-Optical-Infrared Space Astronomy, AURA, http://ngst.gsfc.nasa.gov/project/bin/HST_Beyond.PDF
2. Williams, R., et al. (1996) AJ, **112**, 1335
3. Williams, R., et al. (1998) AAS, **193**, 7501
4. Steidel, C. C. et al. (1996) ApJ, **462**, L17
5. Fontana, A. et al. (1999), Astron. Astrophys,**343**,P L19
6. Seery, B. (1996) http://www.ngst.nasa.gov/cgi- bin/pubdownload?id=21
7. TRW (1998) http://www.ngst.nasa.gov/cgi- bin/pubdownload?id=309-310
8. Ball (1998) http://www.ngst.nasa.gov/cgi- bin/pubdownload?id=307-308
9. Arnaboldi, M. (1999) these proceedings
10. Weymann, R. et al (1999) ApJ, in press, astro-ph/9807208
11. Madau, P. et al. (1998) ApJ, **498**, 106
12. Steidel, C. C. et al. (1999) ApJ, in press, astro-ph/9811399
13. Madau, P. (1999) these proceedings
14. Gardner, J. et al. (1999) private communication
15. Stiavelli, M, (1999) Astron. Astrophys,**343**,P L25
16. ESO website (1999) http://www.eso.org/instruments/

Clusters of Galaxies at High Redshift

Bernard Fort

Institut d'Astrophysique de Paris, 98 bis boulevard Arago, 75014 Paris, France

Abstract. For a long time, the small number of clusters at $z > 0.3$ in the Abell survey catalogue and simulations of the standard CDM formation of large scale structures provided a paradigm where clusters were considered as young merging structures. At earlier times, loose concentrations of galaxy clumps were mostly anticipated. Recent observations broke the taboo. Progressively we became convinced that compact and massive clusters at $z = 1$ or possibly beyond exist and should be searched for.

Large and deep X-ray surveys (AXAF, XMM) give the possibility of finding them up to $z = 1$. New "color redshift" techniques on wide field CCD surveys allow their detection as a volume excess of galaxies, possibly in the surrounding of luminous radio sources. It is also possible to see their deep potential wells directly on gravitational shear maps of background galaxies or on SZ maps.

Clusters at high redshifts are excellent tracers of both the cosmic evolution of mass condensations and the formation of Large Scale Structure. Their abundance can provide strong constraints on the density parameter Ω and their lensing effects may possibly allow the curvature of the Universe to be tested.

Besides, clusters of galaxies are natural places where all the environmental processes that govern the cosmic evolution of galaxies are currently observed. The content and morphology of cluster members depend on the merging and metal enrichment history of the peculiar volume of galaxies from which the cluster has formed in the young Universe. At the same time, the gravitational telescope effect of distant compact clusters magnifies our view of the distant Universe.

Clusters at high redshift will be indeed a hot topic in the coming years. This short summary talk gives observational trends that have emerged for the VLT during the parallel workshop.

1 Introduction

The probability of finding a rich compact cluster at high redshift similar to nearby rich clusters has been considered very low for a long time. Only a small number can possibly form with the standard CDM scenario. Besides, observers were lacking the observational tools to find them. It is only recently, with the improvement of data analysis and spectroscopic observations with large telescopes, that we have had to convince ourselves that they were far more abundant than expected.

During the last two years a handful of clusters has already been found beyond $z = 0.8$ with a typical mass $M(r < 1Mpc) > 2.10^{14} M_\odot$ or a radial velocity dispersion $\sigma > 600$ km/s. Their number is already enough to exclude an Einstein - de Sitter Universe ($\Omega_0 < 0.4$, Bahcall et al. 1998). The motivation of observers, in finding high-redshift clusters, has thus increased considerably to give a strong constraint on the possible upper value of Ω_0. Both with the determination of the acceleration parameter q_0 from the analysis of type Ia supernovae and P(k) constraints from CMB, the measurement of the relative abundance of clusters with a mass $M(> M_0, < R_0)$ can provide a first opportunity to obtain a direct estimation of the range of values for a non null cosmological constant $\Lambda = \lambda/3H_o^2$. The question "Do there exist massive compact clusters at a high redshift and how can they be found?" has indeed represented a major observational topic of the workshop.

The second main topic concerned the possibility of using a peculiar (unbiased) sample of high redshift clusters to try to check (or falsify) the validity of the hierarchical scenario for galaxy formation. A key issue seems to be the confrontation of observations with the fascinating simulations of the formation and cosmic evolution of galaxies within the largest and densest primeval structures of the Universe that collapse into groups and clusters. In this respect, it seems that the X-ray, SZ and weak lensing observational methods have reached their maturity and will now be used in strong conjunction.

The possibility of using rich clusters as giant natural telescopes was a small but remarkable presentation of the workshop. The technique requires almost perfect modeling of cluster lenses. This is now possible for most of the arc clusters that have been observed again with the HST. This technique is paving a new road to probe the young Universe with the NGST and ALMA.

From the introduction above it is natural to report on the workshop by successfully considering the following topics: observational searches for high redshift clusters, cosmological implications of their abundance, cosmic evolution of their content and clusters as gravitational telescopes.

2 High Redshift Clusters, How To Find Them?

A cluster of galaxies at a low redshift is recognized as a galaxy excess within a circular area of about 5 Mpc radius. The galaxies are displayed in groups of various mass still undergoing a hierarchical merging process into a small number of dominant potentials. The merging end of the process is ultimately the formation of a large central cD galaxy with an extended faint (star) halo tracing the geometry of the potential well. As the merging process of galaxies of any type tends to produce ellipticals, a cluster of galaxies is more easily recognized when a concentration of bright ellipticals within a radius of about 500 kpc is already formed. In some cases, like MS2137-23, the detection of an intense and extended X-ray emission reveals a huge, elliptical galaxy that has already swallowed most of the nearby groups, leaving only in its immediate surrounding galaxies several magnitudes fainter than the central one. At large redshift, this massive condensation of mass would not be seen as a galaxy excess. However, at z=2, the simulation shows that such large *dominant* central potentials have not formed yet, especially for an Einstein

- de Sitter universe. The (proto)cluster should rather appear like a fractal pattern of condensation of mass at an early phase of a merging process that took place in the densest part of the large scale structures. It is obvious that for large look-back time we have to agree on the definition of a cluster. In the workshop, the commonest definition adopted to claim the existence of a cluster at redshift beyond $z = 0.8$ was a "central" mass condensation equivalent to an isothermal spherical structure of about 600 km/s or a mass larger than $2.10_14M_\odot$ within a 1 Mpc radius. This definition of a cluster mass is mainly convenient to estimate Ω_0 from their relative abundance at various cosmic times (Section 2). Several methods are considered for finding such clusters at a high redshift.

At optical and near infrared wavelengths, various photometrical methods are used successfully to find a local galaxy excess on the sky. Clearly the galaxy dimming which comes from the reddening effect (K-correction) imposes to extend optical surveys in the I, J and K color bands. The HST observations of a mosaic field of MS1054-0321 from B to I colors beautifully demonstrate that we can have clusters similar to Abell clusters at a redshift z=0.8 (Franx 1999, this conference). The ESO Imaging Survey (Da Costa et al. 1999) or the Dalcanton survey (Zaritsky et al. 1997) with recent IR follow-ups show the road that will be followed to find several cluster candidates/arcminute square at high redshift from multicolor photometry. For redshifts at $z > 1$, the K band seems necessary. The possible cluster candidate PKS1136-282 at $z = 2.3$, presented by G. Miley and coworkers (1999, this conference), is a fascinating example of what could be found in the field of radio sources having a large Faraday rotation. At a low redshift, such radio sources are mostly found in dense environments. This characteristic will probably hold at larger redshifts. So far, since we still do not have panoramic IR detectors to start an all-sky survey of (proto)clusters at $z > 2$, the observation in the K-band of preselected radio source fields is the best way to find them rapidly (Röttgering 1999, this conference).

At redshift up to 1, the EMSS and ROSAT survey demonstrated that AXAF, XMM and their successors will be efficient at finding deep X-ray potentials with an equivalent velocity dispersion larger than 600 km/s. However, the most impressive observational results for the future search of high redshift clusters come from the possibility of using the SZ effect. From the BIMA and OVRO results, assuming an LCDM cosmology, Mohr and coll. (1999) showed that a small and dedicated wide field mm interferometer can start a large cluster survey (10 deg^2/yr, 100 clusters/yr at $z > 1$) for X-ray and lensing follow-up. If such an instrument is operated at the end of 2001 it will give spectacular results on the Universe at large scales (Sunyaev 1999).

Mellier and Schneider (this conference) have demonstrated that it is now possible to measure the cosmic shear with an accuracy of 0.2% on a 5 arc minute scale. This will give a unique opportunity to map the cosmic shear pattern over a large sky area and to detect cluster mass with the aperture mass density method (Schneider & Bartelmann 1997). This gives good hopes for studying the cosmic evolution of mass condensations without any reference to the luminous mass delivery up to a redshift of about 1. Beyond this redshift and with present day instrumentation the exposure time will be too long for an all-sky survey. At this point, the discussion during the workshop has shown that weak lensing is the only method to detect "dark clusters" with no luminous counterpart. So far, very few candidates have been subjected to deep observational follow-ups in the optical or IR (Mellier 1999). They have often revealed possible baryonic counterparts (a faint distant group of IR distant galaxies or a giant isolated elliptical). The discovery of a true dark cluster seems unlikely but will be a big event. Such investigations will be pursued with a high priority even if it appears very difficult to prove their existence due to the rapid dimming of luminous matter with redshift.

The cosmic shear provides such a detailed map of the projected mass spectrum from groups to LSS that it does not provide a natural filter to select the largest compact cluster mass concentration like X-ray and SZ methods do. But this characteristic gives an unchallenged method to directly measure Ω_0 (see Sect. 3)

A direct mass selection of clusters can only be done with cosmic shear measurement on the arc minute scale. In conjunction with other surveys, it will allow the cosmic evolution of the biasing factor to be studied (Van Waerbeke 1999). It is already quite clear that the methods used to find clusters at high redshifts (IR galaxy excess, SZ and shear detection) are complementary and will be urgently implemented on the same wide field.

3 Clusters and Cosmological Tests

To date, about 8 massive clusters have been detected at $z > 0.7$ (MS1137+66 at $z = 0.783$; MS0152-1357 at $z = 0.8$; RXJ1716+67 at $z = 0813$; MS1054-0.3 at $z = 0.83$; probably Q1622+238 at $z = 0.929$; 3C324 at $z = 1.206$ and MS0848+44a,b, respectively at $z = 1.27$ and 1.28). For nearly all of them, the deep potential is detected from X-ray, SZ and/or weak lensing maps. It is also associated with a galaxy excess clearly seen in the near-IR at the cluster center. At a larger redshift the best candidate is the cluster candidate associated with PKS 1138-262 at $z = 2.2$ (Miley and coll., this conference). It needs to be observed extensively to determine its mass.

Several authors claim (Bahcall & Fan Xiao-Hui 1998) that this number of massive clusters at redshift above 0.8 $(M > 8 \times 10^{14} M_\odot).h*3/Mpc*3)$ is already enough to rule out an Einstein - de Sitter universe. Only a value of Ω_0 below 0.3 can be reconciled with such a number density of massive clusters at large redshifts. In fact, the cluster mass function at various redshifts gives an extremely sensitive test of the cosmic rate of mass condensation. Indeed, the measurement of cluster mass at various epochs to derive a measurement of Ω_0 is a vigorous driver for observation of clusters at large redshifts.

The CNOC project has demonstrated that it is possible to use a mapping of the radial velocity dispersion of a large sample of cluster members to obtain the mass profile of a cluster (Ellingson et al. 1998). X-ray surveys have been more extensively used since we have observed a strong correlation at the low redshift between the cluster velocity dispersion $sigma_0$, the X-ray luminosity L_X and the temperature T_X. Using a reasonable prescription like a Press-Schecter formalism, and a typical mass or temperature profile, it is possible to define an Ω_0 test by measuring one cluster X-ray parameter which can be related to the mass at various redshifts. The observers verify that the test is compatible with the sample of low redshift clusters and uses a maximum likelihood method with the high redshift sample to determine an estimation of Ω_0. Clearly almost all results give on average a maximum of probability for a value between 0.1 and 0.4 that also excludes $\Omega_0 = 1$ but the confidence level is still not high enough to get a more accurate value. If we try to understand the limitation of the method one can see that the limitation comes from possible bias in the cluster sample that is used, and approximation in the prescription that describes the geometry and physical characteristics of the cluster content, as well as the normalisation that is used for the power spectrum of initial density fluctuation (Tozzi 1998). However, the detailed modeling of the coherent approach of Cavaliere, Menzy and Tozzi (1998) gives hope that the cluster abundance approach will be improved soon. This group makes a complete modelling of the formation and cosmic evolution of clusters of galaxies including the cosmic history of baryons and galaxy hierarchical merging and interactions to derived scaling law and correlation between mass and gas parameters. As an example, the correlation between L_X and T_X, which is observed with a slope less steep from groups to more massive clusters, becomes a natural throughout the theory that can be used in a more controlled way for cosmological tests (Cavaliere et al. 1997).

Measurement of the cluster abundance or mass function at various redshifts is a method to determine the density parameter. However, it now seems possible to measure the cosmic shear with accuracy of a fraction of one percent (Schneider and Mellier this conference). If the result is regularly obtained when observing a field of view with a size of about 25 square degrees there is no way to compete with the cosmic shear method in this matter: Ω_0 will be determined with accuracy of a few percent from the statistical analysis of the variance and skewness of the shear at various scales up to 20 arc minutes (Mellier 1999). The only limitation of the method will come from the knowledge of the average redshift distribution of the galaxies up to about $z = 1.5$–2 (Bernardeau et al. 1999). Compared to the cluster abundance method the weak lensing approach also tests the growth of mass condensation with time but it is a more powerful and elegant tool because it directly probes the distribution of mass along many lines of sight. It is also important to note here that the possibility of estimating the redshift of background galaxies from their colors allows the measurement of the curvature of the Universe (Lombardi et al.

1999 and Gautret et al. 1998). The methods used a relative comparison of the amplitude of the local lensing effects around distant clusters for various redshifts of background sources.

4 Cluster Content

4.1 Total (Dark) Mass

The mapping of the cluster mass has improved greatly during the last years with strong lensing modeling both from HST images and from weak lensing analysis (Mellier 1999). When a cluster is observed at the end of its collapsing process, the central lensing mass is observed similar to the X-ray mass within the accuracy limitation of the various methods (<30%), as expected. Two striking results were presented which do not concern high redshift clusters yet. Mellier and coll. showed that the mass reconstruction of Abell 1942 (including clumps of dark matter) is very robust and gives a similar result for different instruments and observations. This means that we are now ready to observe the mass function of a fair sample of clusters up to large redshifts (the EIS project). A second interesting result comes from Broadhurst (this conference) who points out the fact that the radial magnification is more sensitive to the slope of the potential than the radial shear. He claims that the observational data on A1699, as well as on Cl0024+1654 data previously obtained by Fort et al. (1997), support the real existence of a universal halo profile emerging from the numerical simulation (Navarro et al. 1997).

4.2 X-Ray Gas

Renzini (1999) summarized an observational analysis of a history of the metal enrichment of the ICM that seems compatible with the hierarchical scenario of galaxy formation with a standard IMF. It results that about 1/3 of the galaxy stars formed at $z>3$ with a primeval abundance of metal about 0.1 Z_0. Type II supernovae will have played a dominant role as compared to type I in the enrichment process. This topic certainly awaits new X-ray observations of distant clusters with a more detailed description of temperature at subarcminute scale and coherent modeling of the formation and cosmic evolution from groups to clusters. In the same way, the merging history of the ICM gas and its dynamics will be better understood with high resolution X-ray images and comparison with lensing maps of the projected potential.

4.3 Galaxy Content

All the major processes that contribute to the formation and evolution of galaxies can be observed in distant clusters in a great number of occurrences (*merging*, *gas stripping*, passive evolution, etc...) for a period of time about two thirds the age of the Universe. Distant clusters are actually perfect templates to study the cosmic evolution of galaxies, the star formation rate, metal enrichment and indeed to compare with simulations of cosmic scenario having a high mass resolution. From presentations during the workshop, it seems that spectrophotometry of cluster members in optical and near IR shall be completed by a good description of their morphology in various colors (HST images). As an example, observations at the high resolution of galaxy members in MS1054-03 (z=0.83) reveal merging red pairs, possibly small spheroidal galaxies without apparent star formation in the immediate surrounding, as well as the classical Butcher-Oemler effect observed at the lower redshift. Besides, Aragon-Salamanca and coll. (1998) pointed out the fact that from intermediate redshifts (z=0.6), bright ellipticals seem to follow passive evolution while their average IR luminosity (mass) increases. All the observations of distant clusters confirm that we already see a younger class of evolving galaxies, some of them possibly older and smaller spheroidal galaxies without gas which also contribute to the general mass increase of central galaxies. We do not have a detailed observational picture of the cosmic evolution of galaxies in clusters yet. We guess that simulations of the cosmic evolution of clusters will be a clue to validate or to falsify the standard hierarchical scenario of galaxy formation and evolution by comparing with the next coming observations.

A question that generated lively discussion concerns the fundamental plan relation that correlates the luminosity of cluster bright ellipticals to their central velocity and surface brightness (Bender 1998, Kelson et al. 1999). Why do observational data have such a small scatter around the fundamental plan? It looks as if most of the brightest ellipticals in distant clusters are formed at the same early time and/or mostly built from the same primeval bricks. The fundamental plane gives such a tight relation at any redshift that it can provide, from the luminosity shift of the fundamental plane with redshift, an estimate of the acceleration parameter q_0 and the epoch of galaxy formation. Clearly the VLT will be the best tool to measure the velocity dispersion of ellipticals in a large sample of distant clusters. In this respect, it is also important to notice that the faint star halo around central giant ellipticals and cDs can now be easily observed with CCD mosaics and VLTs. The halo has implications on the baryonic component of clusters. It can be described with a good spatial resolution and compared with strong lens modeling of the potential. The measurement of the velocity dispersion of stars up to the radius of large arcs could give additional constraints on the dominant elliptical potential of compact clusters, an interesting point to implement new cosmological tests of the curvature of the Universe with clusters of galaxies.

5 Distant Clusters and LSS

Observations of clusters at redshift above 0.6 are still in their infancy. Also the possibility of using them to trace large scale structures, to see possible X-ray filaments that can link merging groups of galaxies or to look at the general magnification bias on background sources (radio sources), lies still in the future. However, Boehringer (this conference) presented a splendid result obtained with the X-ray bright REFLEX clusters sample. It beautifully traces the LSS on a field of about 400 Mpc/h size and gives the power spectrum of the spatial distribution for k ranging from 0.02 to 0.3 h/Mpc. This result gives an idea of what will be done at higher redshifts with the next coming generation of X-ray satellites. Another new topic was the possibility of studying the clustering of Lyman break galaxies at large redshifts. However the clustering history can be compared with cosmological models only if the galaxy mass halo can be measured (Gialavisco, this conference). This programme clearly has a window of opportunity with VIRMOS for deep photometry (I=27+/-1) and 2D-spectroscopy (I=25-26), as well as with NGST photometry (I= 29+/-1).

6 Clusters as Giant Natural Telescopes

The central part of compact clusters of galaxies can strongly magnify background sources when they are small and located on the caustic lines of the lens. It is a way to combat the redshift dimming factor when trying to detect galaxies at z=5 or beyond (Franx et al. 1998). At larger distances from the critical lines in the image plan, say within a radius of several arc minutes, the average magnification is approximately 1.5 magnitudes for the background sources. In this field, the modelling of the lens can be good enough to obtain the magnification factor with accuracy better than 20%, an error largely compatible with the best galaxy modeling of stellar population synthesis. Pello gave a summary of what can be done with the gravitational telescope effect from optical and IR to the mm wavelength (Blain 1998). It is obvious that the magnification allows VLT spectroscopic surveys of field galaxies to be extended down to I=27. However, the field of view of the gravitational telescope effect is small ($<$ 5 arcminutes). The number density of galaxies at $z > 4$ or extremely red galaxies (I–K $>$ 6) is so low that the study of their distribution and evolution requires a fair sample of clusters (n $>$ 36). So the first scientific return of ongoing surveys is the detection, in the IR, of several B, R or I dropout candidates, a few extremely red objects and 50% SCUBA sources identification in the field already observed. However the stretching of the images could be large enough to see primeval merging blocks and to study their dynamics. The detection of such candidates seems very timely since the VLT will soon be equipped with 2D spectrographs. Indeed, scrutinizing the locus of critical lines to find highly magnified primeval objects is a fascinating VLT topic that will later become one of the hottest with the NGST.

7 Conclusion

It is not possible to cover here all the interesting results, ideas and projects that have been presented in the oral presentations and posters. The general impression is that observations of distant clusters are really entering a golden age. All new and powerful observational tools will be in operation within a couple of years (VLT, X-ray observatories, SZ mm interferometers, wide field telescopes for weak lensing measurements, etc.). It is quite certain that new and spectacular results will emerge from the observations. We may have to reconsider our view and timing on the hierarchical scenario of galaxy and cluster formation, with implications for the epoch of first star formation, the history of the metal (dust) enrichment and the dynamics of the merging dark matter and gas. Distant clusters could also be a good statistical tool to place strong constraints on the values of the cosmological parameters. At a redshift about 0.7 they are optimal gravitational telescopes to probe the youngest baryonic structures of the Universe. From all presentations it is strikingly apparent that most programmes need a complete and unbiased clusters sample from X-ray to the mm wavelength. Therefore, it is important to develop the concept of public surveys with HST, X-ray and SZ follow-ups.

References

1. Aragon-Salamanca A., Baugh C., Carlton M., Kauffmann G. (1998) MNRAS **297**, 427
2. Bahcall N., Fan Xiao-Hui. (1998) ApJ **504**, 1
3. Bender R., Saglia R., Zigler B., Belloni P., Greggio L., Hopp U., Bruzual G. (1998) AJ **493**, 529
4. Bernardeau F., Van Waerbeke L., Mellier (1997) A&A **322**, 1
5. Blain A. (1998) AAS 192.6027
6. Cavaliere A., Menci N., Tozzi P., (1999) Astro-ph /9904125, Astroph/9802185
7. Cavaliere A., Menci N., Tozzi P., (1997) ApJ **484**, L21
8. Da Costa L., Scodeggio M., Olsen L., Nonino M., Rengelik R., Bender R., Franx M., Jorgensen H., Renzini A., Rosati P. (1999) A&A **343**, 29
9. Ellingson E., Yee H., Abraham R., Morris S., Carlberg R. (1998) ApJS **116**, 247
10. Fort B., Mellier Y., Dantel-Fort M., (1997) A&A **321**, 353
11. Franx M., Illingworth G., Kelson D., Van Dokkum P., Tran Kim-Vi. (1997) ApJ **486**, L75
12. Gautret L., Fort B., Mellier Y. (1999) Astro-ph /9812388
13. Kelson D., Van Dokkum P., Franx M., Illingworth G., Fabricant D. (1997) ApJ **478**, L13
14. Lombardi M., Bertin G. (1999) A&A **342**, 337
15. Mellier Y. (1999) ARAA **37**, in press
16. Mohr J., Mathiensen B., Evrard G., (1999) Astro-ph /9901281
17. Navarro J., Frank C., White S. (1997) ApJ **490**, 493
18. Olsen L., Scodeggio, M. et al. (1999) A&A **345**, 681

19. Pello R. (1998) Astro-ph /9810390
20. Renzini A. (1998) Astro-ph /9810304
21. Schneider P., Bartelmann M. (1998) MNRAS **286**, 696
22. Sunyaev R.. (1999) Astro-ph /9904125
23. Tozzi P. (1998) Astro-ph /9804123
24. Van Waerbeke L. (1998) A&A **333**, 767
25. Zaritsky D., Nelson A., Dalcanton J., Gonzales A. (1997) ApJ **480**, 91

PARALLEL WORKSHOP 1:

CLUSTERS OF GALAXIES AT HIGH REDSHIFT

Edited by

Bernard Fort and Piero Rosati

Galaxy Populations in High Redshift Clusters

S.A. Stanford

Physics Department, University of California at Davis

Abstract. I review recent research on the galaxy populations in moderate and high redshift clusters. Most studies using the high resolution capability of *HST* find that the fractions of late-type and disturbed galaxies increase with redshift up to $z \sim 0.5$. I present new results on the morphological types determined from *HST* imaging of 4 $z > 1$ galaxy clusters in K-band selected samples. The trends seen at lower redshifts appear to continue past $z \sim 1$, with an increasing number of disturbed galaxies and a decreasing number of early-type galaxies, on average after a statistical field correction, in clusters at $z \sim 1.2$. These results suggest that some elliptical galaxies in high redshift clusters are being constructed at $z < 1.2$.

1 Introduction

Early-type galaxies completely dominate the core region of the nearby Coma cluster in the central ~1 Mpc. There are almost no late-type disk galaxies nor disturbed systems. Naturally the question arises: were clusters were always dominated by ellipticals and S0s? Clearly if we want to study the evolution of galaxy populations in clusters, the only choice for an observer is to compare clusters at a variety of redshifts. Its important to remember that that there may be at least two major problems with this approach. First, clusters in the present epoch may be fundamentally different from the clusters we find at higher redshifts because of the growth in large scale structures on such mass scales. Second, currently we have no complete and well-defined samples over a sufficiently large range in z. Great progress has been made recently to construct well-defined cluster samples, particularly using the Rosat archive (e.g. [6]), but even in these samples much work remains to be done on spectroscopic membership within the individual clusters.

With these caveats in mind, let us consider one of the most distant galaxy clusters known, 3C 324 at $z = 1.206$. This cluster was discovered by Dickinson et al. by virtue of a large number of very red galaxies around the central radio galaxy which have the $R-K$ colors of E/S0s at $z \sim 1.2$ ([3]). The cluster was the target of a deep *HST* WFPC2 exposure, which roughly samples the rest frame U band and covers the central ~1 Mpc, that shows a large number of very peculiar galaxies, at least some of which are no doubt in the cluster. So a simple comparison of the galaxy populations in the cluster around 3C 324 with those in Coma suggests that there have been large scale morphological transformations of cluster galaxies since $z \sim 1$.

2 Moderate-Redshift Galaxy Clusters

Now this is not exactly news to anyone who has been working in the field. With the advent of WFPC2 in *HST*, a number of studies were conducted of the galaxy populations in moderate-redshift clusters, mostly at $z < 0.5$, which have already found similar results regarding a change in the morphological types of galaxies in clusters. I'd like to briefly review these results, which have been the work mostly of various permutations of the MORPHS group involving Dressler, Oemler, Ellis, Couch, and collaborators. Their WFPC2 imaging of the core regions of about 10 clusters at $0.3 < z < 0.6$ has shown a number of very interesting changes occuring in the galaxy populations of clusters. First, the well-known Butcher-Oemler effect is primarily due to the disappearance of normal late-type galaxies in clusters at the present epoch ([5]). A secondary cause of the Butcher-Oemler effect is a decline since $z \sim 0.5$ in the number of disturbed galaxies. Third, the fraction of ellipticals remains constant with redshift, while that of the S0s declines dramatically to the present epoch ([4]). Taken together, these results may be understood by a model for the evolution of cluster galaxies wherein ellipticals formed at $z > 2$ as the core of a cluster, followed by the infall of late-type galaxies from the field which are converted into S0 galaxies by interactions with each other and with the intracluster medium ([1]).

3 High Redshift Clusters

Next let us examine the highest redshift clusters known to see what happens to the trends described above in the morphological distribution of galaxies within clusters. At $z > 1$ very few clusters are known, and even fewer have been imaged by *HST*. So we must work with what we have if we wish to determine the morphological types of galaxies in clusters at these redshifts. I will present here results on four such clusters for which my collaborators and I have obtained *HST* imaging. One of these clusters was found serendipitously in a near-IR field survey, while the other three have been found by targetting known high-z radio galaxies. ClG J0848+4453 was confirmed to be at $z =$ 1.27 by Keck spectroscopy of galaxies in a compact group of very red galaxies found in the imaging survey of Elston et al. Subsequently, it was discovered that a deep Rosat exposure had been obtained of this field and that a faint X-ray source coincides with the spectroscopic members ([7]). Although not appreciably extended in the PSPC data, the X-ray source is probably due to hot gas in the intracluster medium, indicating an $L_x = 1 \times 10^{44}$ ergs s^{-1}. To date, 14 members have been spectroscopically confirmed in this cluster, many of which fall within a well-defined early-type sequence, which, by comparison with the predicted no-evolution color-mag lines based on early-type galaxies in the Coma cluster, indicates color evolution on the order of e.g. 0.5 mag in $R - K$.

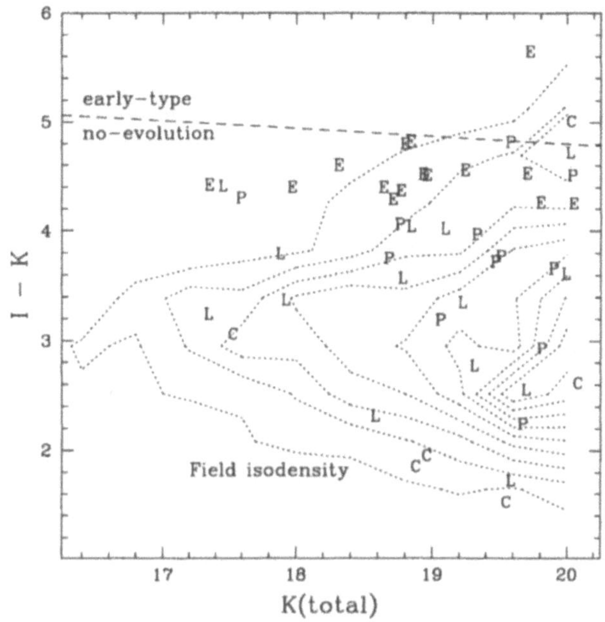

Fig. 1. $I - K$ vs K diagram for ClG J0848+4453 at $z = 1.27$. Each galaxy is represented by the letter corresponding to its morphological type, where E stands for early-types, L for late-types, P for peculiar, and C for compact. The dashed line is the predicted no-evolution colors for early-type galaxies. The dotted lines represent contours of the isodensity of field galaxies.

Although large telescopes such as Keck and now the VLT are capable of obtaining spectroscopic redshifts of galaxies in clusters at $z \sim 1.3$, such work is very difficult and slow. Until such time as complete spectroscopic identification of magnitude-limited samples are available, we can use statistical techniques to correct K-band selected galaxy samples in $z > 1$ clusters for field contamination. We have made use of the Elston et al. survey which covers 100 arcmin^2 down to $K = 21$ in the $BRIzJK$ bands to perform a correction in color-magnitude space rather than just by K magnitude. By calculating the density of field galaxies at any one position in a color-magnitude diagram, we can assign a probability to each galaxy detected in a cluster field as to its likelihood of being a cluster member. The technique is illustrated in Fig. 1 for 0848+4453. Morphological types in a $K < 20.1$ sample were determined from *HST* imaging using the broad categories of early-type (E), late-type (L), peculiar/disturbed (P), and compact (C). Overlaid in the $I - K$ vs K diagram are contours of isodensity of objects in the Elston et al. field survey. Its clear by visual inspection of Fig. 1 that the E types in the red sequence are very likely to be cluster members (which we know to be true in many cases because of the available spectroscopy for these objects at $K < 19$), while the bluer objects at fainter K have a stronger probability of being field galaxies.

Weighting by these probabilities enables us to calculate the morphological fractions of likely cluster members in 0848+4453.

We have performed a similar analysis on the other three $z > 1$ clusters for which we have *HST* imaging. These include 3C 324, 3C 210 at $z = 1.169$, and 3C 267 at $z = 1.140$. The results are shown in Fig. 2, Fig. 3, and Fig. 4. In these color-magnitude diagrams we again see the well-defined early-type galaxy sequence, except for in 3C 267 where very few such galaxies are found. This may be related to its optical appearance as an irregular cluster which may be early in its formation. The morphological fractions for the galaxies in $K < 20$ samples, after weighting by the field probability, were then determined for each of the radio galaxy clusters.

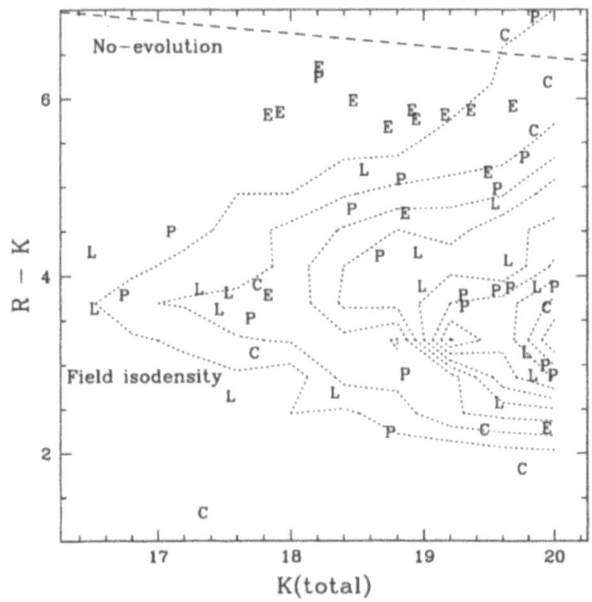

Fig. 2. $R - K$ vs K diagram for 3C 324 at $z = 1.206$, with the same symbols and lines as in Figure 1.

After averaging over all 4 of the $z > 1$ clusters, the morphological fractions for the three main groups are shown in Table 1. We have also included the fractions for $z < 1$ clusters as determined by [5], [4], [2], and [8] from *HST* imaging of the core regions. The trends found at $z < 1$ in the declining numbers of early-type galaxies and increasing numbers of late-type and disturbed/peculiar galaxies appear to continue towards higher redshifts. While we did not attempt to classify ellipticals and S0s separately in the $z > 1$ clusters, it is of interest to attempt to constrain the fraction of ellipticals in the highest redshift clusters. Even if all the early-types are ellipticals in the $z \sim 1.2$ clusters, then the number of ellipticals appears to be increasing with

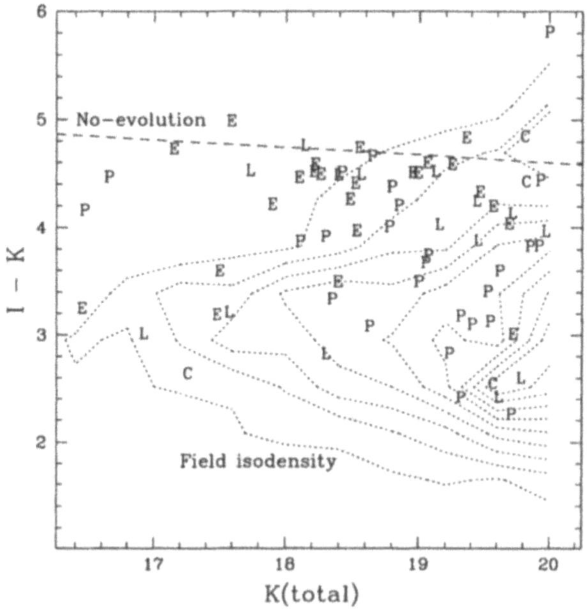

Fig. 3. $I - K$ vs K diagram for 3C 210 at $z = 1.169$, with the same symbols and lines as in Figure 1.

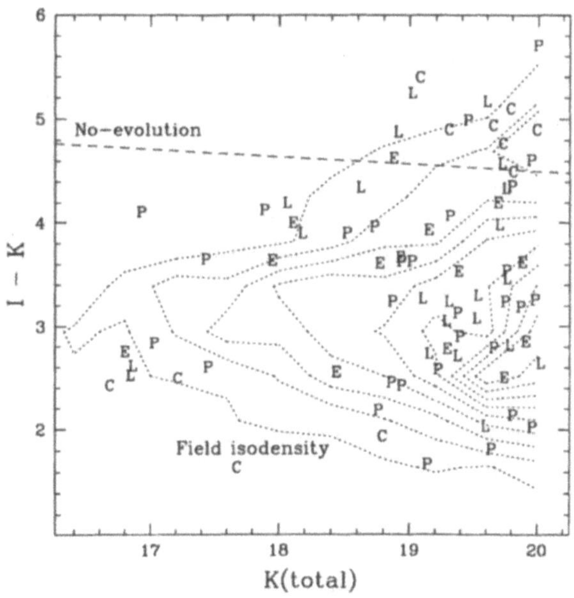

Fig. 4. $I - K$ vs K diagram for 3C 267 at $z = 1.14$, with the same symbols and lines as in Figure 1.

decreasing z, which suggests that ellipticals are still being created in clusters at $z < 1$.

Table 1. Galaxy Morphological Percentages in Clusters

redshift	Early	Late	Disturbed
present	80	16	3
0.3	67	20	13
0.4	51	30	15
0.5	50	36	13
0.8	42	27	20
1.2	31	25	31

4 Summary

The conclusions reached above on the evolution of ellipticals in high redshift clusters must be regarded as tentative. Clearly, a larger sample of clusters at $z > 1$ as well as at $0.5 < z < 1$ must be studied. A number of new facilities will make such studies feasible in the near future, including the VLT and its powerful complement of instruments. The advent of AXAF and XMM will enable more detailed studies to be carried out on high redshift clusters which will help us to understand the way the clusters are forming which no doubt affects their galaxy populations. Finally, in the slightly more distant future, SIRTF will enable the study of the rest frame near-IR luminosity functions in clusters past $z \sim 1$ which will help to constrain the possible merging history of cluster galaxies.

References

1. Couch, W. & Sharples, R. 1987, MNRAS, 229, 423
2. Couch, W. et al. 1998, ApJ, 497, 188
3. Dickinson, M. 1996, astro-ph/9612177
4. Dressler, A. et al. 1997, ApJ, 490, 577
5. Oemler, A., Dressler, A., & Butcher, H. 1997, ApJ, 474, 561
6. Rosati, P., Della Ceca, R., Norman, C., & Giacconi, R. 1998, ApJ, 492, L21
7. Stanford, S.A. et al. 1997, AJ, 114, 2232
8. Stanford, S.A., Eisenhardt, P.R., & Dickinson, M. 1998, ApJ, 492, 461

One of the Highest Redshift X–Ray Selected Clusters of Galaxies

Ingo Lehmann[1], Günther Hasinger[1], Riccardo Giacconi[2], Jim E. Gunn[3], Maarten Schmidt[4], Don P. Schneider[5], Thomas Stanke[1], Joachim Trümper[6], Joachim Wambsganss[1], David Woods[1], and Gianni Zamorani[7]

[1] Astrophysikalisches Institut Potsdam, An der Sternwarte 16, D-14482 Potsdam, Germany
[2] European Southern Observatory, Karl–Schwarzschild–Str. 2, 85748 Garching bei München, Germany
[3] Princeton University, Princeton, NJ 08544, USA
[4] California Institute of Technology, Pasadena, CA 91125, USA
[5] The Pennsylvania State University, University Park, PA 16802, USA
[6] MPI für extraterrestrische Physik, Karl–Schwarzschild–Str. 1, 85748 Garching bei München, Germany
[7] Osservatorio Astronomico, Via Zamboni 33, 40126 Bologna, Italy

Abstract. An unusual double-lobed extended X-ray source with an angular size of 1.7×0.7 arcmin^2 and an X-ray flux of $2 \cdot 10^{-14}$ erg cm^{-2} s^{-1} has been detected in the ROSAT ultra-deep HRI survey of the Lockman Hole. An R-band image from the Keck telescope shows only a marginal excess of galaxies brighter than 24.5 mag and Keck LRIS spectroscopy of 24 galaxies around the X-ray centroid did not yield a significant number of concordant redshifts. The brightest galaxy near the centre of the eastern emission peak is a gravitationally lensed arc at z=2.570, indicating that the X-ray object is a high-redshift lensing cluster of galaxies. This is the first arc, which is optically brighter than any of the galaxies in the lensing cluster. Recently, deep K'-band images of the region revealed a number of bright galaxies (K'=16-17) with very red colours $(R - K') > 6.3 - 6.7$. From Bruzual & Charlot evolution models for elliptical galaxies we have estimated a redshift of z$\sim$1.3. This is therefore one of the most distant X-ray selected clusters of galaxies known. Its X-ray luminosity, temperature and lensing mass are in the same range as for clusters found in the local universe.

1 Introduction

The ROSAT Deep X–ray surveys have resolved a significant fraction of the soft X–ray background into discrete sources [1]. The optical identification of the faint X-ray sources using the Keck telescopes has revealed that the majority of objects are active galactic nuclei (AGN) [2]. Among the AGN is the highest redshift (z=4.45) X–ray selected QSO found to date [3]. The soft X–ray surveys have detected a larger surface density of AGN than seen in any other waveband and find significant evolution in the space density of high-luminosity AGN [4].

The second most abundant class of faint X–ray sources in the deep X–ray surveys are clusters and groups of galaxies [2]. They are the largest bound structures in the universe, composed of dark matter condensations, hot X–ray emitting gas and galaxies. Several soft X–ray surveys have shown in the last few years that X-rays are a powerful tool to detect high–redshift clusters of galaxies [5], [6]. Recently, the highest-redshift X–ray selected cluster of galaxies (z=1.26) was found in the ROSAT Deep Cluster Survey [7]. This cluster is separated only 4.2 arcmin from a cluster at z=1.27 selected in the NIR [8]. In this paper we present preliminary results on a cluster, detected in the ROSAT ultradeep HRI image of the Lockman Hole [1], which contains a a bright gravitationally lensed arc [9] and is possibly at an even higher redshift.

2 The X–Ray Observations

We have detected a bright extended X-ray source (RX J105343+5735) in a 1.32 Msec ROSAT HRI pointing (Ultradeep HRI Survey) in the direction of the Lockman Hole, a region with an exceptionally low HI column density. Together with a series of deep ROSAT PSPC pointings the observations are part of the ROSAT Deep Survey project (RDS) [1]. In Fig. 1 we show the HRI X–ray contours of the double-lobed extended structure superposed on a K'-band image. Both the Eastern and the Western lobes are clearly extended (1.7 arcmin in E-W and 0.6 arcmin in N-S direction). The source was also detected in the ROSAT PSPC pointing, but is not well-resolved due to the large off-axis angle of ~19 arcmin. From the HRI and PSPC data we have derived an observed flux of $\sim 2{\cdot}10^{-14}$ erg cm^{-2} s^{-1} in the 0.5-2.0 keV band. Assuming a redshift of the cluster between 0.3 and 1.2, the estimated temperature range is between 1.8 and 3.5 keV and the X-ray luminosity in the 0.5-2.0 keV band is between 10^{43} and $2.4 \cdot 10^{44}$ erg s^{-1} [9].

3 Optical/Infrared Observations

We have obtained a 300 sec R-band exposure on 1996 April 13 with the Low Resolution Imaging Spectrometer (LRIS; [10]) on the Keck I telescope. Under seeing conditions of 0.75" (FWHM) the image reaches a limiting magnitude of R≈25.5 mag. The image scale of the back-illuminated 2048 × 2048 Tektronik CCD is 0.215" pixel^{-1}.

The R-band image indicates only a marginal excess of galaxies brighter than 24.5 mag [9]. But close to the centroid of the Eastern X-ray lobe there is a bright arc-like feature, which has a length of ~ 7" along the major axis and is not resolved along the minor axis (cf. Fig 2). An optical spectrum of the arc was taken together with 24 galaxies in the field with the multislit mask mode of LRIS on Keck II on 1998 March 19. The seeing was ~ 1". Using a 1.4" slit width for all objects and a 300 line mm^{-1} grating, the wavelength

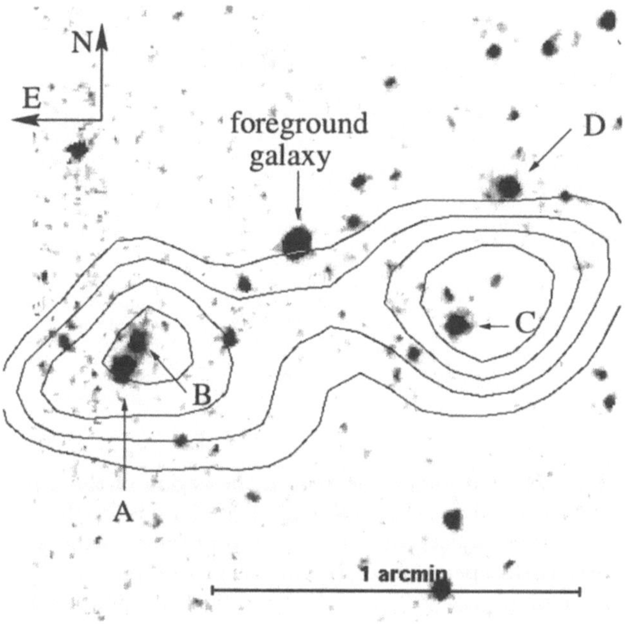

Fig. 1. X-ray contours of the galaxy cluster in the ROSAT HRI ultradeep pointing of the Lockman Hole, superposed on a Calar Alto OMEGA prime K'-band exposure. The K'-band image contains four bright galaxies (marked with A, B, C and D) with $K' = 16.0 - 17.0$ mag in the region of the extended X-ray emission. These galaxies have very red colours ($6.3 \simeq (R - K') \simeq 6.7$ mag). The bright source near the center is a foreground galaxy at z=0.308.

range is between 3200–8200 Å with a resolution of ~ 15 Å. The spectrum of the gravitationally lensed arc (see Fig. 2) shows prominent absorption line features of Lyα and low-ionization metal lines (at z=2.570±0.002) typical of the rest-frame UV emission of starburst galaxies. The apparent magnitude of R=21.4 mag and the redshift are very similar to the gravitationally lensed galaxy in A2218 [11], supporting the gravitational arc interpretation for our object. This is the first arc discovered, which is optically brighter than any of the galaxies in the lensing cluster. Assuming a redshift between 0.3 and 1.2 the lensing mass inside the circle defined by the arc is $1.1 - 1.4 \times 10^{14}$ $M_\odot$ [9].

The spectroscopy of 24 galaxies in the field of the extended X-ray source did not yield a significant number of concordant redshifts. A determination of the spectroscopic redshift of the lensing system was therefore not yet possible.

A K'-band image of the field with a net exposure time of ~ 1.5 hours was taken on 1998 December 4–7 with the OMEGA prime camera at the 3.5m Calar Alto telescope. The OMEGA prime camera has a large field of view (*6.7 × 6.7 arcmin*) and its pixel scale is 0.396" pixel^{-1}. The seeing

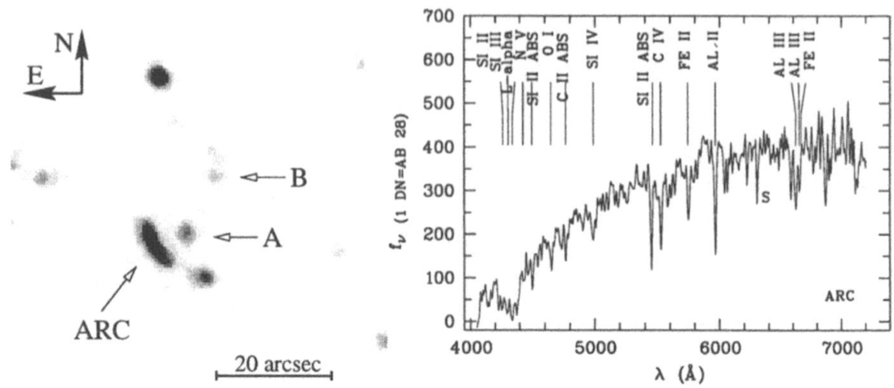

Fig. 2. Left: Keck *R*-band image zoomed on the region of the Eastern lobe. Near the very red objects A and B a gravitationally lensed arc is visible, which is optically brighter than the other objects, but not detected in the K'-band image (cf. Fig. 1). Right: The Keck LRIS spectrum of the arc shows prominent UV metal absorption lines (redshifted to z=2.570), which are typical for high redshift star-forming galaxies. "S" denotes the position of a strong sky line.

conditions during the observations varied from 1.1" to 1.3". The resulting K'-band magnitude limit is around 19.6 mag. The deep K'-band image of the region of the extended X–ray emission (cf. Fig. 1) revealed a number of bright sources with K'-magnitudes between 16.0 and 17.0 and with R-magnitudes of $\sim$ 23.0–24.0. At the center of the Eastern lobe there are two K'-band bright galaxies (A and B). Another bright galaxy C is located near the center of the Western X–ray lobe. In total four galaxies in the field have very red colours $(R-K') \simeq 6.3-6.7$, which are redder than found for cluster members of the highest redshift X–ray selected cluster of galaxies known to date [7].

If we assume that these galaxies are ellipticals we can estimate a photometric redshift of $\sim$ 1.3 from Bruzual & Charlot evolution models [12], using the $R-K'$ colour. Therefore this could be one of the most distant X–ray selected cluster of galaxies known. In the meantime we have taken Keck LRIS and NIRC spectra of red cluster galaxies, but the analysis has not yielded a conclusive redshift.

4 Conclusion

We report the discovery of probably one of the most distant, X-ray selected clusters of galaxies. Its X-ray luminosity, temperature and lensing mass are in the same range as for clusters found in the local universe. Future deep X–ray surveys with XMM and Chandra, in combination with optical/infrared follow-up observations on large telescopes (eg., the VLT and Keck telescopes), will offer a unique opportunity in order to detect and study clusters of galaxies at very high redshifts.

References

1. Hasinger, G., Burg, R., Giacconi, R., et al., 1998, A&A 329, 482
2. Schmidt, M., Hasinger, G., Gunn, J.E., et al., 1998, A&A 329, 495
3. Schneider D.P., Schmidt M., Hasinger G., et al., 1998, AJ 115, 1230
4. Hasinger, G., 1998, AN 319, 37
5. Luppino G.A., Giola I.M., 1995, ApJ 445, L77
6. Rosati P., Della Cella R., Norman C., Giacconi R., 1998, ApJ 492, L21
7. Rosati, P., Standford, S.A., Elston, R., et al., 1999, submitted to AJ
8. Standford, S.A., Elston, R., Eisenhardt P.R., et al., 1997, AJ 114, 2232
9. Hasinger, G., Giacconi, R., Gunn, J.E., et al., 1999, A&A 340, L27
10. Oke J.B., Cohen J.G., Carr M., et al., 1995, PASP 107, 375
11. Ebeling H., Voges W., Böhringer H., et al., 1996, MNRAS 281, 799
12. Bruzual A.G., Charlot S., 1993, ApJ 403, 538

Intracluster Planetary Nebulae

Magda Arnaboldi[1], Massimo Capaccioli[1], Holland Ford[2], Kenneth C. Freeman[3], Rolf Kudritzki[4], Ortwin Gerhard[5], and Roberto Méndez[4]

[1] Osservatorio Astronomico di Capodimonte, Naples, Italy
[2] Dept. of Physics and Astronomy, John Hopkins University, Baltimore, USA
[3] Mt. Stromlo and Siding Spring Observatory, Canberra, Australia
[4] Munich University Observatory, Munich, Germany
[5] Astronomical Institute, University of Basel, Basel, Switzerland

Abstract. We briefly describe the properties of a newly discovered population of intracluster planetary nebulae in the Virgo cluster and the expected developments with the use of the VLT to study these objects in the nearest cluster of Virgo and Fornax.

1 Introduction

Back in 1951, in his introduction on the study of the galaxy luminosity function in the Coma cluster, Zwicky pointed out that "... Vast and often very irregular swarms of stars and other matter exists in the spaces between the conventional spiral, elliptical, and irregular galaxies. It is not at all clear how such swarms should be incorporated into any distribution function of galaxies.." [1]. Therefore the presence of a diffuse intracluster light filling up the regions between galaxies in a given cluster has been known since several decades. Further studies on the presence of diffuse light in clusters were based on photographic technique, and were later on followed by CCD observations in the nineties, but these studied ended with ambiguous results. The best estimate for the contribution of the diffuse light ranged from 10% to 50% of the total light emitted by cluster galaxies in the central region of the cluster itself. Such studies based on the detection of the continuous light above the sky background on large area of sky are hampered by two problems: 1) the typical surface brightness of the intracluster light is less than 1% of the sky brightness and 2) it is difficult to disentangle between diffuse light associated with the halo of the cD galaxy and the diffuse light component of the cluster.

2 Candidates for Intracluster Stars

An alternative to this method for probing intracluster light is through the direct detections of the stars themselves. In 1979 a search was done for intergalactic supernovae in the Coma cluster, but the outcome was negative. In 1981 a supernova Type Ia was discovered in the region between M86 and

M84 in the Virgo cluster. In the eighties several studies argued for the pronounced globular cluster over-abundance of some galaxies at the centre of galaxy clusters to be best explained by assuming the existence of a population of intergalactic globular clusters.

A new interest in these kind of studies was brought by the discovery of intracluster planetary nebulae free flying in the intracluster regions of Virgo and Fornax. In a radial velocity survey of planetary nebulae (PNe) in NGC 4406 (v_{sys} = -227 kms^{-1}), three objects were found with radial velocities $v > 1300$ kms^{-1} [2]. Independent evidence for 10 intracluster PNe candidates were found in the Fornax cluster [3] and in 1998 observations done with the HST of a field near to M87 detected the presence of a diffuse light component via statistical excess of red star counts [4] with respect to similar detections in the HDF. Additional surveys in Virgo ([5], [6], [7]) aiming at detecting point like-emission objects in the [OIII] 5007 line found more than a hundred PNe candidates in the Virgo core region.

3 Cluster Surface Mass Density and the Light Fraction in the Diffuse Component

In the study of NGC 4406, Arnaboldi et al. found 3 PNe in the galaxy field with velocities close to the Virgo cluster mean (1400 kms^{-1}). From the internal kinematics of NGC 4406 stars, one PN is detected for each $8 \cdot 10^9 M_\odot$ of galaxy mass (stellar + dark). 3 intracluster PNe from the survey area (13.5 $arcmin^2$) would imply a surface density of 80 $M_\odot$ pc^{-2} of intracluster matter near NGC 4406. We get a mass density of similar order of magnitude from the kinematics of galaxies in the Virgo cluster itself. In fact the virial surface density for the Virgo cluster at the position of NGC 4406 computed using a King model, $\sigma_{cluster} = 800$ kms^{-1} and $R_{core} = 0.5$ Mpc ([8]) is 390 $M_\odot$ pc^{-2}. The similar order of magnitude of the surface mass density inferred from the presence of intracluster PNe, and from the kinematics of the Virgo cluster itself seems to indicate that much of the mass in the Virgo cluster is in stars plus dark matter as in galaxies themselves.

In 1997, Méndez et al. surveyed a 50 $arcmin^2$ region in the Virgo cluster core far from any galaxies and found 11 PNe intracluster candidates. This photometric selected sample is consistent with the previously spectroscopic detected one. The [OIII] λ5007 measured fluxes of these unresolved sources are consistent with these objects being PNe from an intergalactic population of stars at the distance of the Virgo cluster. In 1998, [6] and [7] discovered 151 intracluster PNe candidates in two $16' \times 16'$ fields in the Virgo cluster core well away from any Virgo galaxies.

How much light is there? The luminosity derived by Méndez et al. 1997 corresponds to a surface brightness of $\mu_b = 28.3$ mag $arcsec^{-2}$ – difficult to detect on the angular scale of the Virgo cluster. In terms of light fraction emitted by the diffuse component vs. the light in galaxies, the Méndez et al. and Feldmeier et al. extended PNe surveys in the Virgo cluster core indicate that the diffuse intracluster population could make up to 50% of the luminosity from all galaxies in the Virgo cluster. Such a high fraction of light in the diffuse component is observed for example also in the Coma cluster from extended CCD surface photometry ([9]). The estimate of the surface brightness in the diffused component inferred from Ferguson et al. HST data is $\mu_b = 31.33$ mag arcsec^{-2} [4], and it disagrees by an order of magnitude in flux from the estimate on PNe candidates number counts. The survey of intracluster PNe is done via the so-called "on-off" technique, i.e. images are taken with an "on-band" filter, centred at the expected redshifted [OIII] emission from the planetary nebulae, and then these images are compared with "off-band" images taken with an adjacent filter. The intracluster PNe candidates are selected as those point-like sources that are visible in the on-band image and are not visible in the off-band one. These photometric samples need to be spectroscopically confirmed, as there are several sources of contaminations. Indeed emission line galaxies at high redshifts (e.g [OII] λ3727 at z = 0.35, or Lyα at z=3.1), or faint stars can simulate spurious counts in the [OIII] images and can be erroneously classified as PNe. So the PNe photometrically selected sample could lead to an overestimation of the light fraction in the diffuse intracluster light. In a very recent spectroscopic run at the Anglo Australian Telescope with the 2 degree field fiber spectrograph (2dF), Freeman et al. 1999 (in preparation) obtained spectroscopic confirmation for 47 PNe out of the 151 PNe candidates in the combined [7] and [5] sample. The surface brightness associated with the spectroscopically confirmed sample is $\mu_B \sim 29.4$ mag arcsec^{-2}, a factor 5 in flux greater than the Ferguson's estimate. Furthermore, recent un-published results from Feldmeier and collaborators indicated that the number counts of PNe candidates in a field including Ferguson HST field east of M87 are consistent with smaller values for the surface brightness luminosity. The evidence gathered so far indicates a clumpy distribution for the Virgo intracluster population of stars, and this population could contribute al least 15% of the total light in galaxies in the Virgo cluster. The presence of the diffuse light, e.g. stars, in the intracluster regions and the hints for its distribution to be clumpy (a clumpiness so far detected only in the projected distribution of the PNe number density) are to be expected if this diffuse stellar population is a remnant from "galaxy harassment" caused by high speed encounters of galaxies in clusters [10]. This cluster-galaxy and galaxy-galaxy interactions can lead to a loss of up to 50% of a galaxy mass in a Hubble time. If tidal interactions and stripping of stars are the origin of this population, one should expect a non-Gaussian distribution of the intracluster PNe radial velocities, with the PNe radial velocities being quite similar to those of the nearby galaxies they originated from.

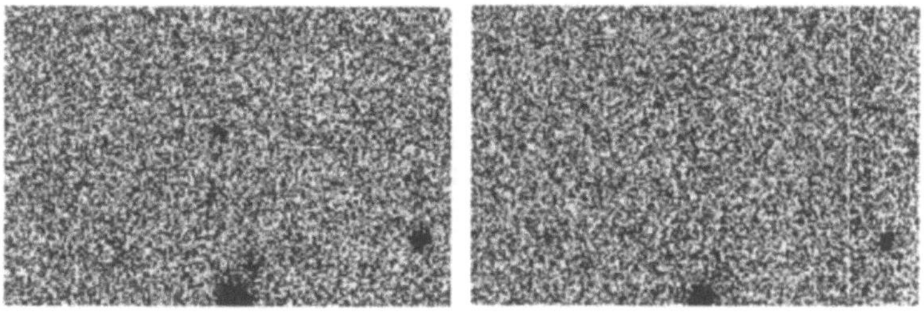

Fig. 1. Image of a PNe candidate in Virgo cluster field centred on the dEs and dS0 number density distribution. The PN candidate is visible in the [OIII] on-band image, left frame, and does not show up in the V band image, right frame. These date were acquired in March 1999 with the wide field imager (WFI) at the ESO MPI 2.2m tel.

4 Intracluster PNe and the VLT

Clues on the origin of the intracluster population can be obtained by studying their radial velocity field: a preliminary study is now undergoing at the AAT with the 2dF, but the VLT will be much more efficient and give much accurate spectra, also for line identifications and metallicity studies. The use of the VLT to test the origin and properties of the intracluster population will make this challenging project within reach in a short time scale of 1-2 yrs: further details on VLT efficiencies, strategies, and limiting fluxes for detections of [OIII] PNe emission lines are presented in detail by R. Méndez's contribution, this volume. We are now carrying out extended survey at the WFI of the ESO MPI 2.2m telescope, to map the extension of this population with respect to the galaxies in the Virgo cluster: preliminary candidates of these PNe are shown in Figure 1.

Not only extended deep surveys of intracluster PNe can shed light on their origins, but also they have implications on the determination of baryonic content of clusters. For the Virgo cluster, the estimate of the $M_{\mathrm{Bary}}/M_{\mathrm{Tot}}$ ratio computed by taking into account stars in galaxies, hot gas and the diffuse stellar population, gives a value $\sim$ 0.2 in 250 kpc from the Virgo centre, pointing towards a $\Omega_0 = 0.2$. In this respect, extended surveys for the intracluster light stars are crucial to establish whether there is a variation of baryonic ratio and therefore a baryonic segregation in the cluster core.

References

1. Zwicky, F. 1951, PASP, 63, 61
2. Arnaboldi, M., et al. 1996, ApJ, 472, 145
3. Theuns, T. Warren, S. 1997, MNRAS, 284, L11
4. Ferguson, H. et al. 1998, Nature 391, 461
5. Méndez, R. et *al. 1997*, ApJL, 491 23

6. Ciardullo, R. et al. 1998, ApJ 492, 62
7. Feldmeier, J.J., et al. 1998, ApJ, 503, 109
8. Binggeli, B., et al. 1987, AJ, 94, 251
9. Bernstein, G.M. et al. 1995, AJ, 110, 1507
10. Moore, B. et al. 1996, Nature 379, 613

Probing Distant Galaxies with Lensing Clusters

Roser Pelló

Observatoire Midi-Pyrénées, LAT, UMR5572, 14, Av. Edouard-Belin, F-31400 Toulouse, France

Abstract. Clusters of galaxies as gravitational lenses allow to study the stellar content and properties of high-z galaxies much fainter than the usual spectroscopic field surveys. We review the recent results obtained on the identification and study of very distant galaxies seen through lensing clusters. Using the gravitational amplification effect in lenses with well known mass distributions, it is possible to build up and study a sample of galaxies with $1 \lesssim z \lesssim 7$. Source candidates are selected close to the critical lines at high-z, through photometric redshifts computed on a large wavelength interval, as well as lens-inversion criteria. This procedure allows to reduce the selection biases in luminosity, and towards active star-forming objects, provided that selection criteria include IR photometry. This is the method presently used in our large collaborative program, aiming to perform the spectroscopic follow up with the VLT of high-z candidates selected from suitable photometric campaigns.

1 The Gravitational Telescope

The identification and study of high-z galaxies is probably one of the most direct methods to constrain the scenarios of galaxy formation and evolution. With respect to large field surveys, cluster lenses can be used to build up an *independent* and complementary sample of distant galaxies, because of their lensing properties. The major bonus is the amplification close to the critical lines, which is $\Delta m \sim 1$ to 3 magnitudes depending on location (see [6] for a review). The expected 2D distribution of arclets (mean redshift and surface density) can be easily obtained for lenses with well known mass-distributions ([2]). These clusters are used here as Gravitational Telescopes .

The first lensed galaxy confirmed at $z \gtrsim 2$ was the spectacular blue arc in Cl2244-02 ([10]). Recent examples of highly magnified galaxies, identified either purposely or serendipitously, strongly encourages this approach: the star-forming source #384 in A2218, at z=2.51 ([4]), the luminous z=2.7 galaxy behind MS1512+36 ([17]), three z $\sim$ 4 galaxies in Cl0939+47 ([16]), a z=4.92 system in Cl1358+62 ([7], [14]), the z=2.72 arc behind MS1512+36

[0] In collaboration with : B. Fort (IAP, F), J.-P. Kneib (OMP, F), J.-F. Le Borgne (OMP, F), Y. Mellier (IAP, F), I. Appenzeller (Obs. Heidelberg, D), R. Bender (Munich Observatory, D), L. Campusano (U. Chile, Chile), M. Dantel-Fort (O. Paris, F), R.S. Ellis (IoA, UK) A. Moorwood (ESO, D), S. Seitz (Munich Observatory, D)

([13]), and the two red galaxies at $z \sim 4$ in A2390 ([8], [12]). We present the method proposed by the Gravitational Telescope Program (hereafter, GT) to access the distant population of galaxies and some illustrative results.

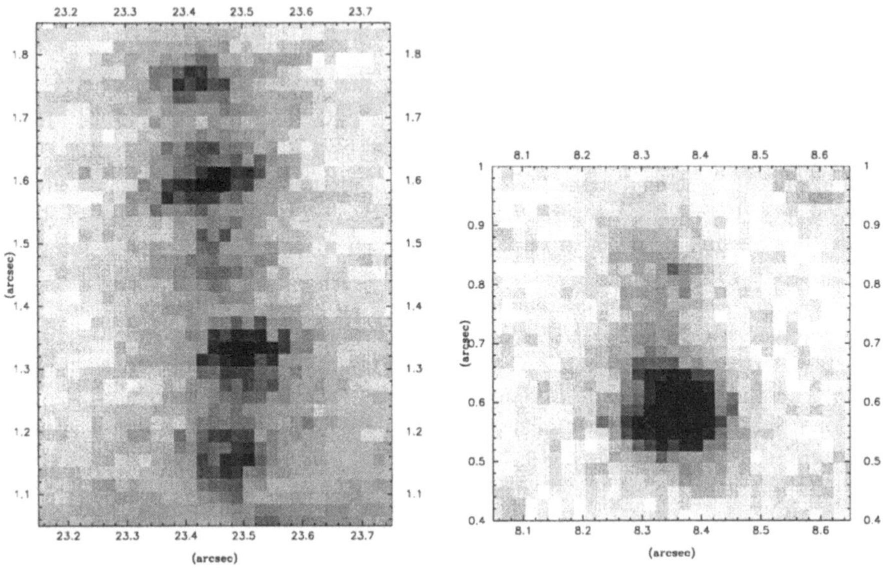

Fig. 1. Images of 2 $z = 4.05$ sources behind A2390: H3 (**left**) and H5 (**right**) ([12]), reconstructed by lens-inversion according to the model by Kneib et al. (1999). Both sources are clumpy, elongated and exhibit the same orientation. The true distance between H3 and H5 is 16″ ($1'' = 9.4(15.7)h_{50}^{-1}$ kpc with $q_0 = 0.5(0.1)$)

2 The Method

Lensed sources in the GT program are selected according to 2 criteria: **1)** they lie close to the high-z critical lines, and **2)** they have a photometric redshift compatible with $z \geq 2$. The former criterion restricts the application to clusters with mass distributions highly constrained by multiple images (revealed by HST or ground multicolor images), where the amplification uncertainties are $\Delta m_{lensing} < 0.3$ mags. Cluster lenses with well constrained mass distributions enable to recover precisely the properties of lensed galaxies (morphology, magnification factor). Fig. 1 gives an example of source reconstruction for two multiple-images at the same $z \sim 4$, behind A2390. The sample of lenses includes clusters with a strong X-ray emission and, as a first priority, those with HST images available. In all cases, deep multicolor photometry has been obtained in the visible and near-IR domains.

Photometric redshifts have been derived using the standard SED fitting method by Miralles & Pelló (1998). The observed SEDs are compared to a set of templates linked to the Bruzual & Charlot code (GISSEL98, [3]). The database includes 255 synthetic spectra, with A_v from 0 to 0.5 mags, and flux decrements in the Lyman forest modelled according to Giallongo & Cristiani (1990; see also Madau 1995). The photometric z criterion can be used advantageously in this case because photometry is available on a large wavelength domain, including near-IR. This selection allows to reduce the biases towards intrinsically bright or strong star-forming sources. Fig. 2 presents an example of likelihood map for one of the $z \sim 4$ sources in Fig. 1.

3 Main Goals

The GT can be used to determine the redshift distribution up to the faintest levels through magnified sources. It is also the natural way to search for primeval galaxies, in order to put strong constraints on the scenarios of galaxy formation. The use of visible and near-IR spectroscopy, combined with photometric z, enables to study the SED of $z \geq 2$ galaxies for a sample which is less biased in luminosity than the field ones. In particular, the present and past SFR history (respectively obtained through the UV/visible/IR and near-IR domains), and the permitted region in the age- metallicity -A_V parameter space. Highly magnified arcs are presently the only way to access the dynamical properties of galaxies at $z \geq 2$, through 2D spectroscopy (Narasimha & Chitre, 1993; Soucail et al. 1998). As shown in Fig.1, lens reconstruction allows to explore the morphology of distant sources with a spatial resolution of $\sim$ 1 kpc. When this technique will be coupled with 2D spectroscopy, the dynamics involved in galaxy formation would be directly observed.

4 First Results on Selected Lenses

H3 and H5 are the first two $z \geq 2$ objects spectroscopically confirmed in A2390, and this is one of the best studied lenses in our sample. $\sim$ 30 sources at $z \geq 2$ have been identified among the numerous arcs and arclets in this field. Their redshift distribution is shown in Fig. 5, and compared to the present spectroscopic survey in A2390. The redshift distribution in other GT targets (A2218, A370, Cl2244-02, AC114) also displays interesting candidates at $1 \lesssim z \lesssim 7$. Once spectroscopic data is available for high-z sources, the SF rates and other stellar properties can be derived, although there is some degeneracy in the SFR-age-metallicity-reddening space. The permitted parameter region can be constrained using the GISSEL98 code when z is well known. Fig. 4 illustrates this point with two examples: H5 and D, the later being one of the brightest ISO sources in A2390 ([9], [1]). H3 and H5 are intrinsically bright star-forming systems ($M_B^* - 2$ to -1 mags, depending on A_v), slightly brighter than the $z \sim 4$ objects found in Cl0939+4713 [16].

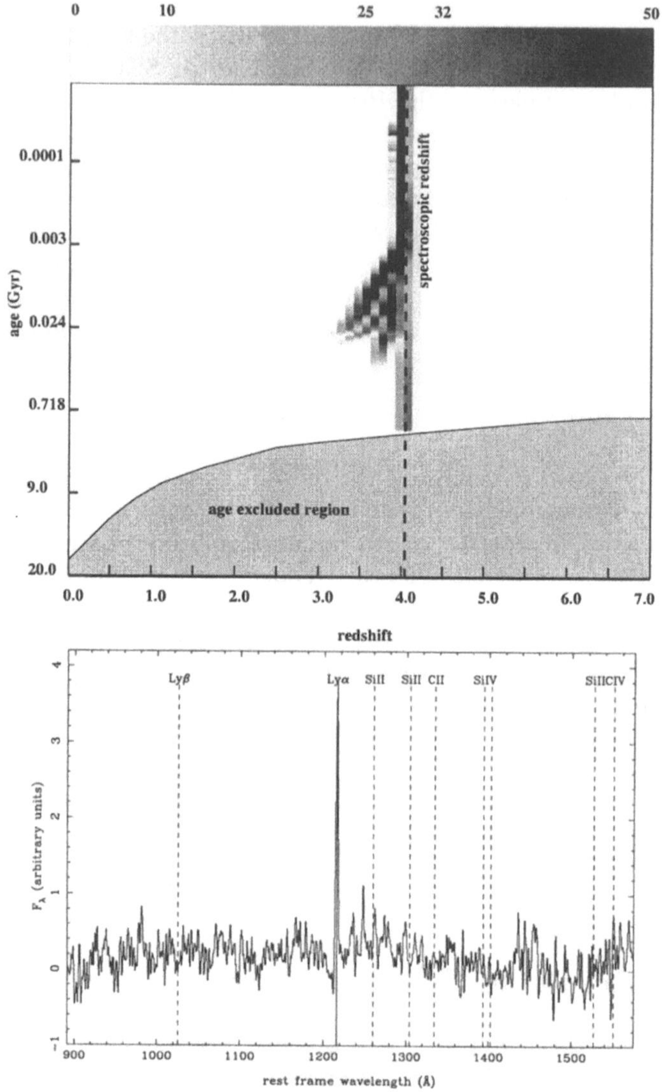

Fig. 2. H5, one of the z=4.05 sources in A2390: **Top)** Photometric z likelihood-map showing the excellent agreement with the spectroscopic z. The shaded region is the limit set for the age of the Universe ($H_0 = 50$ km s^{-1} Mpc^{-1}, $q_0 = 0$). **Bottom)** Averaged spectrum of H5, obtained at WHT, showing a strong Lyα emission line

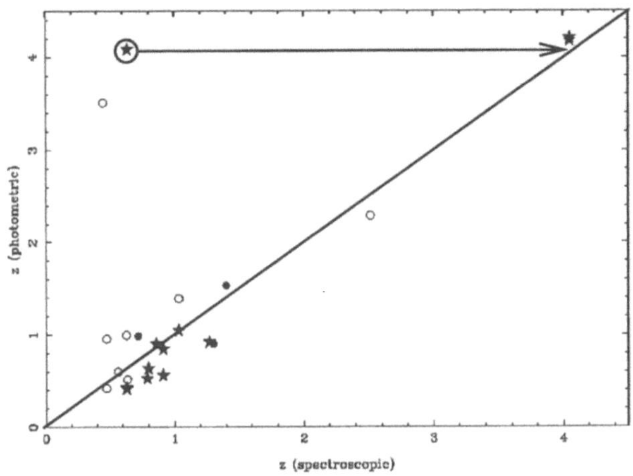

Fig. 3. Spectroscopic versus photometric redshift determination for a sample of multiple images in A370 (black dots), A2390 (black stars) and A2218 (open dots). One of the two catastrophic identifications corresponds to a wrong spectroscopic determination based on a single emission line

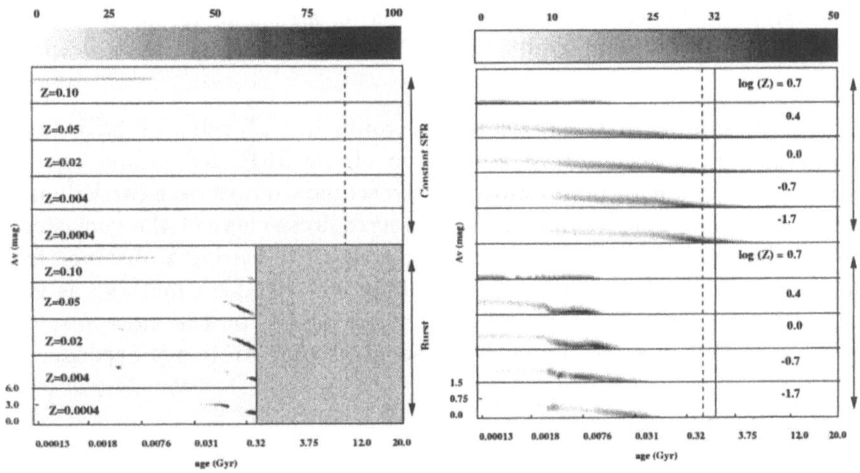

Fig. 4. Likelihood map for two sources in A2390, showing in dark the most probable regions and the degeneracy in the parameter space SFR - age - metallicity - reddening: **left)** Source D (z=0.913): the dotted line gives the age limit of the stellar population ($q_0 = 0$); the hatched region is forbidden according to the optical spectrum, which shows a strong [OII]3727 Å e-line, and evidence for ongoing star-formation. The most likely solutions exhibit high reddening, as expected for a such bright ISO source. **right)** H5 at z=4.05; dotted and solid lines are the cosmic ages ($H_0 = 50$ km s^{-1} Mpc^{-1}, $q_0 = 0.1$ and $q_0 = 0.5$)

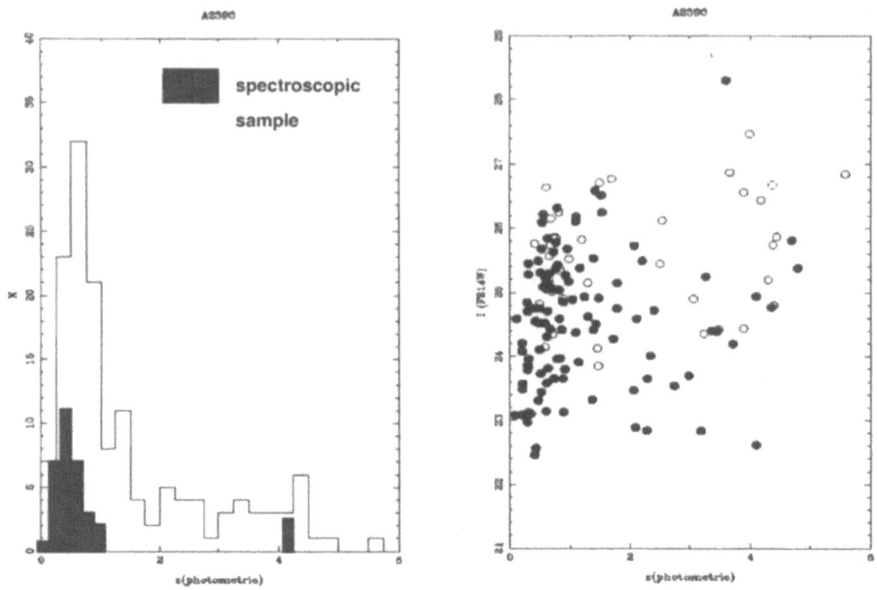

Fig. 5. left) Photometric redshift distribution for arclets in A2390, for the most secured sources (black dots). **right)** I (F814W) magnitude versus photometric z

For a subsample of spectroscopically confirmed objects, we have tested the photometric z accuracy as a function of the SFR, reddening, age and metallicity of the stellar population. Spectroscopic surveys were carried out at CFHT, WHT and ESO (NTT,3.6m). We have cross-checked the consistency between the photometric, the spectroscopic and the lensing z obtained from inversion methods ([5]). The agreement between the three methods is good up to at least $z \lesssim 1.5$. For higher redshifts, the results on the most amplified sources are promising, but an enlarged spectroscopic sample is needed.

5 Future Issues

The uncertainty in the amplification is typically $\Delta m \sim 0.2$ to 0.3 mags. Thus, intrinsic luminosities and SFRs are known to $\sim 30\%$ accuracy using the GT. Only well studied lensing clusters, with fairly well known mass distributions, are actually useful as GTs.

Up to now, all spectroscopically confirmed sources at high-z in our sample are intrinsically bright. In order to constrain their parameter space (metallicity, A_V, ...), spectroscopic information on the UV restframe absorption lines, and near-IR spectroscopy aiming to access the Balmer region are strongly needed. Using arcs in clusters is probably the only way to access dynamics for $z \geq 2$ sources (velocity gradients, 2D velocity fields, ...). This systematic study is a well defined program for 8m telescopes such as the VLT. Combining photometric redshifts with lensing inversion techniques provides with an alternative way to determine the redshift distribution of high-z sources. If $\Delta z \sim 0.1$ is enough for most applications, photometric redshifts allow to go further. Lensing clusters could be used as a tool to check photometric redshifts up to the faintest limits. An Ultra-Deep Photometric Survey of cluster lenses is urgently needed to probe the distant Universe.

References

[1] Altieri, B., et al. (1999) A & A **343**, L65
[2] Bézecourt, J., Pelló, R., Soucail, G. (1998) A & A **330**, 399
[3] Bruzual, G., Charlot, S. (1993) ApJ **405**, 538
[4] Ebbels, T.M.D., et al. (1996) MNRAS **281**, L75
[5] Ebbels, T.M.D., et al. (1998) MNRAS **295**, 75
[6] Fort, B., Mellier, Y. (1994) A & A Review **5**, 239
[7] Franx M., et al. (1997) ApJ **486**, 75
[8] Frye B., Broadhurst T. (1998) ApJ **499**, 115
[9] Lémonon, L., et al. (1998) A & A **334**, L21
[10] Mellier et al. (1991) ApJ **380**, 334
[11] Narasimha, D., Chitre, S.M. (1993) A & A **280**, 57
[12] Pelló, R., et al. (1999) A & A in press (astro-ph/9810390)
[13] Seitz, S., et al. (1998) MNRAS **298**, 945
[14] Soifer, B.T., et al. (1998) ApJ **501**, 171
[15] Soucail, G. et al. (1998) ASP Conference Series, vol. 146, 457
[16] Trager S. C., et al. (1997) ApJ **485**,92
[17] Yee, H.K.C., et al. (1996) AJ **111**, 1783

Gravitational Redshift and Cluster Masses

Tom Broadhurst[1] and Evan Scannapieco[2]

[1] ESO, Garching, Germany
[2] University of California, Berkeley

Abstract. We examine the gravitational redshift of radiation emitted from within the potential of a cluster. Spectral lines from the ICM are redshifted in proportion to the emission-weighted mean potential along the line of sight, amounting to ≈ 50 km/s at a radius of 100 kpc/h for a cluster dispersion of 1200 km/s. We show that the relative redshifts of different ionization states of the gas, provide a unique probe of the 3D distribution. Note that in general the gravitational redshift results in a small overestimate of the recessional velocity of a cluster by an average of ~ 20 km/s.

1 Introduction

The central mass profile of clusters are predicted to show a pronounced flattening in the case that the dark matter is cold and does not dissipate energy. Recent N-body work has revealed that the flattening of the mass profile claimed for simulations of CDM model of structure formation is quite severe inside a radius of of ~ 400 kpc for cluster sized halos, turning over from an outer slope of r^{-2} to r^{-1} (Navarro, Frenk, & White 1996; Moore et al. 1998). This change in slope although distinct is not easily examined with lensing observations. The distortion of images by lensing provides a powerful way of locating dark matter but is insensitive to the gradient of the distribution, particularly in the central region which is subject to a degeneracy (Schneider & Seitz 1995; Kaiser 1995). Thus, the slightly greater distortion generated by a flattening of the slope to r^{-1} cannot be distinguished in practice from an isothermal profile using only lensed distortions (e.g. Clowe et al. 1998). In principle, the magnifications of images are more sensitive to the projected mass profile, being much greater for the flatter NFW profile within r_s or $\approx 5\theta_e$. However, the use of magnifications requires knowledge of both the source redshifts and intrinsic unlensed properties to be reliable (Broadhurst, Taylor, & Peacock 1995). Of course all lensing signals relate only to the projected mass distribution and thus provide only indirect information through the cluster.

Here we evaluate a means of exploring the distribution of matter in clusters that complements and extends the information available from lensing. The gravitation potential of a cluster not only deflects light passing nearby, but imprints a gravitational redshift on emission generated within it. Cluster gas provides a suitable tracer of this redshift effect effect via the wealth of X-ray emission lines. Since the line emission is excited by a two body process, the velocity information is weighted by the square of the gas density along a given sight-line. Furthermore, temperature gradients in the gas cause different lines to be emitted at different distances along the line of sight through the cluster, and may thus be used as a possible three-dimensional probe of the matter profile. Note that in contrast the gravitational redshift of the central starlight of an ordinary isolated galaxy is calculated to be only a few km/s depending on the sharpness of the potential and may be confused with core rotation and other bulk motion of the same order (Stiavelli & Setti 1993).

While this X-ray spectroscopy has been difficult in the past it is advancing to the point now where the velocity dispersion of the gas may be marginally resolved. In the near future, an order of magnitude improvement in resolution and effective collecting area will be achieved using improved techniques, making relevant discussions of detailed spectroscopic effects.

2 Gravitational Redshift

The simplest useful illustration of the gravitational redshift of cluster radiation is an isothermal sphere in which the density $n(r)$ is proportional to r^{-2} for radii less than a truncation radius R_0, and $n(r) = 0$ if $r > R_0$, so that the gravitational potential is given by

$$\phi = \begin{cases} 2\sigma^2[\ln(r/R_0) - 1] & r < R_0 \\ -2\sigma^2 R_0/r & r > R_0, \end{cases} \tag{1}$$

where σ is the velocity dispersion.

The gravitational redshift of light from point sources, such as galaxies, located within this potential are then given simply by

$$\Delta\nu = \nu_0 \frac{2\sigma^2}{c^2}. \tag{2}$$

In a gas with constant temperature, however, spectral line emission is simply proportional to $n_e(r) n_{\text{ion}}(r) \propto n^2(r)$. Thus by averaging the potential weighted by $n^2(r)$ along different lines of sight, we can compute the average redshift of a line at an impact parameter x from the center. In this case, the frequency of the emission line will be shifted by an amount

$$\Delta\nu \equiv \frac{\nu_0}{c^2} \frac{\int_0^{R(x)} dl\, \phi((x^2 + l^2)^{1/2}) n((x^2 + l^2)^{1/2})^2}{\int_0^{R(x)} dl\, n((x^2 + l^2)^{1/2})^2}, \tag{3}$$

where ν_0 rest-frame emission frequency, l is the distance along the line of sight such that $l = 0$ is the point of closest approach to the center of the cluster, $\phi(r)$ is the cluster potential, and $R(x) \equiv \sqrt{R_0^2 - x^2}$.

For a singular isothermal sphere we can further replace $n(r)$ with r^{-2} and $\phi(r)$ with Eq. 1. In this case the denominator becomes

$$\int_0^{R(x)} dl \frac{1}{(x^2 + l^2)^2} = \frac{\theta_x + \sin\theta_x \cos\theta_x}{2x^3}, \tag{4}$$

where $\theta_x \equiv \arctan(R(x)/x) = \arccos(x/R_0)$, and the numerator becomes

$$\int_0^{R(x)} dl \frac{2\sigma^2[\ln((x^2+l^2)^{1/2}/R_0) - 1]}{(x^2+l^2)^2}$$
$$= \frac{2\sigma^2}{x^3}\left[\frac{1}{2}[L(\theta_x) - \theta_x + \theta_x \ln\cos\theta_x] - \frac{\theta_x + \sin\theta_x\cos\theta_x}{4}\right], \quad (5)$$

where $L(\theta_x) \equiv -\int_0^{\theta_x} d\theta \ln(\cos\theta)$ is Lobachevskiy's function. Thus each line is redshifted by an amount

$$z = -\frac{\Delta\nu}{\nu_0} = \frac{2\sigma^2}{c^2}\left[\frac{1}{2} - \frac{L(\theta_x) - \theta_x + \theta_x \ln\cos\theta_x}{\theta_x + \sin\theta_x\cos\theta_x}\right] \approx \frac{2\sigma^2}{c^2}\left[\frac{3}{2} - \ln\left(\frac{2x}{R_0}\right)\right], \quad (6)$$

where the approximation is valid for small impact parameters. The velocity dispersion is only a weak function of cosmology (Eke, Cole, & Frenk 1996) and can for example be estimated as

$$\sigma = 730\, M_{14}^{1/3}(1+z_c)^{1/2}\,\mathrm{km/s}, \quad (7)$$

where M_{14} is the mass of the collapsed halo in units of $10^{14}h^{-1}$ solar masses and z_c is the redshift of collapse. In the singular isothermal case a simple relation exists between this dispersion and the Einstein ring radius θ_{E}

$$\frac{2\sigma^2}{c} = -13.9\,\mathrm{km/s}\,(\theta_e/\mathrm{arcmin}) d_{\mathrm{s}}/d_{\mathrm{ls}}, \quad (8)$$

where d_s is the distance to the source, and d_{ls} is the distance from the source to the lens, which helps to give a rough sense of magnitude of the redshifts in Eqs. 2 and 6. Finally, the relation between total mass and velocity dispersion ($\frac{GM}{R_0} = 2\sigma^2$) gives us the truncation radius $R_0 = 400\, M_{14}^{1/3}(1+z_c)^{-1}$ kpc/h.

In the upper panel Figure 1 we plot the redshift as a function of radius for three clusters of $M_{14} = 2$, 4, and 8, with $z_c = 1.0$. Here we see that while the gravitational redshift of emission lines near the edge of the cluster is quite small. The r^{-4} weighting boosts the effect significantly along lines of sight near the center of the cluster, reaching values of order 10% of the velocity dispersion within the inner 10 – 50 kpc of the cluster.

2.1 Future Prospects and Observational Issues

Thus X-ray spectroscopy approach outlined in §2 and 3 represents the most straightforward method for studying gravitational redshifts. Note that this we require both high spectral resolution to determine the shift of the line centroids, and fair spatial resolution, to obtain radial information. Little information can be learned from a pure spectral analysis, however, as overall cluster-integrated line profile indistinguishable from a thermal profile with a slightly broadened width.

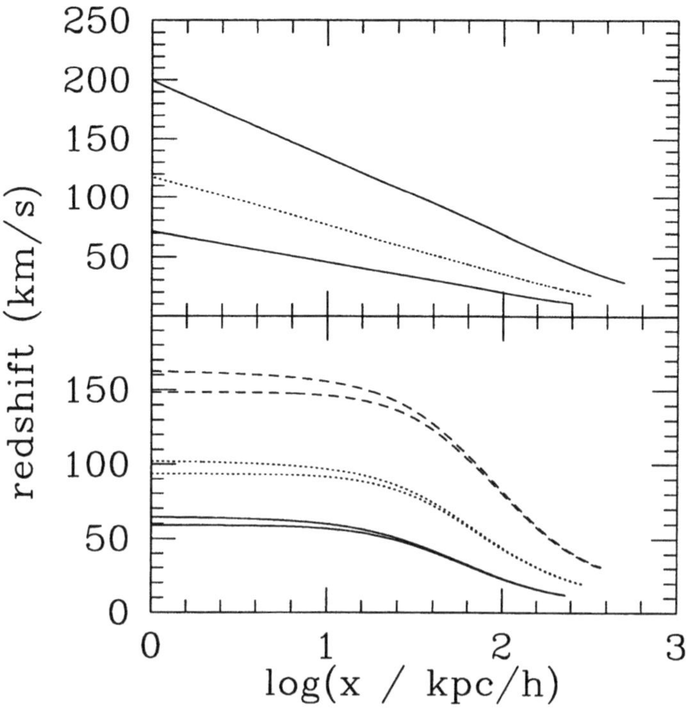

Fig. 1. Upper panel: Redshift as a function of radius for a singular isothermal profile with $M_{14} = 2$ (solid), $M_{14} = 4$ (dotted), $M_{14} = 8$ (dashed), $z_c = 1.0$, and σ as given in Eq. 7. Lower panel: Redshift as a function of radius for a nonsingular isothermal profile with masses as in the upper panel. In each pair of curves the lower curve represents the redshift purely from the profile, while the upper curve includes an additional contribution from a cD galaxy modeled as point mass with 2% of the cluster mass.

This analysis is beyond the abilities of the current generation of X-ray satellites, but only marginally. Relatively high spectral resolution is already achievable in the present X-ray satellites with energy resolution of 300 and lines may now be allowing marginal resolution of gas velocity dispersions in nearby clusters (nice web page at Goddard). The next generation of satellites, however, should

Large scale bulk motion of gas should also be detected from recent group infall in some clusters. The disturbance of cluster plasma by gas infall has claimed to be significant over a period of 3Gyrs in favorable conditions, before shocks are thermalised and bulk flow through the core stops (Roettiger, Burns, & Loken 1996; Roettiger, Loken, & Burns 1997). If infall is common then the relatively small perturbation to the line centroid by gravity proposed here may be made noisy by the continuous process of cluster interaction, or at worst relegated to a correction applied to a larger dynamical disturbance.

However this turns out, it is clear that a careful measurement of frequency structure of X-ray emission lines over the surface of clusters will provide useful information for understanding cluster formation.

References

1. Binney, J., & Tremaine, S. 1987, Galactic Dynamics (Princeton: Princeton University Press)
2. Bird, C. M. 1994, AJ, 107, 1637
3. Bird, C. M. 1995, ApJ, 445, L81
4. Bird, C. M., Davis, D. S., & Beers, T. C. 1995, AJ, 109, 920
5. Bliton, N., Rizza, E., Burns, J. O., Owen, F. N., & Ledlow, M. J. 1998, MNRAS, 301, 609
6. Broadhurst, T., Taylor, A. N., & Peacock, J. A. 1995, ApJ, 438, 49
7. Burns, J. O. et al. 2000 to appear in *Life Cycles of Radio Galaxies* eds J. Biretta et al. astro-ph/9908361
8. Churazov, E., Sunyaev, R., Gilganov, M., Forman, W., & Jones, C. 1998, MNRAS, 297, 1279
9. Clowe, D., Luppino, G. A., Kaiser, N., Henry, J. P., & Gioia, I. M. 1998, ApJ, 497, L61
10. Eke, V. R., Cole, S., & Frenk, C. S. 1996, MNRAS, 282, 263
11. Fabian, A.C. 1994, ARA&A, 32, 277
12. Hill, J. M., Hintzen, P., Oegerle, W. R., Romanshin, W., Lesser, M. P., Eisenhamer, J. D., & Natuski, D. J., 1988, ApJ, 332, L23
13. Hill, J. M., & Oegerle, W. R. 1993, AJ, 106, 831
14. Hill, J. M., & Oegerle, W. R. 1998, AJ, 116, 1529
15. Kaiser, N, 1996, ApJ 439, L1
16. Kriessler, J. R., & Beers, T. C. 1997, ApJ, 481, 775.
17. Moore, B., Governato, F., Quinn, T., Stadel, J., & Lake, G. 1998, ApJ, 499, L5
18. Oegerle, W. R. & Hill, J. M. 1994, AJ, 107, 857
19. Peres, C. B., Fabian, A. C., Edge, A. C., Allen, S. W., Johnstone, R. M., & White, D. A. 1998, MNRAS, 298, 416
20. Navarro, J. F., Frenk C. S., & White, S. D. M. 1996, ApJ, 462, 563
21. Roettiger, K., Burns, J. O. & Loken, C. 1996, ApJ, 473, 651
22. Roettiger, K., Loken, C., & Burns, J. O. 1997, ApJS, 109, 307
23. Schneider, P. & Seitz, C., 1995, A&A, 294, 411
24. Sharples, R.M., Ellis, R.S., Gray, P.M. 1988, MNRAS, 231, 479
25. Stiavelli, M., & Setti, G., 1993, MNRAS, 262, L51
26. Zabludoff, A. I., Huchra, J. P., Geller, M. J., 1990, ApJS, 74, 1
27. Zabludoff, A. I., Geller, M. J., Huchra, J. P., & Ramella, M. 1993, AJ, 106, 1273
28. Zabludoff, A. I., & Zaritsky, D. 1995, ApJ, 447, 21
29. Zombeck, M. V. 1990, *Handbook of Space Astronomy and Astrophysics*, (Cambridge: University Press)

Velocity Dispersion Analysis of CNOC Clusters

Stefano Borgani

INFN - Sezione di Trieste, c/o ICTP, Strada Costiera 11, I-34100 Trieste, Italy

Abstract. I present results of the analysis of the internal velocity dispersions, σ_v, for the sample of distant galaxy clusters ($0.17 \lesssim z \lesssim 0.55$) provided by the Canadian Network for Observational Cosmology (CNOC). Different σ_v estimates are provided, all based on an interlopers removal algorithm, which is different from that originally applied by Carlberg et al. (1996). We find that all such methods provide σ_v estimates which are consistent with $< 10\%$ among themselves and with the original estimates provided by the CNOC collaboration. The resulting distribution of velocity dispersions is used to trace the redshift evolution of the cluster abundance with the aim of constraining the matter density parameter, Ω_m. We find that constrains on Ω_m are very sensitive to the adopted value of $\tilde{\sigma}_8 = \sigma_8 \Omega_m^\alpha$ ($\alpha \simeq$ 0.4–0.5), as obtainable from the local cluster abundance. As $\tilde{\sigma}_8$ varies from 0.5 to 0.6, the best fitting density parameter varies in the range $0.3 \lesssim \Omega_m \lesssim 1.0$.

1 Introduction

Galaxy clusters represent the virialization stage of exceptionally high peaks of initial density perturbations on comoving scales of about $10\,h^{-1}$Mpc (h is the Hubble constant in units of $100\,\mathrm{km\,s^{-1}\,Mpc^{-1}}$). Therefore, their abundance is sensitive to the amplitude of the fluctuation power spectrum on such scales. Standard analytical arguments [1] show that the number density of clusters of a given mass provides a robust constraint on $\tilde{\sigma}_8 = \sigma_8 \Omega_m^\alpha$, where σ_8 is the r.m.s. fluctuation amplitude within a sphere of $8\,h^{-1}$Mpc radius, Ω_m is the matter density parameter and $\alpha \simeq$ 0.4–0.5, weakly dependent on Ω_m, on the presence of a cosmological constant term and on the shape of the power–spectrum. Despite the variety of methods and of data sets employed to trace the local cluster abundance, different analyses consistently indicate that $\tilde{\sigma}_8 \simeq$ 0.5–0.6 [2–4]. Once a model is tuned so as to reproduce the abundance of local clusters, its evolution mainly depends on Ω_m (e.g., ref.[5]). Therefore, having a statistical sample of high–redshift clusters with reliable mass determinations would allow in principle to break the degeneracy between Ω_m and σ_8.

A well established technique to trace the cluster mass function back in redshift is offered by measurements of the velocity dispersions, σ_v, of member galaxies. This technique has been already extensively applied to samples of local clusters (see [6,7], and references therein). Systematic estimates of the velocity dispersions for a statistical sample of distant clusters has been performed for the first time by Carlberg et al. [8]. They [9] analyzed the

Canadian Network for Observational Cosmology (CNOC) sample of clusters sample to trace the evolution of the cluster abundance and concluded that Ω_m values below unity are in general to be preferred (see also ref.[10]).

Here, I present results from a complete re–analysis of the cluster velocity dispersion of the CNOC sample by applying different algorithms, which are similar to that used by [6,11] for local clusters. This allows for the first time to make a homogeneous comparison of the internal dynamics for local and distant clusters. Such estimates of the velocity dispersions are used to estimate the evolution of the cluster abundance. Further details about this analysis are presented in a forthcoming paper [12].

2 The Velocity Dispersion Estimation

A comprehensive description of the CNOC cluster sample is provided in ref.[13]. The CNOC cluster redshift survey includes all the EMSS clusters [14] which have $L_X > 4 \times 10^{44}\,\mathrm{erg\,s^{-1}}$ ($h = 0.5$ and $q_0 = 0.5$) and $f > f_{lim} = 4 \times 10^{-13}\,\mathrm{erg\,s^{-1}cm^{-2}}$ for X–ray luminosities and fluxes in the [0.3, 3.5] keV energy band, $0.17 \le z \le 0.55$ and $-15^\circ < \delta < +55^\circ$. The solid angle covered by EMSS in this declination range is of 587 sq. deg. The CNOC survey provided a fairly large number of (50 to 200) of galaxy redshifts per cluster over a region covering up to $6\,h^{-1}$Mpc projected on the sky.

In order to compute velocity dispersions, σ_v, we apply the same procedure adopted for nearby clusters by F96 and G98. As for the interloper removal, we use the adaptive kernel method [15] to find the significant peaks in velocity distributions and then the "shifting gapper" algorithm to reject galaxies that are too far in velocity space (by $\ge 1000\,\mathrm{km\,s^{-1}}$) from the main body of galaxies at a given distance from the cluster center (within a shifting annulus of $0.4\,h^{-1}$Mpc). From the above member selection analysis only the MS0906+11 cluster, which has been considered as a "binary cluster" by C96, shows two separable peaks, whose distance is $> 3000\,\mathrm{km\,s^{-1}}$ in velocity space.

Velocity dispersions are computed by using the robust estimator [16], σ_{bwt}, as well as three other estimators: (*a*) the usual standard deviation σ_s, (*b*) its weighted version σ_{w-s}, and (*c*) a weighted version of the robust estimate σ_{w-bwt}. As for the latter, the weight associated to each galaxy is a magnitude–dependent geometric weight, which has been introduced in order to account for the geometry of the field surveyed for each cluster and the magnitude completeness of the sample (cf. ref.[13] for further details). We compare in Figure 1 our different estimates of σ_v with that provided by [8]. The median values of the ratio σ_v/σ_{C96} for the different estimators lie in the range 0.96 – 1.04, with $\sigma_{rob}/\sigma_{C96} = 1.04^{+0.06}_{-0.07}$ (at 90% c.l.). This represents a remarkable result, since it indicates that quite different methods to analyze the internal dynamics of galaxy clusters provide consistent results. This supports the reliability of such cluster dynamical studies from optical observations. Unless otherwise specified, we adopt in the following the σ_{bwt} estimator.

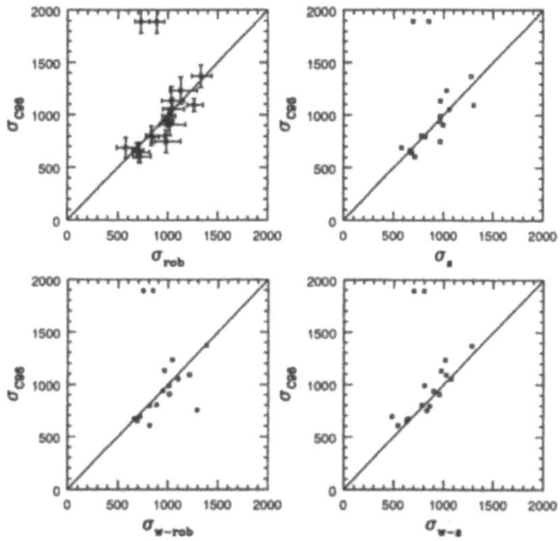

Fig. 1. The comparison between line-of-sight velocity dispersions (in units of $\mathrm{km\,s^{-1}}$) as estimated by Carlberg et al. (1996), σ_{C96}, and the four types of estimates we obtain in this work (see text). The 1σ bootstrap error bars are shown only in the σ_{C96}–σ_{rob} comparison.

3 Constraining Ω_m

The starting point of this analysis is represented by the standard Press–Schechter [1] expression for the comoving number density, at redshift z, of virialized halos of mass M. In order to convert the distribution of cluster masses into that of one–dimensional velocity dispersions, $N(\sigma_v, z)$, we use a virial relation whose amplitude and scatter has been tuned by resorting to high–resolution N–body simulations [12].

In order to constrain cosmological parameters we compare the cluster distribution on the (σ_v, z) plane with model predictions for samples having the same completeness and selection criteria of the CNOC sample. Let $\lambda(\sigma_v, z)$ be the model probability of observing a CNOC cluster with velocity dispersion σ_v at redshift z. The likelihood function $\mathcal{L}$ is defined as the product of the probabilities of observing exactly one cluster in $dz\,d\sigma_v$ at each of the $(\sigma_{v,i}, z_i)$ positions occupied by the CNOC clusters, and of the probabilities of observing zero clusters in all the other differential elements of the (σ_v, z) plane. Assuming Poisson statistics and introducing the quantity $S = -2\ln\mathcal{L}$, it is

$$S = -2\sum_i \ln[\lambda(\sigma_{v,i}, z_i)] + 2\int dz \int d\sigma_v\, \lambda(\sigma_v, z)\,, \qquad (1)$$

where $\lambda(\sigma_v, z) = N(\sigma_v, z)\,(dV(z)/dz) f_{sky}$. Here $dV(z)$ is the comoving volume element in the redshift interval $[z, z + dz]$ and f_{sky} is the flux–dependent EMSS sky–coverage [14].

At a given σ_v, the X–ray luminosity– and flux–limit criteria causes faint clusters to be missed in the CNOC sample. To correct for this σ_v–incompleteness, we select those CNOC clusters which have $\sigma_v > 800\,\mathrm{km\,s^{-1}}$ and $L_{X,bol} > 10^{45}\,\mathrm{erg\,s^{-1}}$. Then we resort to a complete local cluster sample to estimate the fraction $\mathcal{F}_{800}$ of all clusters with $\sigma_v > 800\,\mathrm{km\,s^{-1}}$, which also satisfy the above luminosity selection. Finally, we correct for the σ_v incompleteness by multiplying by $\mathcal{F}_{800}^{-1}$ the sum over the occupied cells in the r.h.s. of eq.(1). To estimate $\mathcal{F}_{800}$, we cross–correlate the ENACS cluster sample [17], which at $z \lesssim 0.1$ is representative of the whole cluster population for $\sigma_v > 800\,\mathrm{km\,s^{-1}}$, with the XBACs [18], which includes all the Abell/ACO clusters identified in the *ROSAT* all–sky survey and is 80% complete for fluxes $f_{[0.1-2.4]} > 5 \times 10^{-12}\,\mathrm{erg\,s^{-1}cm^{-2}}$. Therefore, $\mathcal{F}_{800}$ is given by the fraction of ENACS clusters with $\sigma_v > 800\,\mathrm{km\,s^{-1}}$ and belonging to the XBACs with $L_{X,bol} > 10^{45}\,\mathrm{erg\,s^{-1}}$. The ENACS contains 26 clusters with $\sigma_v > 800\,\mathrm{km\,s^{-1}}$, out of which 6 have $L_{X,bol} > 10^{45}\,\mathrm{erg\,s^{-1}}$. Therefore, we obtain $\mathcal{F}_{800}^{-1} = 4.3 \pm 2.0$, where the uncertainty is the 1σ Poissonian error.

Figure 2 shows constraints on the σ_8–Ω_m plane taking $\Gamma = 0.2$ for the power–spectrum shape Such results are based on *(a)* adopting the unweighted robust estimator for σ_v and *(b)* taking only clusters with $\sigma_v > 800\,\mathrm{km\,s^{-1}}$ and $L_{X,bol} > 10^{45}\,\mathrm{erg\,s^{-1}}$. The plotted iso–likelihood contours corresponds to 1σ, 2σ and 3σ confidence levels, while the dashed curves are the σ_8–Ω_m relation coming from the local cluster abundance and correspond to $\tilde{\sigma}_8 = 0.50$, 0.55 and 0.60, while the shape of the σ_8–Ω_m relation is that provided in ref.[4].

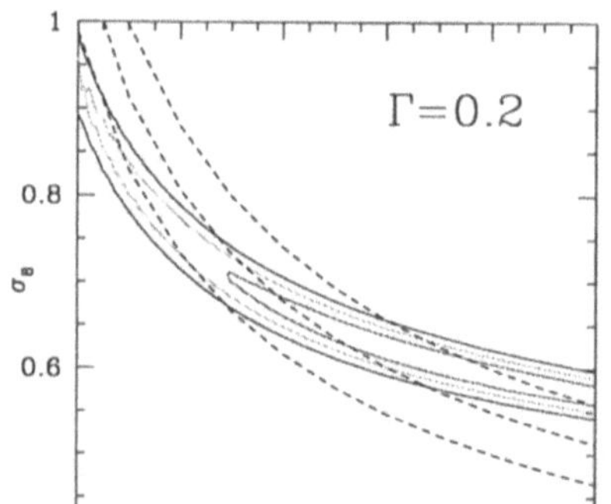

Fig. 2. Iso–likelihood contours on the (σ_8, Ω_m) plane for $\Gamma = 0.2$. Results refer to the reference analysis method, as described in Section 2. Iso–likelihood contours are for $\Delta S = 2.30$,6.17 and 11.8, which corresponds to 1σ, 2σ and 3σ c.l. for two significant fitting parameters.

Based only on the CNOC data, only very loose constraints are allowed for the density parameter, the whole considered range of Ω_m values being allowed within the 2σ confidence level. Both the small number of clusters on which the analysis is based and the limited leverage in redshift allow a rather high level of degeneracy in the σ_8–Ω_m plane. As expected, stronger constraints on Ω_m comes only when we project the likelihood along the σ_8–Ω_m relation from the local cluster abundance. However, even in this case, systematic uncertainties in fixing the local cluster abundance translates into rather large uncertainties in the resulting Ω_m. This is better shown in the panel (*a*) of Figure 3, where we show the variations of ΔS around its minimum for different choices of $\tilde{\sigma}_8$. We find that the best–fitting Ω_m increases from $\simeq 0.3$ to $\simeq 1.0$ as $\tilde{\sigma}_8$ is increased from 0.5 to 0.6. Therefore, although discrepancies among different analyses of the local cluster abundance are believed to be rather small, their impact on the determination of Ω_m is not. This is essentially due to the fact that CNOC clusters are rather rich systems, therefore probing the high–mass tail of the mass function, which is very sensitive to the power spectrum normalization. Therefore, even small variations in the normalization at $z \simeq 0$ translates into sizeable differences in the evolution of the mass function and, therefore, of the inferred Ω_m. While we regard this as a general problem one has to face with, even assuming that the analysis of high–z sample is perfectly under control, further uncertainties in constraining the cosmological parameters may come from other systematics, like the method to estimate σ_v or the way to go from σ_v determinations to the σ_v-distribution.

Among the effects that we checked, it turns out that the least affecting the inference of Ω_m are the choice of the σ_v limit [panel (*b*) in Fig.3] and the method to estimate σ_v [panel (*c*)]. In particular, we find that increasing σ_{lim} from $800\,\mathrm{km\,s^{-1}}$ to $900\,\mathrm{km\,s^{-1}}$ only somewhat increases the uncertainties, due to the decrease of the cluster weights, while providing constraints on Ω_m which are consistent within the 1σ uncertainties. Also, as expected on the ground of the results of Section 2, almost the same constraints come both from our analysis of σ_v and from the explicit background subtraction method by C97. Larger systematics are associated to uncertainties in the correction for the σ_v incompleteness. In panel (*d*) we show the effect of allowing $\mathcal{F}_{800}^{-1}$ to vary within the 1σ Poisson range, due to the limited number of local clusters on which the calibration of $\mathcal{F}_{800}$ is based. The effect of this uncertainty is only slightly smaller than that associated to the $\tilde{\sigma}_8$ uncertainty.

Far from meaning that the evolution of the cluster abundance can hardly be used to place significant constraints on cosmological parameters, the results of this analysis anyway indicate that the situation is more complex than sometimes suggested. Somewhat surprisingly, the availability of cluster mass measurements at high z may be of partial help in increasing the strength of the constraints unless *(i)* systematic uncertainties in the local mass–function are reduced at a $\lesssim 10\%$ level, and *(ii)* the completeness criteria of high–z cluster samples are well under control.

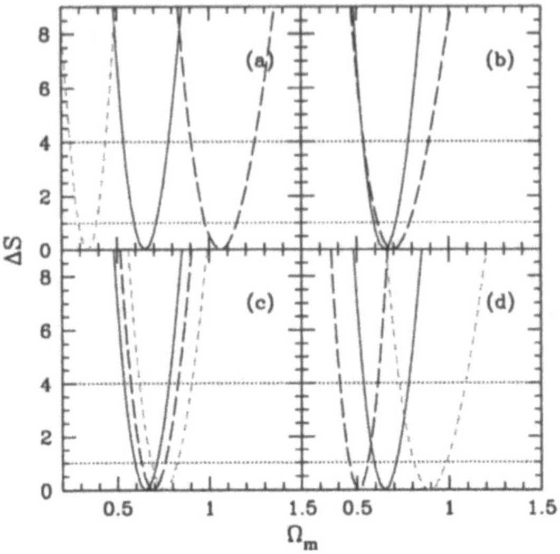

Fig. 3. Variation of the likelihood around its minimum as a function of Ω_m, after marginalizing with respect to σ_8. In all the panels the solid curve correspond to $\sigma_{lim} > 800\,\mathrm{km\,s^{-1}}$ as estimated from the unweighted bi–weight method (see text), $\Gamma = 0.2$ and $\tilde{\sigma}_8 = 0.55$. Panel (a): effect of changing $\tilde{\sigma}_8$; short– and long–dashed curves are for $\tilde{\sigma}_8 = 0.5$ and 0.6, respectively. Panel (b): effect of changing σ_{lim}; the long–dashed line is for $\sigma_{lim} = 900\,\mathrm{km\,s^{-1}}$. Panel (c): effect of changing the σ_v estimator; short– and long–dashed curves are for the explicit background subtraction by C96, and for the weighted bi-weight method (see text), respectively. Panel (d): Effect of the Poissonian uncertainties in the estimate of $\mathcal{F}_{V//}$. The two dotted horizontal lines and the top of each panel correspond to the 1σ, 2σ and 3σ c.l. for one significant fitting parameter.

Acknowledgements

I wish to thank my collaborators in this project: Ray Carlberg, Erica Ellingson, Marisa Girardi, Simon Morris and Howard Yee.

References

1. Press, W.H., Schechter, P. (1974) Astroph. J. **187**, 425–438
2. Eke V.R., Cole S., Frenk, C.S. (1996) Mon. Not. of the royal Astr. Soc. **282**, 263–280
3. Viana P.T.P., Liddle A.R. (1998) Mon. Not. of the royal Astr. Soc. **303**, 535–545
4. Girardi M., Borgani S., Giuricin G., Mardirossian F., Mezzetti, M. (1998) Astrophys. J. **506**, 45–52
5. Oukbir J., Blanchard A. (1992) Astr. Ap. **262**, L21–L24
6. Fadda D., Girardi M., Giuricin G., Mardirossian F., Mezzetti, M. (1996) Astroph. J. **473**, 670
7. Mazure A., et al. (1996) Astr. Ap. **310**, 31–48

8. Carlberg R.G., Yee H.K.C., Ellingson E., Abraham R., Gravel P., Morris S., Pritchet C.J. (1996) Astrophys. J. **462**, 32
9. Carlberg R.G., Morris S.L., Yee H.K.C., Ellingson E. (1997) Astrophys. J. **479**, L19–L23
10. Bahcall N.A., Fan X., Cen R. (1997) Astrophys. J. **485**, L53–L57
11. Girardi M., Giuricin G., Mardirossian F., Mezzetti M., Boschin W. (1998) Astrophys. J. **505**, 74–95
12. Borgani S., Girardi M., Carlberg R.G., Ellingson E., Morris S., Yee H.K.C. (1999) in preparation
13. Yee H.K.C., Ellingson E., Carlberg R.G. (1996) Astrophys. J. Suppl. **102**, 269
14. Henry J.P., Gioia I.M., Maccacaro T., Morris S.L., Stocke J.T., Wolter A. (1992) Astrophys. J. **386**, 408–419
15. Pisani A. (1993) Mon. Not. of the royal Astr. Soc. **265**, 706
16. Beers T.C., Flynn K., Gebhardt K. (1990) Astron. J. **100**, 32–46
17. Katgert P., Mazure A., den Hartog R., Adami C., Biviano A., Perea J. (1998) Astr. Ap., **129**, 399–412
18. Ebeling H., Voges W., Böhringer H., Edge A.C., Huchra J.P., Briel U.G. (1996) Mon. Not. of the royal Astr. Soc. **281**, 799–829

A Sunyaev-Zel'dovich Effect Survey for High Redshift Clusters

J.J. Mohr[1], J.E. Carlstrom[1], G.P. Holder[1], W.L. Holzapfel[2], M.K. Joy[3], E.M. Leitch[1], and E.D. Reese[1]

[1] University of Chicago, Chicago, IL USA
[2] University of California at Berkeley, Berkeley, CA USA
[3] NASA/Marshall Space Flight Center, Huntsville, AL, USA

Abstract. Interferometric observations of the Sunyaev-Zel'dovich Effect (SZE) toward clusters of galaxies provide sensitive cosmological probes. We present results from 1 cm observations (at BIMA and OVRO) of a large, intermediate redshift cluster sample. In addition, we describe a proposed, higher sensitivity array which will enable us to survey large portions of the sky. Simulated observations indicate that we will be able to survey one square degree of sky per month to sufficient depth that we will detect all galaxy clusters more massive than $2 \times 10^{14} h_{50}^{-1} M_\odot$, regardless of their redshift. We describe the cluster yield and resulting cosmological constraints from such a survey.

1 Interferometric Sunyaev-Zel'dovich Effect Imaging

Interactions between cosmic microwave background (CMB) photons and the hot, ionized plasma in galaxy clusters introduce changes in the spectrum of the CMB. At wavelengths of 1 cm we observe this Sunyaev-Zel'dovich Effect (SZE) [1] as a decrease in the intensity of the CMB; this decrement $\Delta T(R)$ at projected cluster radius R can be expressed as

$$\frac{\Delta T(R)}{T_{CMB}} = -2\frac{\sigma_T}{m_e c^2} \int dl\, n_e(l,R) k_B T_e(l,R) \qquad (1)$$

where σ_T is the Thomson cross section, $m_e c^2$ is the electron rest mass, n_e is the electron number density and k_B is the Boltzmann constant. Note that the fractional change in the CMB temperature $\Delta T/T_{CMB}$ is independent of redshift; it depends only on properties of the ICM. Higher order corrections to this expression and a consideration of the effects of cluster peculiar velocities on the CMB spectrum can be found elsewhere (e.g., [2]).

The typical central decrement for a massive $\sim 10^{15}\ M_\odot$ galaxy cluster is ~ 1 mK, and for typical radio telescope system temperatures the SZE decrement is only a few parts in 10^5 of the power detected by the receiver, making SZE observations challenging. Characteristics of radio interferometry make it well suited to meet this challenge. Perhaps the most important of these is that only the correlated portion of the receiver output is signal, while the uncorrelated output serves as noise; differential ground spillover and detector noise are naturally separated from the cluster decrement. Another is that interferometry effectively simultaneously differences over the sky, easily filtering the smoothly distributed ~3 K background from the cluster SZE decrement we seek. Finally, with interferometry one obtains information on many scales simultaneously; by placing some dishes at larger separations it is straightforward to filter out point source contributions.

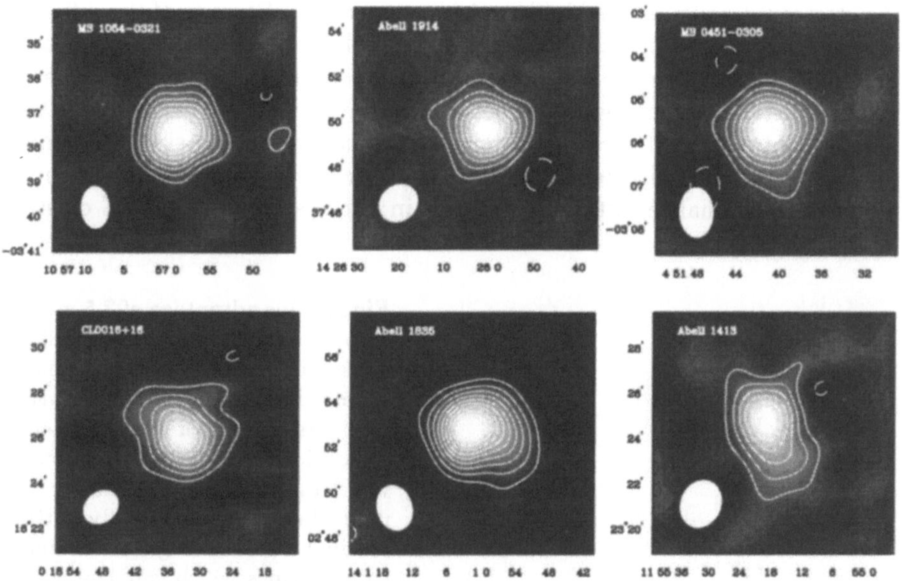

Fig. 1. SZE decrement images of six of the 28 clusters observed with the Sunyaev-Zel'dovich Imaging Experiment at BIMA and OVRO. Contours are spaced by 2σ, and the synthesized beam is represented by the white ellipse in the lower left of each frame.

Over the past several years, Carlstrom, Joy and collaborators have developed sensitive, low noise receivers and used them at the BIMA and OVRO observatories to obtain high quality SZE data on a sample of 28 intermediate redshift clusters [3]. Figure 1 contains greyscale SZE images and signal to noise contour overlays of 6 clusters. The contours are spaced by 2σ, and the white ellipses in the lower left of each frame represent the FWHM of the synthesized beam.

As demonstrated by Figure 1, the interferometric SZE Imaging Experiment does deliver high signal to noise SZE images of intermediate redshift clusters. We use these SZE data to constrain the cluster ICM distribution, and, in combination with published measurements of the ICM temperature, these data enable estimates of cluster ICM masses and mass fractions [4,5]. In combination with X-ray images of galaxy clusters, the SZE data allow one to estimate cluster sizes and measure direct, distance-ladder-independent distances. The X-ray surface brightness $I(R)$ at projected cluster radius R can be expressed as

$$I_X(R) = \frac{1}{4\pi(1+z)^4}\frac{\mu_e}{\mu_H}\int dl\, n_e^2(l,R)\Lambda(T_e) \tag{2}$$

where n_e is the electron number density, $\Lambda(T_e)$ is the temperature dependent cooling coefficient and $n_e = \rho/m_p\mu_e$. Note that $I(R)$ depends on the square of the electron density, and is highly redshift dependent. Because of the different dependence on the electron density, observations of $I(R)$ and $\Delta T/T_{CMB}$ provide a direct measure of the line of sight distance through the cluster. With additional assumptions about the cluster geometry, we can combine this line of sight size with the angular size of the cluster on the sky to estimate the cluster angular diameter distance D_A. We are currently using archival X-ray data to finalize distance estimates in our ensemble of 28 SZE clusters [6].

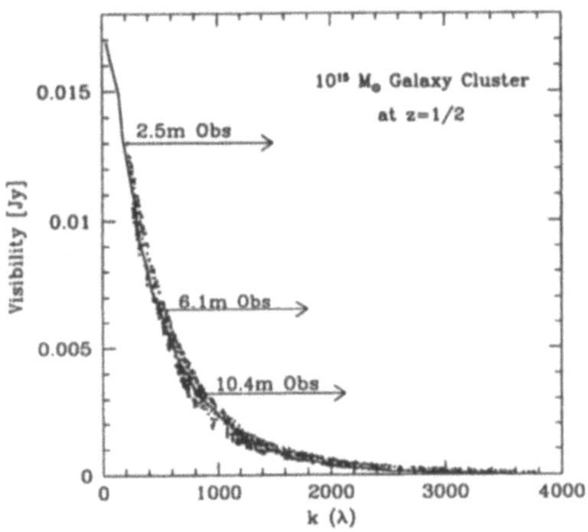

Fig. 2. One advantage of 2.5 m telescopes over the currently available 6.1 m BIMA and 10 m OVRO dishes is clear in this plot of cluster visibility versus k which is the telescope separation in units of the observing wavelength (1 cm). The cluster visibility falls off exponentially with k, yielding significant gains for smaller telescopes which can sample the visibility at smaller separations. The points are visibilities from a simulated galaxy cluster, and the line is the best fit model.

2 The Proposed SZE Interferometric Array

The difficulty of detecting X-ray emission from high redshift clusters– even with the high quality X-ray instruments available today– is made clear by Eqn. 2. Indeed, it is the redshift independent nature of the SZE (see Eqn. 1) that makes it an ideal tool for studying collapsed structures at high redshift. The current best instruments for SZE studies of high redshift clusters do not have the sensitivity required to detect sizeable numbers of clusters in a non-targeted search within reasonable observing times of 1 mth to 1 yr. Therefore, we have proposed to build a new array composed of ten 2.5 m dishes and a wide-band digital correlator. This array, when operating at 1 cm, is optimized for high redshift cluster detection and will be approximately 100 times more sensitive than our current SZE Imaging Experiment on BIMA.

In this era of large telescopes it may be surprising to learn that the best telescope size for SZE cluster studies is 2.5 m. There are two main reasons for this. First, the angular size θ_{BEAM} of the primary beam of a radio telescope scales with the dish diameter D as $\theta_{BEAM} \propto \lambda_{obs}/D$, where λ_{obs} is the observation wavelength. θ_{BEAM} determines the field size in a survey, and cluster signal on scales larger than $\theta_{BEAM}/2$ is lost. For 1 cm observations on a 6.1 m dish, the FWHM of the primary beam is $\theta_{BEAM} \sim 6$ arcmin, whereas for the 2.5 m telescopes $\theta_{BEAM} \sim 15$ arcmin. The primary beam of a 2.5 m dish is better matched to cluster angular scales than a 6.1 m dish. Second, as is shown graphically in Figure 2, interferometry samples the cluster visibility, which is the Fourier transform of the CMB decrement distribution on the sky. Cluster visibilities typically fall off exponentially with distance from the origin k (as opposed to point sources whose Fourier transforms are independent of k). Therefore, with smaller dishes one samples a region of the U-V plane where there is significantly more cluster flux.

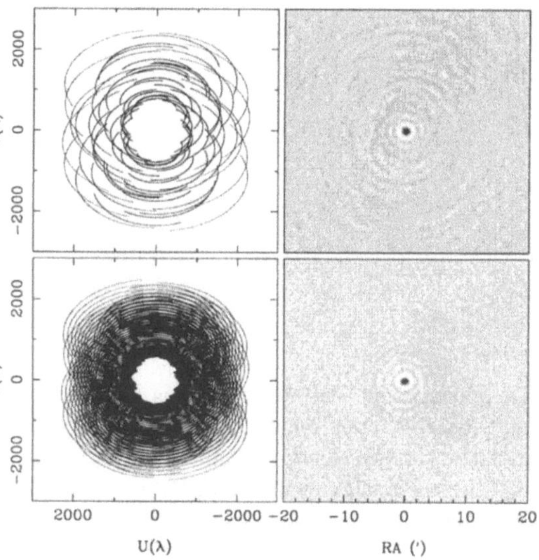

Fig. 3. The wide-band correlator allows us to take advantage of a larger portion of the signal output by the 10 GHz bandwidth receivers. This translates into higher sensitivity and more coverage in the U-V plane. The left panels show the U-V coverage for a 7 hr observation with the current correlator (upper) and the proposed correlator (lower). The right panels show the synthesized beam for each observation. The coverage available with the new correlator leads to lower sidelobes in the synthesized beam and enhanced image quality.

The proposed wide-band digital correlator will enable us to use a larger fraction of the signal output from the 10 GHz bandwidth receivers. Currently, the maximum bandwidth at BIMA is ~800 MHz; thus, the proposed 8 GHz correlator will increase the sensitivity by an order of magnitude. In addition to higher throughput, the wide-band correlator provides enhanced image quality. As is apparent in Figure 3, the fraction of the U-V plane sampled in a 7 hr exposure with the old correlator (top left) is much smaller than the fraction sampled in the same exposure with the new correlator (bottom left). This additional U-V coverage leads to a synthesized beam (right panels) which has smaller sidelobes. This enhanced image quality simplifies structure detection in SZE images.

3 The SZE Galaxy Cluster Survey

X-ray samples of high redshift clusters are disappointingly small, primarily because of the strong redshift dependence of cluster emission (see Eqn. 2); only the most luminous clusters at each redshift tend to be found. However, with this sensitive new SZE array, we will be able to carry out a survey of $\sim 1\ \mathrm{deg}^2$ per month and detect all clusters more massive than $2\times10^{14}h_{50}^{-1}\ M_{\odot}$, independent of their redshift. Interestingly, the cluster selection function from such a survey is essentially a cut in cluster binding mass. So not only will an SZE survey yield a sizeable high redshift cluster sample for the first time, but the survey will also deliver a sample which is much easier to understand than typical X-ray samples at intermediate redshift.

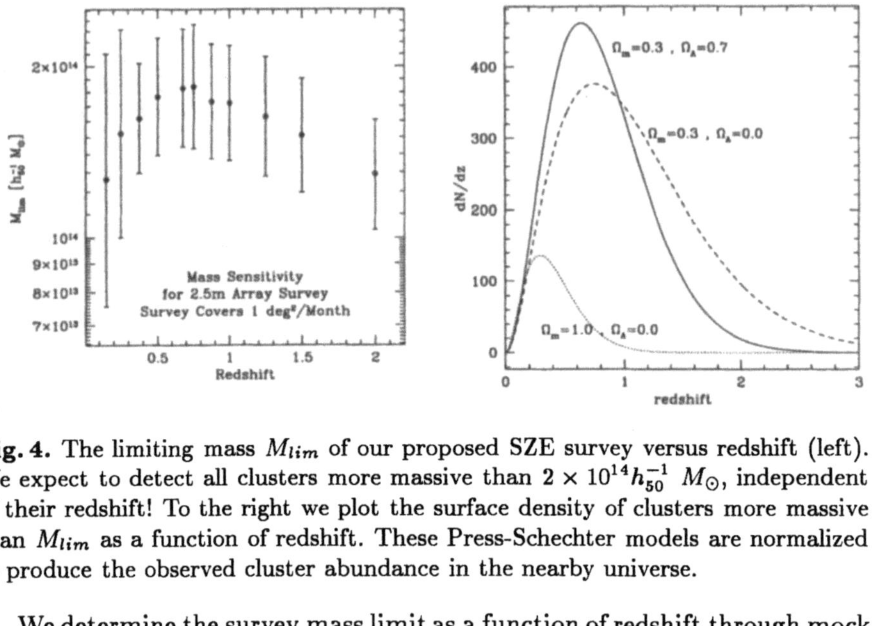

Fig. 4. The limiting mass M_{lim} of our proposed SZE survey versus redshift (left). We expect to detect all clusters more massive than $2 \times 10^{14}h_{50}^{-1}\ M_{\odot}$, independent of their redshift! To the right we plot the surface density of clusters more massive than M_{lim} as a function of redshift. These Press-Schechter models are normalized to produce the observed cluster abundance in the nearby universe.

We determine the survey mass limit as a function of redshift through mock observations of N-body plus SPH simulations of galaxy clusters. These mock SZE observations include noise and have exposures tuned so that we will be able to survey 1 deg^2 in a month. The N-body plus SPH simulations are state of the art simulations sampled from four different cosmological models: (1) SCDM: standard, biased $\Omega = 1$ CDM, (2) OCDM: unbiased $\Omega_m = 0.3$ CDM, (3) LCDM: unbiased $\Omega_m = 0.3$ and $\Omega_\lambda = 0.7$ CDM, and (4) τCDM: $\Omega_m = 1$ but CDM power spectrum with $\Gamma = 0.24$ (see [7,8] for more details). These simulations include only gas dynamics and gravity. We use three particle species; this allows us to follow tidal fields from surrounding large scale structure while still resolving the dynamics in the cluster core. We incorporate no early heating of the ICM from galaxy formation.

To determine the detection limit at each redshift we use outputs of the 48 simulations from the appropriate redshift; with this approach the morphological evolution of galaxy clusters is naturally included. We determine the detection probability in each mock observation by fitting a cluster model and determining the $\Delta\chi^2$ between the best fitting and null models. We set our detection threshold at 5σ.

Figure 4 (left) is a plot of limiting mass versus redshift. The uncertainties on the limiting mass correspond to the scatter about the best fit $\Delta\chi^2$-binding mass relation, determined using mock observations of the entire cluster ensemble at each redshift; this scatter is an indication of the morphological sensitivity of the detection threshold. We find that the limiting mass is independent of cosmological model– at least for the four models tested. Also in Figure 4 (right) is a Press-Schechter estimate of the surface density of clusters more massive than $M_{lim} = 2 \times 10^{14} h_{50}^{-1}\ M_\odot$ as a function of redshift for three cosmological models (noted on the figure). These models are all normalized to produce the correct cluster abundance at the present epoch. Note the cosmological sensitivity of the cluster surface densities; we expect to find very few clusters with our survey if $\Omega = 1$, whereas for more favored, low Ω_0 models we expect hundreds of clusters [10].

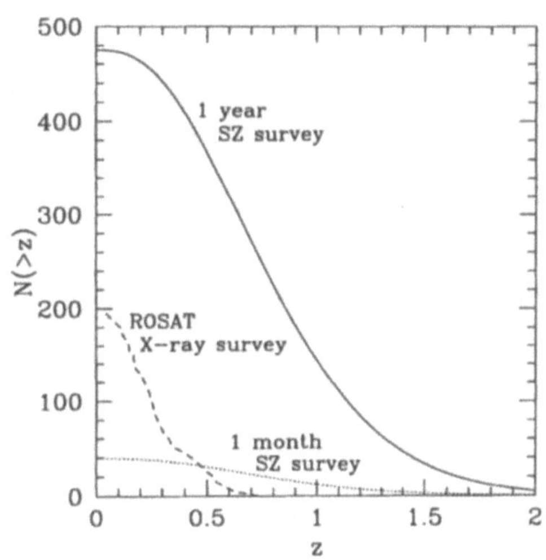

Fig. 5. We plot the expected yield of our proposed SZE survey after one month (dotted) and one year (solid) if $\Omega_0 = 0.3$ and $\Omega_\lambda = 0.7$. For comparison we plot the yield of the ROSAT PSPC deep cluster survey (dashed). After one month we will have more clusters at redshift $z > 0.5$ than the entire ROSAT survey. After one year we will have ~100 clusters at redshifts $z > 1$ and ~400 clusters at redshifts z.0.5.

In Figure 5 we plot the survey yield after one month (dotted line) and one year (solid line), where we have assumed that the currently favored model $\Omega_m = 0.3$ and $\Omega_\lambda = 0.7$ is correct (if $\Omega_\lambda = 0$ the cluster count increases). For comparison we also plot the yield of 200 clusters from a deep ROSAT PSPC X-ray cluster survey [9]. Note that after a one month SZE survey we will have more redshift $z > 0.5$ clusters than the deep ROSAT survey. After one year we expect to have ~100 clusters at redshifts $z > 1$ and almost 400 clusters at redshifts $z > 0.5$ [11]. This highly sensitive probe for massive collapsed structures in the high redshift universe constitutes an exceedingly powerful test of current structure formation models.

The SZE survey itself will yield a list of galaxy clusters and constraints on the cluster surface density. Cluster redshifts, masses, gas distributions and galaxy populations can then be examined through followup observations with X-ray satellites and 8 m class telescopes like the VLT. We anticipate that the scope and number of scientific issues addressed with followup observations of the high redshift clusters from the SZE survey will lead to enormous progress in cosmology.

References

1. Sunyaev, R. A. and Zel'dovich, Y.B., 1972, Comments Astrophys. Space Phys., 4, 173
2. Birkinshaw, M. 1999, *Physics Reports*, 310, num 2-3, 98
3. Carlstrom, J.E., Joy, M.K. & Grego, L. 1996, ApJ, 456, L75
4. Grego, L., Carlstrom, J.E., Joy, M.K., Reese, E.D., Holder, G.P., Patel, S., Cooray, A.R. & Holzapfel, W.L. 1999a, ApJ, submitted
5. Grego, L., Carlstrom, J.E., Joy, M.K., Reese, E.D., Holder, G.P., Patel, S., Cooray, A.R. & Holzapfel, W.L. 1999b, ApJ, submitted
6. Reese, E.D., Mohr, J.J., Carlstrom, J.E., Joy, M.K., Holder, G.P., Patel, S., Grego, L, Holzapfel, W.L., 1999, in preparation
7. Mohr, J.J. & Evrard, A.E. 1997, ApJ, 491, 38
8. Mohr, J.J., Mathiesen, B. & Evrard, A.E. 1999, ApJ, in press, astro-ph/9901281
9. Vikhlinin, A. McNamara, B., Forman, W., Jones, C., Quintana, H., & Hornstrup, A. 1998, ApJ, 502, 558
10. Holder, G.P. & Carlstrom, J.E. 1999, to appear in "Microwave Foregrounds", eds. de Oliveira-Costa & Tegmark (ASP, San Francisco, 1999), astro-ph/9904220
11. Holder, G.P., Carlstrom, J.E. & Mohr, J.J., 1999, in preparation

Probing the Large-Scale Structure with the REFLEX Cluster Survey

H. Böhringer[1], L. Guzzo[2], C.A. Collins[3], P. Schuecker[1], S. Schindler[3], W. Voges[1], D.M. Neumann[4], R. Cruddace[5], S. DeGrandi[2], G. Chincarini[2], T.H. Reiprich[1], A.C. Edge[6], H.T. MacGillivray[7], and P. Shaver[8]

[1] Max-Planck-Institut für extraterr. Physik, D-85740 Garching, Germany
[2] Osservatorio Astronomico di Brera, Milano/Merate, Italy
[3] Liverpool John-Moores University, Liverpool, U.K.
[4] CEA Saclay, Service d'Astrophysique, Gif-sur-Yvette, France
[5] Naval Research Laboratory, Washington, D.C., USA
[6] Durham University, Durham, U.K.
[7] Royal Observatory, Edinburgh, U.K.
[8] European Southern Observatory, Garching, Germany

Abstract. Clusters of galaxies detected in X-rays are ideal probes for the study of the large scale structure in the Universe. Based on the ROSAT All-Sky Survey we have conducted a large redshift survey as an ESO key programme to identify the X-ray brightest clusters found in the southern hemisphere. We present first results from the bright part of the survey comprising 460 identified clusters for an X-ray flux limit of $3 \cdot 10^{-12}$ erg s^{-1} cm^{-2}. We show results for the X-ray luminosity function and discuss the assessment of the large-scale structure be means of the cluster distribution power spectrum on a scale of $400h^{-1}$ Mpc.

1 Introduction

For the study of the structure of the present day Universe on very large scales (≥ 100 Mpc for $H_0 = 50$ km s^{-1} Mpc^{-1}) the use of galaxy clusters as tracers constitutes a very interesting alternative to the conventional galaxy redshift surveys for several reasons. One can study a larger volume with a relatively small number of objects (see e.g. Postman 1998 for optical cluster surveys). The spatial correlation in the cluster distribution is strongly magnified (by up to an order of magnitude) with respect to the galaxy and presumably also with respect to the mass distribution (Kaiser 1984). The biasing factor relating the cluster distribution power spectrum to the mass density fluctuations can be calculated *ab-initio* (see e.g. Bardeen et al. 1986, Bond et al. 1991).

X-ray astronomy has offered a unique tool to efficiently detect and characterize galaxy clusters out to large distances. Originating in the hot intracluster plasma that fills the gravitational potential well of the clusters, the X-ray emission is an equally robust parameter for a first estimate of the size and mass of clusters as the galaxy velocity dispersion. The tight correlation between X-ray luminosity and cluster mass is illustrated in Fig. 1 for a sample of the brightest and best observed X-ray clusters.

With the ROSAT ESO Flux Limited X-ray (REFLEX) Cluster Survey we exploit the unique opportunity provided by the ROSAT All-Sky Survey (RASS, Trümper 1993, Voges et al. 1996), the first complete sky survey conducted with an imaging X-ray telescope, to compile a flux limited cluster sample for cosmological studies in the southern celestial hemisphere. The cluster candidates found in the RASS are subsequently identified and redshifts are measured by optical follow-up observations in the frame of an ESO key programme (Böhringer 1998, Guzzo et al. 1999). In total we are investigating more than 800 very likely cluster candidates.

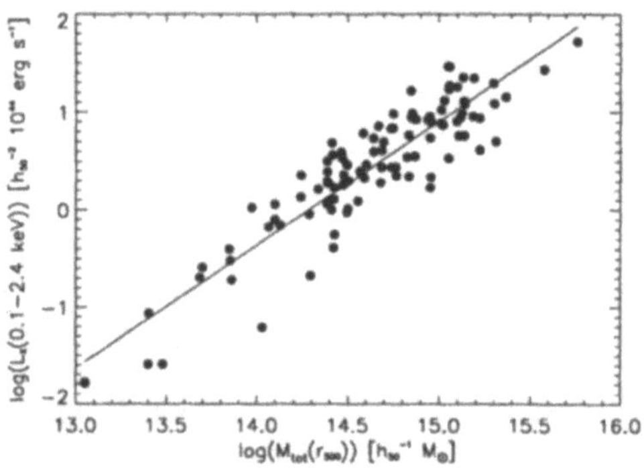

Fig. 1. Relation between the gravitational mass and the X-ray luminosity determined from X-ray observations for the brightest galaxy clusters found by ROSAT (Reiprich 1999).

2 Construction of the REFLEX Cluster Sample

The X-ray sky atlas constructed from the ROSAT All-Sky Survey Mission with its more than 100,000 X-ray sources (Voges et al. 1996) contains thousands of mostly unidentified galaxy clusters. Since the majority of these sources are characterized by few photons, only the brightest, well extended X-ray cluster sources are readily identified. For the selection of the cluster candidate sample we use the COSMOS optical object catalogue (e.g. Heydon-Dumbleton et al. 1989) to search for overdensities in the galaxy density distribution at the position of the X-ray sources (see Böhringer et al. in preparation). Not all the sources flagged by this technique are true galaxy clusters, however. The price paid for aiming at a high completeness in the final catalogue is a contamination of the candidate list by more than 30% non-cluster sources. This contamination is removed from the sample by a critical evaluation of the photographic plates, the detailed X-ray properties, and literature information as well as by the spectroscopic data of the follow-up observations.

These follow-up identifications and the redshift measurements are conducted within an ESO key programme comprising a total of 90 observing nights distributed evenly among the 1.5m, 2.2m, and 3.6m telescopes at La Silla.

The observing program is practically complete with 96% of the cluster redshifts determined and the last missing data will be obtained in an observing run in May 1999. We have thus completed a redshift survey for the cluster candidates down to an X-ray flux-limit of $3 \cdot 10^{-12}$ erg s^{-1} cm^{-2}. The sample of confirmed galaxy clusters comprises 460 objects. 53% of these clusters can be found in the main catalogues of Abell (1958) and Abell, Corwin, & Olowin (1989) and further 10% in the supplementary list. Most of the remaining clusters are previously unknown objects. This result highlights the importance of the selection independent of existing optical catalogues. The high completeness of this sample is demonstrated by a counter-test, in which we searched for X-ray emission at the position of all Abell and ACO clusters in the ROSAT Survey independent of a previous detection of these sources by the survey source identification algorithm. Only 5 clusters ($\sim$ 1%) were found to have been missed by the cluster search based on the COSMOS data.

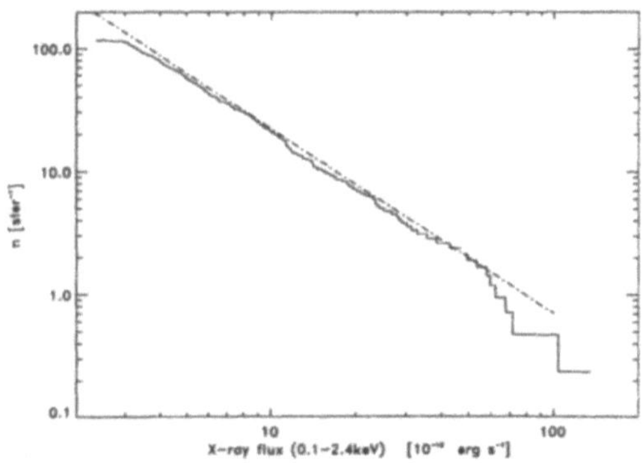

Fig. 2. Cumulative number counts of X-ray clusters as a function of X-ray flux for the REFLEX sample.

3 Properties of the REFLEX Sample

A plot of the number counts of the cluster population as a function of the limiting X-ray flux is shown in Fig. 2. The distribution is very straight and shows no indication for an incompleteness at low flux levels. Fig. 3 shows the distribution in redshift and X-ray luminosity. While most of the clusters detected are not very distant (median redshift is $z = 0.085$), a few very luminous clusters are found out to redshifts beyond $z = 0.3$ with one outstanding object at $z = 0.45$, the most luminous cluster detected so far. Some more high redshift clusters just found in the last data reduction at redshifts between 0.3 and 0.4 fill to gap seen in the figure.

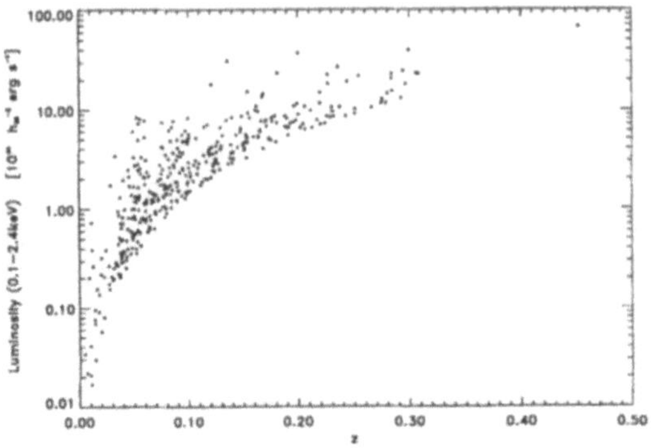

Fig. 3. Distribution of the X-ray luminosities as a function of redshift for the galaxy clusters of the REFLEX sample.

A preliminary version of the X-ray luminosity function of the sample (at a redshift completeness of 80%) is shown in Fig. 4. But in spite of this incompleteness the luminosity function already recovers the densities reached in previous surveys (e.g. Ebeling et al. 1997, DeGrandi 1996, De Grandi et al. 1999) to which the present result is compared in the figure.

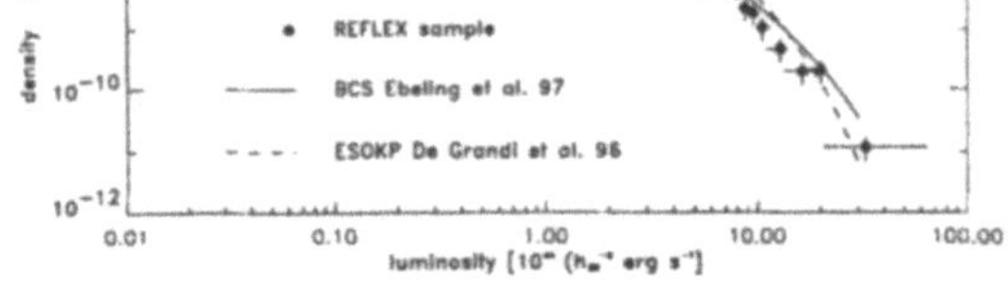

Fig. 4. Differential X-ray luminosity function for the REFLEX sample (points with error bars; for details see Böhringer et al. in preparation) compared to the previous results by DeGrandi (1996) and Ebeling et al. (1997). A value of $H_0 = 50$ km s^{-1} Mpc^{-1} is assumed for the scaling.

The most exciting aim of this programme is to assess the large-scale structure of the galaxy cluster and matter distribution in the Universe. The currently most popular and important statistical function for the characterization of the large-scale structure is the density fluctuation power spectrum. The form of the power spectrum depends on such basic features of our Universe as the mean density, the nature of the dark matter, and the physics of the inflationary process if such an event happened in the early Universe. Its knowledge provides therefore very important constraints on the possible cosmological models. A first preliminary result for this function obtained from 188 clusters of our sample in a box of 400 h_{100}^{-1} Mpc is shown in Fig. 5 (details in Schuecker et al. in preparation). The function is featuring an interesting maximum at a scale of about 100 - 150 h_{100}^{-1} Mpc. The location of this maximum is related to the size of the horizon of the Universe at an epoch when the energy density of matter and radiation was equal and it is therefore a very important calibration point for the theoretical modeling of the evolution of our Universe. Note that the construction of the power spectrum shown in Fig. 5 is based on a luminosity (mass) selection of the clusters which varies with redshift. Therefore the quantitative interpretation of this result is not straightforward and more work is needed to relate this function to the power spectrum of the matter density fluctuation in the Universe.

Also shown in Fig. 5 are the power spectra of the IRAS and Las Campanas galaxy redshift surveys (Fisher et al. 1993, Lin et al. 1996) and the COBE fluctuation spectrum extrapolated to the present (for $\Omega_0 = 1$). We clearly note the high amplification ("biasing") of the cluster power spectrum, as mentioned in the introduction. But both functions have approximately the same shape and maximum position as expected.

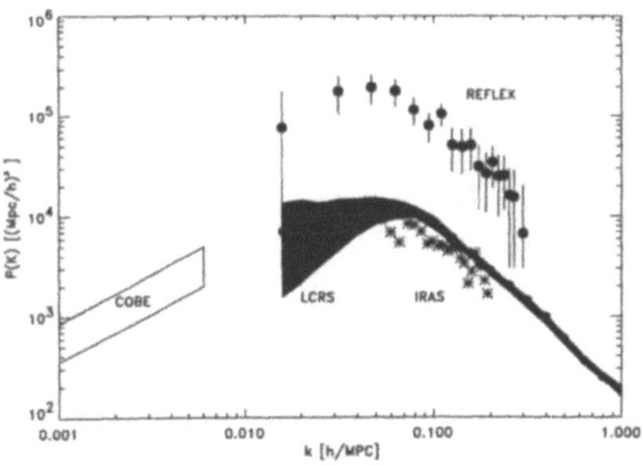

Fig. 5. Power spectrum of the density distribution of the galaxy clusters in the REFLEX sample (upper curve). For the analysis only 188 clusters in a box of 400 Mpc side length was used (Schuecker et al. in preparation). The results are scaled to $H_0 = 100$ km s^{-1} Mpc^{-1}. The power spectra of the galaxy distributions from the IRAS (Fisher et al. 1993) and Las Campanas Redshift Surveys (Lin et al. 1996) as well as the COBE spectrum extrapolated to the present are shown for comparison.

4 Conclusions

It is obvious that these studies of the large-scale structure using clusters as probes require great care in the preparation of the sample of test objects. The first results of the REFLEX Survey presented here and some further tests that we have conducted already demonstrate the high quality of the present data. This is the result of a very homogeneous and highly controlled selection process used to construct the sample. The challenge of the selection work in this survey is to combine the complementary information from X-ray and optical wavelength in the most homogeneous way. Contrary to several earlier studies that we have conducted (Pierre et al. 1994, Romer et al. 1994, Ebeling et al. 1997, De Grandi 1996, De Grandi et al. 1999) in which the identification process was optimized to find as many clusters as possible by using all available sources of information, we have now achieved a highly complete cluster selection by just combining the ROSAT All-Sky Survey data and the COSMOS optical data base in a homogeneous way, completely controlled by automated algorithms. This is an important achievement.

For all the future work with distant cluster surveys e.g. using the Wide-Field Camera and VLT (see the contribution by Da Costa in these proceedings) a precise knowledge of the present day cluster population is required to draw any conclusions on evolutionary effects. The present work is meant to provide the so far best reference for this purpose.

References

1. Abell, G.O., 1958, ApJS, 3, 211
2. Abell, G.O., Corwin, H.G. & Olowin, R.P., 1989, ApJS, 70, 1
3. Bardeen, J.M., Bond, J.R., Kaiser, N. & Szalay, A.S., 1986, ApJ, 304, 15
4. Böhringer, H., et al., 1998, The Messenger, No. 94, 21
5. Bond, J.R., Cole, S., Efstathiou, G., Frenk, C.S., 1991, ApJ, 379, 440
6. DeGrandi, S., 1996, in *Röntgenstrahlung from the Universe*, H.U. Zimmermann, J.E. Trümper, H. Yorke (eds.), MPE Report 263, p. 577
7. De Grandi, S., Böhringer, H., Guzzo, L., Moldendi, S., Chincarini, G., Collins, C.A., Cruddace, R., Neumann, D., Schindler, S., Schuecker, P., Voges, W., 1999, ApJ, 514, 148
8. Edge, A.C., Böhringer, H., Guzzo, L., et al., 1994, A&A, 289, L34.
9. Ebeling, H., Edge, A.C., Fabian, A.C., Allen, S.W., Crawford, C.S., & Böhringer, H., 1997, ApJ, 479, L101.
10. Fisher, K.B., Davis, M., Strauss, M.A., Yahil, A., Huchra, J.P., 1993, ApJ, 402, 42
11. Guzzo, L., et al., 1999, The Messenger, No. 95, 27
12. Heydon-Dumbleton, N.H., Collins, C.A. & MacGillivray, H.T., 1989, MNRAS, 238, 379
13. Kaiser, N., 1984, ApJ, 284, L9
14. Lin, H., Kirshner, R.P., Shectman, S.A., Landy, S.D., Oemler, A., Tucker, D.C., Schechter, P.L., 1996, ApJ, 471, 617

15. Pierre, M., Böhringer, H., Ebeling, H., Voges, W., Schuecker, P., Cruddace, R.G., & H.T. MacGillivray, 1994, A&A, 290, 725
16. Postman, M., 1998, astroph/9810088, to appear in the proceedings of "Evolution of Large-Scale Structure: From Recombination to Garching"
17. Reiprich, T.H., Böhringer, H., 1999, in proceedings of the 4th ASCA Symposium on "Heating and Acceleration in the Universe", Tokyo, March 1999
18. Romer, A.K., Collins, C.A., Böhringer, H., Cruddace, R.G., Ebeling, H., H.T. MacGillivray, & Voges, W., 1994, Nature, 372, 75
19. Schindler, S.C., Guzzo, L., Ebeling, H., et al., 1995, A&A, 299, L9.
20. Schindler, S., Hattori,M., Neumann, D.M., & Böhringer, H., 1997, A&A,317, 66
21. Trümper, J., 1993, Science, 260, 1769.
22. Voges, W., Boller, T., Dennerl, K., et al., 1996, in *Röntgenstrahlung from the Universe*, H.U. Zimmermann, J.E. Trümper, H. Yorke (eds.), MPE Report No. 263, p. 637
23. White, S.D.M., Efstathiou, G., & Frenk, C.S., MNRAS, 1993, 262, 1023.

The Cluster Population in a Complete Sample of Optically Identified RASS X-Ray Sources

Immo Appenzeller[1], Reinhold Kneer[2], Franz-Josef Zickgraf[3], Joachim Krautter[1], and Inge Thiering[3]

[1] Landessternwarte, D–69117 Heidelberg, Germany
[2] Hewlett Packard Corp., Böblingen, Germany
[3] Max-Planck-Institut für Astronomie, D–69117 Heidelberg, Germany

Abstract. Using the 2.1-m telescope of the Guillermo Haro Observatory (Cananea, Mexico) the optical counterparts of a count-rate and area limited complete subsample of the ROSAT All Sky Survey (RASS) have been identified. While the majority of the observed sources were found to be AGN or stars, 78 sources (12% of all identified objects) were found to be galaxy clusters with redshifts between 0.027 and 0.404. Most of these clusters had not been cataloged before. On the other hand, about 2/3 of the Abell clusters in our study areas were not detected as RASS sources. The X-ray bright clusters are on average more distant than the Abell clusters and tend to be dominated by central galaxies (corresponding to Bautz-Morgan types I and II).

1 Introduction

X-ray detected samples of galaxy clusters are becoming increasingly more important for cosmological studies[1], and various large surveys have been carried out to investigate the distribution and evolution of clusters (see e.g. [2], [3], [4]). On the other hand, not all clusters are strong X-ray emitters and X-ray surveys, thus, tend to select distinct clusters types. To help clarify what types of clusters are X-ray bright we compared properties of galaxy clusters identified in a statistically complete sample of ROSAT All Sky Survey (RASS) sources with the properties of the optically selected Abell clusters in the same study area.

2 The Data

Our investigation is based on an optical identification program of all RASS sources in six representative high-galactic latitude regions north of $\delta = -9^0$, comprising a total area of 685 square degrees. The details of the survey, which was carried out with the LFOSC faint object spectrograph at the 2.1-m Telescope of the Guillermo Haro Observatory in Cananea, Mexico, have been described in [5]. The catalog of the resulting identifications has been published in [6]. Of the 674 reliably detected RASS sources of our area and flux limited complete sample 662 sources (98%) could be identified with

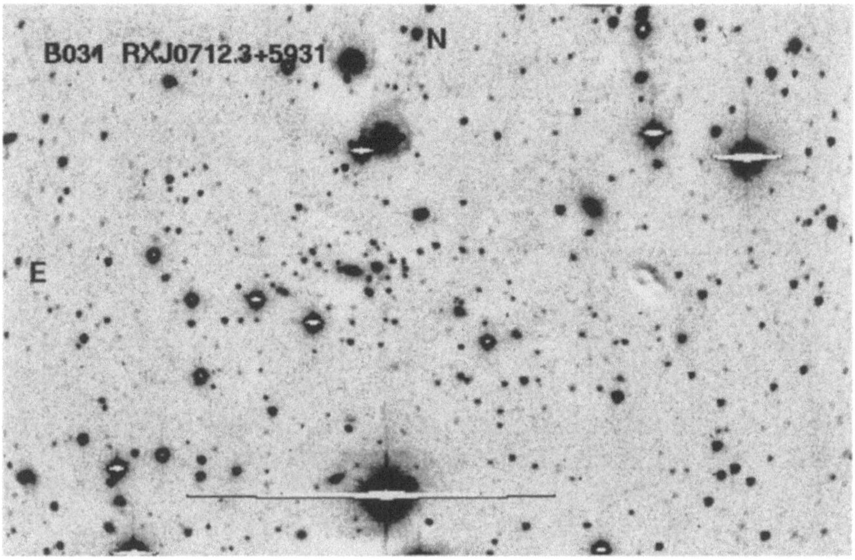

Fig. 1. Example of a new identification of a distant galaxy cluster. The X-ray position and the cluster center are to the left of the image center. The FOV of the 15 minutes R-image is 6×10 *arcmin*^{2}

optical counterparts. Most of these counterparts turned out to be AGN or stars [7]. Only 78 sources (12%) turned out to be galaxy clusters. There were 8 additional sources coinciding with galaxy clusters but with other objects (in most cases AGN) in front of, in, or behind the cluster being a more likely emitter of (at least most of the) the observed X-ray flux. Only 25 (+3) of the 78 (+8) clusters are also listed in the Abell catalog. On the other hand, the total number of the (optically selected) Abell clusters in the study area was found to be 102. Hence 77 Abell clusters (2/3 of the total in our study area) were not detected by RASS.

3 Results

To clarify why only part of the Abell clusters were detected, while other clusters not listed in the Abell catalog were seen by ROSAT (cf. Figure 1), we investigated the distance distribution and other basic properties of the clusters in question. Figure 2 shows the redshift distribution of all clusters in our sample with reliable spectroscopic redshifts. Although Figure 2 contains the majority of our X-ray clusters, 29 objects are missing because no or no reliable redshifts could be measured, usually because the galaxies were too

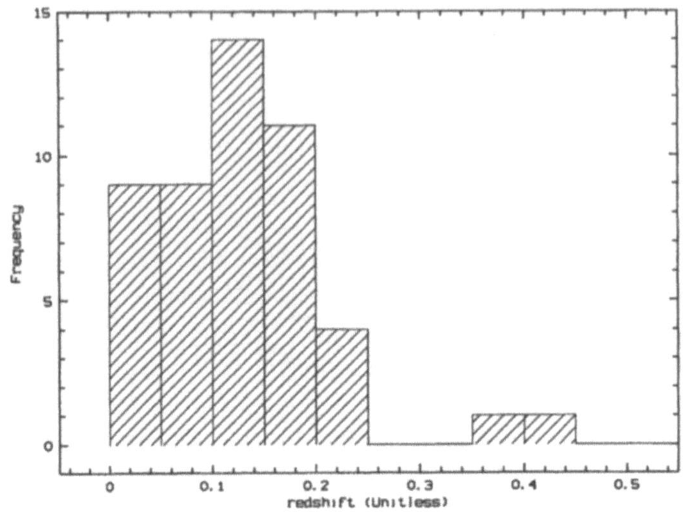

Fig. 2. Redshift distribution of the X-ray clusters with spectroscopic redshifts

faint for spectroscopy with our equipment. Rough estimates of the distance of these clusters were obtained from the galaxy sizes on our images (after calibration with the clusters with known z). The resulting more complete z-distribution indicate that the clusters missing in Figure 2 are on average more distant, but that the maximum of the distribution is shifted only slightly towards higher redshifts. The decline beyond z= 0.2 (probably reflecting the RASS flux limit) remains present. We also determined the redshift distribution of the subsample containing the (RASS-detected) Abell clusters only, but found no significant deviation from Figure 2. In Figure 3 we present the corresponding distribution of the Abell clusters with known redshifts in our test areas which were not detected as RASS sources. The average redshift of this sample is lower, although the difference is significant only when compared to the sample which includes the estimated redshifts. Nevertheless the fact that the Abell clusters are on average less distant than the X-ray clusters rules out that the non-detection of the Abell clusters is due to their distance.

An intrinsic property which seems to distinguish our X-ray bright clusters from the X-ray quiet Abell clusters is the presence of a dominating central galaxy. As demonstrated by Table 1, clusters of Bautz-Morgan type I (i.e. containing a cD galaxy) or II (containing another dominating galaxy) are more frequent among the X-ray bright objects than among the X-ray quiet clusters, which are mostly of type III (i.e. without a dominating central galaxy). Among possible explanations of this correlation could be a deeper potential well or a different evolutionary state of the type I and II clusters.

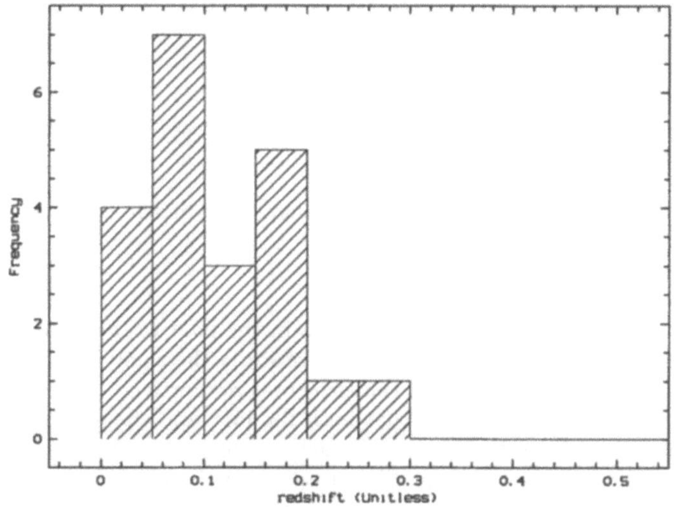

Fig. 3. Redshift distribution of the X-ray quiet Abell clusters with reliable redshifts

Table 1. Distribution of Bautz-Morgan types among the clusters in the study area

Subsample	B-M I&II	B-M III	not classified
All X-ray detected cls.	40%	56%	4%
"New" clusters	50%	47%	3%
X-ray quiet Abell cls.	19%	77%	4%

References

1. Böhringer, H. (1995) Annals of the New York Academy of Sciences **759**, 67–86
2. Collins C.A., Burke, D.J., Romer, A.K., Sharples, R.M., Nichol, R.C., (1997) ApJ **479**, L117
3. Scharf, C.A., Jones, L.R., Ebeling, H., Perlman, E., Malkan, M., Wegner, G. (1997) ApJ **477**, 79
4. Rosati, P., Della Ceca, R., Norman, C., Giacconi, R. (1998) ApJ **492**, L21–L24
5. Zickgraf, F.-J., Thiering, I., Krautter,J., et al. (1997) A&AS **123**, 103
6. Appenzeller, I., Thiering, I., Zickgraf, F.-J. et al., (1998) ApJS **117**, 319–340
7. Krautter, J., Zickgraf, F.-J., Appenzeller, I., et al. (1999) A&A (in press)

The Metal Enrichment History of Clusters of Galaxies

Alvio Renzini

European Southern Observatory, Karl-Schwarzschild-Straße 2,
D-85748 Garching, Germany

Abstract. The ratio of the total iron mass in the ICM to the total optical luminosity of the cluster (the iron mass-to-light-ratio) is the same for all clusters which ICM is hotter than ~ 2 keV, and the elemental proportions (i.e. the [α/Fe] ratio) appear to be solar. This suggests that both the IMF and the relative contributions of SN types are universal. The past SN activity in galaxy clusters is derived, and it is argued that the past average SNIa rate was at least 10 times higher than the present rate in local ellipticals. It is argued that cluster metallicity ($\sim 1/3$ solar) is representative of the low-z universe as a whole. Since the bulk of stars in cluster as well as in field ellipticals and bulges formed at high redshifts ($z \gtrsim 3$), and such stars account for at least $\sim 30\%$ of the baryons now locked into stars, it is concluded that at least 30% of stars and metals formed before $z \simeq 3$. This implies the metallicity of the universe at $z = 3$ to be $\sim 1/10$ solar, which requires the cosmic star formation rate to run at least flat from $z \sim 1$ to ~ 5, in agreement with the most recent determinations of the star formation rate of Lyman-break galaxies.

1 The Chemistry of Galaxy Clusters

If the baryon fraction of rich clusters does not change appreciably in the course of their evolution as theoretical simulations predict (White et al. 1993), then we can then expect within a cluster to find confined in the same place all the dark matter, all the baryons, all the galaxies, hence all the metals, that have participated in the play. Clusters are then ideal *archives* for their past star formation and metal production history. As a whole, clusters behave much like a simple, *closed-box* model of chemical evolution.

Metals in clusters are partly spread through their ICM, partly locked into galaxies and stars. ICM abundances can be obtained from X-ray observations, while optical observations combined to population synthesis models provide estimates for the metallicity of the stellar component of galaxies.

1.1 The Iron Content of the Intracluster Medium

Fig. 1 shows the iron abundance in the ICM of clusters and groups as a function of ICM temperature. Data come from an earlier compilation (Renzini 1997, hereafter R97), with one difference. Clusters and groups in R97 with $kT < 2$ keV have been substituted by a set of 11 groups from Boute (1999).

Earlier estimates gave extremely low iron abundances in groups cooler than $\sim$ 1 keV, which determination comes from the iron-L emission. Such low abundances were regarded with suspicion in R97, being possibly the result of inaccuracies in the complex atomic physics of iron-L. Boute argues that the atomic physics may be correct, but gets substantially higher (probably more realistic) abundances when using 2-temperature fits, as illustrated in Fig. 1. It remains to be established whether the apparent correlation of the iron abundance with temperature is real or instead is an artifact of inadequate diagnostics at low temperatures. Radial gradients in the iron abundance have also been reported for several clusters (e.g. Dupke & White 1999, and references therein), but again such gradients seem to be confined to low-T clusters. I remain highly suspicious of all reported abundances when $kT \lesssim 3 - 4$ keV.

Perhaps even more interesting than the ICM iron abundance is the iron-mass-to-light-ratio (FeM/L) of the ICM, i.e. the ratio $M_{\rm Fe}^{\rm ICM}/L_{\rm B}$ of the total iron mass in the ICM over the total B-band luminosity of the galaxies in the cluster. The FeM/L of clusters and groups is shown in Fig. 2. Again, objects with $kT < 2$ keV in R97 have been omitted, with the exception of 4 objects from Buote (1999) for which a measure of the ICM mass was available from the R97 compilation. The drop of the FeM/L in poor clusters and groups

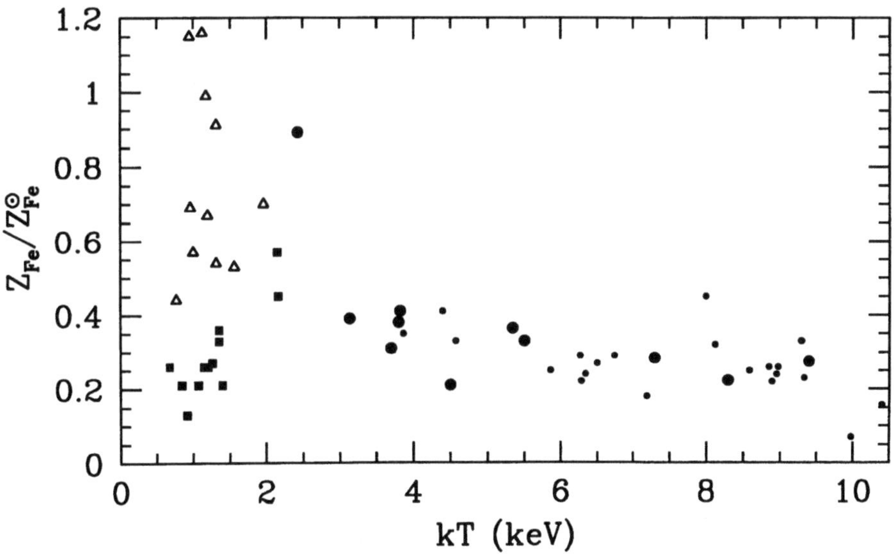

Fig. 1. The iron abundance in the intracluster medium as a function of ICM temperature from the R97 compilation (filled circles). Large circles refer to nearby clusters and small circles to cluster at various redshifts up to $z \simeq 0.5$. Objects with $kT \lesssim 2$ keV are from Buote (1999), with temperatures and abundances determined from one- and two-temperature fits, filled squares and open triangles, respectively. In the latter case the temperature is an emission-measure weighted average.

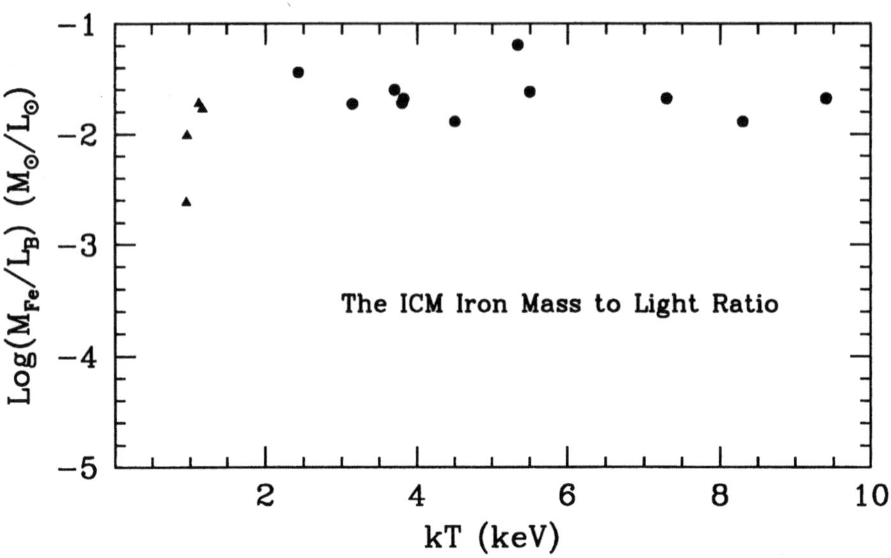

Fig. 2. The iron mass to light ratio of the ICM of clusters and groups (for $H_\circ = 50$) as a function of the ICM temperature (R97).

(i.e. for $kT \lesssim 2$ keV) noticed in R97 can be traced back to a drop in both the iron abundance (which however may not be real) *and* in the ICM mass to light ratio. Such groups appear to be gas poor compared to clusters, which suggest that they may be subject to baryon and metal losses due to strong galactic winds driving much of the ICM out of them (Renzini et al. 1993; R97: Davis, Mulchaey & Mushotzky 1998). For the rest of this paper I will concentrate on clusters with $kT \gtrsim 2-3$ keV.

Fig. 1 and 2 show that both the iron abundance and the FeM/L in rich clusters ($kT \gtrsim 3$ keV) are independent of cluster temperature, hence of cluster richness and optical luminosity. For these clusters one has $Z^{\rm Fe}_{\rm ICM} = 0.3 \pm 0.1$ solar, and $M^{\rm ICM}_{\rm Fe}/L_{\rm B} = (0.02 \pm 0.01)$ for $H_\circ = 50$. The most straightforward interpretation is that clusters did not lose iron (hence baryons), nor differentially acquired pristine baryonic material, and that the conversion of baryonic gas into stars and galaxies has proceeded with the same efficiency and stellar IMF in all clusters (R97). Otherwise, we should observe cluster to cluster variations of the iron abundance and of the FeM/L. The theoretical prediction by White et al. (1993) of the constancy of the baryon fraction in clusters is nicely supported by these evidences.

1.2 The α-Elements in the Intracluster Medium

X-ray observations with *ASCA* have also allowed to measure the ICM abundance of some of the α-elements such as O, Ne, Mg, and Si. A high α-element enhancement, $<[\alpha/{\rm Fe}]> \simeq +0.4$, was initially reported (Mushotzky 1994), an estimate that was later revised down to $<[\alpha/\mathrm{Fe}]>\simeq +0.2$ (Mushotzky et al. 1996). This modest α-element enhancement still suggests an ICM enrichment dominated by SNII products. However, the α-element overabundance vanishes when consistently adopting the "meteoritic" iron abundance for the sun, as opposed to the "photospheric" value (Ishimaru & Arimoto 1997). One can then conclude that there is presently no evidence for an appreciable α-element enhancement in the ICM, with a formal average $<[\alpha/\mathrm{Fe}]>\simeq +0.04 \pm \sim 0.2$ (R97). Clusters of galaxies as a whole are therefore nearly *solar* as far as the elemental ratios are concerned, which argues for stellar nucleosynthesis having proceeded in much the same way in the solar neighborhood as well as at the galaxy cluster scale. In particular, this implies a similar ratio of the number of Type Ia to Type II SNs, as well as a similar IMF (R97). This result suggests that the star formation process (IMF, binary fraction, etc.) is universal, with little or no dependence on the global characteristics of the parent galaxies in which molecular clouds are turned into stars. Perhaps this is just telling us that like it is difficult for us to look into a molecular cloud, it is also difficult for the star forming core of the cloud to have an idea of the galaxy in which it happens to be located.

It is nevertheless worth mentioning that other interpretations have been proposed. For example, bimodal star formation, with most of the metals in the ICM having been produced by a now extinct generation of only massive stars (Elbaz, Arnaud, & Vangioni-Flam 1995; Larson 1998). Decoupling the formation of the stellar population having made the cluster metals from the formation the one now making the cluster light requires fine tuning to account for the constancy of the FeM/L (Fig. 2). This is instead the straightforward consequence of a universal IMF.

1.3 The Iron Content of Galaxies and the Iron Share

An average metal abundance of the stellar component of cluster galaxies can be inferred from integrated spectra coupled to synthetic stellar populations. Much of the stellar mass in clusters is confined to passively evolving spheroids (ellipticals and bulges) for which the iron abundance may range from $\sim 1/3$ solar to a few times solar. Hence, the average iron abundance cannot be much different from solar, even when taking into account the presence of radial gradients (Arimoto et al. 1997), α-element enhancements (Thomas, Greggio & Bender 1999), and the luminosity bias (Greggio 1997).

The global iron abundance of a whole cluster is therefore given by:

$$Z_{\mathrm{CL}}^{\mathrm{Fe}} = \frac{Z_{\mathrm{ICM}}^{\mathrm{Fe}} M_{\mathrm{ICM}} + Z_*^{\mathrm{Fe}} M_*}{M_{\mathrm{ICM}} + M_*} = \frac{5.5 Z_{\mathrm{ICM}}^{\mathrm{Fe}} h^{-5/2} + Z_*^{\mathrm{Fe}} h^{-1}}{5.5 h^{-5/2} + h^{-1}}, \qquad (1)$$

where Z_*^{Fe} is the average abundance of stars in galaxies and M_* is the mass in stars. For the second equality I have assumed as prototypical the Coma cluster values adopted by White et al. (1993). Similarly, the ratio of the iron mass in the ICM to the iron mass locked into stars is:

$$\frac{Z_{\rm ICM}^{\rm Fe} M_{\rm ICM}}{Z_*^{\rm Fe} M_*} \simeq 1.65 h^{-3/2}. \tag{2}$$

With $Z_{\rm ICM}^{\rm Fe} = 0.3$ and $Z_*^{\rm Fe} = 1$ solar, eq. (1) gives $Z_{\rm CL}^{\rm Fe} = 0.34$ and 0.41, and eq. (2) gives 4.6, and 1.65, respectively for $h = 0.5$, and 1. Therefore, most of the cluster iron resides in the ICM, rather than being locked into stars, especially for low values of $H_{\rm o}$ (R97). This must be taken as a fundamental constraint in models for the chemical evolution of galaxies: galaxies do not evolve as a closed box, and outflows clearly play a crucial role.

With the adopted masses and iron abundances for the two baryonic components one can also evaluate the total cluster FeM/L: $(M_{\rm Fe}^{\rm ICM} + M_{\rm Fe}^*)/L_{\rm B} \simeq 1.3 \times 10^{-2}(1.65\,h^{-1/2} + h)$ $(M_\odot/L_\odot)$, or FeM/L=0.037 and 0.034 $M_\odot/L_\odot$, respectively for $h = 0.5$ and 1. The total FeM/L is therefore fairly insensitive to the adopted distance scale.

1.4 Empirical and Theoretical Metal Yields

From the near solar proportions of cluster abundances one obtains the total metal mass to light ratio of a typical cluster as $M_{\rm Z}/L_{\rm B} \simeq 10 \times M_{\rm Fe}/L_{\rm B} \simeq 0.3 \pm 0.1$ $(M_\odot/L_\odot)$. It is worth emphasizing that this is an interesting, fully empirical estimate of the metal yield of stellar populations. As well known, the bulk of stars in galaxy clusters are very old, say ~ 15 Gyr old. Hence, a single burst approximation may not be too rough for some applications. This empirical yield means that ~ 15 Gyr after a burst of star formation there are $\sim 0.3\,M_\odot$ of metals for each $L_\odot$ of blue light from the surviving (low-mass) stellar population. It therefore connects the prompt release of the metals by stars at the top end of the IMF to the luminosity released ~ 15 Gyr later by the stars at the lower end of the IMF of this same stellar population.

Following Tinsley (1981), the metal yield is usually defined *per unit mass* of stars, a quantity which theoretical counterpart depends on the poorly known low mass end of the IMF. The estimate above gives instead the yield *per unit luminosity* of present day (cluster) galaxies, which depends on the IMF only for $M \gtrsim M_\odot$. Theoretical mass-related yields have been recently estimated by Thomas, Greggio, & Bender (1998) based on current models of supernova explosions. Purging these yields of their mass dependence and transforming them into luminosity-related yields one obtains $M_{\rm Z}/L_{\rm B} = 0.08$, 0.24, and 0.33 $M_\odot/L_\odot$, respectively for $x = 1.7$, 1.35, and 1.00 (Renzini 1999b), which compares to $M_{\rm Z}/L_{\rm B} \simeq 0.3 \pm 0.1 M_\odot/L_\odot$ for the empirical cluster value. One can conclude that current stellar yields do not require a very flat IMF to account for the metal content of clusters.

1.5 The Relative Role of Type Ia and Type II Supernovae

Clusters are now dominated by E/S0 galaxies, which produce only Type Ia SNs at a rate of ~ 0.06 SNU (1 SNU $= 10^{-12}$ SNs $\rm yr^{-1} L_{B\odot}^{-1}$) for $h = 0.5$

(Cappellaro et al. 1993). Assuming such rate to have been constant through cosmological times (15 Gyr), the number of SNIa's exploded in a cluster of present-day luminosity $L_{\rm B}$ is therefore $\sim 6\times10^{-14}\times1.5\times10^{10}L_{\rm B}\simeq10^{-3}L_{\rm B}$. With each SNIa producing $\sim 0.7\,M_{\odot}$ of iron, the resulting FeM/L of clusters would be $\sim 7\times10^{-4}$, which falls short by a factor ~ 50 compared to the observed cluster FeM/L. The straightforward conclusion is that either SNIa's did not play any significant role in manufacturing iron in clusters, or their rate in what are now E/S0 galaxies had to be much higher in the past. This argues for a strong evolution of the SNIa rate in E/S0 galaxies and bulges, with the past *average* being some 10 to 50 times higher than the present rate (R97). The direct observation of SNIa's at high redshift may soon check whether this is the case.

We believe to have a fairly precise knowledge of the amount of iron released by each SNIa event, but it is difficult for the theory to predict the evolution of the SNIa rate past a burst of star formation (e.g. Greggio 1996). The case of Type II SN's is quite the opposite: their rate is just the death rate of massive stars ($M\gtrsim8M_{\odot}$), but a great uncertainty affects the mass of iron $M_{\rm Fe}^{\rm II}$ produced by each SNII as a function of its progenitor's mass. There are empirical reasons to believe that this is not a strong function of initial mass, and that assuming $0.07\,M_{\odot}$ of iron per event (as in SN 1987A) cannot be too far from reality. The total number of SNIIs $-N_{\rm SNII}-$ is obtained integrating the stellar IMF from e.g. 8 to 100 $M_{\odot}$, with the IMF being $\psi(M)=3.0L_{\rm B}\,M^{-(1+x)}$,. Thus, integrating the IMF one gets:

$$\left(\frac{M_{\rm Fe}}{L_{\rm B}}\right)_{\rm SNII}=\frac{M_{\rm Fe}^{\rm II}N_{\rm SNII}}{L_{\rm B}}\simeq\begin{cases}0.003 & \text{for } x=1.7\\ 0.009 & \text{for } x=1.35\\ 0.035 & \text{for } x=0.9.\end{cases}\tag{3}$$

This means that if the Galactic IMF slope ($x=1.7$, Scalo 1986) applies also to ellipticals in rich clusters, then SNII's underproduce iron by about a factor of 10 compared to the observed FeM/L. Making all cluster iron by SNII's requires a very flat IMF ($x\simeq0.9$). On the other hand, with an IMF slope somewhere between Scalo's and Salpeter's SNIIs make about 1/4 to 1/3 of the total iron, while the rest has to be made by SNIa's, just as in the so-called *standard chemical model* for the chemical evolution of the Milky Way (cf. Renzini et al. 1993, R97).

2 Clusters as Fair Samples of the Local Universe

According to current estimates, with $H_{\rm o}=50$ the present stellar mass density parameter is $\Omega_*=0.0036$, and the baryon mass density parameter $\Omega_{\rm b}=0.05$ (e.g. Madau et al. 1996). Therefore, $\sim 7\%$ of the baryons are now locked into stars, which compares to $\sim 1/(1+5.5h^{-3/2})$ in clusters, or $\sim 6\%$ and $\sim 10\%$, respectively for $h=0.5$ and 0.75. Fukugita, Hogan & Peebles estimate $(\Omega_*/\Omega_{\rm b})_{\rm o}=0.13$ for $h=0.5$, which is a factor of 2 higher than the cluster value, but they may have overestimated the stellar mass in spheroids. Therefore, it appears that the global, universal efficiency of baryon conversion into galaxies and stars – $(\Omega_*/\Omega_b)_o$ – is nearly the same as that observed in local clusters, which supports the notion of clusters being representative of the low-z universe as a whole.

If so, then no major difference in metal enrichment exists between field and clusters. Therefore, the global metallicity of the present-day universe is likely to be $\sim 1/3$ solar, as that of the only place where we can thoroughly measure it: in galaxy clusters. As a consequence, the metal share of the IGM to galaxies should also be nearly the same as the cluster ICM to galaxies metal share, as given by eq. (2), with most of the metals now residing in the IGM rather than within field galaxies (R97).

2.1 The Prompt Initial Enrichment of the Universe

The metallicity of the low redshift universe is the result of the integrated star formation history, from *first stars* to the present. However, our current knowledge of stellar ages can tell us something more about the actual history of metal production. Over the last several years compelling evidence has been gathered for the bulk of stars in galactic spheroids (i.e. ellipticals and bulges) – in clusters as well as in the general field – being very old, likely to have formed at $z \gtrsim 3$ (for a comprehensive review see Renzini 1999a).

With spheroids containing at least 30% of all stellar mass in the local universe (Schechter & Dressler 1987; Persic & Salucci 1992), an perhaps as much as $\sim 75\%$ (Fukugita, Hogan & Peebles 1998), one infers that at least 30% of all stars *and metals* have formed at $z \gtrsim 3$ (Renzini 1998, 1999b). This expectation from the *fossil* evidence is in good agreement with the most recent direct estimates from Lyman-break galaxies (Steidel et al. 1999), where the cosmic SFR runs flat from $z \sim 1$ all the way to $z \sim 5$, as indeed in one of the plausible options offered by Madau et al. (1998).

With $\sim 30\%$ of all stars having formed at $z \gtrsim 3$, and the metallicity of the $z = 0$ universe being $\sim 1/3$ solar, it is straightforward to conclude that the global metallicity of the $z = 3$ universe had to be $\sim 1/3 \times 1/3 \sim 1/10$ solar, or more (Renzini 1998). The metallicity of DLA absorbers at $z = 3$ appears to be $\sim 1/20$ solar (Pettini et al. 1997), just a factor of 2 below the predicted value from the *fossil evidence.* However, the Ly_α forest is much more metal poor, with $Z \simeq 10^{-3} Z_\odot$ at $z = 3$ (e.g. Songaila 1997). On the other hand, DLA and Ly_α-forest systems provide a vision of the early universe that is biased in favor of cold, metal-poor gas that may have been only marginally affected by star formation and metal pollution. Metal rich objects such as giant starbursts that would be dust obscured, the metal rich passively evolving spheroids, and the hot ICM/IGM obviously do not enlist among QSO absorbers, and the metallicity of high-z QSOs themselves appear to be several times solar (e.g. Hamann & Ferland 1999). There is indeed ample evidence (from the Ly_α forest to the broad emission line region of QSOs) that the universe was already extremely inhomogeneous at $z \sim 3$, which makes it quite difficult to measure its mass-averaged metal abundance. In any event, the prediction is: 1/3 solar.

References

1. Arimoto, N., Matsushita, K., Ishimaru, Y., Ohashi, T., & Renzini, 1997, ApJ, 477, 128
2. Arnaud, M., Rothenflug, R., Boulade, O., Vigroux, L., & Vangioni-Flan, E. 1992, A&A, 254, 49
3. Cappellaro, E., Turatto, M., Benetti, S., Tsvetkov, D.Yu., Bartunov, O.S., & Makarova, I.N. 1993, A&A, 268, 472
4. Davis, D.S., Mulchaey, J.S., & Mushotzky, R.F. 1998, astro-ph/9808085
5. Dupke, R.A., & White, R.E.III. 1999, astro-ph/9902112
6. Elbaz, D., Arnaud, M., & Vangioni-Flam, M. 1995, A&A, 303, 345
7. Fukugita, M., Hogan, C.J., & Peebles, P.J.E. 1998, ApJ, 503, 518
8. Greggio, L. 1996, The Interplay between Massive Star Formation, the ISM and Galaxy Evolution, ed.s D. Kunth et al. (Gif sur Yvettes, Edition Frontieres), p. 89
9. Greggio, L. 1997, MNRAS, 285, 151
10. Hamann, F., & Ferland, G. 1999, ARA&A, in press, astro-ph/9904223
11. Ishimaru, Y., & Arimoto, N. 1997, PASJ, 49, 1
12. Larson, R.B. 1998, MNRAS, 301, 569
13. Madau, P., Ferguson, H.C., Dickinson, M.E., Giavalisco, M., Steidel, C.C., & Fruchter, A. 1996, MNRAS, 283, 1388
14. Madau, P., Pozzetti, L., & Dickinson, M. 1998, ApJ, 498, 106
15. Mushotzky, R.F. 1994, in Clusters of Galaxies, ed. F. Durret, A. Mazure, & J. Tran Thanh Van (Gif-sur-Yvette: Editions Frontières), p. 167
16. Mushotzky, R.F., et al. 1996, ApJ, 466, 686
17. Persic, M., & Salucci, P. 1992, MNRAS, 258, 14p
18. Pettini, M., Smith, L.J., King, D.L., & Hunstead, R.W. 1997, ApJ, 486, 665
19. Renzini, A. 1997, ApJ, 488, 35 (R97)
20. Renzini, A. 1998, ASP Conf. Ser. 146, 298
21. Renzini, A. 1999a, astro-ph/9902108
22. Renzini, A. 1999b, astro-ph/9902361
23. Renzini, A., Ciotti, L., D'Ercole, A., & Pellegrini, S. 1993, ApJ, 419, 52
24. Scalo, J.M. 1986, Fundam. Cosmic Phys, 11, 1
25. Schechter, P.L., & Dressler, A. 1987, AJ, 94, 563
26. Songaila, A. 1997, ApJ, 490, L1
27. Steidel, C.C., Adelberger, K.L., Giavalisco, M., Dickinson, M., & Pettini, M. 1999, ApJ, 519, 1 460, 408
28. Thomas, D., Greggio, L., & Bender, R. 1998, MNRAS, 206, 119
29. Thomas, D., Greggio, L., & Bender, R. 1999, MNRAS, 302, 537
30. Tinsley, B.M. 1981, Fund. Cosmic Phys. 5, 287
31. White, S.D.M., Navarro, J.F., Evrard, A.E., & Frenk, C.S. 1993, Nat, 366, 429

The Connection of X-Ray Clusters of Galaxies with the Early Star Formation Histories

Nicola Menci[1] and Alfonso Cavaliere[2]

[1] Osservatorio Astronomico di Roma, via di Frascati 33, I-00040 Monteporzio (Roma), Italy
[2] Dip. Fisica, II Universitá di Roma, via della Ricerca Scientifica 1, I-00133 Roma, Italy

Abstract. Using current models of structure and star formation based on hierarchical clustering, we connect the cosmic star formation rate (SFR) to the X-ray emission properties of groups and clusters of galaxies. We show that when the baryons cool down from a hot phase to condense into stars with a SFR flat for $z \geq 2$, then such hot phase yields deep X-ray source counts about 10 times larger than in the case of a SFR peaked at $z \approx 1.5$. We also discuss the effect of the SFR on the shape and evolution of the X-ray $L - T$ relation.

1 Introduction

In current models of galaxy formation the SFR of the Universe is tightly related to the amount of cold and hot gas (m_c and m_h, respectively) contained in the dark matter (DM) halos where the galaxies are located. As a consequence, we expect the history of the SFR to be related to the X-ray properties of groups and clusters of galaxies, since these host most of the hot baryons emitting by thermal bremsstrahlung luminosities $L \sim 10^{44}$ erg/s. In particular, the process of star formation is expected to affect the X-ray properties of *small* galactic systems, since the "stellar" temperature $kT_* = E_* \, m_H / 3 m_h \sim 0.2$ keV (E_* being the Supernova energy) which measures the thermal energy provided by the Supernovae (SN) is close to the virial temperature $kT \approx G\,M/10\,R \propto M^{2/3}$ of small galaxy groups (DM masses $M \sim 10^{13}\, M_{\odot}$). Thus, in X-rays the connection with the SFR should show up mainly at faint luminosities.

Since the behaviour of the cosmic SFR at $z > 2$ is still plagued by observational uncertainties in the O-IR bands, we explore here the above connection with the X-ray data, and show that such observations are within the reach of forthcoming X-ray observatories like AXAF. These observations can discriminate between a declining SFR for $z > 2$ (as initially proposed [1]) and a flat SFR for $z > 2$ (as indicated by recent observations; see, e.g., [2], [3]).

The result may have relevant implications for all sides of the current theories of galaxy formation, including the cosmogony of DM halos. In fact in these theories the SFR is strongly coupled to the dynamical history of the DM halos, whose hierarchical growth is thought to proceed through merging and accretion events.

2 Star Formation Models

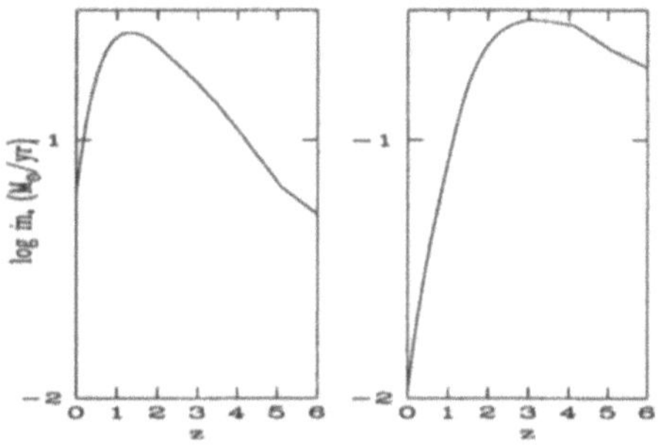

Fig. 1. The SF histories in Model A (left panel) and model B (right panel). Critical Universe with $H_o = 50$ km/s Mpc and tilted CDM spectrum of perturbations.

The baryonic processes inside dynamically evolving DM halos are usually treated using semi-analytic models (SAM) in the framework of hierarchical clustering (for a comprehensive account see [4]). According to the SAMs, the baryons gravitationally bound to a DM halo are assumed to be exchanged among three phases: the *cold* phase produced by radiative cooling; the resulting *condensed* phase into stars; the *hot* phase constituted by gas preheated by SN explosions and further raised to the virial temperature of the potential well. Besides, a fraction of the hot gas may be expelled by SN winds; this depends on the key parameter ϵ_o, defining the fraction of SN energy going into kinetic energy. The SAMs also include coalescence of baryons following halo merging and the luminosity-colour evolution due to rise and fall of successive star generations. In the end, the statistics of the baryonic phases are related to the statistics of the DM halos, usually characterized in terms of their circular velocity v_c.

We developed an analytic version of the SAMs (based on the Extended Press & Schechter Theory, see [5]), to describe the above baryonic phases and the exchanges among them (we refer to [6] for an extended description of the model and of its results). The parametrizations we adopt are the same as in the SAMs. The star formation time scale and the SN feedback are related to v_c by scaling laws depending on a set of four global parameters which, as described in [7], are related to the detailed parameter ϵ_o. We focus on two particular sets of parameters. The first (Model A) is characterized by a relatively large value of $\epsilon_o = 0.1$. With this, the SN winds are effective in expelling hot gas from small halos. Correspondingly, the resulting SFR we find (see fig. 1) is peaked at $z \approx 1.5$ and then declines considerably for larger z. The second set (Model B) is characterized by a small $\epsilon_o = 0.01$; in this case the SN are not effective in expelling baryons from the halos, but they heat them to somewhat larger temperatures compared with model A.

3 X-Ray Observables Related to SFR

The different conditions described by Models A and B lead to different amounts and physical states for the hot baryons which are contained in the DM potential wells and emit in X-rays. The X-ray luminosity of groups or clusters may be written in the form [8]

$$L \propto \rho^{1/2}(z) \left(\frac{m_h}{M}\right)^2 I(M,z)\, G^2(M)\, T^2(M) \; . \tag{1}$$

Here the normalization is adjusted (as usually) so as to match the height of

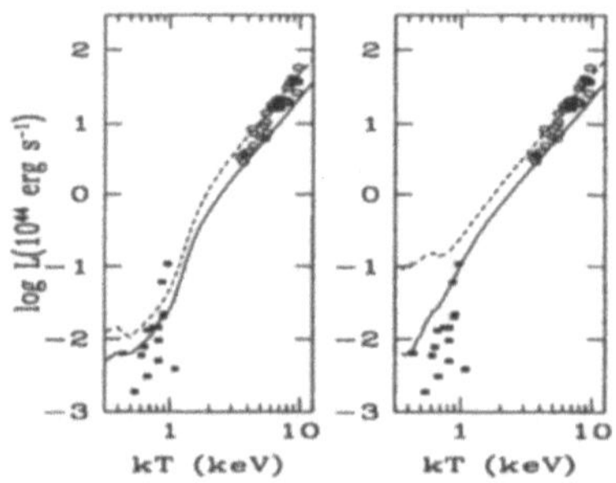

Fig. 2. The $L - T$ correlation at $z = 0$ (solid line) and $z = 1$ (dashed) in Model A (left) and Model B (right). The data are taken from [9] and [10]

the observed *local* $L - T$ relation at $T = 4$ keV; $\rho(z)$ is the DM density of the halo; the shape factor I describes the internal ICM distribution. In addition, G denotes the density jump across the shock induced by the infalling gas, averaged over the merging histories, that is, over the probability to merge clumps M' onto M; G is governed by T/T', where T' is the temperature of infalling gas. The overall effect of G^2 is to depress the $L - T$ relation at the scale of groups, bending it down from the self-similar behaviour $L \propto T^2$, which applies to very rich clusters [11].

The different amounts of hot baryons retained in small halos in Model A and Model B affect the z-evolution of the $L - T$ relation (see fig. 2); the considerable *increase* with z at the faint end of L found in Model B reflects the larger amount of emitting baryons in the smaller halos, which dominate the statistics at $z > 1$. In addition, the larger amount of expelled diffuse baryons marking Model A, together with the corresponding lower SN-induced temperature, make the shocks more effective in this case. Correspondingly, the factor G^2 in eq. 1 is more effective in bending down the $L - T$ relation from its self-similar shape.

Such effects of different SFR on the X-ray emission at the groups scale interestingly affect the deep source counts, as shown in fig. 3. Note that at fluxes $F \approx 10^{-15}$ erg/s cm^2 (reachable, e.g., by AXAF) a flat SFR implies counts larger by a factor ~ 10 with respect to a SFR declining for $z > 2$.

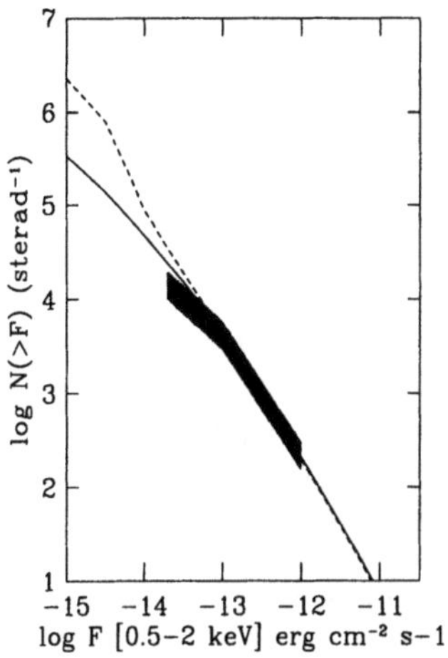

Fig. 3. The X-ray source counts in the energy band 0.5-2 keV for the SFRs derived from Model A (solid line) and from Model B (dashed line). The assumed cosmological/cosmogonical parameters are as given for fig. 1. The shaded region corresponds to the ROSAT cluster counts observed by [12].

References

1. Madau, P., Ferguson, H.C., Dickinson, M.E., Giavalisco, M., Steidel, C.C., & Fruchter, A. (1996), Monthly Not. Roy. Astron. Soc., **283**, 1388
2. Pettini, M. et al. (1997), *The Spectra of Star Forming Galaxies at High Redshift*, in *The Ultraviolet Universe at Low and High Redshift: Probing the Process of Galaxy Evolution*, ed. Waller W. et al., AIP Conf. Proc. 408, 279
3. Meurer, G. Heckman, T.M., Lehnert, M.D., Leitherer, C., & Lowenthal, J. (1997), Astron. Journ., **111**, 54
4. Sommerville, R.S., & Primack, J.R. (1998), preprint [astro-ph/9802269]
5. Lacey, C., & Cole, S. (1993), Monthly Not. Roy. Astron. Soc., **262**, 627
6. Menci, N., & Cavaliere, A. (1999), preprint
7. Cole, S., Aragon-Salamanca, A., Frenk, C.S., Navarro, J.F., & Zepf, S.E. (1994), Monthly Not. Roy. Astron. Soc., **271**, 781
8. Cavaliere, A., Menci, N., Tozzi, P. (1998), in press [astro-ph9810498]
9. Ponman, T.J., Bourner, P.D.J., Ebeling, H., Böhringer, H. 1996, Monthly Not. Roy. Astron Soc., **283**, 690
10. Markevitch, M. (1998), Astrophys. Journ., **504**, 27
11. Allen, S.M, & Fabian, A.C. (1998), Monthly Not. Roy. Astron. Soc., 297, L57
12. Rosati, P., Della Ceca, R., Norman, C., & Giacconi, R. (1998), Astrophys. Journ., **492**, 21

The Connection Between X-Ray Clusters and Star Formation

Paolo Tozzi and Colin Norman

The Johns Hopkins University, 3400 N. Charles, 21218, Baltimore, MD

Abstract. The properties of X-ray clusters of galaxies can be well understood in terms of a competition between shock heating and adiabatic compression. Strong shocks are expected to be important for massive clusters, while adiabatic compression is dominant for small clusters and groups. The scale of the shock/adiabatic transition is marked by a change of slope of the L–T relation and in the global properties of the emitting plasma. This scale is connected to star formation processes. Two quantities are crucial: the *average energy* injected in the IGM from stars and SNe, and the *epoch* of the energy injection. We show how these quantities can be synthesized in terms of specific entropy, which ultimately determines the X-ray emission from groups and clusters.

1 The Properties of X-Ray Clusters of Galaxies

Clusters of galaxies are objects of great cosmological relevance. In fact, they are believed to constitute a well–defined population of X–ray sources with simple evolutionary behaviour (see, e.g., [2] and references therein). However, their X–ray emission properties are not fully understood: the simple self–similar scaling predicts a relation between the X–ray luminosity and the temperature, $L \propto T^2$ [5], which is at variance with the observed relation $L \propto T^3$. There is further steepening of the L–T at temperatures around 1 keV [7]. Moreover, a sharp transition in the global properties of the emitting plasma is evident below 1 keV [9]. This behaviour has been related to the presence of an entropy minimum in the intergalactic medium (IGM) which, in turn, can be adiabatically compressed in the cluster potential [4][6][1] or accreted through shocks, as suggested by numerical simulations and analytical models [3]. Here we present a simplified argument to show how shock and adiabatic compression compete in a manner regulated by the entropy level of the IGM. The entropy production (before accretion) is due to non–gravitational heating by stars and SNe, and is determined by the average energy and the epoch when such energy is released [8].

2 The L–T Relation in Terms of Entropy

Here we present a simple argument to understand the L–T relation in terms of entropy production. The minimum entropy of the diffuse IGM is due to non–gravitational energy injection kT_* at the epoch z_*, when the gas density was

$\rho_* \propto (1+z_*)^3$. The specific entropy initially is then $S \propto \log(K_*)$, where $K_* \equiv kT_*/\mu m_H \rho_*^{2/3}$. If, in the following, the IGM is only adiabatically compressed, the final relation between density and temperature is $\rho \propto (T/K_*)^{3/2}$. The X–ray emissivity due to bremsstrahlung is $\epsilon_{ff} \propto \rho^2 T^{1/2}$. The central total luminosity of the cluster can be written:

$$L \propto \epsilon_{ff} V \propto T^{7/2} K_*^{-3} M \propto T^5 K_*^{-3}, \quad (1)$$

where V is the emitting volume. Then, in the limit of pure adiabatic compression, the L–T relation is naturally expected to be steeper than the self similar case (see also [1]).

On the other hand, for a generic IGM distribution we can write K as:

$$K = T/\rho^{2/3} = (T/T_*)(\rho/\rho_*)^{-2/3} K_*. \quad (2)$$

If we neglect further adiabatic compression, the quantity K is determined as follows. In the case in which all the gas is accreted through a strong shock, the ratio (ρ/ρ_*) is simply the maximum compression factor allowed, $(\rho/\rho_*) \to 4$. If the gas is isothermal, we have $K \propto K_* T$. Substituting the actual value of K for K_* in equation 1, we obtain $L \propto T^2$. Thus, in the limit of strong shocks the self similar scaling is recovered.

Of course it is too simplistic to describe the observed L–T relation in terms of two power–laws. In the real world, the transition between the adiabatic and the shock regime is gradual. However, the scale of the onset of the shocks is an important quantity, and, from observations, is expected to occur around 1 keV. In the following section we show how this scale depends on the entropy, which requires two pieces of information: *how much* energy is supplied, and *when*.

3 The Synergy of X–Ray and Optical Observations

The mass scale at which shocks start to appear, is the scale for which the infalling velocity of the accreting baryons is larger than the sound speed $v_s = (\gamma kT/m_H \mu)^{1/2}$, where $\gamma = 5/3$ and $\mu = 0.59$ for a primordial IGM.

If the potential well is deep enough, the infall velocity is essentially the free fall velocity v_{ff} (i.e., the baryons follows the dark matter). However part of the work done by the gravitational field goes into compression of the gas. As a result, the baryons are delayed with respect to the dark matter and the infall velocity v_i of the gas is smaller than v_{ff}. If we approximate the adiabatic external flow with a Bernoulli equation [10], the condition for the formation of the shock is $v_i > v_s$. This occurs when the gravitational term (or in other words, the free fall velocity) is dominant with respect to the pressure term. At the virial radius, where the gas is accreted and reaches the maximum velocity, this condition can be well approximated by:

$$\frac{v_{ff}^2}{c_s^2} > \left(\frac{\gamma+1}{\gamma-1}\right) - \frac{2}{\gamma-1}\left(\frac{\rho_{ta}}{\rho_s}\right)^{\gamma-1} \quad (3)$$

where ρ_s and ρ_{ta} are the densities of the gas at the shock and at the turn–around. Assuming the shock occurs at the virial radius and using mass conservation, we have $\rho_s = f_B \rho_{dm} v_{ff}/v_i$. Expanding in ρ_{ta}/ρ_s, at the first order we have $v_{ff} \geq (1.87 - 1.95)c_s$, when $\rho_{ta}/\rho_s \simeq 1/8$ and $1/30$.

If we assume that the entropy minimum $\propto \log(K_*)$ is produced at a single epoch when the IGM was uniform (before the formation of clusters and groups), we have $K_* = kT_*/\mu m_H \rho_o^{2/3}(1+z_*)^2$. Since $c_s^2 = \gamma K_* \rho_s^{2/3}$, the mass scale of the shock/adiabatic transition is determined by K_*, as shown in the panel a) of figure 1. The two horizontal lines bracket the mass range roughly corresponding to $0.5 - 2$ keV, with an associated set of values for K_* in the range $0.2 - 1 \times 10^{34}$ ergs g$^{-5/3}$cm^2. Here we use an average M–T relation and take into account the fact that a fraction $\leq 1/2$ of the mass is responsible for more than 80 % of the X–ray emission in groups. This argument is analogous to the one presented in [8], based on the entropy core in clusters.

In panel b) and c) we show the behaviour in terms of the two variables T_* and z_* independently. In panel b) we see that the transition scale occurs in the same mass interval for a range of initial temperature $kT_* = (0.1 - 0.5)$ keV. If the same energy is injected later the entropy and the predicted transition mass scale are higher. This is clear in panel c), where the dependence on z_* for a given energy input is shown. These results hold under the assumption that the gas was uniform at the epoch of entropy production. If, instead, the gas was at a contrast δ_* with respect to the background density, the relation between energy and epoch is $T_* \propto (1+z_*)^2\delta_*^{2/3}$ for a given K_*. This indicates why, after the collapse, a larger energy input is needed to obtain the same effect (i.e., the same K_*).

Our discussion shows that the entropy is the key parameter, and demonstrates how both energy and time are equally important in shaping the X–ray emission from large scale structure. Such considerations are mandatory to calculate the energy budget that is necessary to reproduce the observed L–T relation.

3.1 What is Still Missing?

The description of the X-ray properties in terms of entropy can be much more detailed. In particular, the complete density and temperature profiles, including shock heating and adiabatic compression, can be computed in different regimes and cosmological frameworks [10]. However, to define statistically the population of local and high z clusters of galaxies, we still need some important pieces of information.

The injection of T_* is a continuous process, eventually peaking at redshift $z \simeq 2 - 3$. The emitting properties of the ICM will result from the competition between the timescales for dynamical evolution and entropy production. Since there is a large scatter in the formation history of massive halos in

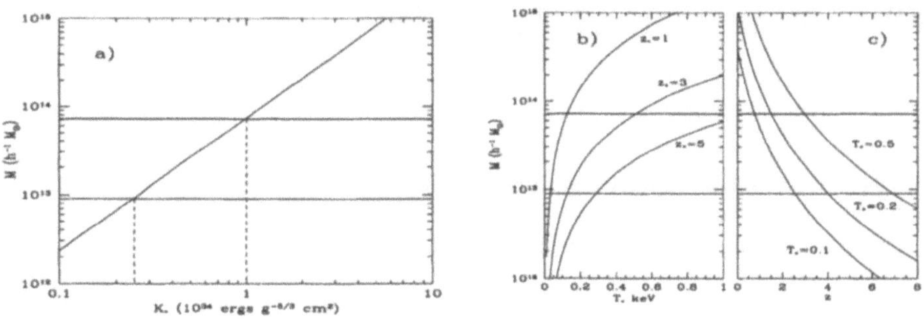

Fig. 1. Panel a): the mass scale of the adiabatic/shock transition as a function of K_*; b): the same as a function of the non gravitational energy scale T_* for three different redshifts z_*; c): the same as a function of redshift for three different T_*

all the hierarchical cosmologies, the properties fluctuate with the epoch of observation and the mass scale. Such intrinsic scatter is one of the most relevant characteristics of the observed L–T relation. The accretion of already virialized, small halos is expected to occur with the free fall velocity, since the bound, clumped gas is not slowed by pressure effect. Consequently, part of the ICM will be always shocked, in an amount depending on the presence of small scale structure. Cooling flows are not included in the present treatment; on the other hand, they are expected to be important, not only for mass deposition in the center, but even for their contribution to the X-ray properties.

In summary, the physics of the X–ray emitting plasma in clusters of galaxies is well understood. The most meaningful parameter is the entropy, which is directly connected to the energy and the temporal scales set by star formation processes. These, in turn, can be directly observed in the optical and infrared bands from ground and space based telescopes. In the near future, the synergy between the X–ray facilities and optical telescopes like the VLT, is expected to give powerful insights in the formation and evolution of cosmic structures.

References

1. Balogh, M.L., Babul, A., & Patton, D.R. 1998, preprint [astro/ph9809159]
2. Borgani, S., Rosati, P., Tozzi, P., & Norman, C. 1999, ApJ, in press
3. Cavaliere, A., Menci, N., Tozzi, P. 1997, ApJL 484, 1
4. Evrard, A.E., & Henry, J.P. 1991, ApJ, 383, 95
5. Kaiser, N. 1986, MNRAS, 222, 323
6. Kaiser, N. 1991, ApJ, 383, 104
7. Ponman, T.J., Bourner, P., Ebeling, H., Bohringer, H. 1996, MNRAS, 283, 690
8. Ponman, T. J., Cannon, D. B., Navarro, J. F. 1999, Nature, 397, 135
9. Renzini, A. 1997, ApJ, 488, 35
10. Tozzi, P., & Norman, C. 1999, in preparation

Clustering at High Redshift

Stefano Cristiani[1,2]

[1] Space Telescope-European Coordinating Facility,
European Southern Observatory
[2] Dipartimento di Astronomia dell'Università di Padova

Abstract. Together with the CMB, the three sources of information that astronomers have at high redshift as probes of the formation and evolution of the LSS are QSOs, galaxies and absorbers observed in the spectrum of distant background objects. In this contribution I try to give a hint of historical perspective, following how the technological advances have driven the emphasis from one class to another, in order to show what are the likely forthcoming milestones.

1 QSOs

QSOs have been the first class of sources used to obtain direct information about clustering at high redshift. In the 80's they were the only available high-z objects bright enough to be discovered on photographic plates and observed in relatively large quantities with the existing spectrographic facilities.

Systematic searches began with the pioneering work of Osmer (1981) and the detection of a clustering signal on scales ~ 10 Mpc was achieved first using inhomogeneous QSO catalogs [24] and then the statistically-well-defined samples used to study the QSO luminosity function (LF) [23].

QSOs display a number of appealing properties when compared to galaxies as cosmological probes of the intermediate-linear regime of clustering: they have a rather flat redshift distribution, their point-like images are less prone to the surface-brightness biases typical of galaxies and they sparse-sample the environment. In recent times complete samples totaling about 2000 QSOs have been assembled, providing a $\simeq 5\sigma$ detection of the clustering with an amplitude of $6h^{-1}$ comoving Mpc [2,17,8], consistent with or slightly larger than what is observed for present-day galaxies and definitely lower than the clustering of clusters.

The standard objection with regard to the information content in surveys of QSOs and radio galaxies is that, since they are exotic objects, one risks learning possibly more about the formation and evolution of super-massive black holes in galactic nuclei, which is something interesting *per se*, than about cosmology. In fact when the statistics on the clustering became sufficient to address its evolution, and people started plotting amplitudes of the correlation function (CF) as a function of redshift, surprisingly the trends did not follow any of the canonical patterns of constant, stable or collapsing clustering, in terms of the parameterization $\xi(r,z) = \xi(r, z=0)(1+z)^{-(3+\epsilon-\gamma)}$, where ϵ models the gravitational evolution of the structures.

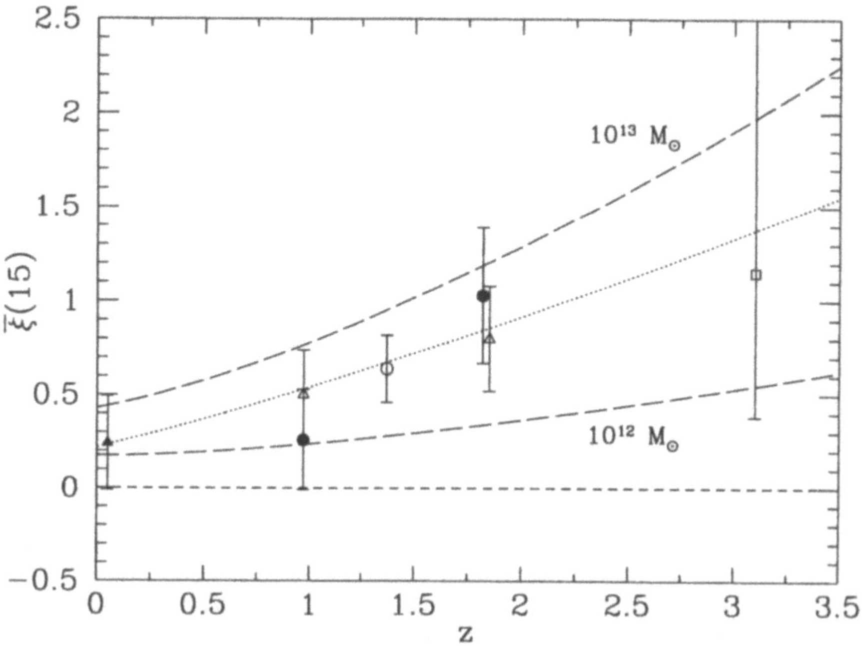

Fig. 1. The amplitude of the average two-point correlation function (TPCF) $\bar{\xi}(15\ h^{-1}$ Mpc) as a function of redshift (from La Franca, Andreani and Cristiani 1998). The dotted line is the $\epsilon = -2.5$ clustering evolution, the dashed lines show the evolution of the clustering for a minimum halo mass of 10^{12} and 10^{13} $M_\odot$ h^{-1} according to the transient model of Matarrese et al. (1997).

Evidence is found for an *increase* of the clustering with increasing redshift [14]. Not unreasonably, after all, since the observed clustering is the result of an interplay between the clustering of mass, decreasing with increasing redshift, and the bias, increasing with redshift, convolved with the redshift distribution of the objects [19,4] and is critically linked to the physics of the QSO formation and evolution. Let us model, following [5], the rise and fall of the QSO LF as the effect of two components: the newly formed BH, which are dominant at $z > 3$ and the reactivated BH, which dominate at $z < 3$. The reactivation is triggered by interactions taking place preferentially in groups of galaxies. In this way the clustering properties of QSOs are related to those of transient, short-lived, objects [16]. At high redshifts QSOs correspond to larger (rarer) mass over-densities collapsing early and cluster very strongly. Then the clustering amplitude decreases until the mass scale typical of a QSO reaches the value of the average collapsing peaks, after which clustering may grow again.

If we think of QSOs as objects sparsely sampling halos with $M > M_{\rm min}$, we can see from Fig. 1 that an $M_{\rm min} = 10^{12} - 10^{13}\ h^{-1} M_\odot$ would provide at the same time both the desired amount *and* evolution of clustering. $5 \cdot 10^{12}$ $M_\odot$

is also the typical halo mass of the groups of galaxies, in which interactions are most effective, that fits correctly the evolution of the luminosity function.

The idea that QSOs reside in small to moderate groups of galaxies is corroborated by the study of the environments of (radio-quiet) QSOs, measuring QSO-galaxy clustering [11] or associated absorption in close line-of-sight QSO pairs. A recent result based on the WENSS and Green Bank radio surveys [22] also confirms the general tendency for radio sources (QSOs+radio galaxies) to become increasingly biased tracers of the matter distribution with increasing redshift.

The cosmogonic occurrence of the QSO phenomenon is beginning to be understood. It is still arduous to turn to cosmology, but it is also easy to predict that the SLOAN and 2DF QSO surveys will bring dramatic progress in this direction.

2 Clustering of Galaxies

The type of rapid progress that in the meantime has taken place for galaxies: CCDs have become larger and better, the HDF and other deep fields have cleaned away some misconceptions about confusion limits. Simple color criteria show remarkable efficiency, making it possible to select high-z galaxies at an industrial rate. It is sufficient to obtain good multicolor imaging of deep fields down to $I \simeq 25$ to get about one high-z galaxy per sq.arcmin. The success rate is so good that one can measure clustering at high-z almost without spectroscopic redshifts. The basic idea is that the redshifts estimated on the basis of the photometry are sufficiently good to subdivide the sample in redshift bins and compute the angular correlation function (ACF) within the individual bins. This is not the poor man's approach to the high-z universe, but a technique which makes it possible to derive redshifts (and clustering) for galaxies which are about two magnitudes fainter than the deepest limits for spectroscopic surveys (even with 10m class telescopes).

We [3] have built a photometric redshift code based on the comparison of the observed colours of galaxies with those expected from template SEDs derived from the GISSEL library of models. Spectroscopic redshifts are still important, to check that the code works well and that the estimated uncertainties correspond to the observed ones, with σ_z increasing from 0.1 to 0.2 with increasing redshift. As a matter of fact it turns out that spectroscopic redshifts are not always more reliable than the photometric ones: in a few cases discrepancies have been found to be due to wrong spectroscopic estimates published in the literature.

Monte Carlo simulations have been carried out to infer the uncertainties also in the domain inaccessible to spectroscopy ($I_{AB} \geq 26$) and a comparison with other photometric methods has been performed to test the stability as a function of the adopted templates. This makes it possible to derive a map of the contamination effects due to the uncertainties on the photometric

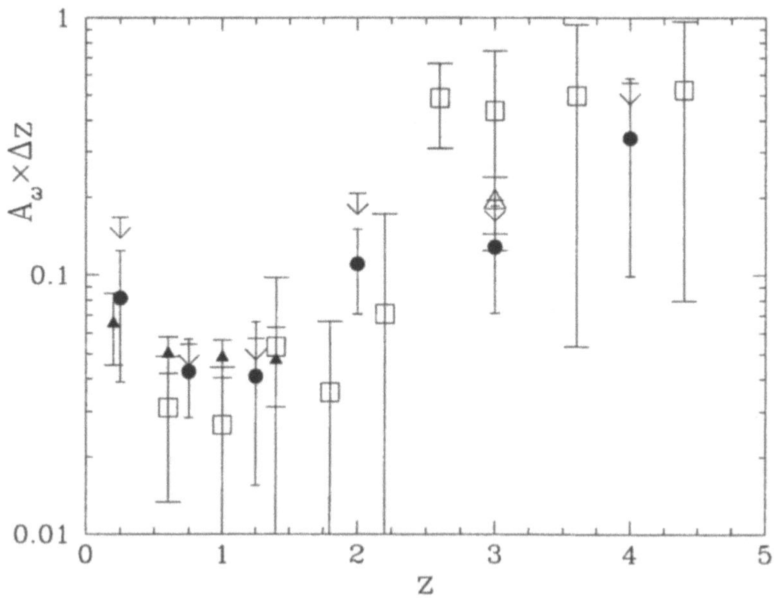

Fig. 2. The amplitude of the angular correlation at 10 arcsec (A_ω) as a function of redshift. Filled circles show the values obtained by Arnouts et al. 1999 (the arrows are upper limits estimated on the basis of a contamination correction), filled triangles the estimates obtained by Connolly et al. (1998). The open triangle is the value for a LBG sample of Giavalisco et al. (1998) and open squares refer to the values obtained by Magliocchetti & Maddox (1998).

redshifts and guided our choice of optimal redshift bins to minimize the effects of the errors, shot-noise and small field of view (as a note in passing, also spectroscopic redshifts often end their lives in bins either to compute a LF or a global star formation rate or a CF).

The evolution of the ACF amplitude shows again what is familiar from the case of QSOs: an initial decrease followed by an increase. As a consequence of the bias dependence on the redshift and on the selection criteria of the samples, the behaviour of the galaxy clustering cannot provide a straightforward indication of the evolution of the underlying matter clustering. For this reason, the parametric form mentioned in Sect.1 cannot correctly describe the observations for any value of ϵ.

We have compared our results with the theoretical predictions of a set of different cosmological models belonging to the class of CDM scenario. The halo masses required to match the observations depend on the adopted background cosmology. For Einstein de Sitter (EdS) universes, the SCDM model reproduces the observed measurements if a typical minimum mass of $10^{11}h^{-1}\ M_\odot$ is used, while the τCDM and TCDM models require a lower typical mass of $10^{10}-10^{10.5}h^{-1}\ M_\odot$. For OCDM and ΛCDM models, the mass is a *function of redshift*, with $M_{\rm min} \leq 10^{10}h^{-1}\ M_\odot$ at $z \leq 1$, $10^{11}h^{-1}\ M_\odot$

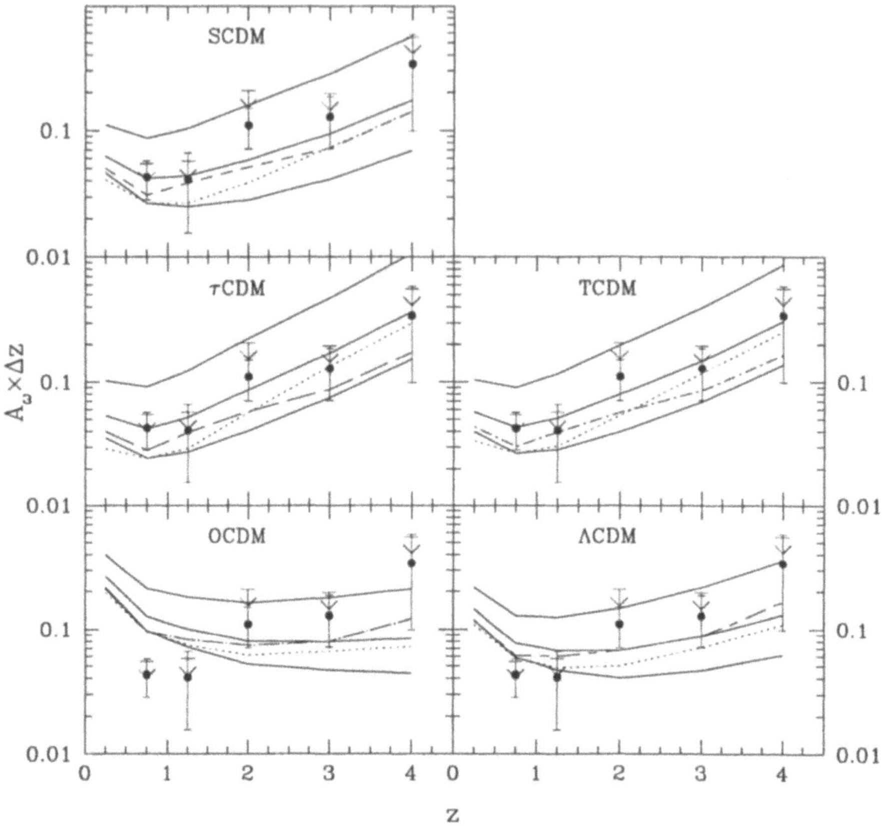

Fig. 3. Comparison of the observed A_ω (filled circles for $I_{AB} \leq 28.5$, arrows for the upper limits estimates) with the prediction of the various theoretical models. The solid lines show the measurements expected when a minimum mass $M_{\rm min} = 10^{10}$, 10^{11} and $10^{12} h^{-1} M_\odot$ is assumed; the lower curves refer to smaller masses. The dotted lines show the prediction obtained by using the median masses at any redshift estimated with GISSEL models. The dashed curves correspond to models where the masses necessary to reproduce the observed density of objects in each redshift bin are used.

between $1 \leq z \leq 3$ and $10^{12} h^{-1}\ M_\odot$ at $z \simeq 4$. The higher masses required at high z to reproduce the clustering strength for these models are a consequence of the smaller bias they predict at high redshift compared to the EdS models. At $z \simeq 3$, the clustering strength and the observed density of galaxies are in good agreement with the theoretical predictions for any fashionable cosmological model. At $z \simeq 4$, the present analysis seems to be more discriminant. Due to the remarkably high correlation strength, for some models the observed density of galaxies starts to be inconsistent with the required theoretical halo density. The difference in the predicted masses (a factor $\simeq$ 15 to 30 at $z \simeq 3$ and 4) between EdS and non-EdS universe models is also highly discriminant and in principle testable in terms of measured velocity dispersions.

A prediction of the hierarchical models is the dependence of the clustering strength on the limiting magnitude of the samples. At $z \simeq 3$ the density of galaxies in the HDF is approximately 65 times higher than the spectroscopic samples of LBGs [1,12]. Comparing our measurements with the latter a clear trend of decreasing clustering strength with increasing density is detected. This result is in excellent agreement (both qualitatively and quantitatively) with the predictions of hierarchical models [18] and, as noted by Adelberger et al. (1998), suggests the existence of a strong relation between the halo mass and the absolute UV luminosity: more massive haloes host the brighter galaxies.

Finally, we have compared our σ_8^{gal} to that of the mass predicted for three cosmologies to estimate the bias. For all cases, we found that the bias is an increasing function of redshift with $b(z \simeq 0) \simeq 1$ and $b(z \simeq 4) \simeq 4.5$ (for EdS universes), and $b(z \simeq 0) \simeq 0.5$ and $b(z \simeq 4) \simeq 3$ (for open and Λ universes). This result confirms and extends in redshift the results obtained for LBGs at $z \simeq 3$ [1,12], suggesting that these high-redshift galaxies are located preferentially in the rarer and denser peaks of the underlying matter density field.

What are the descendants of the galaxies observed at $z \simeq 3-4$ in the HDF? In a simple scenario assuming that only one galaxy is hosted by the descendants the resulting local bias implies that the descendants in the case of OCDM and ΛCDM models can be found among the brightest and more massive galaxies inside clusters, while for EdS universes they are field or normal bright galaxies.

The present results have been obtained in a relatively small field for which the effects of cosmic variance can be important[25]. Nevertheless they show a possibility of challenging cosmological parameters which becomes particularly exciting in view of the rapidly growing wealth of multi-wavelength photometric databases in various deep fields and availability of 10m-class telescopes for spectroscopic follow-up in the optical and near infrared.

3 QSOs Strike Back: Clustering of Absorbers

After all galaxies are exotic objects too, in the sense that they are also biased tracers of mass. There is, however, a way to follow more "normal" matter at high redshifts: the absorption spectra of QSOs, which sensitively probe the gaseous baryonic component of the universe. Sensitivity is the key word: with the present instrumentation it is possible to detect neutral HI column densities down to 10^{12} cm^{-2}, while for example 21 cm radio observations which roughly correspond to the visible extent of a galaxy are limited in the best cases to column densities that are 7 orders of magnitude larger. In this way observations of the Lyman-α forest reveal with an enormous dynamic range very different structures, ranging from fluctuations of the diffuse intergalactic medium to the interstellar medium in protogalactic disks.

In recent times a large database of high-resolution ($\sim$ 10 km/s), high-S/N spectra of the Lyman forest has become available, allowing a detailed investigation of the clustering of Lyman-α lines, especially at high redshift, where the density of lines provides particular sensitivity. Cristiani et al (1997) have used more than 1600 Lyman-α 's to detect a weak but significant signal, with $\xi \simeq 0.2$ on scales of 100 km s^{-1} at a 4.6σ level. Exploring the variations of the clustering as a function of the column density a trend of increasing amplitude of the TPCF with increasing column density is apparent, showing again that bias is at work.

Metal systems show stronger clustering: the CIV CF at $z \simeq 2 - 2.5$ is consistent with a canonical power-law form with an $r_o = 3.5$ Mpc and $\gamma \simeq 1.8$ with evidence for an increase of the clustering strength with increasing column density [9,21]. The relatively large correlation scale indicates that the CIV absorbers are biased tracers of relatively high density regions.

To get a rough idea of the (over- and under-)densities corresponding to Lyman-α absorbers it is useful to estimate their size, on the basis of the statistics of coincidences and anti-coincidences of absorptions in close lines-of-sight to QSO pairs and groups [10], from which one can infer the ionization and the relationship between the observed optical depths and the total density. It turns out that these structures are big (a few hundred kpc), highly ionized and at the lower column densities ($\log N_{HI} \lesssim 14$) probe weakly biased or even anti-biased regions of the Universe. It makes sense then to compute the power spectrum (PS) of mass density fluctuations and assume that it corresponds to the linear regime. The recovery of the 3-D PS is a complex procedure and various recipes have been proposed by a number of authors [7,13,20]. The present results indicate that the PS amplitude is consistent with some scale-invariant, COBE-normalized CDM models (an open model with $\Omega_0 = 0.4$ OCDM, variants of the SCDM) and inconsistent with others (the CCDM model, with $\Omega_0 = 1$, $h = 0.5$, $\sigma_8 = 1.2$)[7]. Even with limited dynamic range and substantial statistical uncertainty, a measurement of the PS that has no unknown "bias factors" offers many opportunities for testing theories of structure formation and constraining cosmological parameters.

In a similar fashion, it has been proposed to apply the Alcock-Paczinsky test to QSO groups to investigate the cosmological geometry [13]. This test, based on the comparison between the clustering properties of the (Lyman-α) absorbers, namely the TPCF, observed along the line of sight and the corresponding estimate in the transverse direction, is especially sensitive to Ω_Λ and should be able to discriminate between an ($\Omega_m = 0.3$, $\Lambda = 0$) and an ($\Omega_m = 0.7$, $\Lambda = 0.3$) universe at a 4σ level by observing $\sim$ 25 QSO pairs.

Future prospects look really exciting. Starting at the end of 1999, the echelle spectrograph UVES will be available on UT2 of the VLT, extending, with respect to HIRES at Keck, the possibility of high-resolution ($\Re \sim 50000$)

observations down to at least $V \sim 20$ and to a shorter wavelength range (i.e. to more numerous, lower-redshift QSOs). Then (by 2001) the FLAMES facility (see `http://http.hq.eso.org/instruments/flames/`) will allow observation with UVES of up to 8 QSOs at the same time in a field of 25 arcmin diameter and/or 135 targets in the same area with an intermediate/high resolution spectrograph, GIRAFFE. The Alcock-Paczinsky test and other cosmologically discriminant measures could be carried out in a few nights.

Acknowledgments

I wish to thank P. Andreani, S. Arnouts, V. D'Odorico, S. D'Odorico, A. Fontana, E. Giallongo, F. La Franca, F. Lucchin, S. Matarrese, L. Moscardini with whom I have been collaborating and P. Bristow for carefully reading the manuscript.

References

1. Adelberger K.L., Steidel C.C., Giavalisco M., Dickinson M.E., Pettini M., Kellogg M., 1998, ApJ **505**, 18
2. Andreani P., Cristiani S. 1992, ApJ **398**, L13
3. Arnouts S., Cristiani S., Moscardini L., Matarrese L., Lucchin F., Fontana A., Giallongo E. astro-ph/9902290
4. Bagla J.S., 1998, MNRAS **297**, 251
5. Cavaliere A., Perri M., Vittorini V., 1997, Mem.S.A.It. **68**, 27
6. Connolly A.J., Szalay A.S., Brunner R.J., 1998, ApJ, 499, L125
7. Croft R.A.C., et al., 1998, astro-ph/9809401
8. Croom S.M., Shanks T., 1996, MNRAS **281**, 89
9. D'Odorico V., Cristiani S., D'Odorico S., Fontana A., Giallongo E., 1998a, A&AS **127** 217
10. D'Odorico V., Cristiani S., D'Odorico S., Fontana A., Giallongo E., Shaver P., 1998b, A&A **339**, 678
11. Fisher K.B., Bahcall J.N., Kirhakos S., Schneider D. P., 1996, ApJ **468**, 469
12. Giavalisco M., Steidel C.C., Adelberger K.L., Dickinson M.E., Pettini M., Kellogg M., 1998, ApJ **503**, 543
13. Hui L., Stebbins A., Burles S., 1999, ApJ **511**, L5
14. La Franca F., Andreani P., Cristiani S. 1998, ApJ **497**, 529
15. Magliocchetti M., Maddox S.J., 1998, astro-ph/9811320
16. Matarrese S., Coles P., Lucchin F., Moscardini, L., 1997, MNRAS **286**, 115
17. Mo H.J., Fang L.-Z., 1993, ApJ **410**, 493
18. Mo H.J., Mao S., White S.D.M., 1999, MNRAS **304**, 175
19. Moscardini L., Coles P., Lucchin F., Matarrese S., 1998, MNRAS **299**, 95
20. Nusser A., Haehnelt M., 1999, MNRAS **303**, 179
21. Quashnock J.M., Vanden Berk D.E., 1998, ApJ **500**, 28
22. Rengelink R., 1999, Ph.D. Thesis
23. Shanks T., Fong R., Boyle B., Peterson B.A., 1987, MNRAS **227**, 739
24. Shaver P.A., 1984, A&A **136**, L9
25. Steidel C., 1998, astro-ph/9811400

EIS – An Imaging Survey for VLT Science

Luiz da Costa

European Southern Observatory, Garching bei München, D-85748, Germany

Abstract. The imaging data assembled by the recently completed ESO Imaging Survey (EIS) are reviewed and their scientific value briefly assessed. Among the various applications, the imaging data has been used to build a large sample of candidate distant clusters of galaxies in the Southern Hemisphere to be used for follow-up observations with the VLT as well as other space and ground-based facilities. Preliminary results from ongoing work to confirm these candidates are reported and the future prospects discussed.

1 Introduction

The ESO Imaging Survey (EIS) was a joint effort of ESO and its community to carry out a public imaging survey and to prepare target lists for different scientific applications in preparation for the beginning of regular operation of the VLT. The primary scientific goals were conceived by a Working Group (WG) composed of experts selected from the community at large and the survey was conducted by a team formed primarily by astronomers from the community. Among the different goals of EIS, one of the top priorities was the construction of a catalog of distant cluster of galaxies candidates over an extended redshift baseline from which targets could be drawn for subsequent follow-up work. Such a sample was constructed by the EIS team and it is the main focus of this presentation.

The importance of a large sample of confirmed distant clusters is vast, as emphasized throughout this meeting, ranging from the study of galaxy and large-scale structure formation and evolution to estimates of cosmological parameters. This has motivated several recent attempts to find and investigate intermediate and high-redshift clusters using X–ray, optically- and infrared-selected samples. Even though X–ray samples are in many respects superior to those selected in optical or infrared, the number of confirmed distant clusters in the Southern Hemisphere is still quite small. Currently, the best samples of distant X–ray clusters are all based on deep pointings of the ROSAT PSPC [11]. In particular, the RDCS [10] includes about 30 $z > 0.5$ spectroscopically confirmed clusters, the majority in the Northern Hemisphere. Moreover, considering the currently foreseen timetable of future X-ray missions it is clear that no new X-ray based high-redshift cluster sample will be available in the next few years. Clearly, if VLT is to play a role in the study of the distant clusters it will have to rely on optically- or infrared–selected cluster samples, at least in the near future. It is worth pointing out that while some investigations require well-defined complete samples, usually difficult to asses from optically-selected ones, others can benefit solely by a statistically large sample.

The need for cluster targets was foreseen by the EIS WG which recommended a moderately deep, wide-angle I-band survey to search for distant clusters, despite the inadequacy of the available detectors for wide-field imaging at the beginning of the survey. Still, EIS is currently the largest imaging survey available (17 square degrees) in the Southern Hemisphere to a depth ($I \lesssim 23$) suitable for blind searches of distant clusters.

2 General Results from EIS

EIS was conceived as an attempt to optimize the use of the NTT by carrying out a coordinated imaging survey in the one-year period between the re-commissioning of the NTT and the commissioning of the first unit of the VLT project. The main objectives of EIS were: 1) to conduct a public optical-IR imaging survey in preparation for VLT; 2) to produce target lists for VLT before the start of the Science Verification period. The survey was designed to satisfy the requirements of a broad range of applications. Following the recommendations of the EIS WG the survey consisted of two parts: **1)** An optical, moderately deep, multi-passband, wide-angle survey (EIS-WIDE) covering four patches distributed in the right ascension range $22^h < \alpha < 10^h$ to produce targets almost year-round. The survey consisted of a mosaic of overlapping EMMI-NTT frames with each position on the sky being sampled twice for a total integration time of 300 sec. The observations were carried out in the period July 97- March 98, with all of the data, including pixel maps and derived catalogs, being publicly released in March and July 1998; **2)** Deep optical/infrared observations of the HDF-S and AXAF Deep Field (EIS-DEEP). These observations were obtained using SUSI2 and SOFI at the NTT in the period August-November 1998, with all of the data, including derived catalogs, being publicly released December 10-12 1998 [3] [9].

The accumulated imaging data provided the following data sets: **1)** $\sim$ 17 square degrees in I-band ($I_{AB} \sim 23.5$); **2)** $\sim$ 3 square degrees in VI; **3)** $\sim$ 2 square degrees in BVI; **4)** $\sim$ 125 square arcmin in JKs ($K_{AB} \sim 23.5$); **5)** $\sim$ 80 square arcmin in $UBVRI$ ($I_{AB} \sim 26$). In addition, combined, the optical/infrared deep observations provide a coverage of 15 and 25 square arcmin in eight and seven passbands, respectively [3] [9]. Similar data will soon become available for the HDF-S STIS field.

From the available photometric data several target lists were extracted for immediate use in follow-up observations at different facilities and for Science Verification observations. Among them are candidate galaxy clusters, quasars, high-redshift galaxies as well as galactic objects such as white dwarf and brown dwarf candidates [17]. Some of these targets have already been used for follow-up observations. Quasar candidates identified in patch B have been observed with 2dF at AAT, leading to the confirmation of about 50%

of the candidates. Distant EIS cluster candidates and high-redshift galaxies, identified using dropout and photometric redshift techniques, have been successfully confirmed from VLT observations using the VLT test camera [2] and FORS (see ESO PR 12/99), respectively, as part of the Science Verification of UT1.

3 EIS Cluster of Galaxies Candidates

As part of the main effort of the EIS team, the galaxy catalogs derived from the I-band images were used to search for candidate clusters using the matched–filter algorithm [7] to detect clusters and to estimate redshifts. The final EIS candidate clusters sample consists of 304 objects in the redshift range $0.2 \lesssim z \lesssim 1.3$ [5] [6] [14], with a median redshift of $z \sim 0.5$. This EIS cluster sample is the largest such a catalog presently available in the Southern Hemisphere and it is arguably one of the main results of the EIS-WIDE effort.

An extensive effort is now underway to validate these candidates at different redshift intervals either by direct spectroscopy, using 4-m class telescopes for the low-redshift ($z \lesssim 0.5$) cluster candidates, or by using optical/infrared color-magnitude relations to identify the red-sequence of early-type cluster galaxies known to exist over the redshift range considered. The latter has been carried out by complementing the I-band EIS data with suitably deep optical data ($z \lesssim 0.7$) and by infrared observations of high-redshift candidates ($z > 0.7$). The identification of the red-sequence is a necessary step for a preliminary "confirmation" and for the selection of targets for subsequent spectroscopic observations with 8-m class telescopes.

These follow-up observations include: **1)** 2dF@AAT test runs of EIS cluster candidates in patch B and D [13]; **2)** EFOSC2@3.6m observations of seven $z \sim 0.5$ clusters [8]; **3)** JKs SOFI@NTT observations of fifteen clusters with $z > 0.6$ [4] [8]; **4)** deep optical observations of two high-z candidates with the VLT-TC [2]; **5)** deep optical observations (VI) of four clusters with the Danish 1.5m telescope [4]; and **6)** deep optical observations (V) of thirteen clusters with the Nordic Optical Telescope [4]. In addition, deep $I-$band imaging of $\sim$ 10 clusters with $z \lesssim 0.5$ will be carried out in the near future for weak lensing studies [15]. Altogether this implies that about one-third of the EIS cluster candidates will have already been followed-up by the end of 1999.

Gathering the information already available from the various groups, the results from these preliminary confirmation efforts can be summarized as follows: **1)** the combination of new spectroscopic measurements, redshift data from the literature, and the presence of color-magnitude relations [6] provide either direct or indirect evidence that most EIS clusters with $z < 0.3$ are likely to be real. Particularly promising are the preliminary results of the 2dF observations for candidates with $z < 0.5$. Test observations (3 hours long), conducted under poor seeing conditions, showed that at least 50%

of the observed candidates have two or more concordant redshifts and are very likely real associations; **2)** new imaging and spectroscopic data for ten candidates with $0.4 < z < 0.6$ provide strong evidence that nine of them are real associations; **3)** optical and infrared color information for fifteen candidate clusters with $z > 0.6$ suggest that twelve of them are real, including five at $z \gtrsim 0.9$.

These preliminary results are extremely encouraging, suggesting that a confirmation rate of about 70% can be achieved. This clearly justifies vigorously pursuing the confirmation effort. The data already obtained have also shown ways to improve the confirmation process.

4 Future Prospects

4.1 Clusters

The work on the EIS clusters is expected to proceed in several fronts which include imaging and spectroscopic confirmation, velocity dispersion measurements, deep imaging for weak lensing studies, and X−ray observations with XMM. Such a program requires a variety of observations that can be conducted at different facilities. For instance, the detection of the early-type red sequence of intermediate redshift candidates ($z \lesssim 0.7$) is possible from optical observations with 2-m class telescopes, while for candidates at higher redshifts infrared imaging is needed. Spectroscopic confirmation of low-redshift EIS candidates ($z \lesssim 0.5$) can be successfully carried out using 2dF@AAT. In fact, by the appropriate selection of galaxies using color information, or whenever possible photometric redshifts, as well as further constraints on the positioning of fibers will certainly increase the success rate of confirmation, and can lead not only to accurate redshift measurements but may also provide a robust determination of the velocity dispersion.

Beyond $z > 0.5$ 8-m class telescopes are required if one expects to build a large sample of confirmed clusters in a timely fashion. To confirm the reality of these candidates requires the observation of at least a few galaxies and the best candidates are those lying along the red-sequence. Even though FORS1 is not the ideal spectrograph, given the geometry of the slits, the observations can be made more efficient by using photometric redshift techniques. Using the optical/infrared data accumulated in the process of determining the red-sequence and those available from the public surveys one can select galaxies over the field-of-view of FORS1 within a small estimated redshift interval about the nominal cluster redshift. The effectiveness of the method has been demonstrated for the two high-redshift candidates observed with the VLT-TC for which $BVIJKs$ color catalogs are available [1]. By selecting galaxies in a photometric redshift bin of 0.2 around the nominal cluster redshift the surface density of galaxies in the field is reduced by a factor of more than five, thereby significantly increasing the chances of targeting other cluster members well-beyond the cluster core and with different SEDs.

It remains to be seen in practice how large the success rate of measuring redshifts for cluster members will be using this method. However, these observations should provide enough data to confirm the reality of the cluster and, perhaps, to obtain a first estimate of the velocity dispersion. More accurate velocity dispersion measurements will require the next generation of VLT instruments (FORS2 and especially the integral field unit of VIMOS), which would greatly benefit from having already a sample of confirmed clusters. It is worth mentioning that the infrared data used to detect red-sequences will also be used to search for higher redshift clusters. Preliminary analysis of the JKs data available, covering about 500 square arcmin, have led to the discovery of at least seven concentrations of red objects with estimated redshifts $1.1 < z < 1.7$, reminiscent of recently confirmed clusters at $z \sim 1.2$ [12] [16]. Finally, deep I-band imaging is required to estimate weak lensing masses. For low-redshift clusters this will require observations with the WFI@2.2, while at higher redshift VLT observations are needed.

If the program outlined above is successful it will naturally provide a unique data set of confirmed high-z clusters, with virial and weak lensing masses, for follow-up observations with XMM in time for the second announcement of opportunities. This concerted study of clusters over such an extended look-back time will provide a unique database to investigate the evolution of clusters and their galaxy population.

4.2 Public Surveys

Recently, the WG for public surveys endorsed the continuation of such programs using the recently commissioned wide-field imager (WFI@2.2) and the NTT for infrared observations. These surveys are expected to be conducted by both the EIS team and outside groups under the same general guidelines adopted for EIS. The two-year plan recommended by the WG envisions a deep five-passband multicolor survey of three square-degrees, complemented by deep infrared observations (JKs) of an area of 1000 square arcmin. It also calls for shallower observation of selected stellar fields in preparation for follow-up observations with the FLAMES spectrograph. In the short-term the WG has also reaffirmed its support for completing the EIS-wide survey in two additional passbands (BV) of the 17 square degrees area covered in $I-$band by EIS.

5 Summary

The initiative taken by ESO and the Member States to carry out a coordinated public imaging survey in a one-year time-scale has been successful providing a vast set of imaging data, for a wide-range of applications, in time for the Science Verification period of UT1 and for the first year of VLT operation. As a by-product the EIS experiment has led to the development of a powerful data reduction pipeline prototype and a survey environment, both essential ingredients for the continuation of other Public Surveys using the WFI@2.2 imager. The target lists prepared by EIS have already been used in a number of follow-up works, in preparation for VLT programs, and for the Science Verification period, thus meeting the primary requirement of the survey. In particular, a significant effort has been made by different groups to investigate the EIS candidate cluster catalog leading to encouraging results which, if vigorously pursued, could lead to the construction of a large sample of high-redshift clusters for observations with the VLT as well as other facilities such as HST and XMM.

The response of the community to EIS has been good, judging by the interest manifested by the number of accesses to the EIS web pages, the amount of data retrieved, the number of follow-up programs undertaken over the past six months and the desire of the WG to continue public surveys. However, a significant effort must still be made to cope with the large volume of data expected from the wide-field imager.

Acknowledgments

I would like to thank all of the past and present EIS team members for their enormous effort and dedication in making it possible to meet the deadlines of the survey. I would also like to thank M. Colless, H. Jørgensen, M. Ramella, R. Saglia and S. Cristiani for reporting their preliminary results. Special thanks to M. Scodeggio, L. Olsen, R. Rengelink and M. Nonino for their contribution to the cluster work reported here and A. Renzini for his support throughout this program. Finally, I thank R. Giacconi for providing the coordinates for the AXAF Deep Field.

References

1. Arnouts, S. & Benoist, C., 1999, *private communication*
2. da Costa *et al.* , 1999, A&A, 343, L29
3. da Costa *et al.* , 1999, A&A, submitted (astro-ph/9812105)
4. Jørgensen *et al.* , 1999, *private communication*
5. Olsen, L.F., *et al.* , 1999a, A&A, in press (astro-ph/9803338)
6. Olsen, L.F., *et al.* . 1999b, A&A, in press (astro-ph/9807156)
7. Postman, M. *et al.* , 1996, AJ, 111, 615
8. Ramella *et al.* , 1999, *private communication*
9. Rengelink *et al.* 1999, A&A, submitted (astro-ph/9812190)
10. Rosati, P. *et al.* . 1998, ApJ, 492, L21
11. Rosati, P. 1998, astro-ph/9810054
12. Rosati, P. *et al.* . 1999, astro-ph/9903381
13. Saglia, R. & Colless, M., 1999, *private communication*
14. Scodeggio, M., *et al.* . 1999, A&A, in press (astro-ph/9807336)
15. Seitz *et al.* , 1999, *private communication*
16. Stanford, S.A. *et al.* . 1997, AJ, 114, 2232
17. Zaggia *et al.* , 1999, A&A, *in press* (astro-ph/9807152)

The Space Missions WAXS/WFXT and SWIFT: New Targets for the VLT

Guido Chincarini[1,2]

[1] Osservatorio Astronomico di Brera, via E. Bianchi 46, I-23807 Merate (LC), Italy

[2] Università degli Studi di Milano, via Celoria 6, I-20133 Milano, Italy

Abstract. At OAB we were, during the year 1998, deeply involved in planning two important space missions for which very large ground based telescopes, VLT in particular, would play a very large and important role in the optical followup. The study of the first mission, Wide Angle X-ray Survey using a Wide Field X-ray Telescope, was coordinated by the Observatory of Brera, involved mainly Italian industries and resulted in a proposal to the Italian Space Agency under the Small Satellite Program. The second mission, SWIFT, has been coordinated and directed by the Goddard Space Flight Center and resulted in the submission of a proposal to NASA under the MIDEX program. The science goal of this mission is the detection and study of the Gamma-ray bursts.

1 Background of the WAXS/WFXT Mission

It is very well known that the clue to the understanding the formation and evolution of Clusters of Galaxies resides in the detection and detailed study of clusters at redshifts larger than 0.5 and up to a redshift z$\sim$ 2 assuming they still exist and are recognizable. It is also well known that the detection of distant clusters by optical surveys is very limited because of the confusion of the galaxy background at faint magnitudes and the fading in that of the member galaxies. Indeed that had been clearly discussed some time ago by Cappi et al. (1989) and more recently evidenced by the galaxy surveys carried out by Postman et al. (1996) and by the ESO Survey (EIS), these proceedings. On the other hand while the X-ray cluster detection is a much cleaner method due to the net separation at any redshift between an extended emission and a point like source over a rather faint X-ray background, a good resolution of the optics is a must. Possibly the X-ray telescope should allow a resolution better than 15 arcsec Half Energy Width (HEW). A good sensitivity is also needed since the surface brightness of even bright clusters dims considerably with distance. It has been shown, especially by Bahcall and Soneira (1983) but see also the most recent results obtained with the REFLEX project, Guzzo et al. (1999), that clusters are unique tools in describing the statistical characteristics of the Large Scale Structure. While shortly the Power Spectrum for the local Universe will be quite nicely measured by REFLEX, it remains fundamental to verify if the theoretical prediction are confirmed at redshifts z$\sim$ 1 and larger. Statistical information on much higher redshifts could be obtained only by a deep AGN sample at z> 2.5. To create a statistical significant catalogue we need to survey a very large and fairly deep area of the sky. Following ROSAT observations and hydrodynamics simulations, furthermore, it has been shown that groups of clusters are interconnected by reasonably hot matter and galaxies which may extend, at lower temperature however, in the Intergalactic Medium forming an interconnected net spreading about the Universe in the form of filaments (see Chincarini and Rood 1980, and references therein, for an early discussion of this concept). Clusters, during their lifetime, will be accreting matter, gas and galaxies, from these filaments and from the surrounding environment and this will characterize their properties at various redshifts. It becomes very important, therefore, to have a rather large field of view in order to observe simultaneously, and with comparable good resolution, the clusters and their close environment. To optimize this research sometime ago we started to design, and plan, an X-ray space mission capable of optimizing the achievement of the science goals. The requirements, derived directly by what has been briefly mentioned above, were: 1) X-ray optics with good resolution over a large field of view, 2) reasonable good sensitivity over the energy range 0.2 – 2.5 keV with some sensitivity up to about 7 – 8 keV and 3) high accuracy in the image reconstruction (better than 2 arcsec) and absolute astrometry better than 2 – 3 arcsec, this being very important for AGN especially. This idea, originally triggered by Riccardo Giacconi, grew interest in the international astronomical community and we started long ago to refine the details and study the spacecraft and instrument requirements. The Astronomical Observatory of Brera got deeply involved not only in the definition of the Science but, with the support of ASI and under the technical direction of Citterio, started to develop and refine the design of a wide field optics and its construction. More recently the effort of a rather large international collaboration came up with a detailed study of phase A submitted to ASI by the end of 1998. I refer for a detailed discussion of the science and of the project to the proposal, and references therein, we submitted (about 250 pages) and to the paper presented by Chincarini (1999) on behalf of the collaboration. In this brief communication I will limit myself to describe the very recent results obtained with the last optics we made and measured at the Marshall Space Flight Center X-ray facility since this, by itself, is a major result in the fabrication of high quality X-ray optics. This is a break through equivalent to what occurred when we discovered the design, and had the capability of making, the Ritchey Chretien mirrors in place of the standard parabola and hyperbole.

2 The Performance of the X-Ray Optics

The results we are going to illustrate have been obtained with a 60 cm wide shell made with a carrier of Silicon Carbide. The optical surface is achieved by filling the gap between the super-polished mandrin and the carrier with epoxy resin. The reflecting surface is made out of gold. The fastest way to illustrate the result is by Figure 1. The two horizontal lines indicate the minimum acceptable resolution for the mission ($\leq$ 20 arc sec) and the goal ($\leq$ 15 arc sec). The two continuous lines at the bottom of the Figure refer to the optical design of the mirror. The design of the optics is obtained by optimizing the resolution over about 1 degree field of view and that is the input for the construction of the super-polished mandrin whose resolution, after construction, is given by the continuous line running between 6 and 8 arc sec HEW. Obviously no replica can be better than the master mandrin. To illustrate the performance of a classical Wolter I optics we dashed the resolution as a function of the distance from the optical axis for the XMM telescope. As it is well known while on axis and within 10 arc minutes the resolution is good, it deteriorates drastically at larger distance from the optical axis. That is over a field of 1 square degree the ratio (Area with good resolution)/(Area with poor resolution) $\sim$ 0.16. Above 20 arc min off center the resolution is not even acceptable. The WFXT optics, the various replica have been obtained using always the same mandrin, have been tested at the PANTER facility of MPE (Germany) and at the Marshall facility (USA). The filled points refer to the first shell at 1.5 keV energy while the filled square refer to the tests obtained at 0.5 keV. This is already a great result! It shows, but we had already some tests showing this, that we are very close to reaching the target goal and that we are capable of making optics having good resolution over a large field of view. We gain, compared to the Wolter I optics above, 84% of the field! Excellency, however, was reached testing a new shell, Sic2. In the making of this we obviously used of the manufacturing experience we gained with the previous ones. Here at 0.1 keV (we can measure at this low energy only at Marshall) we measured an HEW of about 10 arc sec, stars in Figure 1 (due to a minor miss-alignment the measurement at -30 arc min is actually more than 30' off axis). This test was very significant for an other aspect. The quality of the images deteriorates somewhat going to higher energy due to the development of wings in the profile which are due to scattering. By reducing the number of replica per mandrin and by caring somewhat more in the process, that can be strongly reduced.

With this result and with the phase A study on the focal plane (SAO and Leicester), telescope and spacecraft we have shown that we are ready for a fantastic mission because of the goals we can achieve in cosmology and in other astronomical studies and because of readiness. Whether it will fly or not will depend on the budget and on the vision, or strong limitations and cultural bias, of those who will evaluate it.

3 SWIFT - A Panchromatic Gamma Ray Burst MIDEX Mission

In the limited time, and space, at my disposal I can not make justice to such an excellent mission. I feel however obliged to mention it, more information

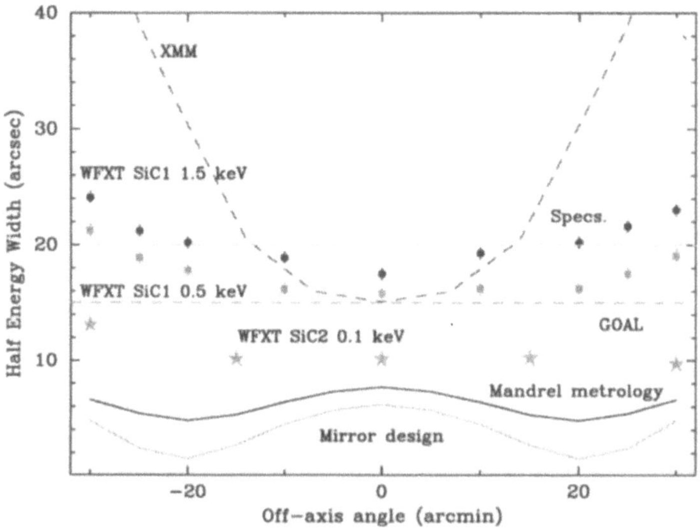

Fig. 1. Resolution of the X-ray optic vs. off-axis angle

can be obtained from *http://swift.gfsc.nasa.gov/*, because in the coming years I have no doubt that the VLT will play a fundamental role in the follow up studies of these very energetic events. A second motivation is that during this meeting, Gamma ray bursts were not mentioned at all.

The discovery by BeppoSAX of afterglow from Gamma-ray bursts has revolutionized our understanding of these events and opened an extremely interesting field of endeavor which will keep busy the science community for years to come. We are dealing with extreme energetic events, 10^{51} - 10^{54} ergs, which occur, in a very short time. The understanding of the phenomenon requires detailed observations of the burst itself and, whenever possible, of the faint hosting galaxy. It is of paramount importance to observe at various wavelengths soon after the burst has been detected. SWIFT satisfies to this need. The best way to illustrate the capability of the planned mission is to mention briefly the mode of operation. The spacecraft host three instruments. A Burst Alert Telescope (BAT) sensitive in the range 10 - 150 keV and capable of determining the GRB location to about 4 arc minutes. It has a detecting area of 5200 cm^2 and a Field of View of about 2 sr. An X-Ray telescope (XRT) sensitive in the range 0.2-10keV and capable of locating the afterglow with an accuracy of $\sim$ 2 arc sec. The effective area at 1.5 keV is 110 cm^2. An Ultra Violet Optical telescope (UVOT) which can observe in 6 colors (from 170 nm to 650 nm). The aperture of the telescope is 30 cm with a pixel scale of 0.5 arc sec.

The winning strategy is fast response and communication. Soon after the burst (10 sec) the slew begins and the BAT location is transmitted to Earth. After about 55 sec in addition to having the BAT observation we plan to have the XRT location and image which, within 10 sec, is send to Earth together with the BAT light curve. If visible by UVOT this instrument will gather data in parallel with the other instrument and send a 2×2 arc minutes^2 finding card to the ground. It will take about 1200 sec to get the XRT spectra and 7200 sec to have a set of UVOT observations in the different filters. Etc. The strategy and the mission design satisfy the two main requirements mentioned above: immediate accurate pointing of the spacecraft after the detection of the burst and immediate alert to the ground to enable the medium size and large telescopes to initiate the follow-up observations. In this role the VLT, and other ground-based telescopes, will play a fundamental role in measuring those quantities that allow the understanding and modeling of the phenomenon. Since these objects are at cosmological distances, indeed it may be a way to know about the Universe at very large redshifts, the observations of the host galaxy, when detected, is a must. These observations must be carried out with 4 m and 8 m class telescopes, however they could be planned and will not be part of a TOO program. The SWIFT mission has been selected for the Phase A Study and the final evaluation is due in September 1999. Once selected the mission will fly in 2003.

4 Acknowledgments

I would like to thank the OAB staff who largely contributed to the continuous development of the X-ray optics at OAB: Conconi, Ghigo, Mazzoleni, Bergamini, Cantu', Casiraghi, Crimi, Garignani, Salini and Valtolina. The Institutes involved in the hardware study of phase A for **WAXS/WFXT** were: Brera Astronomical Observatory (OAB), Smithsonian Astrophysical Observatory (SAO), University of Leicester (UL), Max-Planck-Institute für Extraterrestrische Physik (MPE). The Italian Industries which participated: ALENIA, MEDIA-LARIO, TELESPAZIO, OFFICINE GALILEO. P.I: G. Chincarini (OAB); co-PI:S. Murray (SAO); Co-I J. Trümper (MPE), A. Wells (UL), Telescope PI: O. Citterio (OAB); PM: G. Tagliaferri (OAB); PS: S. Sciortino (OAPA). Six panels headed by S. Sciortino coordinated all the work for the science proposal. The panels were chaired by: Colafrancesco (LSS), Chincarini (Clusters of galaxies), Zamorani (AGN), Forman (Galaxies) Pallavicini (Stars) and Watson (Compact Objects). Contribution to this work came from Antonuccio-Delogu, Arnaboldi, Bandiera, Bardelli, Böhringer, Bonometto, Borgani, Campana, Catalano, Cavaliere, Conconi, Covino, Della Ceca, De Grandi, De Martino, Fiore, Garilli, Ghisellini, Giacconi, Giommi, Girardi, Giuricin, Governato, Guillout, Guzzo (who revised the final version of the science proposal), Iovino, Israel, Lazzati, Le Fevre, Longo, Maccacaro, Maccagni, Mardirossian, Matarrese, Micela, Molendi, Molinari, Moscardini, Murray, Norman, Perola, Osborne, Pye, Ramella, Randich, Robba, Rosati, Scaramella, Stella, Stewart, Tagliaferri, Tozzi, Trinchieri, Vikhlinin, Vittorio, Wolter. Members of the following Institution expressed direct interest in the mission: ASI (SDC), Bologna (OA), Catania (OA), Firenze (OA), Milano (IFCTR, UNI), Napoli (OA), Padova (UNI, OA), Palermo (OA, UNI), Perugia (INFN), Roma (OA-UNI), Trieste (OA), Marsiglia (CNRS, LAS), Baltimore (IHU), Cambridge (MIT), Princeton (UNI), Munich (ESO, MPE), Copenhagen (DSRI).

Our involvement in the **SWIFT** mission came following the proposal made by Nick White to use the spare optics of the JETX Telescope (SRG Mission) for a MIDEX proposal. The mission benefits of the extensive experience of scientists at GSFC, Penn State University, Los Alamos National Laboratory, Berkeley, Mullard Space Science Laboratory (UK), Leicester University, Brera Observatory, Italian Space Agency (ASI for the Malindi Station). The SWIFT industrial partner Spectrum Astro. Neil Gehrels (Goddard) is the Principal Investigator, Tim Gehringer (Goddard) the Project Manager. At present participate to the mission the following co-Is/Lead representatives: N. White, J. Nousek, L. Angelini, S. Barthelemy, T. Cline, R. Corbet, K. Jahoda, F. Marshall, R. Mushotxky, J. Norris, D. Palmer, A. Parsons, A. Smale, J. Tueller, W. Zhang, L. Whitlock, D. Burrows, G. Garmire, S. Horner, P. Mezaros, E. Feigelson, A. Wells, M. Turner, M. Ward, R. Willingale, K. Mason, M. Cropper, O. Citterio, G. Chincarini, G. Tagliaferri, P. Caraveo, L. Stella, M. Vietri, E. Fenimore, K. Hurley, F. Lebrun, J. Paul, B. Paczynski, F. Cordova, T. Sasseen, L. Cominsky; and the following Associate Scientists (Follow-up Team): M. Boer, M. Busby, A. Cimatti, M. della Valle, A. Filippenko, D. Frail, G. Ghisellini, P. Madau, B. Margon, M. Metzger, H. Pederson, P. Saracco, B. Schaefer, D. Schneider, I. Smith, C. Stubbs, C. Thompson, F. Vrba.

We refer to the proposal and web information for the various roles of the participants to this fantastic mission. This is designed to serve all of the international community, the data will be public and working groups will be created for the follow-up research, interested in this new and fascinating field of endeavor.

References

1. Bahacall, N., & Soneira (1983) Ap.J., **270**, 20
2. Cappi,A., Chincarini, G., Conconi, P., Vettolani, P., 1989, A&A, **223**, 1
3. Chincarini, G. (1999) astro-ph 9902184. To appear in the proceedings of "Observational Cosmology, The development of galaxy systems".
4. Chincarini, G. and Rood, H. (1980) Sky & Telescope, **59**,364
5. Guzzo,L., Boehringer, H., Schuecker,P., Collins, C.A., Schindler, S., Neumann, D.M., De Grandi, S., Cruddace, R., Chincarini, G., Edge, A.C., Shaver, P.A., Voges, W. (1999) "The ESO Messenger", **95**, 27.
6. Postman, M.,Lubin,L.M., Gunn,J.E., Oke,J.B., Hoessel, J.C. et al. (1996) A.J., **111**, 615

http://www.merate.mi.astro.it/wfxt.html
http://swift.gsfc.nasa.gov/

The VST – VLT Survey Telescope

Magda Arnaboldi[1], Massimo Capaccioli[1], Dario Mancini[1], Roberto Scaramella[2], Giorgio Sedmak[3], and Richard Kurz[4]

[1] Osservatorio Astronomico di Capodimonte, Napoli, Italy
[2] Osservatorio Astronomico di Roma, Monteporzio, Italy,
[3] Università di Trieste, Trieste, Italy
[4] European Southern Observatory, Garching bei München, Germany

Abstract. We describe the telescope concept of the VST – the VLT survey telescope – and the scientific and instrumental background which lead to the kick-off of the project for the realization of a 2.6 m telescope for wide field to be placed at Paranal, on the VLT plateau within the year 2001. The project for the camera and the data handling and archiving, plus the analysis pipeline are outlined, with particular emphasis on the possible scientific surveys and stand alone projects. We will briefly review some of the scientific projects for which the VST was conceived, and its impact on the VLT science.

1 Introduction

The VST – VLT survey telescope – will be a 2.6 m reflector devoted to wide field imaging. It is part of a joint venture between the Astronomical Observatory of Capodimonte (OAC) and the European Southern Observatory. This telescope is designed for Cassegrain operations, with a corrected field–of–view (FOV) diameter of 1.47 degree, to be matched with a 16k$\times$16k CCD mosaic camera, with a 15 μm pixel of equivalent size of $0.21''$. This instrument represents a change of philosophy with respect to the general multipurpose telescopes and is more flexible than single purpose telescopes such as in the SLOAN project. The VST is a very focussed experiment, so that the design is optimised for the wide field imaging, with a fast realization track (3 years) and a given operational time span ($\approx$ 10 years).

The VST main use will be as an ancillary telescope to the VLT and, together with the secondary units for interferometry, it will be the only additional telescope to be placed on the Paranal plateau. The VST will be capable of yielding significant scientific returns on its own, thanks to unparallelled (optical + site) quality $\times$ image size ([1],[2]). It will also pose the astronomical community with a challenging data flow, which will require very specific procedures and data centres.

2 Telescope Description

2.1 Mechanics

The telescope has an Alt-Az mounting, which allows a high mechanical stiffness and a compact overall structure. The telescope enclosure will be designed in collaboration with ESO. The final design of the enclosure will be compliant with a number of nights lost due to wind of about 10% during a year of operation, based on the wind statistics of the Paranal site. The overall tracking accuracy will be very closed to 0.05″ rms and it will be improved by means of a field stabilisation device. The VST telescope control architecture follows the concepts developed for the VLT by ESO; in doing so, the system coherence is increased, an easier maintenance program is possible, and the integration into the existing ESO hardware and software environments is ensured.

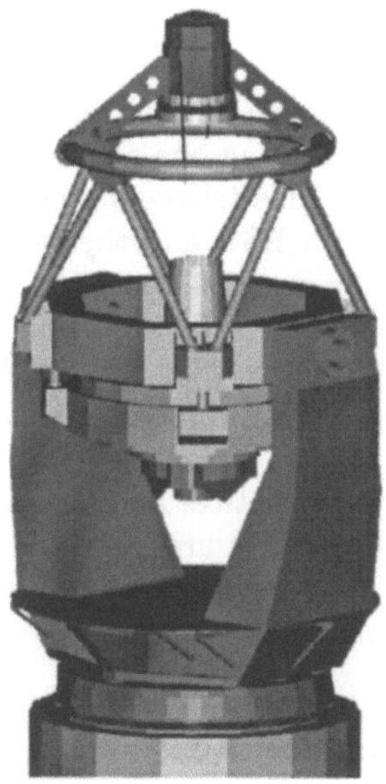

Fig. 1. CAD 3-D view of the VLT survey telescope

2.2 Optics

On March 1998, Zeiss Jena was awarded the realization of the VST optics. The telescope design is a modified Ritchey-Chretien, with a f/5.5 focal ratio. The circular diameter of the corrected field of view is 1.47 degree, envisaging the use of a one square degree detector. The wide-field corrector is of a new design: it will cover the whole visual wavelength range, from U to I, with an encircled energy of 80% in a 1.7 pixel or better, and a maximum distortion of 0.01%, which is much smaller that the median seeing at Paranal of 0.7″, i.e. more than 3 pixels. The mirror surfaces and the lens for the wide field correctors will be polished to a precision 2nm rms.

The new design for the corrector foresees two modes of operations: the first for nearly zenithal observations (e.g. in U band) uses two lenses, the second uses a lens plus an atmospheric dispersion corrector (ADC) when observing in the B to I bands at large zenithal distances. The primary and secondary mirrors will be active. A Shack-Hartman wavefront sensor on line will provide the data to an active optics corrector. The M1 cell will support the M1 mirror via 80 active actuators. The motion of the secondary mirror will be controlled by a hexapode system. The primary duties of this system is to correct all secondary mirror de-centring, de-focusing and coma-tilting phenomena during telescope operations.

2.3 VST Camera

On April the 4th, 1998, ESO issued a call for tender for the realization of the 16k×16k CCD camera, covering one square degree, to be placed at the Cassegrain focus of the VST. This telescope needs a peculiar camera of new design. OAC collaborated with ESO and the Max Planck Institute für Astronomie in Heidelberg in the construction of the 8k×8k camera just placed at the ESO 2.2 m telescope in La Silla, and, thanks to special funding from the Italian Consortium of Observatories (C.N.A.A.), is a member, together with the Padua Observatory, of a European Consortium which will build the 16k×16k camera for VST.

The VST is conceived as an optical and also a near infrared (IR) survey telescope: some European institutes are now studying an IR camera feasibility, notwithstanding the major obstacle at present time, which is given by the high price of the IR arrays.

2.4 VST Data Handling and Analysis

Although conceptually simple, the operation of VST is made complex from the huge amount of information collected every clear night. Indeed each frame will consists of 256 Mpix, with roughly 50 to 100 frames per night (included calibrations), and therefore one has to be prepared for a huge data flow of some tera–pixels per year. In order to do so, on the OAC side, a national project financed by the Italian Ministry of University and Research has started the development and implementation of the data reduction and analysis pipelines, and very recently an official collaboration has started with a similar French project, TERAPIX, based at the Institute d'Astrophysique in Paris.

3 VST Main Scientific Goals

3.1 VST Performances

The exploitation of the VLT requires catalogues of objects and supporting observations obtained at other smaller telescopes because of costs and efficiency reasons. Indeed, the typical one hour spectroscopic limit at low resolution of an 8 m telescope, R $\approx$ 25, can be well reached by VST in a similar exposure time.

Indeed, notwithstanding a collecting area $\sim$ 10 times smaller than VLT, for several imaging projects the VST gain is due to its large image area: for instance the first VLT instrument, FORS1, has an imaging field of $6' \times 6'$, resulting in a hundred time smaller area. For survey work, the relative gain of VST is a factor 10, at a much cheaper cost per photon.

Moreover, excellent image quality and sampling is essential for object separations, cross identifications, and astrometric solutions which are needed at very high accuracy in order to exploit the spectroscopic capabilities of the VLT. Placing the VST at Paranal and the strong requirements on both the VST optics and the camera are such that the above goals will be fulfilled.

The telescope performances expected for the VST are based on the optical specifications and on the efficiency curve for the CCDs used for the 8k$\times$8k camera at the ESO 2.2m telescope and a detector area of 1 deg^2. We give here some examples of the telescope performances.

The study of large scale structures in the universe and galaxy evolution requires galaxy catalogues extended over fairly large areas, which can be obtained through multicolour imaging. A very wide survey (several hundred square degrees) with a limiting magnitude in V of 25.5 AB mag arcsec^{-2}, in at least two bands, can fully exploit the capabilities of the VST. In 30 min exposures one reaches $Vw_{lim} \sim 25.5$ AB mag arcsec^{-2}, which could translate to $Vw_{lim} \approx 24$ for galaxies (one also has $Iw_{lim} \sim 24$ AB mag arcsec^{-2}, so $Iw_{lim} \approx 22.5$ for galaxies). So, a reasonable estimate can be $\approx$ 50 hrs of exposures per band per 100 deg^2. If one has about 6 effective hours per night, then we need 17 nights to do accurate photometry in two bands over 100 deg^2.

Furthermore, for the typical magnitude limit of wide area imaging ($R \simeq 24$), one would measure the photometric parameters and colours for several million galaxies, including a large number of low surface brightness galaxies whose properties are still largely unknown. Thanks to the very high throughput of VST in the UV (twice that of the ESO 2.2m), large samples of high-z candidate galaxies can be selected through the Lyman break technique. If 1/3 of a year observing dark time can be devoted to deep multicolour imaging (5 bands, $Iw_{lim} \approx 26$ AB mag arcsec^{-2}), one could cover at least 5 deg^2 and obtain a sample of $\sim$ 8000 candidates with z larger than 3.

High performances are expected also for narrow band imaging with the VST. As an example, the limiting [OIII] 5007 flux at the VST for an emission line stellar object like planetary nebulae (PN) is 4×10^{-18} erg cm^{-2} s^{-1} for 6×2000s exposures (at the 3σ level): this flux is 10 times smaller that the flux we can detect today with old generation 4 m telescopes and detectors for the same exposure time. The limiting [OIII] 5007 flux of 6×2000s exposures at the VLT plus the multi-object spectrograph VIMOS (R=2500) is $\sim 5 \times 10^{-18}$ erg cm^{-2} s^{-1}: by comparison with the limiting flux in this wavelength range at the VST, we can conclude that such a facility is the ideal instrument to produce catalogues for extragalactic PN candidates in the Virgo and Fornax cluster for spectroscopic follow-up at the VLT.

As an example of a self-contained project for the VST, we mention the monitoring of the amplification of a star light due to microlensing events. With a continuous monitoring for a few months of the events towards the Galactic bulge one hopes to find signatures for the presence of extra-solar-planets. Again wide area and excellent image quality make VST a unique instrument for this purpose.

4 Conclusions

The VST quality parameters and the excellent site, the Paranal plateau, make today this telescope the most competitive project for WFI in the Southern hemisphere. The need for an ESO WFI facility such as the VST is very strong primarily as a complementary tool for the VLT, but also as an independent survey tool. The ESO community needs a survey facility such as the VST in order to produce the databases which are essential to achieve excellence in the science of the VLT era.

References

1. Arnaboldi, M., et al. 1998, The Messenger, 93, 30
2. Arnaboldi, M., et al. 1998, Wide Field Surveys in Cosmology, 14th IAP meeting held May 26-30, 1998, Paris. Publisher: Editions Frontieres. ISBN: 2-8 6332-241-9, p. 343.

PARALLEL WORKSHOP 2:

STAR-WAY TO THE UNIVERSE

Edited by

Rolf Kudritzki and Luca Pasquini

KUEYEN: A Unique Machine for Stellar Studies

Luca Pasquini

European Southern Observatory, Karl Schwarzschild Strasse 2,
D-85748 Garching Bei München, Germany

Abstract. The VLT telescope KUEYEN (formerly UT2) will be a unique facility for stellar study thanks to the coupling of high resolution spectroscopy and multiplex capabilities with an 8 meter telescope.

FORS2, UVES and FLAMES will produce data which will in principle be able to solve several current outstanding problems in stellar astrophysics. On the other hand, when considering both the quantity and the quality of the data produced, the full exploitation of the facilities offered at this telescope requires a novel and well planned approach by the users.

1 Introduction

8 meter class telescopes and the VLT in particular represent a major step in the advancement of observational astrophysics. While emphasis is often given to the impact of these telescopes and their instrumentation on observational cosmology and extragalactic astronomy (cfr. the contributions to the plenary session of this meeting), little doubt exists that the VLT will have a tremendous impact also on stellar astrophysics. In particular the instrumentation of KUEYEN (formely UT2) will be especially suitable to stellar studies, thanks to the combination of 4 key characteristics:

1. High Resolution Spectroscopy
2. High Multiplex
3. High Performance
4. Fed by the VLT!

The scheme of the instrumentation on KUEYEN foresees three instruments. On Nasmyth B the UVES high resolution spectrograph will be offered from April 2000, at the same time as the Cassegrain instrument FORS2. The Nasmyth A platform will host the FLAMES facility, which will initially feed the GIRAFFE and UVES spectrographs and it will have the capability of being upgraded to feed a third (TBD) spectrograph. FLAMES will be offered in the second *half of 2001.*

2 Instruments Summary

FORS2 is very similar to FORS1, the focal reducer and low resolution spectrograph successfully installed at ANTU (Appenzeller et al. 1999); the main differences between these two instruments being that FORS2 will be equipped with an echelle which will allow to obtain a resolving power of 5000 and that it will have a multislit (MOS) capability to acquire up to ~50 spectra simultaneously, compared to the maximum of 19 slits with FORS1.

UVES is a high resolution spectrograph, with two (Blue and Red) optimized arms and with a resolving power - aperture product $R \times \alpha = 40000$. A resolving power of up to 120000 can be obtained in the Red arm by using image slicers. Observations can be carried out in the Red or in the Blue arm, or simultaneously in both arms if the dichroic mode is chosen.

Thanks to the optimization of the components of the two arms (optics and detector), UVES is very efficient over the whole 300-1000 nm spectral range. This high efficiency will allow to gather a S/N of 10 on stars of V~20 or a S/N of 100 on stars of V~16 in a 2 hours observation.

FLAMES is a fibre facility composed of a Nasmyth corrector, a fibre positioner and the fibre links to the GIRAFFE and UVES spectrographs. The main characteristics of FLAMES are summarized below:

- Field of View: 25 arcminutes diameter
- Capability of positioning 130 fibres for next observation in less than 20 minutes while simultaneously observing
- Minimum separation between objects: 10 Arcseconds
- 8 one arcsecond fibres feed the Red arm of UVES, with a R=40000
- 132 fibres feed GIRAFFE in Medusa mode, each with a 1.2 arcseconds aperture; with the two GIRAFFE gratings R=15000 and R=5000 can be obtained
- 15 deployable Integral Field Units feed GIRAFFE in IFU mode. Each IFU is composed of 20 microlenses, each of 0.5 arcseconds aperture for a total sky coverage of ~ 2×3 arcseconds, and R=22000 and R=9000
- 1 GIRAFFE Integral field unit (ARGUS): 300 microlenses will provide a field of 7×11 arcseconds and R=22000 and R=9000 with a 0.5 arcsec sampling. A focal enlarger will also allow a 0.3 arcsecond sampling.
- Simultaneous GIRAFFE and UVES observations will be possible.
- FLAMES will be offered to users in Sept 2001 in UVES and GIRAFFE Medusa mode, two months later with the GIRAFFE Argus and IFUs modes.

Although it is not possible to easily summarize the capabilities of the three KUEYEN instruments, Table 1 gives an overview of the multiplex and highest resolution combination available.

Table 1. In this table some characteristics of the KUEYEN instruments are summarized. In column 2 the multiplexity degree is indicated, in Column 3 the aperture on the sky corresponding to the maximum resolving power (given in Column 4). Note that the highest resolving power on UVES Red is obtained through the use of an image slicer. In the last column the approximative wavelength coverage per setting is given.

Inst/Conf	Mult.	Slit (Arcsec)	R_{Max}	Coverage (nm)
UVES Blue	1	0.5	80000	80
UVES Red	1	0.3*	120000	200 or 400
UVES Dichr	1	0.5	80000	280 or 480
FORS2 Slits	19	0.5	5000	
FORS2 MOS	50	0.5	5000	
GIRAFFE Medusa	132	1.2	15000-5000	$\lambda/19$ - $\lambda/7$
GIRAFFE IFUs	15	0.52	22000-9000	$\lambda/19$ - $\lambda/7$
'3D'	20 each	2×3		
GIRAFFE Argus	1	0.52/0.3	22000-9000	$\lambda/19$- $\lambda/7$
'3D'	300	7×11/4×6.3		
FLAMES+UVES Red	8	1	40000	200-400

3 Scientific Applications

Thanks to the combination of the characteristics summarized in Table 1 a number of outstanding observational problems will not only be tackled, but solved by KUEYEN.

By 'solved' in this context is meant that the observational data will be of sufficiently good quality and quantity that no further data should be required to answer the astrophysical questions as they are presently formulated; surely more questions will arise as a result of these observations.

A non-exhaustive list of these astrophysical topics is given below, and, since most of them are discussed by other authors in this volume, the reader may refer to these contributions for a detailed discussion.

- Bulge Chemical Composition
- Light Element Abundances (Be, Li, Li^6/Li^7)
- Abundances of the most metal poor stars
- Chemical composition and dynamics of galactic Globular and Open Clusters
- Magellanic Clouds: Age and chemical composition of Field and Clusters, Star Formation History (SFH), Metallicity dependence of RR Lyr P-M relationship.
- Metallicity, SFH of the Sagittarius dwarf galaxy
- Local Group galaxies: Chemical composition and dynamics
- Accurate distances of clusters and MCs with Eclipsing Binaries

- Evolution of O-A stars in Local Group galaxies; abundances and mass losses along the tracks
- Stellar parameters, mass losses of central stars of planetary nebulae
- Evolution of massive stars as function of metallicity, dating starburst at high Z
- Rotation of white dwarfs from narrow NLTE core of Balmer lines
- Physics of Cepheids, calibration of Baade-Wesselink method.
- Hot stars beyond the local group: chemical composition, distances

3.1 Science Case: 47 Tuc

In order to illustrate the possible observational strategies and the potentiality of KUEYEN, it is instructive to briefly analyze a well established science case.

47 Tuc is one of the best studied Globular Clusters and it has very interesting characteristics: for instance it probably belongs to the old cluster generation, even if it has a quite metal rich (Fe/H ~-0.7) population (see i.e. Hesser et al. 1987). The turnoff is located at V~17.5, which implies that main sequence stars are well reachable with UVES and GIRAFFE.

By studying a large number of stars with high resolution and high S/N ratio observations, it will be possible to resolve many outstanding problems which affect this cluster and have deep implications in many branches of modern astrophysics. For instance, in 47 Tuc evidence exists of chemical differences (CH and CN) among giants *and* among main sequence stars (see i.e. Cannon et al. 1998). What is producing the observed spread? Do other chemical elements show these differences? Does the chemical composition correlate with the spatial distribution of the stars ?

Also, thanks to its combination of old age and high metallicity, 47 Tuc is a very interesting object for the study of the Li evolution and big bang nucleosynthesis (Pasquini and Molaro 1997); what is the Li abundance in turnoff stars? Is that higher than the Li plateau (Spite and Spite 1982) indicating Li production in the early galaxy? Does a dispersion in Li abundance exist among main sequence stars of 47 Tuc with similar temperatures?

Also, after Hipparcos, several authors have investigated the ages of globular clusters calibrating them with nearby subdwarfs (Gratton et al. 1997, Pont et al. 1998). This analysis is based on the assumption that the derived chemical abundances do not suffer from systematic effects when derived for main sequence and giant stars (cfr. i.e. Pont these proceedings). Do systematic effects exist in deriving metal abundances in dwarfs and giants ? An homogeneous data set covering from the RGB to the TO of a Globular Cluster will provide the ideal test.

In order to answer these questions a set of high resolution, high S/N ratio spectra for a large number of stars belonging to the cluster and spatially distributed is required. In the following I assume that 450 stars, divided in several magnitude bins from the tip of the RGB to 0.5 magnitudes below the TO are observed. I also assume that 3 GIRAFFE settings are enough to derive the most relevant elemental abundances (Li, Fe, Ca, C, N, Mg, Na).

- Sample: 450 stars with $14 < V < 18$: 100 stars $14<V<15$; 100 stars $15<V<16$; 100 stars $16<V<17$, and 150 with $17<V<18$.
- 110 stars in GIRAFFE MEDUSA mode are observed simultaneously (leaving 22 fibres for sky recording), with a minimum of S/N=60 per resolution element.
  This translates into approximately:
  6×0.5 Hour for the $V<16$ 200 stars
  3×1 Hours for the 100 $16< V <17$ stars
  6×3 Hours for the 150 $17< V <18$ stars
  for a total of 24 observing hours, or less than 4 VLT UT2 nights and a total of more than 2000 spectra.
- In addition, simultaneously $\sim$ 120 UVES Red Spectra at higher resolution can be obtained. These spectra can be used to determine a more detailed and thorough analysis and/or to validate the results obtained from the GIRAFFE lower resolution spectra.
- By distributing in time the observations, at least 3 spectra at different epochs are obtained, thus providing a good statistics on binary stars and dynamics of this globular cluster as an additional bonus. If, finally, information on the dynamics of the core is required, this could be obtained with one frame of GIRAFFE in Argus mode, or an additional 0.5 hour of observing time.

What clearly emerges from the previous discussion is that with a reasonable number of KUEYEN nights it will be possible to observationally solve a number of outstanding unsolved questions.

What is also relevant, is that when combining the number of targets, the S/N ratio and resolution required, this research seems impossible with any telescope other than KUEYEN.

4 Implications for Astronomers

The example in the previous section has shown how in a few nights KUEYEN will be able to produce thousands of spectra.

Since in typical stellar high resolution work, on the other hand, a detailed analysis is required in order to retrieve the astrophysical information, this large amount of spectra calls for an effort towards the automatization not only of the data reduction (which will be provided by ESO), but also of the data analysis.

In Table 2 the amount of STELLAR spectra produced by KUEYEN in the typical 10 years lifetime of the instruments is estimated. It is assumed that the KUEYEN nights dedicated to stellar work are of the order of 160/year and that a mean of 6 science exposures per night are acquired.

Table 2. Number of stellar spectra produced by KUEYEN instruments in 10 years of use; for assumptions see text. The UVES spectra in FLAMES mode are considered to be acquired simultaneously to the GIRAFFE spectra.

Inst/Conf	Mult.	Nspectra/Night	Nights/Yr	NYr	Total
UVES	1	6	80	10	4800
FORS2 SLIT	19	6	10	10	11400
FORS2 MOS	50	6	10	10	30000
GIRAFFE Medusa	120	6	50	10	360000
GIRAFFE IFUs	15	6	10	10	9000
FLAMES+UVES Red	8	6	60	10	28800
			160	10	444000

Even if this simple prediction may be off by a factor 2, it clearly shows that it will be possible to obtain with KUEYEN $\sim$ 0.5 million stellar spectra in 10 years.

In addition FLAMES will require accurate relative astrometry (0.2 arcseconds) for targets, guide and fiducial stars, and this has to be provided by the observers. One very positive note in this sense is that the 2.2m WFI recently installed (Baade et al. 1999) at the ESO-MPA 2.2m telescope at the La Silla observatory has a field of view well matching the one FLAMES, and a request for a survey of FLAMES stellar field has been recently submitted to the Working Group for Public Surveys (The Messenger **94**, 31).

Nevertheless, it is clear that both the observation preparation and analysis of such an amount of high quality data will require an unprecedented effort by the ESO stellar community;

while KUEYEN will represent a quantum leap in our capability of understanding many stellar astrophysical problems, we need to adjust ourselves to KUEYEN.

References

1. Appenzeller, I., Fricke, K. et al. (1998) The Messenger **94**, 1
2. Baade, D., Meisenheimer, K. et al. (1998) The Messenger **93**, 13
3. Cannon, R.D., Crocke, B.F.W, Bell, R.A., Hesser, J.E., Stathakis, R.A. (1998) MNRAS **298**, 601
4. Gratton, R.G., Fusi Pecci, F. et al. (1997) ApJ **491**, 749
5. Hesser, J.E., Harris, W.E. et al. (1987) PASP **99**, 739
6. Pasquini, L., Molaro, P. (1997) A&A **322** , 109
7. Pont, F., Mayor, M., Turon, C., Vandenberg, D.A. (1998) A&A **329**, 87
8. Spite, F., Spite, M. (1982) A&A **115**, 357

Improvement of Cepheids as Distance Indicators

Wolfgang P. Gieren[1], Pascal Fouqué[2], and Jesper Storm[3]

[1] Universidad de Concepción, Departamento de Física, Astronomy Group, Casilla 160-C, Concepción, Chile
[2] Observatoire de Paris, Section de Meudon, DESPA, F-92195 Meudon Cedex, France; and European Southern Observatory, Casilla 19001, Santiago 19, Chile
[3] Astrophysikalisches Institut Potsdam, An der Sternwarte 16, D-14482 Potsdam, Germany

Abstract. The current status in the use of Cepheids as standard candles is reviewed. Sources of systematic error in HST-determined Cepheid distances are discussed, and observational programs are suggested which will lead to a significant improvement in the accuracy of Cepheid-based distances to galaxies out to the Virgo and Fornax clusters.

1 Introduction

Cepheid variables continue to be our most important primary calibrators of extragalactic distances. Currently we are able to detect and use Cepheids out to $\sim$ 25 Mpc to determine primary distances of their host galaxies, as done over the last years by the HST Key Project teams. The success of Cepheid variables as distance indicators is due to several crucial properties: Cepheids are bright, as compared to most other stellar standard candles; they are easily identified from the characteristic shapes of their light curves; they are sufficiently numerous in late-type galaxies to provide good statistics in distance determinations; and, perhaps most importantly, the physical property used for the determination of luminosity and hence distance, the famous period-luminosity relationship, is well understood from basic physical principles. Since, however, current instrumental capabilities do not permit to observe Cepheids out to regions of unperturbed Hubble flow, the principal use of Cepheids is to provide zero point calibrations of a variety of secondary methods of distance measurement which do reach out to $\sim$ 100 Mpc or beyond; from such secondary methods, an accurate value of the Hubble constant will eventually be established.

There has been substantial progress on Cepheid variables over the last years regarding their usefulness as distance indicators. The various microlensing surveys have returned large samples of Cepheids in LMC and SMC which have led to a better understanding of both the physical properties of individual Cepheids, and of the global properties of large Cepheid ensembles. Hipparcos parallaxes of Galactic Cepheids have become available. Cepheid samples useful for distance determination have now been discovered by the HST Key Project teams in nearly 20 galaxies out to the Virgo and Fornax clusters. Infrared Baade-Wesselink-type techniques have been developed which hold great promise for very accurate direct distance determination of Cepheid variables out to several Mpc. Most of these issues have recently been reviewed in some detail by [22].

The most difficult problem in using Cepheids for extragalactic distance determination is the calibration of local distances to these variables. One could choose to adopt a distance modulus to the LMC and tie the absolute calibration of the PL relation zero point to this adopted LMC distance, as done by the HST Key Project team; however, current distance determinations for the LMC fall into the broad range of 18.2 to 18.7 (see [23], and [25]), introducing a large systematic uncertainty into this approach. A better approach might be to calibrate Cepheid distances locally (within the Galaxy). In the following, we very briefly review the status of the different techniques to derive distances of Galactic Cepheids. We then discuss the typical accuracy of Cepheid distances to external galaxies as determined with the HST method. Finally, we discuss several programs feasible on 8 m-class telescopes for science which will further improve the accuracy of Cepheid distances to nearby galaxies to a level which promises to make our dream of a Hubble constant accurate to better than 10 percent come true within a very few years from now.

2 The Current Status of Milky Way Calibrations of Cepheid Distances

Distances to Galactic Cepheid variables, and a corresponding calibration of the zero points of the period-luminosity (PL) relation in different photometric bands, can currently be obtained principally from the following three independent techniques: (a) Hipparcos parallaxes, (b) Cepheids in open clusters, and (c) infrared Baade-Wesselink (or surface brightness) techniques (which have superseded former techniques calibrated at optical wavelengths). The slopes of the PL relations in different bands are conveniently derived from samples of LMC Cepheids which are all at the same distance, to first order, due to the small intrinsic depth of the LMC in the line of sight.

2.1 Cepheid Distances from Hipparcos Parallaxes

Following the original work of [4] which led to a 18.70 value for the LMC distance modulus, several authors have reanalyzed the Hipparcos parallaxes of Cepheids. The problem is the high relative errors on the parallaxes which make the proper treatment of statistical biases a crucial issue in the analysis. To date, differences in the PL zero point mounting up to an astonishing 0.38 mag ([16]) have emerged from different analyses of the same set of parallax data. A recent attempt to harmonize the situation has been made by [17] who has used Monte Carlo simulations and synthetic sample techniques. His analysis shows that the values derived by different groups might converge to a common PL zero point if the data analysis techniques are properly employed, but it seems clear that the PL zero point cannot be determined with an accuracy better than ±0.2 mag from current Hipparcos Cepheid parallaxes. Future space missions like SIM or GAIA have to be awaited to derive Cepheid distances which can compete in accuracy with, and hopefully supersede, other available techniques.

2.2 Cepheids in Open Clusters with Main-Sequence Fitted Distances

This has been the standard method for several decades. Its main sources of systematic error have been discussed, among others, by [6]. Problems include the small number of Cepheids found in Galactic open clusters, sparse clusters which make main-sequence fitting difficult, and usually high and patchy foreground reddening to these clusters which are located at low galactic latitudes. The most serious current source of systematic uncertainty is the evidence from Hipparcos parallaxes of a set of nearby open clusters that the zero age main sequence location in clusters may vary by as much as 0.6 mag ([19], and references therein). While it had formerly been thought that a determination of the PL zero point to about ±0.2 mag or even better is possible with the cluster method, the systematic uncertainty is clearly higher now, as long as the correct interpretation of the Hipparcos parallaxes of open clusters remains unclear.

2.3 Cepheid Distances from the Infrared Surface Brightness Technique

This technique was introduced by [5], extending preliminary work of [24]. The use of infrared colors and magnitudes instead of optical colors has resulted in a dramatic increase in the accuracy of distance determination to Cepheids (see Fig. 1). A thorough analysis of the known sources of systematic uncertainty affecting the method has shown that Cepheid distances accurate to ±3 percent can be obtained ([7]). The method has the further advantage of being practically independent of assumed absorption corrections and the metallicity of the target Cepheids. [8] have applied the technique to a sample of Galactic Cepheids and have obtained absolute calibrations of the PL relation in the VIJHK bands which show smaller dispersions than any former calibration from distances of Galactic Cepheids, demonstrating the high intrinsic accuracy of the method. Comparison with the observational distance results for Cepheids from the infrared maximum likelihood technique of [14] suggests agreement at the ±0.1 mag level. Also, there is excellent agreement with recent theoretical PL relations in the V and K bands based on consistent evolutionary and pulsational calculations (unaffected by the treatment of convection) as presented by [1], a fact which presents further evidence that systematic errors in the infrared surface brightness technique of distance determination are indeed as low as suggested by [7]. As a consequence of these findings, we conclude that this technique is currently able to provide the most accurate calibration of the absolute zero points of the Cepheid PL relation from Galactic stars. Comparison of the Galactic PL relations to the apparent relations defined by samples of LMC Cepheids has produced a LMC distance modulus of 18.46 ± 0.02 ([8]). However, this value may be affected to some extent by a possible metallicity dependence of Cepheid absolute magnitudes; such a dependence can be avoided by a direct, one-step application of the method to LMC Cepheids (see Section 4).

3 The Accuracy of HST-Determined Cepheid Distances to Nearby Galaxies

Although of course a fantastic achievement in itself, the HST method of deriving distances to late-type galaxies from samples of Cepheid variables is affected by several sources of systematic error. The most relevant ones include the following:

(a) LMC distance (HST Cepheid distances are derived relative to the LMC)
(b) incompleteness bias
(c) metallicity dependence of the PL zero points in the V and I bandpasses used by HST
(d) effects of stellar evolution on the PL relation
(e) separation of reddening due to dust from the effect of metallicity on Cepheid colors

Recent comprehensive reviews of the status of distance determinations to the LMC have been given by [23], and [25]. The main conclusion of both authors is that at the present time, reasonably well-established and apparently well-calibrated methods do produce distance moduli discrepant by nearly a half magnitude, in the range 18.2 to 18.7. The accepted value depends strongly on the adopted weighting which is very difficult to quantify scientifically. It is therefore clear that in spite of the Hipparcos mission, and the enormous improvement in the LMC variable star data base due to the various microlensing surveys, the LMC distance modulus is still uncertain by at least ± 0.2 mag.

A recent detailed discussion of the effect of incompleteness bias on Cepheid distance moduli has been given by [15]. The origin of the bias is that a sample of Cepheids in a given distant galaxy is complete only at the bright end of the PL relation (e.g. for long periods) because the observations are limited to

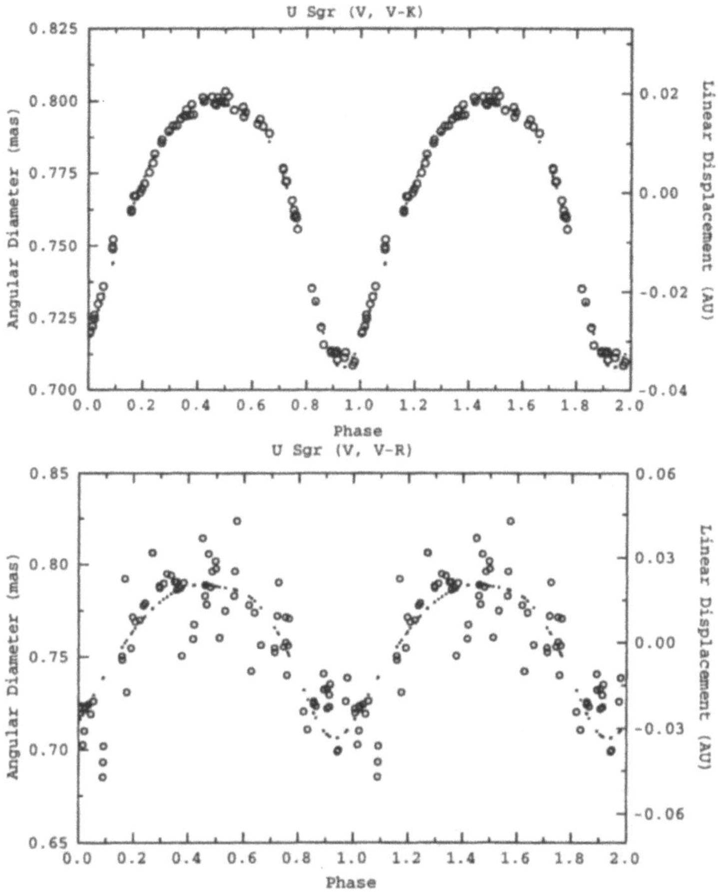

Fig. 1. (top) The angular diameters (open circles, left scale) of the Cepheid U Sgr calculated from V, V-K photometry using the infrared surface brightness method, plotted against phase. Overplotted (dots, right scale) is the linear displacement variation of the Cepheid. (bottom) The same adjustment from the V, V-R_J photometry.

a given apparent limiting magnitude. [15] find that this bias always causes an underestimate of the true distance modulus (even if the correct slope of the PL relation is used) which can amount to 0.25 mag at a distance modulus of 32. While the use of a conservative low period cutoff is the best way to minimize the effect of this bias, the correction is difficult to make as long as sample completeness is not perfectly well established. Systematic errors in the order of 0.1 mag might be present due to this effect in typical HST Cepheid distance determinations.

In spite of several efforts with different approaches over the past few years, the question of what effect metallicity variations may have on PL relations (in different wavelength bands) is not yet satisfactorily answered.

While the studies of [18] (comparison of EROS observations of LMC and SMC Cepheid samples), [12] (simultaneous distance solutions for 17 galaxies), [11] (comparison of HST Cepheid observations in inner and outer fields in M 101), and [21] (Baade-Wesselink analysis of SMC Cepheids) all yield at least the same sign of the effect (e.g. more metallicity-poor Cepheids are intrinsically fainter, thus their distances tend to be overestimated), and basically agree on a value of ~ 0.4 mag/dex in the O/H abundance, the uncertainties associated to each of these determinations are large. Other empirical determinations find a much larger effect of metallicity on the PL relation at optical and near-infrared wavelengths ([10]; [20]). From a theoretical point of view, current results are contradictory: while [1] find a small effect in the sense described above, [3] have found an effect in the opposite sense. We conclude that the effect of metallicity on the PL relation is not yet well determined, neither from observation nor from theory. The amount of systematic error made in the HST method when dealing with galaxies whose metallicities are quite different from the LMC is currently unclear but could amount to several tenths of a magnitude.

Regarding the effect of stellar evolution on the PL relation, we expect a steepening of the slope of the relation at low periods due to the reduction of blue loops during core helium burning. The blue loop near the minimum mass of a Cepheid does not completely penetrate the Cepheid instability strip. This predicted effect has recently been observed by [2] for Cepheids with periods less than 2–3 days. For extragalactic distance determinations, this effect is of no concern since one is usually dealing with Cepheids of periods larger than 10 days. For nearby galaxies in which very short-period Cepheids can be detected, these objects should be excluded from the distance determination (an additional reason for this is the large contamination of Cepheid samples with overtone pulsators at the short-period end of the Cepheid instability strip).

Our conclusion from the remarks given in this section is that there are several sources of systematic error (apart from calibration problems with the data) in the HST process of distance determination from Cepheids which are not well controlled presently. Their combined effect can affect a typical HST Cepheid distance by at least 0.3 mag, and possibly more than that. The largest single source of systematic uncertainty appears to be the distance to the LMC. It is hard to imagine real progress on the determination of H_0 as long as we cannot agree on the LMC distance at a level of better than five percent.

4 Observational Programs to Improve Cepheids as Distance Indicators

To improve the accuracy of galaxy distances measured via their Cepheid variables, various programs should be carried out which will lead to significant progress along at least three lines:

(a) reduction of systematic errors in the HST method
(b) measurement of accurate direct, one-step Cepheid distances to a sample of nearby galaxies which can be used to calibrate secondary distance indicators of long distance range with higher accuracy than currently possible
(c) shift of extragalactic Cepheid observations to the near-infrared, with its distinct advantages over observations in optical passbands

Regarding (a), a determination of the LMC distance accurate to about 3 percent is now feasible by using the infrared surface brightness technique directly on LMC Cepheids. The ideal targets are Cepheid-rich, young clusters which also serve as an ultimate test on the capabilities of the method. A program on NGC 1866 is underway ([9]), as well as on NGC 2031 which has the advantage of being less crowded than NGC 1866 and is closer to the LMC bar. A definitive empirical calibration of the metallicity effect on Cepheid absolute magnitudes can be done in nearby, nearly face-on spiral galaxies like NGC 300 for which accurate, ground-based photometry of Cepheid variables can be obtained with the high resolution imaging capabilities now existent, as well as an accurate determination of the radial metallicity gradient from spectroscopy of A supergiants ([13]). The results of these programs will make HST-determined Cepheid distances of galaxies out to Virgo and Fornax accurate to better than 10 percent.

With 8 m-class telescopes like VLT, a number of nearby, late-type galaxies are in reach for a direct application of the infrared surface brightness technique, including galaxies in nearby groups like Sculptor. Exploiting the full potential of the method, accurate one-step distances will be measured which will be practically unaffected by the reddening and metallicity of the target galaxies. After detecting Cepheids in these galaxies with Wide Field Imagers, follow-up VLT observations with ISAAC and UVES will provide the infrared light curves and radial velocity curves of selected variables. From these direct distances improved zero point calibrations will be obtained for a variety of methods, including the Planetary Nebulae Luminosity Function and the Tully-Fisher method.

Observing extragalactic Cepheids at infrared wavelengths has clear advantages: infrared PL relations are steeper than optical ones and show less intrinsic dispersion; there are less problems with dust absorption; the effect of metallicity on Cepheid infrared colors is negligible; and, infrared mean magnitudes of Cepheids are determined more accurately from a few observations than *blue* or visual magnitudes due to the more symmetric and smaller-amplitude infrared light curves. Re-observation of Cepheid samples in near-infrared bands in HST-observed galaxies will therefore be extremely rewarding and ultimately lead to the most accurate Cepheid distance results for galaxies beyond the reach of direct techniques. Such programs are now clearly feasible with the new large aperture, high resolution ground-based telescopes coming into operation. Combining the results of all these efforts, a Hubble constant accurate to 10 percent should, finally, become a reality in a few years from now.

WPG was supported by Fondecyt grant No. 1971076 which is gratefully acknowledged.

References

1. Alibert Y., Baraffe I., Hauschildt P., Allard F. (1999) A&A submitted, astro-ph/9901294
2. Bauer F. et al. (1998) A&A submitted, astro-ph/9807094
3. Bono G., Caputo F., Marconi M. (1998) ApJ **497**, L43
4. Feast M. W., Catchpole R. M. (1997) MNRAS **286**, L1
5. Fouqué P., Gieren W. P. (1997) A&A **320**, 799
6. Gieren W. P., Fouqué P. (1993) AJ **106**, 734
7. Gieren W. P., Fouqué P., Gómez M. (1997) ApJ **488**, 74
8. Gieren W. P., Fouqué P., Gómez M. (1998) ApJ **496**, 17
9. Gieren W. P., Fouqué P., Storm J. (1999) in preparation
10. Gould A. (1994) ApJ **426**, 542
11. Kennicutt R. C. et al. (1998) ApJ **498**, 181
12. Kochanek C. S. (1997) ApJ **491**, 13
13. Kudritzki R. P. (1998) in Stellar Astrophysics for the Local Group, eds. A. Aparicio, A. Herrero, F. Sanchez, Cambridge: Cambridge University Press, p. 149
14. Laney C.D. & Stobie R.S. (1994) MNRAS **266**, 441
15. Lanoix P., Paturel G., Garnier R. (1999) ApJ submitted, astro-ph/9810148
16. Luri X., Gomez A. E., Torra J., Figueras F., Mennessier M. O. (1998) A&A **335**, L81
17. Pont F. (1999) in Harmonizing Cosmic Distance Scales in a Post-Hipparcos Era, eds. D. Egret, A. Heck, ASP Conf. Ser., in press, astro-ph/9812074
18. Sasselov D. D. et al. (1997) A&A **324**, 471
19. Seggewiss W. (1999) in Harmonizing Cosmic Distance Scales in a Post-Hipparcos Era, eds. D. Egret, A. Heck, ASP Conf. Ser., in press
20. Sekiguchi M., Fukugita M. (1997) Observatory **118**, 73
21. Storm J., Carney B. W., Fry A. M. (1999) in Harmonizing Cosmic Distance Scales in a Post-Hipparcos Era, eds. D. Egret, A. Heck, ASP Conf. Ser., in press, astro-ph/9811376
22. Tanvir N. R. (1999) in Harmonizing Cosmic Distance Scales in a Post-Hipparcos Era, eds. D. Egret, A. Heck, ASP Conf. Ser., in press, astro-ph/9812356
23. Walker A. R. (1999) in Post-Hipparcos Cosmic Candles, Dordrecht: Kluwer, in press
24. Welch D. L. (1994) AJ **108**, 1421
25. Westerlund B. E. (1997) in The Magellanic Clouds, Cambridge: Cambridge University Press

Eclipsing Binaries as Precise Distance Indicators

Jens Viggo Clausen

Niels Bohr Institute for Astronomy, Physics and Geophysics,
Astronomical Observatory, Copenhagen University, Denmark

Abstract. Double-lined eclipsing binaries provide the most direct distance determinations available for the LMC, SMC and other Local Group galaxies. Distance moduli accurate to ±0.10 mag. or better can be reached, but results from extragalactic binaries are still scarce. Complete CCD light curves are, however, available for several MC systems, and the needed accurate radial-velocity curves can be obtained at VLT UT2 plus UVES and/or FORS2 in the near future. In addition, a large number of interesting new MC binaries have recently been discovered from microlensing searches, and better M31 candidates have also become available. See http://www.astro.ku.dk/~jvc for further information.

1 Introduction

Double-lined eclipsing binaries provide, apart from objects where purely geometrical methods can be applied, the most direct distance determinations available. The method includes only one scale dependent step, and from complete, accurate observations and careful analyses of well-detached systems, distances moduli better than ±0.10 mag. can be obtained. Cases where binary distances are especially important include nearby systems, cluster members, and not at least extragalactic binaries on which I will concentrate in the following. The reader is also referred to reviews on the topic by Guinan et al. [17], Hilditch [21] and Paczyński [33], and to the paper by Pritchard & Niemela in these proceedings.

With present day facilities, including VLT class telescopes, precise binary studies are limited to Local Group (LG) galaxies, primarily the Magellanic Clouds (MC). Besides distances, the double-lined detached eclipsing binaries provide accurate masses, radii, and luminosities of the components, important for tests of theoretical models of single stars, and they define the empirical mass-luminosity relations. For young, massive stars of typical LMC or SMC composition, such information is not accessible from studies of Milky Way objects.

Whereas precise dimensions – and distances – are known for about 100 galactic binary components [2,3], results for extragalactic systems are scarce. In fact, only for the LMC binary HV2274 has a high-quality distance been published [19,46], see Table 2. This situation is primarily caused by lack of accurate radial-velocity curves; complete CCD light curves exist for several

MC binaries. With the availability of VLT UT2 plus UVES and FORS2 in the near future, the prospects for a real break through are, however, excellent.

Precise distances from many binaries are of high importance, first of all for refining the distance determination of the LMC (see e.g. [51,7,48] for the significant scatter even in modern results), well-known as a critical step on the extragalactic distance ladder and important for e.g. accurate determinations of H_0 [24].

2 Eclipsing Binaries in Local Group Galaxies

Eclipsing binaries in external galaxies were first discovered in the beginning of this century by Leavitt [27] in her study of MC variable stars. A summary of photographic and photoelectric work on MC systems is given in [22], and Table 1 of [21] lists the about 180 eclipsing binaries discovered in 10 nearby galaxies before 1990 from (mostly) photographic surveys. Since then this number has drastically increased, primarily due to the intensive CCD photometry carried out for microlensing searches.

An up-to-date list of eclipsing binaries discovered in CCD surveys since 1990 is presented in Table 1, but many more systems can soon be added. Only a tiny fraction of the EROS material has been published, about 2600 LMC systems are expected from OGLE observations of 11 central fields (Udalski, private communication), the DIRECT results cover only three out of six M31 fields and none of the M33 fields observed, and finally new systems in LG dwarf galaxies should become available. From the many known binaries a more than sufficient sample of systems ideal for MC (and possibly M31) distance determinations can easily be selected; see e.g. the classification scheme and the preliminary light curve analyses in [1]. Even the new CCD discovery light curves are, however, not suitable for detailed analyses.

Table 1. Eclipsing binaries discovered in CCD surveys.

Project	Ref.	Galaxy	Ecl. bin.	Bands	Mag. limit	P(days)
EROS	[16]	LMC	79	B,R	15-18	0.7-46
MACHO	[1]	LMC	611	V,R	R $\leq$ 18	$<$ 156
OGLE II	[47]	SMC	1459	B,V,I	I $\leq$ 20	0.3-250
DIRECT	[23,39,40]	M31	39	B,V,I	V 18-21.5	0.2-12

The first accurate CCD observations of MC binaries were obtained at ESO by [22], inspired by [15], and until now accurate light curves for about 15 systems in each of the LMC and the SMC have been either published, obtained or are in progress from ongoing projects at ESO and Mt. John Univ. Obs., NZ. Ultraviolet spectrophotometry from IUE and/or HST is available for about 10 of them, see Table 1 of [18]. Low-medium resolution spectroscopy exist for some, primarily obtained by Niemela and collaborators, but radial-velocity curves matching the quality of the photometric data are missing. Further details on existing observations can be seen at http://www.astro.ku.dk/~jvc.

Most of the systems studied so far are early B-type binaries discovered from the photographic Harvard surveys. Observations of a new sample, selected among the EROS, MACHO and OGLE discoveries and including interesting less-massive late B-type systems, have been initiated at ESO 1999; see Pritchard & Niemela (these proceedings) for further information.

3 The Method and its Accuracy

Among the various types of distance indicators [45] binaries may be classified as 'meter sticks', and the principles for establishing distances from eclipsing systems seem to go back to Stebbins [41,42]. Statistical procedures have been applied on binary samples, sometimes leading to apparently very accurate LMC and SMC distances [13,14,10], but these results are of historical interest only. Detailed studies of individual double-lined eclipsing binaries are needed and must be based on complete light and radial-velocity curves, standard photometry, and possibly ultraviolet and visual spectrophotometry, plus preferably high resolution spectra.

The distance modulus for each of the components of a double-lined eclipsing binary is traditionally determined from the following simple relation:

$$(m_V - M_V)_0 = m_V - A_V - M_{bol\,\odot} + 5\log(R/R_\odot) + 10\log(T_e/T_{e\,\odot}) + BC. \quad (1)$$

The strength of the method is the fact that the stellar radius R is obtained from *purely scale independent* quantities,

$$R/R_\odot = r\, 1.9757\, 10^{-2}\, (K_1 + K_2)\, P\, (1 - e^2)^{0.5} / \sin(i)\ , \quad (2)$$

where r is the relative radius in units of the semimajor axis of the orbit, K_1 and K_2 are the radial-velocity semiamplitudes of the components in km/s, P is the orbital period in days, and e and i are the eccentricity and inclination of the orbit, respectively. All these elements are determined from *combined* radial-velocity and light curve analyses, and from accurate and complete observations of well-detached systems, a typical rms error of $\log(R/R_\odot)$ of ± 0.008 can be reached. Strategies to avoid systematic errors are well known and discussed in e.g. [4,2].

m_V (and individual indices) is obtained from standard photometry of the binary plus the luminosity ratio(s) between the components, known from the light curve analyses, and typically an accuracy of $\pm 0.01 - 0.02$ can be reached. When CCD photometry in crowded fields is applied it is critically important to secure well-established zero-points. The interstellar absorption A_V (foreground plus intrinsic) is obtained from the reddening, e.g. $E(b-y)$ which for early-type stars is determined from the standard $(b-y)_0$ - c_0 relation

[8]. Individual absorption results accurate to about ± 0.04 can be established, whereas assuming average values for e.g. LMC or SMC systems is far from accurate enough [26,46].

The only scale dependent part of the method is the determination of effective temperatures and bolometric corrections of the components. They are normally derived from reddening free, suitable standard indices, like for early-type stars [u-b] or c_0, and empirical calibrations [29,9]. Bolometric corrections may also be derived through the corresponding temperatures [12]. The observational uncertainties of the indices translate to typically $\pm 0.008 - 0.013$ for $\log(T_e/T_{e\odot})$ and ± 0.05 for BC.

Summing up directly, the various contributions yield a total accuracy of $\pm 0.10 - 0.15$ for the distance modulus, but this result is slightly too pessimistic. As seen, the temperature term in (1) enters with a factor of 10, but there is a fortunate compensation through the corresponding bolometric correction, since for early-type stars [12] $BC \approx -5.42 \log(T_e) + 21.39$, so the effective factor becomes less than 5. A more realistic accuracy is therefore $\pm 0.07 - 0.09$ *per component of a binary*, excluding possible systematic errors.

The primary sources of possible systematic errors are the effective temperature and BC terms. Firstly, the available empirical scales may contain systematic errors, which can only partly be studied from nearby binaries [36] and by comparing different calibrations. Secondly, they are based on galactic stars of solar composition or similar and not a priori valid for stars of typical LMC or SMC compositions. However, colours and bolometric corrections calculated from theoretical stellar atmospheres show that for early-type stars only small differences should occur, and also that the temperature and BC shifts to some extent compensate each other like expressed above. Further details can be seen at http://www.astro.ku.dk/~jvc.

An alternative to (1) has been applied in some recent analyses where UV spectrophotometry has been available [19,34,35],

$$f_\lambda = (R_1/d)^2[F_{\lambda,1} + (R_2/R_1)^2\ F_{\lambda,2}]10^{-0.4[E(\lambda-V)+A_V]} \tag{3}$$

f is the combined observed flux and F_i the surface flux of component i, and (3) is solved simultaneously for stellar parameters (temperatures and abundance; $\log g$ is known from the masses and radii), interstellar absorption, and distance d [11]. Theoretical atmosphere models are used, and the method is quite complex, but very precise temperatures are reported – in the case of HV2274 about 1% or ± 180 K [19].

Finally, spectral line analyses, free of complications from interstellar absorption, can be used. In recent years effective temperatures for early-type single stars, plus surface gravities (known for binary components) and abundances, have successfully been derived from line analyses based on moderately high-resolution, high S/N spectra and NLTE atmosphere models, see e.g. [25]. Disentangling of the combined spectrum has been applied for galactic early-type binaries, leading to accurate spectroscopic orbits as well as temperatures and abundances for the components [37,43,38]. These methods are promising for analyses of LMC and SMC systems and should be used when high-quality spectroscopy becomes available.

In conclusion – distance moduli accurate to ±0.10 or better *per binary component* can be reached. The method includes only one scale dependent step, the determination of temperatures and bolometric corrections or surface fluxes from observations, and the effect of possible systematic errors from it can be investigated and minimized by applying the different approaches described above. Paczyński [33] is even more optimistic and see no fundamental reason why ≈ 1% could not be achieved for LG galaxies in the future.

4 Results from LMC and SMC Binaries

As mentioned in Sect. 2, sufficient photometric observations are available for several LMC and SMC eclipsing binaries, but due to the lack of high-quality spectroscopy, distances and absolute dimensions have been published for just the few systems listed in Table 2. HV2274 is detached and represents the most accurate LMC distance (18.30 ± 0.07 for the center of the LMC [19]) yet obtained from eclipsing binaries. The four other Harvard variables are semi-detached, but strongly improved distance determinations should be possible, at least for some of them, if better mass, temperature, and reddening determinations become available. The massive and close system Sk-67^o105 is not optimal for the purpose.

Preliminary results from the promising detached LMC system HV982 [35] are presented by Pritchard & Niemela (these proceedings). Analyses based on published light curves [22,50,28,44] and NTT/EMMI radial-velocity observations are in progress for HV12634 and HV12484 in the LMC, as well as an independent study of HV2274 (Clausen et al.).

Table 2. Absolute dimensions and distances for LMC and SMC binaries

System	$M/M_\odot$	$R/R_\odot$	T_e	$(m-M)_0$	References
HV2241	34±2:	16±1:	27000±3000	18.5	[34,30]
(LMC)	17±1:	14±1:	20200±1500	±.5	
HV2274	12.1±.4	9.84±.24	23000±180	18.35	[49,19,46]
(LMC)	11.4±.4	9.03±.20	23110±180	±.07	
HV5936	7.2±.9	4.6±.3	29000±2000	18.1	[22,6]
(LMC)	3.3±.5	5.7±.3	19200±2000	±.3	
Sk−67°105	47.2±.8	15.6±.1	44600±1300	?	[20,31,32]
(LMC)	29..8±.6	12.2±.3	41000 (fixed)		
HV1620	21±3:	6±1:	33000±4500	18.6	[34,30]
(SMC)	16±2:	11±1:	24400±3500	±.5	
HV2226	9.3±.7	5.6±.5	28000±2000	18.64	[22,5]
(SMC)	5.6±.5	5.3±.2	19400±2000	±.27	

5 Conclusions

Definitive, precise distance determinations for the LMC, SMC and other LG galaxies are still lacking. Double-lined eclipsing binaries provide the most direct method available, and an accuracy of ± 0.10 mag. or better can be reached per binary component. Independent ways exist for handling the the only scale dependent step of the method, determination of temperatures and bolometric corrections or surface fluxes from observations, which is the primary source of possible systematic errors. The distances should be based on several individual binary analyses, e.g. 10–15 per galaxy. A sufficient number of suitable candidates is known, including many new and fainter ones discovered from microlensing searches. Complete light curves and other photometric data are available for several MC systems, and more observations are in progress. The required accurate radial-velocity curves can be secured in the near future at VLT UT2 plus UVES and/or FORS2. As a 'by product' the binaries will establish empirical mass-luminosity relations and provide accurate stellar dimensions, important for tests of theoretical models of e.g. young, massive stars of typical MC compositions.

Acknowledgements

I would like to thank John Pritchard for many fruitful discussions and Virpi Niemela for generously making unpublished material available to me. This investigation was supported by the Danish Natural Science Research Council.

References

1. Alcock C., Allsman R.A., Alves D. et al. (1997) AJ **114**, 326
2. Andersen J. (1991), A&AR **3**, 91
3. Andersen J. (1997) in *Fundamental stellar properties: The interaction between observation and theory*, IAU Symp. No. **189**, Eds. T.R. Bedding, A.J. Booth and J. Davis, Kluwer, Dordrecht, 99
4. Andersen J., Clausen J.V., Nordström B. (1980) in *Close Binary Stars. Observations and Interpretation*, IAU Symp. No. 88, Eds. M.J. Plavec, D.M. Popper and R.K. Ulrich, Reidel, Dordrecht, 81
5. Bell S.A., Hill G., Hilditch R.W., Clausen J.V., Reynolds A.P. (1991) MNRAS **250**, 119
6. Bell S.A., Hill G., Hilditch R.W., Clausen J.V., Reynolds A.P. (1993) MNRAS **265**, 1047
7. Cole A.A. (1998) ApJ **500**, L137
8. Crawford D.L. (1978) AJ **83**, 48
9. Davis J., Shobbrook R.R. (1977) MNRAS **178**, 651
10. Dworak T.Z. (1974) Acta Cosmologica **2**, 13
11. Fitzpatrick E.L., Massa D. (1999), in prep.
12. Flower P.J. (1996) ApJ **469**, 355

13. Gaposhkin S. (1968) PASP **80**, 556
14. Gaposhkin S. (1970) Inf. Bull. Var. Stars No. 496
15. Graham J.A. (1983) IAU Highlights Astron. **6**, 209
16. Grison P., Beaulieu J-P., Pritchard J.D. et al. (1995) A&AS **109**, 447
17. Guinan E.F., Bradstreet D.H., DeWarf L.E. (1996) in *The origin, evolution, and destinies of binary stars in clusters*, ASP Conf. Ser. Vol. **90**, Eds. E.F. Milone and J.-C. Mermilliod, 196
18. Guinan E.F., Ribas I., Fitzpatrick E.L., Pritchard J.D. (1998) in *Ultraviolet Astrophysics beyond the IUE Final Archive*, Eds. W. Wamsteker, R.R. Gonzales-Riestra and R. Harris, ESA SP-413, Noordwijk, 315
19. Guinan E.F., Fitzpatrick E.L., DeWarf L.E. et al. (1998) ApJ **509**, L21
20. Haefner R., Simon K.P., Fiedler A. (1994) A&A **288**, L9
21. Hilditch R.W. (1996) in *The origin, evolution, and destinies of binary stars in clusters*, ASP Conf. Ser. Vol. **90**, Eds. E.F. Milone and J.-C. Mermilliod, 206
22. Jensen K.S., Clausen J.V., Giménez A. (1988) A&AS **74**, 331
23. Kaluzny J., Stanek K.Z., Krockenberger M. et al. (1998) AJ **115**, 1016
24. Kennicutt Jr. R.C., Freedmann W.L., Mould J.R. (1995) AJ **110**, 1476
25. Kilian J. (1991) A&A **244**, 419
26. Larsen S.S., Clausen J.V., Storm J. (1999) A&A, submitted
27. Leavitt H.S. (1908) Harvard Annals **60**, 87
28. Loudon M., Budding E. (1993) Ap&SS **201**, 287
29. Napiwotzki R., Schönberner D., Wenske V. (1993) A&A **268**, 653
30. Niemela V.S., Bassino L.P. (1994) ApJ **437**, 332
31. Niemela V.S., Morrell N.I. (1986) ApJ **310**, 715
32. Niemela V.S., Morrell N.I. (1999) in *New Views of the Magellanic Clouds*, IAU Symp. No. 190, in press
33. Paczyński B. (1997) in *The extragalactic distance scale*, STSI Symp. Ser.10, Eds. M. Livio, M. Donahue and N. Panagia, Cambridge Univ. Press, 273
34. Pritchard J.D., Tobin W., Clark M., Guinan E.F. (1998) MNRAS **297**, 278
35. Pritchard J.D., Tobin W., Clark M., Guinan E.F. (1998) MNRAS **299**, 1087
36. Ribas I, Giménez A., Torra J. et al. (1998) A&A **330**, 600
37. Simon K.P., Sturm E. (1994) A&A **281**, 286
38. Simon K.P., Sturm E., Fiedler A. (1994) A&A **292**, 507
39. Stanek K.Z., Kaluzny J., Krockenberger M. et al. (1998) AJ **115**, 1894
40. Stanek K.Z., Kaluzny J., Krockenberger M. et al. (1999) AJ, in press
41. Stebbins J. (1910), ApJ **32**, 185
42. Stebbins J. (1915), ApJ **42**, 133
43. Sturm E., Simon K.P. (1994) A&A **282**, 93
44. Tobin W., Duncan S.P.R., West S.R.D., Gilmore A.C. (1993) MNRAS **260**, 777
45. Trimble V. (1997) in *The extragalactic distance scale*, STSI Symp. Ser.10, Eds. M. Livio, M. Donahue and N. Panagia, Cambridge Univ. Press, 313
46. Udalski A., Pietrzyński G., Woźniak P. et al. (1998) ApJ **509**, L25
47. Udalski A., Soszyński I., Szymański M. et al. (1998) Acta Astron. **48**, 563
48. Walker A.R. (1999) in *Post Hipparcos Cosmic Candles*, Eds. A. Heck and F. Caputo, Kluwer, Dordrecht, 125
49. Watson R.D., West S.R.D., Tobin W., Gilmore A.C. (1992) MNRAS **258**, 527
50. West S.R.D., Tobin W., Gilmore A.C. (1992) MNRAS **254**, 419
51. Westerlund B.E. (1997) *The Magellanic Clouds*, Cambridge Univ. Press

Magellanic Cloud Eclipsing Binaries: Distances and Dimensions

John Pritchard[1] and Virpi Niemela[2]

[1] Mount John University Observatory and Department of Physics & Astronomy, University of Canterbury, Private Bag 4800, Christchurch, New Zealand.
[2] La Plata Observatory, Paseo del Bosque, 1900 La Plata, Argentina. Member of Carrera del Investigador, CIC, Prov. Buenos Aires, Argentina. Visiting Astronomer, CASLEO, operated under agreement between CONICET, SeCyT, and the Universities of La Plata, Córdoba and San Juan, Argentina.

Abstract. In this contribution we analyse previously unpublished radial velocity curves in conjunction with other already published light curves, Strömgren *uvby* photometry and *IUE* spectrophotometry.

We also discuss the selection of candidates for a new effort in this program to study systems which accurately sample a much larger parameter space of the HR diagram than has ever been studied before.

1 Introduction

The reasons for studying Magellanic Cloud Eclipsing Binaries (MCEBs) are well documented elsewhere, in particular see the article by Clausen in these proceedings. Let us simply reiterate the primary goals of this research, namely the determination of accurate distances and dimensions.

In section 2 we discuss the analysis of the Magellanic Cloud eclipsing binary system HV 982 based on previously unpublished radial velocity curves in conjunction with other already published light curves, Strömgren *uvby* photometry and *IUE* spectrophotometry.

Catalogues recently published by the MACHO and OGLE groups [1,6] based on multi-bandpass CCD photometry permit the selection of targets which are ideally suited for the accurate determination of component dimensions and system distances – a process that was somewhat more 'hit and miss' based on the photographic surveys of the past. These new surveys also provide somewhat fainter systems which thus have components with much smaller masses than have been studied in the Magellanic Clouds to date, an important extension to this ongoing programme. In section 3 we discuss the selection of target candidates for a new observational programme with these thoughts very much in mind.

2 Harvard Variable 982

The well detached, eccentric, 5.33 day period early B-type eclipsing binary HV 982 in the LMC is an exciting prospect. High quality CCD light curves with excellent coverage in u, V and I_C (as well as sparser coverage in vby) have been obtained with the McLellan 1-m telescope at the Mount John University Observatory in New Zealand [5]. Radial velocity curves have been made by Niemela with the 2.15-m telescope at CASLEO, San Juan, Argentina. These previously unpublished radial velocities are tabulated in table 1. Additional photometric and spectrophotometric data including Strömgren *uvby* CCD light curves and calibrated photometry obtained at the Danish 1.54-m telescope at ESO, La Silla and *IUE*–SWP and more recently *HST*–FOS uv-spectrophotometry have been obtained, and are under analysis.

Table 1. Radial velocity measurements for HV 982. For each velocity measurement the number of lines the measurement is based (n) on and the mean error (m.e.) are also given. The Remarks indicate the nature of the lines in the spectra used for determination of the radial velocities. Spectra where lines are clearly seen to be double are included with weight 1.0 in the analysis, while measurements based on poorly resolved lines ('double?') are included with weight 0.5. Spectra where only single lines are evident are excluded from the analysis.

	Component A			Component B			
HJD −2450000.0	V_A km/s	n_A	$m.e._A$ km/s	V_B km/s	n_B	$m.e._B$ km/s	Remarks
432.783	+457	6	28	+117	6	29	double?
433.777				+242	9	11	single lines
434.726				+322	7	6	single lines
435.718	+45	6	8	+431	8	13	double lines
435.810	+57	4	10	+475	5	18	double lines
436.740	+284	8	13				single lines
437.718	+354	3	28	+112	3	33	double?
806.118	+424	9	29	+94	9	17	double?
807.656				+252	11	6	single lines
808.753	+135	6	26	+490	5	35	double lines
809.688	+62	5	19	+430	5	16	double lines
810.692	+280	7	8				single lines
811.730	+473	8	18	+66	7	16	double?
812.707	+243	7	10				single lines
862.538	+64	4	14	+449	5	17	double lines

Analysis of the radial-velocity curves yield a mass ratio close to unity but the uncertainty is of the order of 0.1. This uncertainty in the mass ratio leads to correspondingly large uncertainties in the remaining directly related parameters of the system, in particular the derived masses and radii are $14{\pm}2\,M_{\odot}$ and $8.0{\pm}0.5\,R_{\odot}$ respectively. A preliminary estimate of the $T_{\rm eff}$s for the components (the individual $T_{\rm eff}$s differ by only a few percent) based on the Danish-1.54m Strömgren photometry is $24\,000{\pm}1000$ K and this implies $(m-M)_0 = 18.7 \pm 0.15$ [*Clausen*, private communication]. It is known that radial velocities based on lower resolution spectra and poorly resolved lines tend to underestimate the physical size of eclipsing binaries [2]. If this applies in this case we might expect HV 982 to be even more massive and consequently even further away in contrast to other recent results indicating the LMC is possibly rather closer, e.g. [4] and references therein. Higher accuracy radial-velocity curves are fundamental to resolving this problem. VLT + UVES will be an ideal instrument for obtaining the required observations for this system (and many others). Exposures of approximately one hour should yield high resolution ($R = 55\,000$), high signal to noise ($S/N = 100$) spectra, according to the exposure calculator. The all but complete data set for HV 982 thus makes it a potentially very fruitful object for study with the VLT in the future.

3 Moving Down the HR Diagram

The faintest eclipsing systems listed in the Harvard College Observatory catalogues of variable stars in the Magellanic Clouds [3] are no fainter than about $m_{PG}{\sim}16.8$ mag. The recently published catalogues by the MACHO group for the LMC [1] and by the OGLE group for the SMC [6] provide systems at least 2 magnitudes fainter. Furthermore the Harvard catalogues lists only 108 systems in total in the two galaxies, whereas the MACHO catalogue lists 611 systems and the OGLE catalogue lists 1459 systems. The new catalogue also provide many more well detached systems in comparison to the Harvard catalogues, simply due to observational bias – non well detached systems are continually varying and thus much easier to detect than the relatively short periods of variability of well detached systems. This is an important development since well detached systems are much better suited to the purposes of the studies discussed here providing the most accurate source of fundamental data relevant to 'normal' single stars.

Because of these limitations of the Harvard catalogues, which until the publication of the new microlensing project catalogues were the only useful source of potential targets, studies of well detached systems have been limited to a rather confined region of the Hertzsprung-Russell (HR) Diagram namely B0 to B1.5 or perhaps B2.

The new catalogues however provide potentially excellent targets whose components should be as late as A0. A new initiative has therefore been undertaken to observe (initially) twelve new systems in each of the Magellanic Clouds in the range B2–A0. The chosen systems are shown in the HR diagrams in figure 1. The systems were chosen on the basis of their magnitudes and colours, the shape of their light curves and the crowdedness of their immediate surroundings.

Acknowledgements. We would like to thank Jens Viggo Clausen for many interesting and useful comments and discussions. JP wishes to thank the

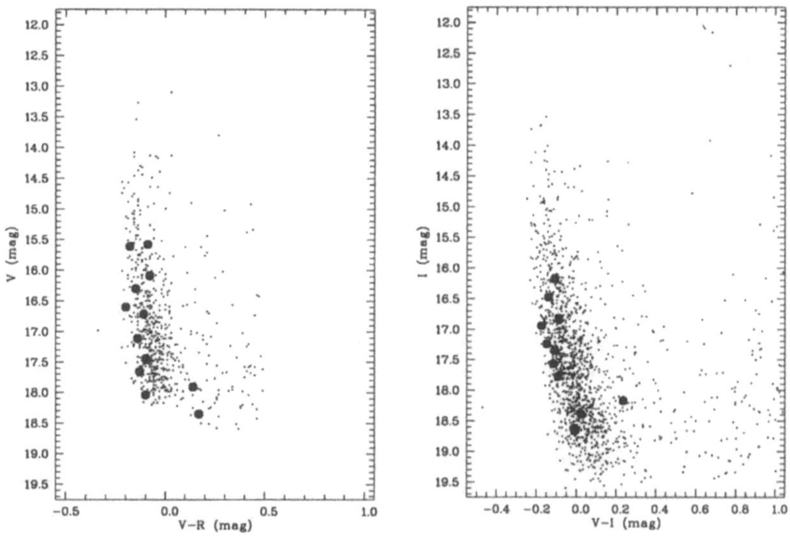

Fig. 1. HR diagrams for the LMC and SMC. The data are taken from the MACHO (LMC) and OGLE (SMC) catalogues of eclipsing binaries. The small crosses show all the systems from the respective catalogues while the large filled circles show the twelve systems in each galaxy chosen for a new project to investigate Magellanic Cloud stars in the spectral range B2–A0.

Marsden Fund of New Zealand for financial support. VN would like to thank the director and staff of CASLEO for the use of their facilities.

References

1. Alcock C., Allsman R. A. et al. (1997) The Macho Project LMC Variable Star Inventory.V.Classification and Orbits of 611 Eclipsing Binary Stars. AJ 114:326
2. Andersen J., Clausen J.V., Nordström B. (1980) Determination of absolute dimensions of main sequence binaries. In: Plavek M.J., Popper D.M., Ulrich R.K. (Eds.) Close binary stars: Observations and interpretation, Proc. IAU Symp. 88, Reidel, Dordrecht, 81
3. Gaposhkin S.I. (1977) One hundred and eight eclipsing variables in the Magellanic Clouds. Research in Space Science SAO Special Report No. 380
4. Guinan E.F., Fitzpatrick E.L. et al. (1998) The Distance to the Large Magellanic Cloud from the Eclipsing Binary HV 2274. AJ,509:L21
5. Pritchard J.D., Tobin W. et al. (1998) CCD photometry of variable stars in the Magellanic Clouds - VI. The eclipsing binary HV 982. MNRAS 299:1087
6. Udalski A., Soszynski I. et al. (1998) The Optical Gravitational Lensing Experiment. Eclipsing Binary Stars in the Small Magellanic Cloud. Acta Astronomica, 48:563

Blue Supergiants – Starway to Virgo and Fornax

R.P. Kudritzki

Munich University Observatory, Scheinerstr.1, 81679 Munich, Germany

Abstract. The mechanical momenta contained in the radiation driven winds of Blue Supergiants are tightly correlated with stellar luminosities. This **Wind Momentum - Luminosity Relationship** provides a unique opportunity to determine extragalactic distances by purely spectroscopic means unaffected by uncertainties in reddening and metallicity. Extrapolating from most recent work on Blue Supergiants in Local Group galaxies it is concluded that with medium resolution, multi-object spectroscopy carried out at 8m class telescopes (for instance FORS at the VLT) distance moduli of spiral and irregular galaxies can be obtained with an accuracy of 0.1 mag out to the Virgo and Fornax clusters of galaxies.

1 Introduction

It has long been a dream of stellar astronomers to study individual stellar objects in distant galaxies by spectroscopy

- to disentangle the kinematics of galaxies and to determine the gravitational potential,
- to determine abundances of the stellar populations and to obtain detailed constraints on the chemodynamical evolution and
- to determine the physical properties of the brightest objects for an estimate of the distances, in this way improving the accuracy of the extragalactic distance scale.

With the new generation of 8m class telescopes this dream will become true. We are approaching the era of **extragalactic stellar astronomy**.

Two classes of stellar objects (apart from Supernovae) are ideally suited for such an investigation because of their intrinsic brightness and spectral characteristics: Planetary Nebulae (irradiated by their hot Central Stars), and Blue Supergiants. During this workshop PNs will be discussed by Roberto Mendez. Therefore, this paper concentrates on Blue Supergiants and their potential as extragalactic distance indicators (Artemio Herrero and collaborators will deal with abundance determinations).

Blue Supergiants have strong stellar winds leading to emission lines all over the spectrum. Stellar astronomers look at these winds from two sides. At the one hand, they are simply sources of trouble. They corrupt the nice photospheric spectra and introduce a substantial complication of the model atmosphere theory. On the other hand, stellar winds are a gift of nature, since they lead to broad and significant spectral features easily to detect. Stellar wind lines allow quantitative spectroscopic studies of the most luminous stellar objects in distant galaxies and, thus, enable us to obtain unique quantitative information about their host galaxies. It is this second aspect, which makes Blue Supergiants so important for extragalactic distance determinations.

Since the winds of Blue Supergiants are a result of radiation pressure, we expect their strengths to reflect the stellar luminosities. Indeed, as we will show in the next section, mechanical stellar wind momenta and stellar luminosities are tightly correlated. This Wind Momentum - Luminosity Relationship provides a unique opportunity to determine extragalactic distances from stellar wind lines of individual supergiants by purely spectroscopic means. The second part of this paper will discuss this aspect in detail.

2 The Wind Momentum – Luminosity Relationship

The theory of radiation driven winds ([1], [12], [5]) predicts a tight relationship between the total mechanical momentum flow $\dot{M} v_\infty$ contained in the stellar wind outflow and the luminosity L of the mass-loosing star

$$\dot{M} v_\infty \propto R_*^{-1/2} L^{1/\alpha_{\rm eff}}. \tag{1}$$

In Eq. 1 $\dot{M}$ is the mass-loss rate, v_∞ the terminal velocity of the stellar wind, and R_* the stellar radius. $\alpha_{\rm eff}$ is a dimensionless number of the order of 2/3 to 1/2 (depending on T_{eff} and ionization) and represents the power law exponent of the distribution function of line strengths of the thousands of spectral lines driving the wind. (For further discussion, see [7], or [8]).

Thus, by measuring the rate of mass-loss and the terminal velocity directly from the spectrum we are – in principle – able to determine the luminosity of a blue supergiant. This is an exciting perspective, because it would give us a purely spectroscopic tool to determine stellar distances. Quantitative spectroscopy would yield T_{eff}, gravity, abundances, intrinsic colours, reddening, extinction, $\dot{M}$ and v_∞. With the luminosity from the above relation one could then compare with the dereddened apparent magnitude to derive a distance.

It is, of course, challenging to compare the prediction of radiation driven wind theory with the observations to see whether an observed WLR does exist. This requires the determination of v_∞ and $\dot{M}$. While the measurement of the former is easy and straightforward by means of the blue edges of the P–Cygni profiles of UV resonance lines (for O-, B-, and A-supergiants) or H_α (for A-supergiants with winds of significant strengths; for examples and references, see for instance [7]), the determination of the latter with sufficient precision used to be problematic. However, the development of hydrodynamic NLTE Unified Model Atmospheres with spherical extension and stellar winds made it possible to use the strength of H_α as a wind line to determine $\dot{M}$ very

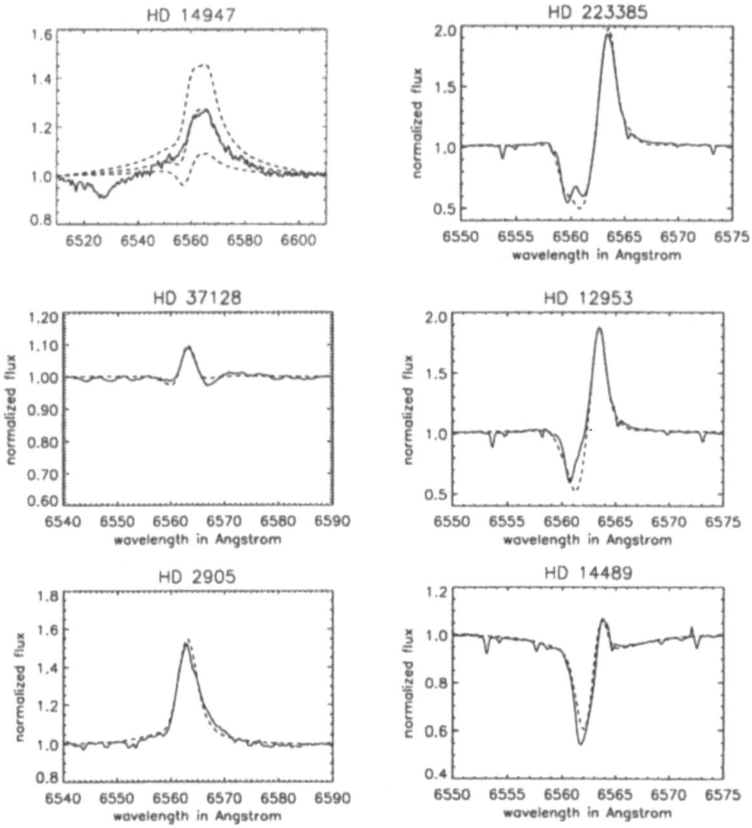

Fig. 1. H_α line profile fits of the galactic supergiants HD 14947 (O5 Ia), HD 37128 (B0 Ia), HD 2905 (B0.7 Ia), HD 223385 (A3 Ia), HD12953 (A1 Ia), HD 14489 (A2 Ia). The fits of HD 14947 are taken from [13] and the three different curves indicate adopted changes in the model calculations of 25 percent in $\dot{M}$. The fits of all other objects are from [9], who used an improved version of the unified model atmosphere code developed by [15] for the line profile fits.

precisely (see again [7] for an overview). In this way, [13] obtained accurate mass-loss rates for a large sample of galactic O-stars and provided a detailed discussion of the H_α line formation process and the possible error sources. This paper also contains an analytical description of the H_α line formation process in stellar winds together with universal curves of growths.

[9] extended this spectroscopic work investigating the winds of the brightest galactic B- and A-supergiants with well determined distances. Fig. 1 indicates how accurately the observed H_α stellar wind line profiles can be reproduced by the theoretical models yielding mass-loss rates with a precision better than 0.15 dex.

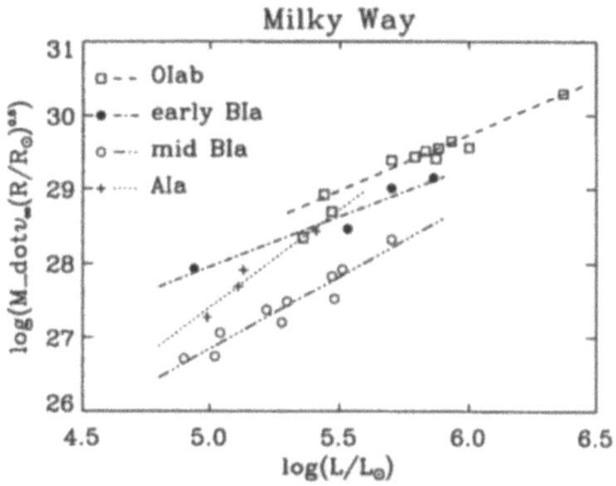

Fig. 2. Wind momenta as function of luminosities for galactic supergiants of spectral types O, B (early B: B0 to B1; mid B: B1.5 to B3) and A. The straight lines for each group of spectral types are the corresponding regression curves. (From [9]).

Fig. 2 combines the results from [13] and [9] with regard to the observed wind momentum and reveals the clear existence of a WLR which, however, appears to vary as function of spectral type. From the viewpoint of the theory this variation of the WLR in slope and height between the spectral types is expected not only because of ionization changes but also because of the different spectral locations of the photon momentum absorbing lines with regard to the flux maximum and absorption edges such as the hydrogen Lyman - and Balmer - edges. Most recent theoretical work ([14]), [6], [10]) indicates agreement with the observed trends.

3 The WLR and Extragalactic Distances

Evidently, the existence of the WLR provides a unique opportunity to obtain information about absolute magnitudes from a purely spectroscopic analysis of the stellar wind lines. The idea is to carry out multi-object spectroscopy with 8m-class telescopes combined with HST photometry to obtain stellar parameters ($T_{\rm eff}$, $\log g$, abundances) and reddening in a first step. Then determination of $\dot{M}$ from H_α and the application of a properly calibrated WLR would yield the absolute magnitude. Comparison with the de-reddened apparent magnitude would, finally, allow to determine the distance. The best spectroscopic targets at large distances are certainly A-supergiants such as shown in Fig. 3, where the results obtained by [11] with Keck HIRES are included. Since massive stars evolve at almost constant luminosity towards the red, A-supergiants are the optically brightest "normal" stellar objects because of the effects of Wien's law on the bolometric correction. In addition, we can determine the wind momenta of these objects solely by optical spectroscopy at H_α (without need of the UV). This means that we can use

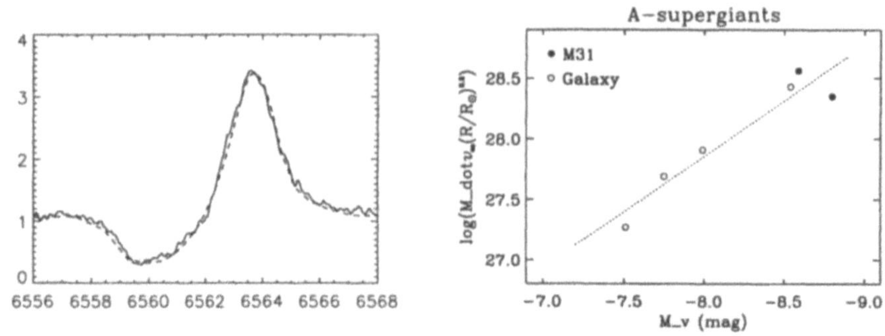

Fig. 3. Left: H_α fit of the Keck HIRES spectrum of the M31 A2 Ia-O supergiant 41-3654 yielding a precise determination of $\dot{M}$ and v_∞. Right: Wind momentum of galactic and M31 A-supergiants as function of absolute visual magnitude. The data of the M31 objects are from [11]. The dashed curve is the linear regression obtained from all objects.

ground-based telescopes of the 8m class for spectroscopy rather than the "tiny" HST (which is then needed for accurate photometry only).

From Fig. 3 we see that the brightest A-supergiants have absolute magnitudes between -9^m and -8^m. Such objects would be of apparent magnitude between 20^m and 21^m in galaxies 6 Mpc away, certainly not a problem for medium (2 Å) resolution spectroscopy with 8 m class telescopes. Even in a galaxy like M100 at a distance of 16 Mpc ([4], [2]) these objects would still be accessible at magnitudes around 22.5 and would yield wind momentum distances. Indeed, the HST colour magnitude diagram published by [3] may show the presence of such objects in M100.

One might ask, of course, whether a medium resolution of 2 Å would still allow a sufficiently accurate determination of effective temperature, surface gravity, abundances, and wind momentum. We note that the rotational velocities of A-supergiants are typically on the order of 40 km s^{-1}, a broadening which is matched by a spectral resolution of roughly 1.4 Å at H_α. Therefore we expect that the medium resolution situation will not be dramatically worse, provided sufficient S/N can be achieved and accurate sky-, galaxy-, and HII region background subtraction can be performed at very faint magnitudes. Experiments with observed spectra degraded with regard to S/N, resolution and sky emission indicate that such observations are challenging but feasible.

[11] and [9] conclude from the dispersion of the observed WLR that with ten to twenty objects per galaxy, distance moduli as accurate as 0.1 mag should be achievable. These uncertainties in WLR distance moduli appear to be comparable to those obtainable from Cepheids in galaxies. The advantage of the WLR-method, however, is that individual reddening (and therefore extinction) as well as metallicity can be derived directly from the spectrum of every object. Moreover, it is a new independent primary method for distance

determination and can contribute to the investigation of systematic errors of extragalactic distances.

Acknowledgements

This work is a collaborative effort of many colleagues at different institutions. I wish to thank Danny Lennon and Stephen Smartt (ING, La Palma), Artemio Herrero and Ilu Monteverde (IAC, Tenerife), Jim McCarthy (Caltech), Kim Venn (Macalester, Minnesota) and Joachim Puls, Johannes Reetz, Norbert Przybilla, Oliver Knörndel, Uwe Springmann (USM Munich) for their continuous contributions to our joint "extragalactic stellar astronomy project". In addition, I wish to express my gratitude to all my colleagues from the **FORS TEAM**. Their incredible team spirit, work morale, devotion and relaxed humour also in critical situations has been an experience which I will never forget.
The funding through the "Verbundforschung Astronomie" by the Bundesminister für Forschung und Bildung and the DLR under grants 05 MU104 and 50 R93040 is gratefully acknowledged.

References

1. Castor, J.I., Abbott, D.C.,& Klein, R.I. 1975, ApJ 195, 157
2. Ferrarese, L. *et al.* 1996, ApJ **464**, 568
3. Freedman, W., *et al.* 1994a, ApJ **435**, L31
4. Freedman, W. *et al.* 1994b, Nature **371**, 757
5. Kudritzki, R.P., Pauldrach, A.W.A.,, Puls,J., & Abbott, D.C. 1989, A&A **219**, 205
6. Kudritzki, R.P., Springmann, U., Puls, J., Pauldrach, A.W.A., Lennon, M. 1998a, ASP Conf. Series Vol. **131**, p. 299 - 308
7. Kudritzki, R.P. 1998, "Quantitative Spectroscopy of the Brightest Blue Supergiant Stars in Galaxies" in Proc. of "Stellar Astrophysics for the Local Group", VIII Canary Island Winterschool for Astrophysics, eds. A. Aparicio et al., Cambridge University Press, p. 149-262, invited lectures
8. Kudritzki, R.P., 1999a, in Proc. of STScI May 1998 Symposium "Unsolved Problems of Stellar Evolution", ed. M. Livio, invited paper, in press
9. Kudritzki, R.P., Puls, J., Lennon, D.J., Venn, K.A., Reetz, J., Herrero, A., & McCarthy, J.K. 1999, A&A , submitted
10. Kudritzki, R.P., Springmann, U., Puls, J., Pauldrach, A.W.A., Lennon, M. 1999b, A&A , in preparation.
11. McCarthy, J.K., Kudritzki, R.P., Lennon, D.J., Venn, K.A., Puls, J. 1997, ApJ **482**, 757
12. Pauldrach, A.W.A., Puls, J., & Kudritzki, R.P. 1986, A&A 164, 86
13. Puls, J., Kudritzki, R.P., Herrero, A., Pauldrach, A., Haser, S.M., Lennon, D.J., Gabler, R., Voels, S.A., Vilchez, J.M., Wachter, S., & Feldmeier, A. 1996, A&A **305**, 171
14. Puls, J., Springmann, U., Lennon, M. 1999, A&A, submitted
15. Santolaya-Rey, E., Puls, J., & Herrero, A. 1997, A&A **323**, 488

Extragalactic Planetary Nebulae and What to Do with Them

Roberto H. Méndez

Munich University Observatory, Scheinerstr. 1, 81679 Munich, Germany

Abstract. More than 2500 planetary nebulae (PNs) have been discovered in galaxies other than our own. Observational efforts involving them are justified because they are very useful in a variety of astrophysical problems. Their observed luminosity function is an important tool for extragalactic distance determinations. Their detectability in the outskirts of elliptical galaxies, and their easily measured radial velocities, make PNs ideal test particles to study the angular momentum and dark matter distribution in such galaxies. The PNs are also efficient tracers of diffuse intracluster stellar populations in clusters like Virgo and Fornax. PN abundances are valuable tools for studies of stellar and galactic evolution. I present a summary of these subjects and describe the expected impact of VLT on them.

1 Introduction

Planetary nebulae (PNs) are among the most easily detectable individual objects in any galaxy. This is due partly to the high stellar luminosities at the tip of the AGB, but mostly to the fact that, when the central star becomes sufficiently hot, the previously ejected envelope, now ionized, transforms a lot of stellar UV radiation into visible light: the PN becomes much brighter than its central star. In fact, sometimes we do not see the star, because it is overwhelmed by the brightness of the PN. In many cases the [O III] emission line $\lambda5007$, which is frequently the brightest emission line in the visible PN spectrum, can radiate away more than 10% of all the stellar luminosity.

In principle, PNs in other galaxies can be identified by comparing two CCD images: one taken through a narrow "on-band" filter (FWHM$\sim 30-50$ Å) centered at the redshifted wavelength of the [O III] $\lambda5007$ nebular emission line, and another one taken through an "off-band" filter at a nearby wavelength ($\sim$ 5300 Å). The PNs, which are unresolved point sources at typical extragalactic distances, are detected by blinking the on-band vs. the off-band image. In practice, combinations of several on- and off-band images are used, in order to detect fainter objects and eliminate CCD defects and cosmic ray events. In cases of bright and/or complex background, the off-band image can be scaled to, and then subtracted from, the on-band image, in such a way that the background continuum is minimized. Photometric measurements and calibration (accomplished using images of standard stars) give the nebular fluxes emitted in the $\lambda5007$ line. The fluxes can be transformed into magnitudes using the following expression:

$$m(5007) = -2.5 \log I(5007) - 13.74 \quad (1)$$

(Jacoby 1989), where $I(5007)$ is the nebular flux, expressed in erg cm^{-2} s^{-1}, and the constant 13.74 is derived from the monochromatic flux of a star with V=0, and from the equivalent width of the Johnson V filter. For example, the brightest PN measured by Ciardullo et al. (1989) in M 31 has $m(5007) = 20.28$. For a distance of 770 kpc ($m - M = 24.42$; Freedman and Madore 1990) and an extinction correction of 0.28 mag, we obtain the corresponding absolute magnitude $M(5007) = -4.42$.
In sections 2 to 6 the following uses of extragalactic PNs are described:

- Distance determinations using the PN luminosity function (PNLF) as a standard candle.
- Dynamical studies in elliptical galaxies using PN radial velocities; this provides strong constraints on dark matter distribution.
- Detection of the diffuse intracluster stellar population in clusters of galaxies (e.g. Virgo, Fornax). The PN radial velocities and abundances may give clues about the origin of this diffuse intracluster stellar population.
- PN abundances are a tool for studies of metallicity and metallicity gradients in elliptical galaxies (similar role as H II regions in spirals and irregulars).
- The shape of the PNLF offers a possible constraint on post-AGB evolution theory: H burners vs. He burners.

Section 7 is a summary of the expected impact of VLT+FORS in this field.

2 The PN Luminosity Function: Distances

Knowing the apparent magnitudes we can build the planetary nebula luminosity function (PNLF). The [O III] λ5007 PNLF describes how many PNs there are at each value of $m(5007)$. Sometimes it is better to use the **cumulative** PNLF, which gives the number of PNs **brighter** than a given magnitude. When we run out of PNs, the PNLF drops to zero, while the cumulative PNLF becomes a horizontal line.

The PNLF method of distance determination is based on the assumption that the "bright end" of the PNLF has the same shape and absolute brightness everywhere in the universe. This universal bright end is represented (Ciardullo et al. 1989) by the following expression for the number of PNs at magnitude M:

$$N(M) = c_1 \exp(c_2 M)\ [1 - \exp(3(M_* - M))] \quad (2)$$

which can be described as an "exponential with cutoff", where c_1 is a suitable *positive constant*, c_2 = *0.307*, and M_* is the $M(5007)$ of the cutoff. This function is calibrated by fitting it to the observed PNLF of the M 31 bulge. If we adopt the distance to M 31 mentioned above, $M_* = -4.54$. Now we can fit any other observed PNLF to the calibrated function, using for example the method of maximum likelihood. For details about this procedure please refer to Ciardullo et al. (1989). The maximum likelihood fit gives a distance modulus and information about the total number density of PNs per sampled solar luminosity, through the constant c_1. In order to obtain an unambiguous fit, the observed PNLF must show a recognizable change of slope, i.e. going from the cutoff zone into the exponential zone. In practice this is the most delicate step in the method: to ascertain that the change of slope is real and not attributable to the onset of incompleteness effects. For this purpose it is advisable to observe the PNLF down to $M(5007) = -3.0$.

The PNLF method shows very good internal consistency, in the sense that all galaxies belonging to a cluster give similar distances. This statement is true even when applied to galaxies of different Hubble types (see e.g. Jacoby 1997 and references therein). There is also excellent agreement with cepheid distances, although here we have a minor problem: the direct comparison with cepheid distances involves only 7 galaxies at rather small distances, up to 10 Mpc. In the cases of the Virgo and Fornax clusters the galaxies with PNLF distances are not the same as those with cepheid distances. This produces some uncertainty from depth effects within each cluster. Even so, the agreement between cepheid and PNLF distances is satisfactory (Jacoby 1997, Méndez 1999). The PNLF method appears to be one of the most reliable and thoroughly tested methods for extragalactic distance determinations. VLT+FORS will surely add many PNLF distance estimates to the $\sim$ 30 already known.

3 PN Kinematics

If a PN has been detected, it is not difficult to obtain a moderate resolution spectrogram and measure a reliable radial velocity, because all the detected photons belong to the [O III] λ5007 emission line. Because of this, in addition to their usefulness for distance determinations, PNs are ideal test particles for kinematic studies.

Most of the information we have about dark matter associated with individual galaxies comes from observations of the rotation curves of spiral galaxies; the presence of high rotation velocities at large distances from the center of the galaxy forces us to conclude that there must be a lot of matter beyond the optical disk of the galaxy. It is not possible to make this kind of studies for elliptical galaxies, because they lack neutral H (which is needed to produce the 21 cm H I emission used to measure velocities). One way to proceed in this case is to obtain integrated spectrograms of the stellar population, and to derive information from the absorption line profiles. This requires a very good signal to noise ratio, which forces us to work not too far from the centers of elliptical galaxies, where the surface brightness is sufficiently high.

This explains why PNs are so useful: they are preferentially detected in the outskirts of elliptical galaxies, where the stellar background is fainter. Therefore they can be used to probe regions where absorption line studies are impossible. These are precisely the regions where the effects of dark matter are expected to be most important. Efforts are currently under way to map the dark matter distribution in elliptical galaxies using PNs. One example of this kind of work is the paper by Hui et al. (1995) on 433 PNs in the galaxy NGC 5128.

One aspect to consider is that in order to get significant dynamical constraints we need many PNs (several hundreds). The cumulative PNLF can be used to estimate how faint we need to go if we want to collect so many PNs. In the case of a giant elliptical galaxy, we need to detect $M(5007) = -1.5$.

There is yet another problem: we need an efficient method to measure the radial velocities. The FORS slitlets allow to measure only 19 objects at a time. In December 1998 we have used NTT+EMMI at La Silla to test another technique, based on taking direct images and grism slitless spectrograms, both through the on-band filter at the redshifted λ5007 in order to keep the background light to a minimum. This technique has given promising results, and we expect to be able to extract the radial velocities of many more PNs per exposure than possible with the FORS slitlets.

4 Intracluster PNs

The first few intracluster PNs in the Virgo cluster were serendipitously identified by Arnaboldi et al. (1996). Many more (including a few candidates in the Fornax cluster) were found soon thereafter (Méndez et al. 1997, Theuns and Warren 1997, Feldmeier et al. 1998). These PNs are ideal tracers of the intracluster stellar population, and their properties may offer vital clues on the origin of this population. For more details please refer to the paper by Arnaboldi in these proceedings.

5 Abundances of Extragalactic PNs

PN abundances are a tool for studies of metallicity, metallicity dispersion and gradients in galaxies without H II regions. Besides, there is a clear correlation between the PN formation rate (derived from PNLF studies) and the color of the galaxy (e.g. Peimbert 1990; Ciardullo et al. 1991), which may offer clues on the effect of metallicity on the AGB mass loss rates. Therefore the direct determination of PN abundances in many elliptical galaxies would be valuable for several aspects of stellar and galactic evolution.

PN abundance determinations at large distances are a very ambitious task because some essential diagnostic lines are much fainter than [O III]

5007. For example, the intensity of [O III] 4363 relative to 5007 is a sensitive indicator of electron temperature. The knowledge of the ratio 4363/5007 is a prerequisite for accurate abundances from collisionally excited lines in the visible spectrum. For typical electron temperatures between 10000 and 20000 K, the ratio 4363/5007 lies between 0.007 and 0.036, respectively. At the distance of the Virgo cluster, this implies that we need to detect fluxes $\sim 10^{-18}$ erg cm^{-2} s^{-1}. This is probably at the limit of what VLT+FORS can do.

However, there is no lack of nearer targets; up to now the only galaxies where a substantial number of PN abundances have been determined are the Magellanic Clouds (Dopita et al. 1997) and M 31 (Richer et al 1999, Jacoby and Ciardullo 1999).

6 On the Shape of the PNLF

The empirical formula (section 2, Eq. 2) used by Ciardullo et al. (1989) has no firm theoretical basis. Or rather, its theoretical basis is an oversimplified view of post-AGB evolution, which involves uniformly expanding, homogeneous spherical PNs ionized by non-evolving central stars. The numerical simulations of the PNLF (e.g. Méndez and Soffner 1997) lead to expect a somewhat different shape. Eq. (2) gives an ever increasing PNLF towards fainter magnitudes (the exponential part), while the numerical simulations show a roughly constant, or even decreasing, PNLF between $M(5007) = -3.5$ and -2.3. The simulated PNLF only starts increasing again for mags fainter than -2.3 (of course the PNLF must drop again as we go to sufficiently faint magnitudes, but here we are not concerned about that). This deficiency of PNs near -2.5 can be considered a “prediction” of the simulation procedure. It is almost certainly attributable to the accelerated evolution of H-burners as they enter the white dwarf cooling tracks; central stars in this comparatively depleted section of the log L - log $T_{\rm eff}$ diagram (see Méndez and Soffner 1997 or Méndez 1999) tend to produce PNs near $M(5007) = -2.5$. If we could solve the problem of generating a full set of post-AGB tracks for He-burners, the shape of the PNLF might become a way to put constraints on the ratio of post-AGB H-burners to He-burners.

The observed PNLF of the bulge of M 31 appears to agree better with the numerical simulation than with Eq. (2); see Méndez (1999). It would be very interesting to obtain statistically complete PN samples reaching similarly faint magnitudes in a few other galaxies, in order to test the universality of the PNLF shape. This would allow to test how reliable is the Monte Carlo procedure to generate a simulated PNLF, eventually indicating if some adjustments are needed. A successful test would give us more confidence about using the simulated PNLF for distance determinations, and might even teach us something about post-AGB evolution.

7 What can be done with FORS

Let us estimate the FORS performance limits for work on extragalactic PNs. Assume a total exposure time of 6 hours, a seeing of 0.5 arc sec, a monochromatic sky brightness of 22.7 mag per sq arc sec at 5000 Å, and a CCD quantum efficiency of 95%. In these conditions the faintest detectable m(5007) is expected to be 31, which is equivalent to a flux 1.27×10^{-18} erg cm^{-2} s^{-1}. If we can detect such fluxes we can do the following:

- Detection of several of the brightest intracluster PNs ($M(5007) = -4$) at distances up to 100 Mpc; for example, in the Coma cluster.
- PNLF distances and PN formation rates at 60 Mpc (we need to detect $M(5007) = -3$).
- Dynamical studies in giant elliptical galaxies at 30 Mpc (PNs as kinematic test particles; about 500 PNs are required; the cumulative PNLF indicates that we need to detect $M(5007) = -1.5$).
- PN abundances at 15 Mpc (we need to detect [O III] 4363, which is roughly a factor 100 fainter than 5007).

Acknowledgements

This work has been supported by the Deutsche Forschungsgemeinschaft through Grant SFB (Sonderforschungsbereich) 375.

References

1. Arnaboldi M., Freeman K.C., Méndez R.H. et al. 1996, ApJ 472, 145
2. Ciardullo R., Jacoby G.H., Ford H.C., Neill J.D. 1989, ApJ 339, 53
3. Ciardullo R., Jacoby G.H., Harris W.E. 1991, ApJ 383, 487
4. Dopita M.A., Vassiliadis E., Wood P.R. et al. 1997, ApJ 474, 188
5. Feldmeier J.J., Ciardullo R., Jacoby G.H. 1998, ApJ 503, 109
6. Freedman W.L., Madore B.F. 1990, ApJ 365, 186
7. Hui X., Ford H.C., Freeman K.C., Dopita M.A. 1995, ApJ 449, 592
8. Jacoby G.H. 1989, ApJ 339, 39
9. Jacoby G.H. 1997, in: Habing H.J., Lamers H.J.G.L.M. (eds) Proc. IAU Symp. 180, Planetary Nebulae, Kluwer, p. 448
10. Jacoby G.H., Ciardullo R. 1999, ApJ in press
11. Méndez R.H. 1999, in: Heck A., Caputo F. (eds) "Post-Hipparcos Cosmic Candles", Kluwer, Ap. Sp. Sci. Library 237, 161
12. Méndez R.H., Guerrero M.A., Freeman K.C. et al. 1997, ApJ 491, L23
13. Méndez R.H., Soffner T. 1997, A&A 321, 898
14. Peimbert M. 1990, Rev. Mexicana Astron. Astrofis. 20, 119
15. Richer M.G., Stasinska G., McCall M.L. 1999, A&AS, in press
16. Theuns T., Warren S.J. 1997, MNRAS 284, L11

Towards Understanding the Planetary Nebula Luminosity Function

Paola Marigo[1,2], Achim Weiss[2], Martin A.T. Groenewegen[2], Léo Girardi[1,2]

[1] Dipartimento di Astronomia, Vicolo dell'Osservatorio 5, I-35122 Padova, Italy
[2] Max-Planck-Institut für Astrophysik, Karl-Schwarzschild-Str. 1, D-85740 Garching bei München, Deutschland

Abstract. We report on a theoretical work in progress about the planetary nebula luminosity function. A new synthetic model is developed to follow the evolution of a planetary nebula and its central star, starting from the onset of AGB phase up to the white dwarf cooling sequence. Both H- and He-burning Post-AGB tracks are used. Basic properties of the circumstellar nebula are predicted, namely: the dynamics and chemical composition of the stellar ejecta, the photoionisation of H and He by the central star, the formation of a few relevant optical emission lines (e.g. Hβ; HeIIλ4686; [OIII]λ5007). As a first application, the planetary nebula luminosity function in the LMC is analysed adopting a population synthesis approach. It turns out that a reasonable reproduction of the observed bright end in the [OIII]λ5007 luminosity cannot be obtained assuming that all central stars are H-burners, but rather considering that the observed population of PNe is excited by a mixture of H- and He-burner central stars.

1 Introduction

The bright end of the planetary nebula luminosity function (PNLF) has been suggested to be used as an extragalactic distance indicator, given its apparent nearly invariance with respect to age population and metallicity (see, for instance, Mendez & Soffner 1997 and references therein; cf. also Mendez, this conference).

In this perspective, an improved synthetic model for PN evolution is being developed (Marigo et al., in preparation) in order to investigate the sensitiveness of the bright cut-off of the PNLF to various possible factors, such as: metallicity of the progenitor stars, mass-loss during the AGB phase, initial-final mass relation, transition time from the tip of the AGB to the onset of ionisation, dynamical expansion of the nebula, evolutionary properties - luminosities and time-scales - of the central star depending on either H- or He-burning nuclear regimes, star formation history of the parent galaxy.

2 A New Synthetic Model of PN Evolution

The main ingredients of the model are summarised in the following. The AGB phase is described by detailed synthetic models covering the evolution of low-/intermediate-mass stars from the first thermal pulse up to the complete ejection of the envelope (Marigo 1998; Marigo et al. 1996, 1998, 1999). These models provide the mass of the central star as a function of the initial mass and metallicity of the progenitor, the density and the chemical composition profiles of the ejected material as a function of time (assuming spherical symmetry), and the possible dynamical interaction of AGB shells ejected at different instants with different velocities.

Post-AGB tracks are taken from Vassiliadis & Wood (1994) for both H- and He-burning central stars over the relevant range of stellar masses and metallicities.

The prescription for the fast wind during the Post-AGB phase is derived from Pauldrach et al. (1988). The dynamical evolution of a PN is described adopting the prescriptions of Kahn & Breitschwerdt (1990) for the momentum-driven phase and the interacting stellar winds model by Volk & Kwok (1985) for the energy-driven phase. The dynamical effect of the shell thickening caused by ionisation is included following Kahn (1983). The fluxes

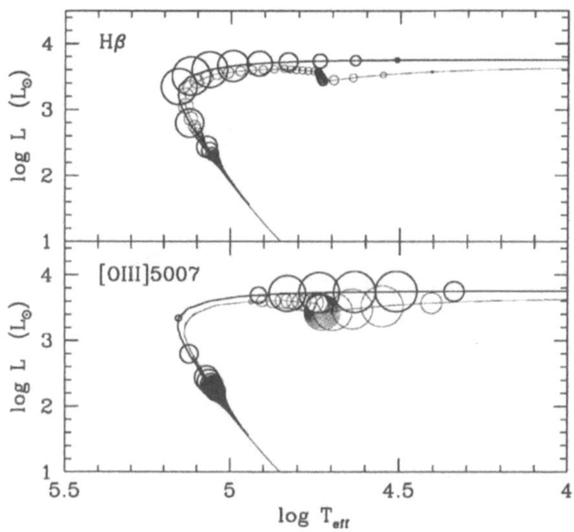

Fig. 1. Representation of the detection probability (circles) of a PN ionised by either a H- or a He-burner central star (thick and thin line, respectively) in the luminosity emitted in the Hβ line (top panel) and [OIII]λ5007 line (bottom panel). See text for further explanation.

emitted by the central star for the ionisation of both H and He^+ are specified as a function of the effective temperature and surface gravity, according to non-LTE model atmospheres (Clegg & Middlemass 1987). The intensities of few relevant optical emission lines (Hβ, HeIIλ4686, [OIII]λ5007) are calculated assuming a constant electron temperature $T_{\rm el} = 10^4 K$. This approximation will be improved with the adoption of the Cloudy photoionisation code.

3 H-Burner versus He-Burner Central Stars

The detection probability of a PN depends on whether its central star is H- or He-burner. The circles along the Post-AGB tracks shown in Fig. 1 are equally spaced in time (10^3 yr) and have a radius proportional to the logarithm of the luminosity emitted by the nebula both in the Hβ line (top panel) and in the [OIII]λ5007 line (bottom panel). In the case of Hβ emission, the detection probability increases as the H-burner central star heats up and then reduces as both bolometric luminosity and effective temperature decrease. The distribution of circles along the He-burning track reflects its systematically lower bolometric luminosity and the longer evolutionary time-scales. In the case of [OIII]λ5007 emission a notable feature is the nearly absence of this line as the central star passes through the hottest knee of the track, since at these temperatures the bulk of oxygen is expected to be three (and four) times ionised.

4 The Bright Cut-Off of the PNLF in the LMC

The theoretical PNLF is constructed combining PN models (corresponding to central stars in the mass range $0.6M_{\odot} - 1.1M_{\odot}$ and initial metallicity $Z = 0.008$), with an initial-final mass relation consistently derived from AGB calculations, and simple prescriptions for the history of the star formation rate (constant) and the initial mass function (Salpeter). The results are shown in Fig. 2. It turns out that assuming all central stars of PNe in the LMC are H-burners, the observed bright end is not reproduced. A steady decrease of the distribution towards high luminosities (rather than a sharp cut-off) is predicted, and the brightest end is fainter than the observed one (top right-hand side panel). Part of the discrepancy is removed when using He-burning Post-AGB tracks to construct the PNLF (bottom left-hand side panel), and a satisfactory agreement is finally obtained assuming a mixture of H- and He-burning central stars in the proportions suggested by Dopita et al. (1996, bottom right-hand side panel). The main reason for this resides in the longer evolutionary time-scales characterising the He-burning tracks, which results in a higher probability of detecting PNe in the bright bins of the [OIII]λ5007 luminosity distribution.

Acknowledgments P. Marigo is grateful to the MPA (Garching) where most of this work has been carried out, and to ESO for financial support to attend the workshop.

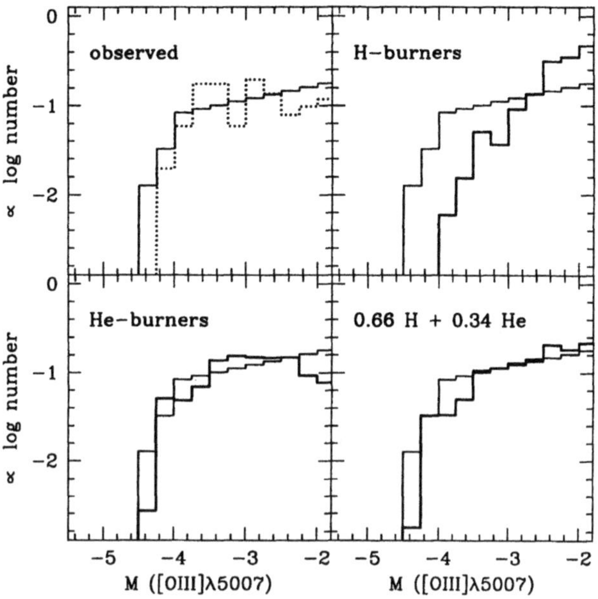

Fig. 2. Bright cut-off of the PNLF in the LMC. Top left-hand side panel: observed distribution according to the original data sample from Jacoby et al. (1990; dashed line), and including the completeness correction suggested by Ciardullo (1995; thin solid line). The absolute magnitude M([OIII]λ5007) is calculated according to the definition by Jacoby (1989). Normalisation is applied for M([OIII]λ5007)≤ -2. Other panels: comparison between observed (thin solid line) and theoretical (thick solid line) PNLFs. See text for more details.

References

1. Ciardullo R., (1995) Highlights of Astron. **10**, 507
2. Clegg R. E. S., Middlemass D. (1987) Mont. Not. Roy. Astr. Soc **228**, 759–778
3. Dopita M. A., Vassiliadis E., Meatheringham S. J. et al. (1996) Astrophys. Journ. **460**, 320–341
4. Jacoby G. H. (1989) Astrophys. Journ. **339**, 39–52
5. Jacoby G. H., Walker A. R., Ciardullo R. Astrophys. Journ. **365**, 471–477
6. Kahn F. D. (1983) Planetary nebulae IAU Symp. **103**, Dordrecht, D. Reidel Publishing Co., 305–315
7. Kahn F. D., Breitschwerdt D. (1990) Mont. Not. Roy. Astr. Soc **242**, 505–511
8. Marigo P. (1998) Astron. Astrophys. **340**, 463–475
9. Marigo P., Bressan A., Chiosi C. (1996) Astron. Astrophys. **313**, 545–564
10. Marigo P., Bressan A., Chiosi C. (1998) Astron. Astrophys. **331**, 564–580
11. Marigo P., Girardi L., Bressan A. (1999) Astron. Astrophys. **344**, 123–142
12. Mendez R. H., Soffner T. (1997) Astron. Astrophys. **312**, 898–906
13. Pauldrach A., Puls J., Kudritzki R. P. (1988) Astron. Astrophys. **207**, 123–131
14. Vassiliadis E., Wood P. R. (1994), Astrophys. Journ. Suppl. Ser. **92**, 125–144
15. Volk K., Kwok S. (1985) Astron. Astrophys. **153**, 79–90

HII Galaxies as Cosmological Probes

Jorge Melnick[1], Roberto Terlevich[2], Elena Terlevich[2], and Fernando Selman[1]

[1] European Southern Observatory, La Silla, Chile
[2] Institute of Astronomy, Cambridge, UK,
[3] Instituto Nacional de Astronomía, Optica y Electrónica, Puebla, México

Abstract. Owing to their large luminosities per unit mass HII galaxies allow us to investigate in considerable detail the process of high mass star formation out to redshifts of cosmological importance. In this contribution we discuss the use of the correlation between the Hβ luminosity and the emission line-profile widths of HII galaxies as a distance indicator. We show that HII galaxies can be reliably used as distance indicators out to redshifts $z > 3$ provided that their line widths, fluxes, extinctions, and metallicities can be accurately measured. FORS, ISAAC, and FLAMES are the ideal instruments for this challenging investigation.

1 Introduction

Observations of distant type Ia Supernovae (SNIa) provide credible evidence that the Universe may well be accelerating (Riess *et al.* 1998; Perlmutter *et al.* 1998). This discovery opens fascinating new perspectives in cosmology, and in particular, is turning Einstein's biggest blunder – the Cosmological Constant Λ – into one of his greatest hits. When the results from the SNIa Hubble diagram are combined with other cosmological tests, such as the CMWB anisotropy, large-scale structure, or others (see eg. Turner, 1999 for a comprehensive review) the rather robust result emerges that Λ contains up to 80% of the total energy density of the Universe. But an energy density of vacuum is not the only possibility to explain the acceleration and several authors have explored the possible existence of a quasi-homogeneous, time-varying scalar field with negative pressure known as 'Quintessence' (because it would be the fifth essential component of the Universe together with photons, baryons, dark-matter, and neutrinos; Turner, 1999).

A very interesting feature of cosmic models with Λ is the so called *focusing effect* mentioned by Refsdal *et al.* (1967) and recently discussed in detail by Melnick, Terlevich, and Terlevich (1999; MTT99). Focusing is also there for Quintessence. Figure 1 presents the Hubble diagram in FRW metric with two extreme forms of Quintessence parametrized by the equation of state $\omega = P/\rho$ (pressure over density). $\omega = -1$ corresponds to vacuum energy. The plots show world lines for three representative values of matter density Ω_M, and five different values of energy density of Quintessence, Ω_Λ.

The focusing effect is clearly seen in these figures as the point where, for a given $\Omega_M (> 0.2)$ the different Ω_Λ curves cross. The redshift of the

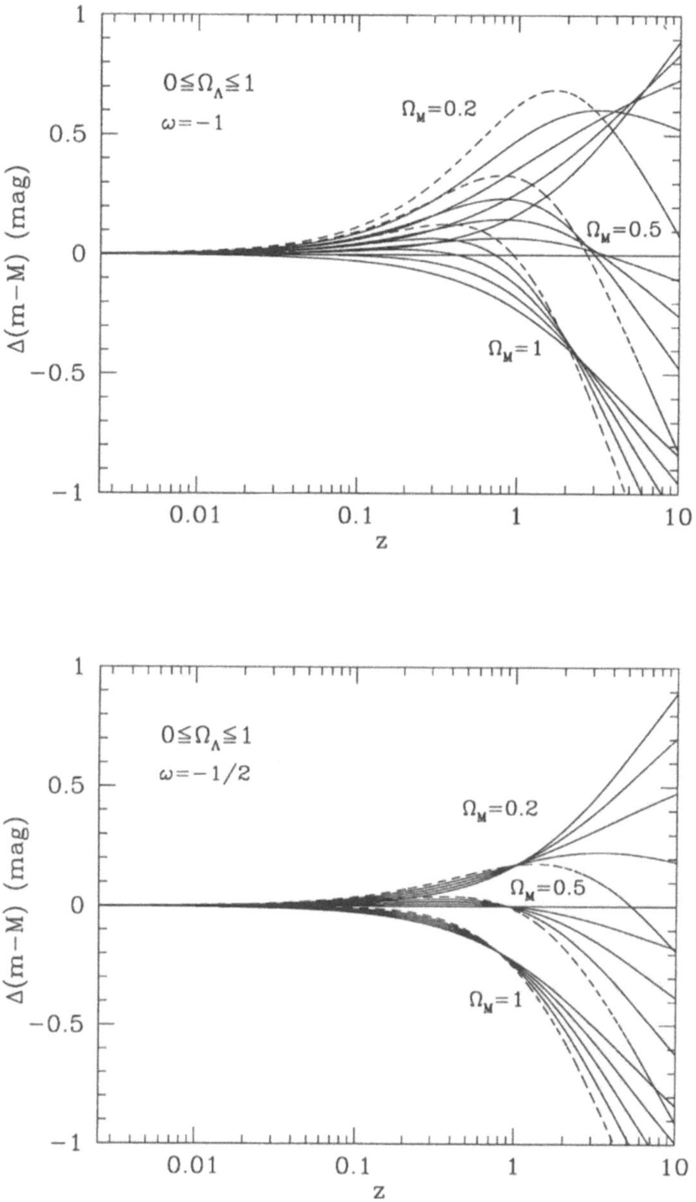

Fig. 1. Normalized Hubble diagrams for models with two different kinds of Quintessence parametrized by the equation of state $\omega = P/\rho$. Each plot shows curves for three different values of matter energy density (Ω_M) and five different values for the energy density of Quintessence (Ω_Λ) ranging from 0 to 1 in steps of 0.25. The *dashed lines in each plot* correspond to $\Omega_\Lambda = 1$.

focusing depends on the equation of state. For $\omega = -1/2$, Ω_Λ is very poorly constrained at $z < 1$, but Ω_M can be constrained very well at z=1. For $\omega = -1$ (Vacuum) the focusing occurs at z=2-3 where there is a very good discrimination of Ω_M of about 1 mag, while models with different Ω_Λ separate out very nicely at $z \sim 1$.

These plots clearly illustrate the cosmological importance of having a distance indicator that can be used over a wide range of redshifts. In this contribution we discuss the use of HII galaxies as distance indicators out to at least z=3.

2 The L(Hβ) $-$ σ Relation for HII Galaxies

Terlevich and Melnick (1981; TM81) showed that there is a very strong correlation between line-width and Hβ luminosity for HII galaxies. Based on the similarity between this correlation and the L(Hβ) $-$ σ relation for ellipticals and spiral bulges, they argued that the widths of the emission lines in HII galaxies could be used to measure their total masses. The correlation was calibrated as a distance indicator by Melnick *et al.* (1988; MTM88) who showed that, in fact, the correlation is a plane with oxygen abundance (O/H) as third parameter.

Figure 2 shows the L(Hβ) $-$ σ relation plotted using the data of MTM88 for local objects plus galaxies at intermediate and high redshifts from the literature (Koo *et al.*, 1995; Guzman *et al.*, 1997; Pettini *et al.*, 1998). The solid line shows the MTM88 fit to the local objects. High-z HII galaxies have luminosities about one order of magnitude larger than the range covered by their local and intermediate redshift counterparts. Therefore, a solid understanding of the underlying astrophysics is required to apply the correlation to calculate distances to high-z HII galaxies.

3 HII Galaxies as Distance Indicators

A detailed discussion of the use of the the L(Hβ) $-$ σ relation as distance indicators has recently been presented by MTT99 and will not be repeated here for reasons of space. Instead I would like to examine the nature of the correlation between L(Hβ) and σ. Of course, understanding the properties of galaxy-sized starbursts is important for cosmology per-se because the majority of galaxies at $z \geq 3$ appear to be young starbursting systems (see contribution by M. Pettini in this conference).

In his thesis, Edu Telles (1995) measured radii and Blue luminosities for a sample of nearby HII galaxies in order to explore in detail the similarities between the dynamics of normal Ellipticals and HII galaxies. Figure 3 summarized his results. The correlations between L_B and σ and between R and σ are very similar both in slope and zero point. For Ellipticals these correlations are in fact two different projections of the so-called Fundamental

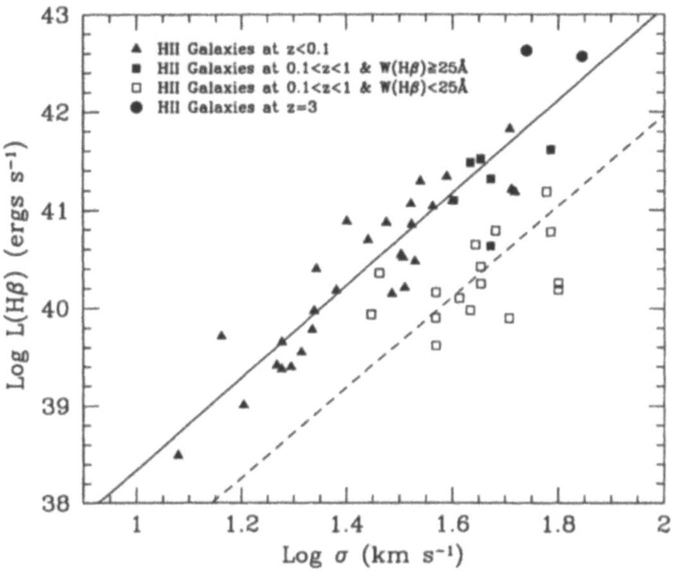

Fig. 2. The $L(H\beta) - \sigma$ correlation for HII galaxies over a wide range of redshifts. The solid line shows the Maximum-Likelihood fit to the $z < 0.1$ galaxies. The dashed line shows the predicted correlation for evolved galaxies (open squares).

Plane. Telles went on to show that HII galaxies also exhibit a well defined fundamental plane and that their Blue luminosity scales approximately as $R\sigma^2$.

These results lend strong support to the hypothesis of TM81 that the ionized gas in HII galaxies is in Virial equilibrium with the total mass of the young component. Extinction and metallicity corrections to the emission line luminosities are not a major concern because they can be directly measured from the spectrum (MTM88, MTT99). If we restrict the sample to objects with very strong ($W(H\beta) > 25$Å) and very narrow ($\sigma < 65 kms^{-1}$) lines we effectively select the youngest objects for which σ is dominated by the young stellar component (MTM88). In this case, the $L(H\beta) - \sigma$ relation is the ratio between total mass and mass of ionizing stars and should be a power-law if the IMF is a power-law of constant slope.

3.1 The IMF of Starburst Clusters

There have been many claims in the literature about variations of the IMF for different stellar systems and different environments (see e.g. Scalo, 1998 for a review). Using new NTT data for the 30 Doradus cluster we have uncovered a new selection effect that we call the *Magnitude Limit effect,* that has been *neglected in most* determinations of the IMF of young clusters (Selman *et*

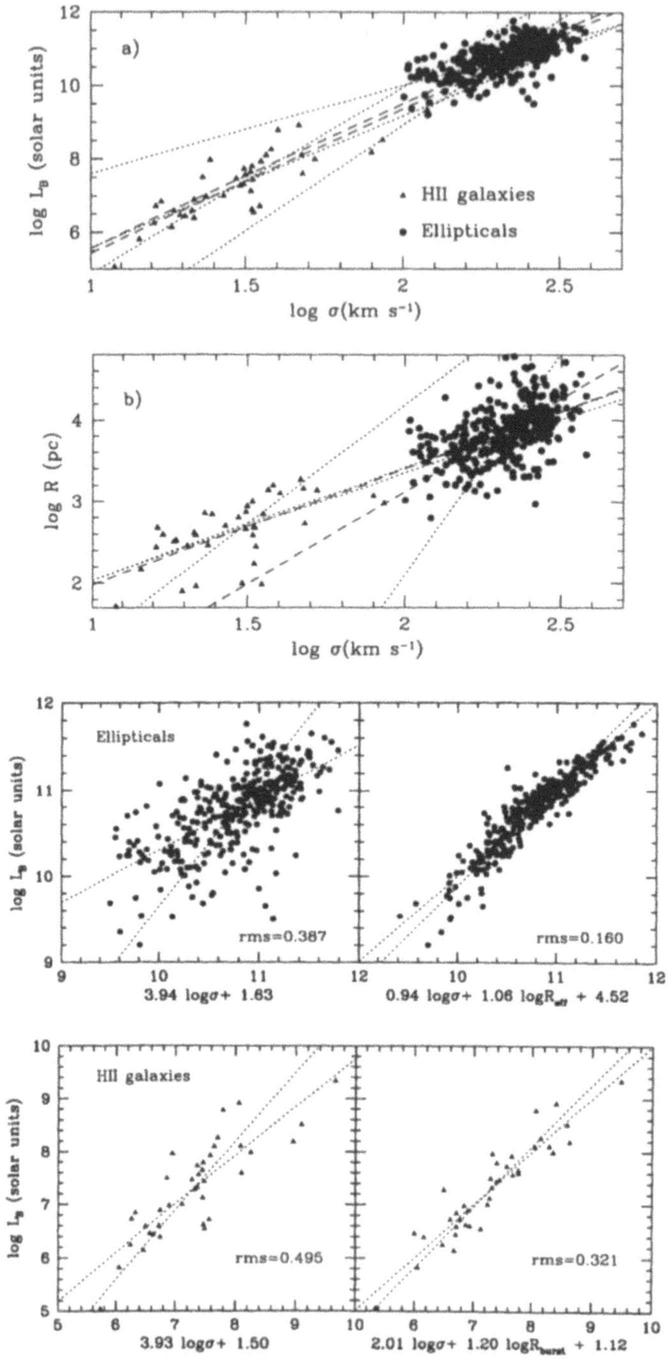

Fig. 3. Comparison between the dynamics of HII galaxies and normal Ellipticals (from Telles 1995).

al. 1999). Figure 4 illustrates this effect. The filled triangles present the raw counts that exhibit the well known turnover due to incompleteness. The filled squares show the data corrected for incompleteness using the standard and tedious artificial star experiments. Since our photometry is complete to about $V \sim 19$, or $M \sim 3\,M_\odot$ if there is no reddening, the turnover of the corrected points could mistakenly be interpreted as a break in the IMF at $M \sim 8\,M_\odot$. However, because young starbursts are dusty, the internal extinction is significant and varies from star to star, so, a well defined magnitude limit does not correspond to a well defined mass limit. In the case of 30 Dor the dispersion in internal reddening is about $\Delta E(B-V) = 1$ mag (peak to peak) so the magnitude limit effect is in fact critical. The open circles show the data corrected for this effect and the solid line shows a fit of slope $\alpha = -2.37 \pm 0.08$ to the fully corrected data, very close to the Salpeter slope $\alpha = -2.35$. Essentially all the claims of departures from a Salpeter law, or high-mass cut-offs we have examined can be ascribed to the magnitude-limit effect. We conclude, therefore, that there is good observational (and theoretical, e.g. Elmegreen, 1999) evidence that the IMF for intermediate and high-mass stars is a universal power-law of Salpeter slope.

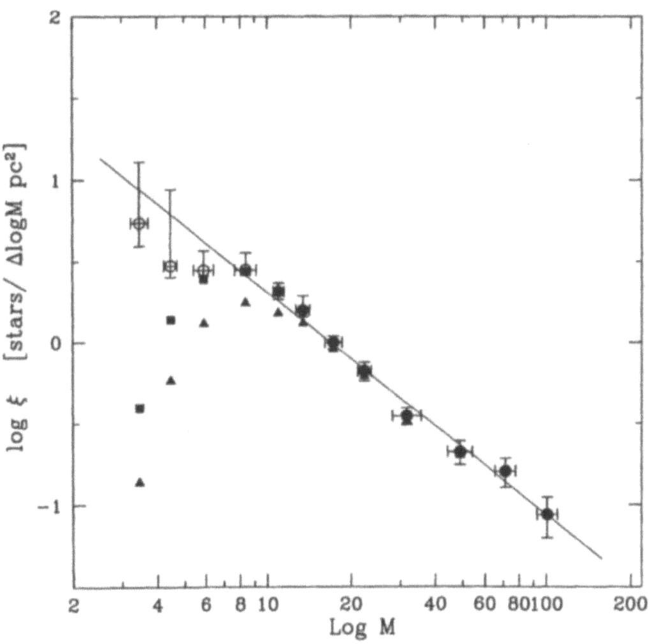

Fig. 4. The IMF of 30 Doradus. The different symbols represent the raw data (triangles), the data corrected for incompleteness (squares) and the data corrected for magnitude limit effect (open circles). From Selman *et al.* 1999.

4 Future VLT Observations

Probably the majority of $z \sim 3$ Lyman Break Galaxies (LBG) are starburst galaxies and even HII galaxies (Pettini *et al.* 1998). Since they are the youngest galaxies we know, they are by definition powerful cosmological probes. We have argued that using their rest-frame emission line spectra we can measure their distances with sufficient accuracy so as to use them to learn about the energy contents of the Universe. Assuming $\omega = -1$ (cosmological constant), this requires galaxies at z=0 (to determine the zero point), z=1 to determine Ω_Λ, and z=2-3 to determine Ω_M.

Large samples of objects at z=3-4 and $z \sim 0.5$ are beginning to accumulate through the Lyman break technique. At intermediate redshifts the densities are several per square arc-minute, so the ideal instrument to obtain the line-widths will be FLAMES, while FORS1 will be needed for the spectrophotometry. For the high redshift samples, ISAAC is required.

References

1. Guzmán, *et al.*, 1997, ApJLett., 427, L9.
2. Koo, D.C., *et al.*, 1995, ApJLett., 440, L49.
3. Koo, etal, 1996, ApJ, 469, 535.
4. Melnick, J., Terlevich, R. & Moles, M., 1988, MNRAS, 235, 297 (MTM88).
5. Melnick, J., Terlevich, R., Terlevich, E., 1999, MNRAS, in press (MTT99).
6. Perlmutter, S., *et al.*, 1998, Ap.J., in press (astro-ph/9812133).
7. Pettini, M., *et al.*, ApJ, 506, 539.
8. Refsdal, S., Stabell, R. & de Lange, F.G. 1967, Memoirs RAS, 71, 143.
9. Riess, A.G., *et al.*, 1998, A.J., 116, 1009.
10. Scalo, J., 1998, in 38th Herstmonceux Conference, eds. Gilmore & Howell, ASP Conf. Series, 143, 201.
11. Selman, F., Melnick, J., Terlevich, R., Bosch, G., 1999, A&A, in press.
12. Telles, E., 1995, Ph.D. Thesis, Cambridge University.
13. Terlevich, R. & Melnick, J., 1981, MNRAS, 195, 839 (TM81).
14. Turner, M.J., 1999, astro-ph, 9901109.

The Hubble Constant from the Fornax Cluster Distance

Tom Richtler[1], Georg Drenkhahn[1,2], Matías Gómez[1,3], and Wilhelm Seggewiss[1]

[1] Sternwarte der Universität Bonn, Auf dem Hügel 71, D-53121 Bonn, Germany,
[2] Max-Planck-Institut für Astrophysik, D-85740 Garching, Germany
[3] Departamento de Astronomía y Astrofísica, PUC, Santiago de Chile

Abstract. Type Ia supernovae are the best cosmological standard candles available. The intrinsic scatter of their decline-rate- and colour-corrected peak brightnesses in the Hubble diagram is within observational error limits, corresponding to an uncertainty of only $3\,\mathrm{km\,s^{-1}\,Mpc^{-1}}$ of the Hubble constant. Any additional uncertainty, resulting from peak-brightness calibration, must be kept small by measuring distances to nearby host galaxies most precisely.

A number of different distance determinations of the Fornax cluster of galaxies agree well in a distance modulus of 31.36 ± 0.04 mag (18.7 ± 0.3 Mpc). This leads to new absolute magnitudes of the well-observed Fornax type Ia SNe SN 1980N and SN 1992D and finally to a Hubble constant of $H_0 = 70 \pm 6\,\mathrm{km\,s^{-1}\,Mpc^{-1}}$.

1 The Hubble Diagram of Type Ia SNe

Supernovae (SNe) of type Ia are known to be excellent standard candles. In order to construct a reliable Hubble diagram it needs more than just to plot the data from the standard SNe catalogue [1] but requires highly uniform spectroscopic and, especially, photometric observations as are now available from the Calán-Tololo survey [2] or the CfA campaign [3].

Fig. 1 shows the Hubble diagram for 29 classified Ia SNe from the Calán-Tololo survey [2]. The intersection Z of the straight line (slope 5!) with the y-axis is related to the Hubble constant H_0 and the absolute magnitude M of the standard candles by the well-known equation:

$$\log H_0 = 0.2 \cdot (M - Z) + 5 \tag{1}$$

For the knowledge of an accurate Hubble constant, one would of course aim at determining the distances to as many SNe as possible.

The quality of the SNe Ia Hubble diagram can be improved further if one corrects the brightness at maximum for the decline rate Δm_{15} of the B lightcurve (e.g. [2]) and the colour $(B_{\mathrm{max}} - V_{\mathrm{max}})$ [4]:

$$B_{\mathrm{corr}} = B_{\mathrm{max}} + b(\Delta m_{15} - 1.1) + R(B_{\mathrm{max}} - V_{\mathrm{max}}) \tag{2}$$

A new analysis of the Calán-Tololo SNe sample has been made recently by *Drenkhahn & Richtler* [5,6]. The derived correction terms are quoted in the

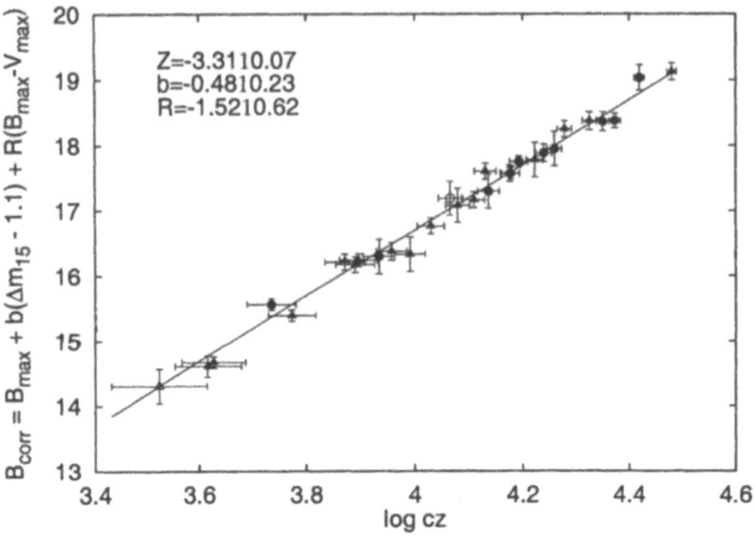

Fig. 1. The Hubble diagram of corrected SN Ia magnitudes in B based on the Cálan-Tololo sample [2]. b and R are the correction coefficients for decline rate and intrinsic colour, resp., and Z is the intersection of the straight line with the y-axis. No systematic differences are seen between SNe Ia's in early-type galaxies (circles), late-type galaxies (triangles), and "red" SNe (open symbols).

upper part of Fig. 1. As a result of the correction process the scatter in the Hubble diagram is reduced to 0.14 mag (even smaller in V and I), which is already consistent with the observational errors. Therefore, the intrinsic scatter must be distinctly smaller than 0.1 mag. This fixes the zero-point Z with high precision and the accuracy of H_0 in (1) depends only on the absolute magnitude M_{corr}. Determining M_{corr} within ± 0.1 mag corresponds to measuring H_0 with an uncertainty of only $\pm 3\,\mathrm{km\,s^{-1}\,Mpc^{-1}}$.

2 The Distance of the Fornax Cluster of Galaxies

With respect to the accurate calibration of the absolute magnitudes of Ia SNe, the Fornax cluster of galaxies is one of the most suitable sites. It hosts the type Ia SNe SN 1992A (in NGC 1380), SN 1980N, and SN 1981D (both in NGC 1316). In contrast to the Virgo cluster, which is the permanent but precarious target of distance measurements, the Fornax cluster has a compact structure without distinct substructures. Promising attempts to tighten the distance to Fornax have been made [8,10,15], but have been criticised when hunting for a small value of the Hubble constant [7]. The last year has brought convincing new results for the Fornax distance (for the following discussion refer to Table 1).

Planetary nebula luminosity function (PNLF). Distances to Fornax galaxies using the PNLF have been deduced by Jacoby and co-workers [8,9]. Their zero-point is based on M31 with a distance of 710 kpc (or distance modulus $m - M = 24.26$ mag) and a reddening of $E_{B-V} = 0.11$ mag. Changing to the more recent distance modulus of $m - M = 24.44$ mag [17] but keeping the reddening one arrives at a Fornax modulus of 31.33 ± 0.08 mag.

Globular cluster luminosity function (GCLF). The turn-over magnitude (TOM) of the GCLF of four Fornax galaxies has been derived by Kohle et al. [10] to be $V_{\mathrm{TO}} = 23.80 \pm 0.06$ mag. Applying a more recent calibration of the TOM of the Galactic globular cluster system: $M_{V,\mathrm{TO}} = -7.61 \pm 0.08$ mag [5,7], one gets a Fornax distance modulus of 31.41 ± 0.10 mag. The GCLF of NGC 1380 has also been analysed using ESO NTT data [11], which result in a modulus of 31.35 ± 0.16 mag. Presently, we are investigating the GCLF of the Fornax galaxy NGC 1316 [12]. The preliminary result of 31.32 ± 0.15 mag agrees well with the other values.

Cepheid distances. Now there is also a Cepheid distance of one Fornax galaxy, NGC 1365, available (HST Key Project [13]). The value of 31.35 ± 0.07 mag agrees very well with the results of the previously mentioned methods.

Surface brightness fluctuations (SBF). Jensen et al. [14] report new infrared SBF measurements for five Fornax galaxies. They find a distance modulus of 31.37 ± 0.12 mag.

Tully-Fisher relation (T-F). Finally, a Tully-Fisher distance from 21 cm data for 18 Fornax galaxies can be derived: when combining published 21 cm line widths ΔV_{20} [15] with dereddened integrated B magnitudes B^0_{tot} [16] one finds the relation: $B^0_{\mathrm{tot}} = -6.990(\pm 0.426) \log \Delta V_{20} + 29.312(\pm 0.365)$ (N.B.: The galaxy ESO 357-G25 which has an extremely large error in the line width has been skipped from the sample [15, their Table 1].) We then use the absolute magnitudes M_B of all non-Fornax calibrating galaxies from the work of Federspiel et al. [17, their Table 2] and obtain 31.43 ± 0.12 mag.

Table 1 suggests that the distance of the Fornax cluster is now very well determined. The mean value of the distance modulus is 31.36 mag with a scatter of only ± 0.04 mag (linear distance 18.7 ± 0.3 Mpc).

3 The Absolute Magnitudes of Type Ia SNe in Fornax and the Hubble Constant

Table 2 lists the parameters of the three Fornax Ia events. The photometry is taken from [2]. For shortness, only the brightness B is given. SN 1992A in NGC 1380 is one of the best ever observed SNe. After correction for decline rate and colour and with the new distance modulus of the Fornax cluster the absolute magnitudes M^B_{corr} have been calculated. Eq. (1) and the corresponding data in the B, V, and I bands for the listed SNe yield a Hubble constant of $H_0 = 70 \pm 6\,\mathrm{km\,s^{-1}\,Mpc^{-1}}$.

Table 1. Distance modulus determinations of the Fornax cluster

Galaxy NGC	$m - M$ mag	σ mag	Method[a]	Reference
1316, 1399, 1404	31.33	0.08	PNLF	[8], [9]
1374, 1379, 1399, 1427	31.41	0.10	GCLF	[10]
1380	31.35	0.16	GCLF	[11]
1316	31.32	0.15	GCLF	[12]
1365	31.35	0.07	Ceph	[13]
1339, 1344, 1379, 1399, 1404	31.37	0.12	SBF	[14]
18 galaxies	31.43	0.12	T-F	[15], [16], [17]
mean value	31.36	0.04		

[a] PNLF: planetary nebulae luminosity function
GCLF: globular cluster luminosity function
Ceph: Cepheid distance; HST Key Project
SBF: surface brightness fluctuations
T-F: Tully-Fisher relation

Table 2. Type Ia SNe in the Fornax cluster of galaxies

SN	Host Galaxy	B_{max}	σ	Δm_{15}	$B_{max} - V_{max}$	B_{corr}	M^B_{corr}
SN 1992A	NGC 1380	12.57	0.03	1.47	0.02	12.36	−19.00
SN 1980N	NGC 1316	12.49	0.03	1.28	0.05	12.33	−19.03
SN 1981D	NGC 1316	12.66	0.06	1.20	0.24	12.25	−19.11

Parameters of SN 1981D from own light-curve fit

References

1. Barbon, R., Cappellaro, E., Turatto, M. (1989) A&AS **81**, 421, `http://athena.pd.astro.it/~supern/`
2. Hamuy, M., Phillips, M. M., Suntzeff, N. B., Schommer, R. A., Maza, J., Avilés, R. (1996) AJ **112**, 2398
3. Riess, A. G., Kirshner, R. P., Schmidt, B. P. et mult. al. (1999) AJ **117**, 707
4. Tripp, R. (1998) A&A **331**, 815
5. Drenkhahn, G., Richtler, T. (1999) A&A submitted
6. Richtler, T., Drenkhahn, G. (1999) in: Kundt, W., van de Bruck, C. (Eds.) Cosmology and Astrophysics: A Collection of Critical Thoughts. Lecture Notes in Physics, Springer, Berlin, in press
7. Tammann, G. A., Sandage, A. (1999) in: Egret, D., Heck, A. (Eds.) Harmonizing Cosmic Distance Scales in a Post-Hipparcos Era. Obs. Strasbourg, in press

8. McMillan, R., Ciardullo, R., Jacoby, G. H. (1993) ApJ **416**, 62
9. Jacoby, G. H. (1997) in: Livio, M., Donahue, M., Panagia, N. (Eds.) The Extragalactic Distance Scale. University Press, Cambridge, p. 197
10. Kohle, S., Kissler-Patig, M., Hilker, M., Richtler, T., Infante, L., Quintana, H. (1996) A&A **309**, L39
11. Della Valle, M., Kissler-Patig, M., Danziger, J., Storm, J. (1998) MNRAS **299**, 267
12. Gómez, M., Richtler, T., Infante, L. (1999) in preparation
13. Madore, B. F., Freedman, W. L., Silbermann, N. et al. (1998) Nature **395**, 47
14. Jensen, J.B., Tonry, J.L., Luppino, G.A. (1998) ApJ **505**, 111
15. Bureau, M., Mould, J. R., Staveley-Smith, L. (1996) ApJ **463**, 60
16. de Vaucouleurs, G., de Vaucouleurs, A., Corwin, H. G., Buta, R. J., Paturel, G., Fouqué, G. (1991) Third Ref. Cat. of Bright Galaxies. Springer, New York
17. Federspiel, M., Tammann, G. A., Sandage, A. (1998) ApJ **495**, 115

Lunar Occultations, Setting the Stage for VLTI: The Case Study of CW-Leo (aka IRC+10216)

Hans Ulrich Käufl[1], Bringfried Stecklum[2], Steffen Richter[2], and Andrea Richichi[3]

[1] European Southern Observatory, D-85747 Garching b. München, Germany
[2] Thüringer Landessternwarte, Sternwarte 5, D-07778 Tautenburg, Germany
[3] Osservatorio Astrofisico di Arcetri, Largo E. Fermi 5, I-50125 Firenze, Italy

Abstract. Lunar occultation allows for a ***sneak preview*** of what the VLTI will observe, both with comparable angular resolution and sensitivity. In the thermal infrared ($\lambda \approx 10\mu m$, angular resolution $\leq 0.03''$) the technique has been pioneered with TIMMI on La Silla. Using this technique several dust shells around Asymptotic Giant Branch stars have been resolved. For the Carbon star CW-Leo (IRC+10 216) high S/N scans will allow for '$1\frac{1}{2}$-dimensional' imaging of the source. At the present state of data reduction the light curves already provide for a very convincing proof of theories on the milli-arcsec scale. In combination with VLTI the technique allows for checks of the visibility calibration and related issues. Moreover, in the *(u,v)*-plane both techniques are extremely complementary, so that a merging of the data sets appear highly desirable. At La Silla and Paranal ESO a suite of instruments which can be (*ab*)used for this project is under construction.

1 Lunar Occultations in the Mid-Infrared

Lunar occultation (LO) is the application of Fresnel diffraction of a half plane in astronomy, whereby the edge of the Moon provides for the half plane ('*knife edge*'). The theory is described in optics text books (e.g. [1]). The astronomical context is given e.g. in [2] and references therein. The diffraction pattern is characterized by a main diffraction fringe whose width Θ_F is approximately $\sqrt{\lambda/D}$ where λ is the wavelength and D is the distance to the occulting object. For the Moon at $\lambda = 10\mu m$ one gets a Θ_F of order of $0.033''$. For theoretical curves see fig. 1. Ground based infrared astronomy in the $10\mu m$ window is characterized by a very strong background radiation. Therefore, to avoid saturation, infrared arrays have to be read quite fast (typically once every $0.005 - 0.050s$). So these instruments are *free of charge* high speed photometers allowing for Nyquist sampling of the Fresnel diffraction pattern. The data presented here have been collected with TIMMI([3]), ESO's now historic cryogenic focal reducer on the 3.6m telescope on La Silla. The follow-up instrument, TIMMI2 ([4]) and the corresponding VLT-instrument VISIR ([5]) will have special provisions, allowing to record light curves with much more technical sophistication (e.g. in long-slit spectroscopy mode or with a pupil slicer to enhance the robustness of the data sets).

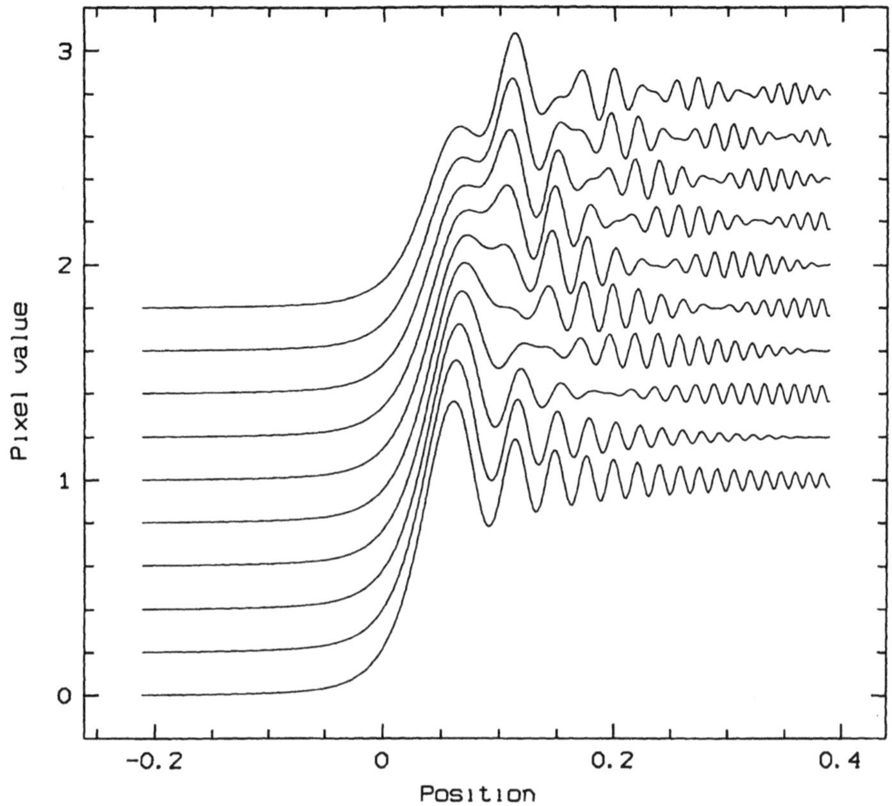

Fig. 1. Synthetic lunar occultation light curves: intensity vs. the angular separation star to lunar limb ["]. The figure shows light curves for a binary star with equal components from zero to 0.054" separation in steps of 0.006". The trace starting at abscissa value 1 corresponds to 0.030" i.e. $\approx \Theta_F$ at $\lambda = 10\mu m$. The traces show, that the higher order fringes carry information on spatial scales $\leq \Theta_F$.

2 Some Geometrical Background Information

Within 9 years the Moon covers an area of 3700 square-degrees, i.e. nearly 10% of the celestial sphere, in a band of $\pm 5.5^o$ along the ecliptic. This band intersects the galactic bulge South of and close to the galactic center, i.e. in a region of very high AGB-star[1] density. An occultation is characterized by the contact angle Φ_c which is the angle between the lunar orbit projected on the sky and a line from the Moon's center to the contact point of the lunar limb where the occultation occurs. The light curve corresponds to a scan with a

[1] Asymptotic Giant Branch (AGB) stars represent a short, but very enigmatic evolutionary phase preceding the planetary nebula formation. All stars with $M_* \leq 8M_\odot$ pass this stage and expel up to 75% of their original main sequence mass in a very short time, forming huge (several AU) dusty envelopes.

half-plane along the direction of this vector. In that sense lunar occultation contains in this direction spatial frequencies up to $\approx 1/\Theta_F$ whereas perpendicular to that it contains the information of normal imaging. For a detailed discussion and examples projected into the Fourier spatial frequency plane see [2]. As the Moon is very close there is a substantial horizontal parallax. For that reason Φ_c is markedly different (up to $\pm 15^o$) when observing the same event in parallel from ESO's two observatories La Silla and Paranal and hence one gets access to different areas of the *(u,v)*-plane. In that sense lunar occultations provides for quasi-imaging capabilities.

3 The Link with Interferometry

At $\lambda \approx 10\mu m$ VISIR in lunar occultation mode and interferometry with ESO's VLTI [6] equipped with suitable instrumentation [7] have approximately the same flux limits and the same spatial resolution [8]. Both techniques are very complementary in *(u,v)*-plane coverage. The LO technique allows to determine phases whereas interferometry with two telescopes only is restricted to visibility measurements. It was therefore proposed to merge data sets obtained with both techniques for one object[2]. This approach offers ultimately imaging with an effective point spread function of 30 mas.

Another important aspect is the calibration of visibilities. A sizable fraction of potential visibility calibrators is known or can be expected to be extended or to have extended envelopes on a scale of 30-50 mas (most are late type stars in some state of mass loss). LO light curves can be used to assess the true spatial properties of such objects.

4 The Preview

Until the VLTI is operational, LO offers the only method to spatially resolve objects in the mid-infrared ($\lambda \approx 10\mu m$). So far circumstellar shells around the semi-regular variable star Z Cnc, the OH-IR star IRAS 18020-1820 and the prototype mass loosing Carbon star CW Leo have been resolved. The data set of Z Cnc is presently being analyzed in full detail (Richter, thesis, in preparation). In combination with a near-infrared LO data set ([9],[10]) a radiative transfer model for the optically thin shell around this star is constructed. Additional constraints are derived from the IRAS Low Resolution Spectroscopy database and from near infrared photometry.

Particularly interesting, however, was the occultation event of Nov. 11, 1998 of CW Leo. For technical details etc. see [11]. Figure 2 shows a preliminary result of the brightness profile. CW Leo, a Mira-variable star with carbon chemistry and a period of 650d, is particularly well suited to study the mass-loss phenomenon of Asymptotic Giant Branch Stars; due to its exceptional brightness and its vicinity to the solar system this is the best studied object

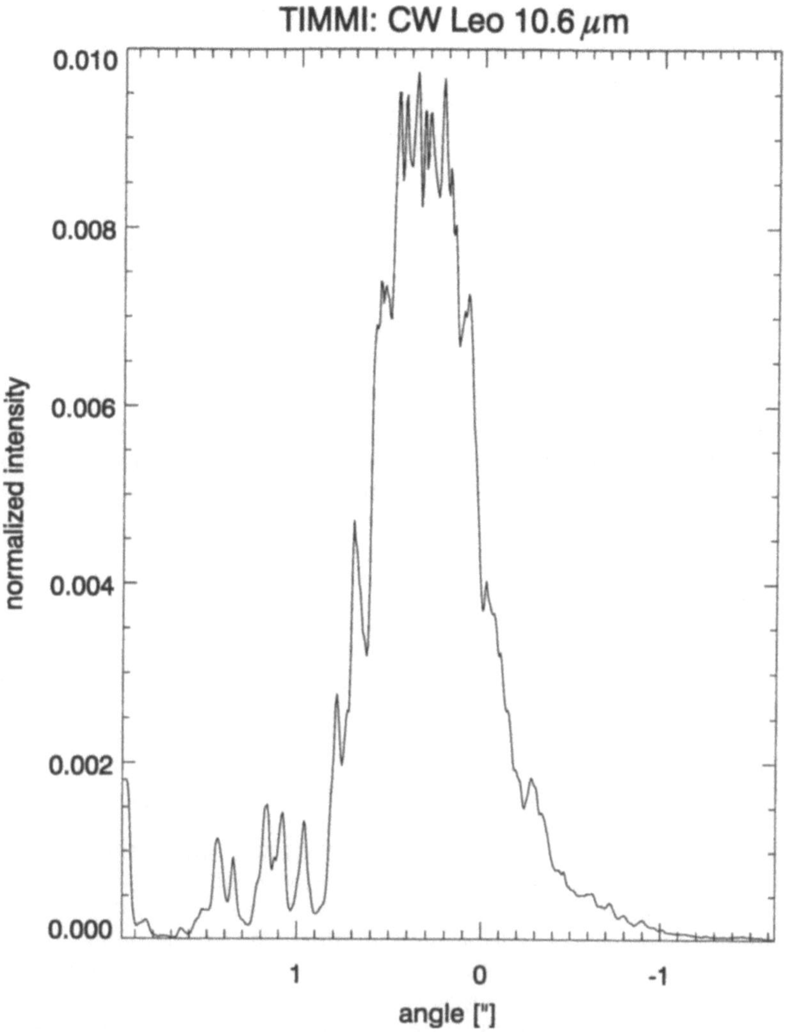

Fig. 2. Spatial Brightness Profile at $\lambda \approx 10.6\mu m$ measured for CW Leo. The central plateau of the source profile corresponds to the innermost dust shell (note: even at this quite long wavelength the dust shell is not optically thin and the direct view on to the photosphere of this star is obscured.). At least three symmetric rings indicative of *onion*-like shells can be seen which are interpreted as the signature of the recent mass loss history of this Mira variable star.

of its class. Also, modeling of dust formation and mass-loss around this object is extremely advanced. Taking the canonical values for CW Leo ($d \approx 120pc$, $v_{exp} \approx 30\frac{km}{s}$) one would expect to see concentric shells with a separation of $\approx 0.1''$ if the mass-loss would follow the period of the luminosity. According to a recent model for mass-loss of bright carbon stars [12] the ejection occurs semi-regularly with sub-harmonic frequencies of the Mira-pulsation. As can be seen in fig.2, the separation of the *onion shells* in fact is of order of $0.3 - 0.4''$, thus nicely reproducing the theoretical prediction.

5 Outlook and Future Work

At this point the lunar occultation data are still being processed. Until the VLTI comes online LO is the only means to achieve milli-arcsec imaging information of astronomical objects. Once the VLTI is operational, reconstruction algorithms will be practically tested and used to image circumstellar shells around AGB stars.

References

1. Sommerfeld A. (1950) Vorlesungen über Theoretische Physik, Band IV Optik, Dietrich'sche Verlagsbuchhandlung, Wiesbaden, 250–252
2. Käufl, H. U., Stecklum, B., Richichi, A. (1998) *Astronomical Interferometry*, Robert D. Reasenberg *Editor*, 267–274
3. Käufl, H. U., Jouan, R., Lagage, P.O., Masse, P., Mestreau, P., Tarrius, A. (1994) *Infrared Phys. Technol.* **35**, 203–210
4. Reimann, H.-G., Weinert, U., & Wagner, S. (1998) Proc. SPIE **3354**, *Infrared Astronomical Instrumentation*, Albert M. Fowler *Editor*, 865–876
5. Y. Rio, Lagage, P.O., Dubreuil, D., Durand, G., et al. (1998) Proc. SPIE **3354**, *Infrared Astronomical Instrumentation*, Albert M. Fowler *Editor*, 615–626
6. Mariotti, J.-M., Denise, C., Dèrie, F. et al. (1998) *Astronomical Interferometry*, Robert D. Reasenberg *Editor*, 800–806
7. Leinert, C., Graser, U. (1998) *Astronomical Interferometry*, Robert D. Reasenberg *Editor*, 389–392
8. Stecklum, B., Käufl, H.U., Richichi, A., (1997) Proc. ESO workshop *Science with the VLT Interferometer*, F. Paresce *Editor* 153–160
9. Richichi, A., Ragland, S., Stecklum, B., Leinert, Ch. (1998) A&A **338** 527–534
10. Richichi, A., Fabbroni, L., Ragland, S., Scholz, M., (1999) A&A **344** 511–520
11. Stecklum, B., Käufl, H.U., Richichi, A. (1999) *The Messenger* **95** 25–27
12. Schröder, K.-P., Winters, J.M., Arndt, T.U., Sedlmayr, E. (1998) A&A **335** L9–L12

Absolute Ages of Globular Clusters and the VLT

Frédéric Pont

Universidad de Chile, Dep. de Astronomia, Casilla 36-D, Santiago, Chile
fpont@das.uchile.cl

Abstract. In the past two years, the ages derived for the oldest globular clusters have significantly decreased, and our confidence in them increased, due on one hand to the arrival of a new generation of evolution models, on the other hand to precise Hipparcos parallaxes for field subdwarfs. The age range inferred, 11–14 Gyr, is now compatible with most cosmological models, alleviating the famous "cosmic age problem". We here examine what could be the contributions of the VLT to further increasing the accuracy of the age determinations. Two possibilities are considered in some detail: (1) deep imaging of the white dwarf cooling sequence, with the possibility of obtaining an independent distance and a -small- probability of reaching the end of the sequence and obtaining the white dwarf cooling age, (2) high-resolution, low S/N spectroscopy of main-sequence globular cluster stars, to ensure the correspondence of the cluster and subdwarf metallicity scales.

1 Introduction

For the past three decades, the most reliable lower limit to the age of the Universe has been derived from the metal-poor globular clusters (GC). This field has seen spectacular progress in the last couple of years, triggered by the apparition of Hipparcos parallax data for local subdwarfs, and of a new generation of evolution models.

Until then, the ages derived for the oldest globular clusters of the Galaxy were in open conflict with the age derived for the Universe by measuring the Hubble constant, an apparent contradiction often called the "cosmic age problem". While cosmological models yielded ages in the 8–10 Gyr range, typical ages of 15–18 Gyr were obtained for the GC. The uncertainties on GC ages were of the order of 20%, mainly due to the uncertainty on the subdwarf distance scale (VandenBerg et al. 1996).

The Hipparcos astrometric satellite has measured precise parallaxes for a large set of local subdwarfs, allowing the distance scale for metal-poor clusters to calibrated empirically. Previous estimates were found to be ~0.1 mag too faint at [Fe/H]~ -1.3 dex. This modification in the distance scale caused the inferred ages to decrease by 1–2 Gyr. However, the largest change in the derived ages was due to the new generation of models, giving ages 2–3 Gyr younger, all other things being equal. As a result, the ages derived for the oldest GC are now in the range 11–14 Gyr (Reid 1997, Gratton et al. 1997, Pont et al. 1997).

When these lower age are compared to the increased expansion age, obtained with current values of H_0 and non-vanishing cosmological constant (as seems to be indicated by the high-z supernovae data, see Perlmutter et al. 1998, Schmidt et al. 1998), the "cosmic age problem" is no longer an open contradiction, and any age for the Universe between 11 and 14 Gyr could satisfy both constraints.

Now, if we want the globular cluster ages to be a really constraining parameter for cosmology, the global uncertainty should be brought down to ~1 Gyr. This requires a finer tuning of the distance scale, the metallicity scale and the evolution models. In this paper, I attempt to consider how the VLT could contribute to the two first points. I also examine another independent approach to globular cluster ages: the white dwarf cooling age.

Metallicities: Clusters, Subdwarfs and Models

The most reliable absolute ages for globular clusters are obtained by comparing the observed magnitude of the main-sequence turnoff with model prediction. The turnoff magnitude essentially expresses the main-sequence lifetime of stars of about 1 M_0 (for the ages typical of globular cluster), and as such is a very basic and robust prediction of stellar evolution models. However, for the magnitude of the turnoff to be known, the distance to the cluster must be accurately determined. This can be done by direct fit of the cluster sequence to local subdwarfs of known parallax. The distance is determined by fitting the cluster sequence to field subdwarfs of similar metallicities. As the metallicities are not exactly equal, colour shift are applied to compensate for the difference. The cluster sequence is then confronted to model isochrones for the same metallicity.

As stated above, precise parallaxes from the Hipparcos mission have put the GC distance scale on much firmer ground. At present, the dominant sources of uncertainty are associated with metallicity. Three metallicity scale must be coherently known for the derived ages to be correct: the calibrating subdwarf scale, the cluster scale and the evolution model scale. The well-known problem of α-element overabundances has shown that the correspondence between the models, where [O/H] plays an important role, and the clusters, for which only [Fe/H] is usually known, must be studied with care. The latest generation of models use α-enhanced mixtures, and it is useful to check the detailed abundance mix of the GCs themselves by *in situ* observations.

Another problem is the correspondence between the GC scale and the calibrating subdwarf scale. In recent years Gratton et al. (1997) and Clementini et al. (1999) have proposed significant revisions of both scales compared to previous studies (e.g. Zinn & West 1984 for the clusters and Carney et al. 1994 for the subdwarfs). But while the calibrating subdwarfs are dwarfs or at most, as in Pont et al. (1997), subgiants, the stars used to derive the metallicity of clusters are giants, and a mismatch between dwarf and giant metallicity scales could exist, and would be reflected in the age derived. This suspicion was given substance by a recent study by King et al. (1998). They obtained high-resolution spectra for six subgiants in M 92 with the Keck telescope, and derive metallicities that are 0.3–0.5 dex lower than those measured on giants. Should it be confirmed, such a metallicity shift would have an important impact in the age scale, and it is crucial to check it by measuring metallicities for subgiant and main-sequence stars in globular clusters (section 3).

2 Deep Imaging (FORS)

The White Dwarf Cooling Sequence

Detecting and defining the position of the white dwarf cooling sequence for the nearest old globular cluster allows an independent distance determination, by comparison with the sequence of local field white dwarfs with precise parallaxes. There are now enough of these calibrators (Bergeron et al. 1997) that, if the white dwarf sequence is measured deep enough in a cluster, an accurate distance can be determined by direct fitting of the two sequences.

Moreover, white dwarfs could contribute to the determination of globular cluster ages in a much more decisive way: if the *end* of the white dwarf sequence can be reached, comparison with theoretical models will directly yield the cooling age of the system (recent white dwarf cooling sequence calculations include Hansen 1998, Saumon & Jacobson 1998, Chabrier 1999). This methods gives ages completely independent of the main-sequence turnoff estimate.

Measuring the white dwarf sequence in even the nearest globular cluster is a tough observational challenge. The top of the sequence has now been seen in 6 clusters with the HST (M15: De Marchi & Paresce 1995; ω Cen: Elson et al. 1995; 47 Tuc: Paresce et al. 1995; NGC 6752: Renzini et al. 1996; NGC 6397: Cool et al. 1996; M4: Richer et al. 1997). In M4 and NGC 6397, the sequence was beautifully defined over several magnitudes, down to $M_V \sim 14$ ($m_V \sim 26.5$). The whole sequence has even been delimited in the old open cluster M67 by Richter et al. (1998), yielding a cooling age of 4 Gyr.

Imaging the oldest globular clusters deep enough to reach the end of the sequence, however, is another matter. Depending on the age of the Universe, and on the set of models chosen, the end of the sequence is expected between $m_V = 28^{th}$ and 30th magnitude. In the first case – the oldest GC $\sim$11 Gyr old as suggested by Reid (1997) or Cassisi et al. (1998) and models by Chabrier et al. (1999) – we may hope to reach the end of the sequence with VLT. In the second – a Universe 14 Gyr old as suggested by Pont et al. (1997) and the models by Hansen (1998) – the sequence will vanish into the faint galaxy background, and its detection will be a hard task even for the future NGST.

The first point to be addressed will be the following: does VLT allow an improvement over HST in the measurement of white dwarf sequences? The VLT has the advantage of collecting surface, allowing FORS1 to nominally

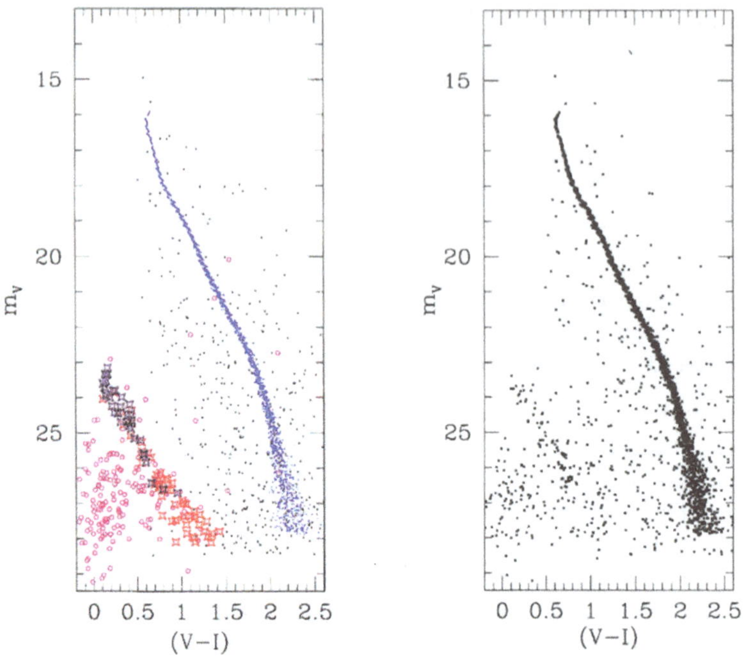

Fig. 1. Simulated deep Colour-Magnitude Diagram for the old cluster NGC 6397. On the left, dots indicate the cluster sequence and field stars. Open circles are unresolved field galaxies, and stars represent the white dwarfs. On the right, same kind of simulation without the identifying symbols, to give an idea of the expected CMD crowdedness. (Galaxy statistics taken from the Hubble Deep Field, cluster statistics from King et al. 1998b).

reach magnitudes of 28.5 in V in a one-night exposure. However, the seeing will be much higher, increasing the crowdedness of the field and the difficulty of sorting out background galaxies, and the sky will be brighter. At $V \sim 28$, background galaxies also become very numerous. Crowding in real space will be severe, as well as crowding in CMD space. Fig. 1 shows a simulation of what a very deep globular cluster sequence could look like. When different symbols are used (left), a clear definition of the white dwarf cooling sequence looks feasible despite the background galaxies, but once identical symbols are used, the difficulty of isolating a clear white dwarf sequence becomes apparent.

The Lower Main Sequence

Deep imaging with the VLT could in principle allow us to reach 1–2 magnitudes deeper than HST along the lower main sequence. However, the lower main sequence does not contain crucial information as far as the age is concerned. Moreover, crucial improvement over HST imaging is less likely here than for the white dwarf sequence because contamination by foreground/background stars and background galaxies will be very severe below $m_V \sim 27$ and will probably prevent a clear definition of the cluster sequence.

3 High-Resolution Spectroscopy (UVES, GIRAFFE)

As detailed above, the metallicity correspondence between the clusters, the field subdwarfs used in the distance calibration, and the evolution models, probably constitute the dominant source of uncertainty at present on globular cluster ages. High resolution spectroscopy with instruments such as UVES and GIRAFFE on the VLT will allow detailed abundance analysis in several cluster. With these, we will be able to check whether the mix of elements used in the models is appropriate, and whether the cluster stars and calibrating subdwarfs share similar element ratios at a given [Fe/H]. Detailed element analyses will almost certainly provide a few surprises, and allow finer tuning of the age determination.

The correspondence between giant and dwarf metallicities, questioned by the study of King et al. (1998), will be harder to ascertain. Main sequence stars for nearby globulars are extremely faint for high-resolution spectroscopy, even with 8-m class telescopes. In the King et al. study, 4 nights with the Keck telescope were necessary to reach subgiants of $m_V \simeq 18$, still 1.5 mag above the main sequence. The red main sequence may be reached with really long exposure times. Alternatively, we can attempt to extract the information that we need from high-resolution spectra of lower signal-to-noise ratio:

Metallicities from Low S/N High-Resolution Spectra Using Cross Correlation

The *cross-correlation function* (CCF) of a star's spectra with a numerical template, routinely used for radial velocity determinations of cool stars, also contains metallicity information. With a template of thousand of small metallic lines carefully chosen to be mostly unsaturated, the **area** of the CCF is a very sensitive indicator of [Fe/H]. Deriving metallicities from the CCF area was done for GCs in Fornax and the LMC by Dubath et al. (1992, 1993), for subdwarfs in Pont et al. (1997), and for Cepheids in Pont 1997, demonstrating the efficiency of the method. A preliminary calibration was obtained for globular cluster giants by Dubath (priv. comm.).

The use of the cross-correlation technique could enable one to ensure the coherence of the GC and local subdwarf metallicity scales in a very direct way. CCF calculated with the same template can be compared directly for local subdwarfs and subgiants and their equivalent in chosen GCs. No intervention from models is necessary, and the method has the advantage of enabling a standard, automatic reduction. Compared to methods using spectral indices, such as the Calcium triplet, it is much neared to a full, line-by-line analysis, in that it actually samples the shape of thousand of small metallic lines, instead of relying on a single saturated feature.

4 Conclusions

In the coming years, we may aim to reduce the uncertainties on absolute globular cluster ages from the present 2–3 Gyr to $\sim$ 1 Gyr. Why 1 Gyr ? That is the accuracy needed for the value to be really "cosmologically constraining", in the sense that it would impose a narrow range of combination of the cosmological parameters. It is also the level of precision needed to give a clear indication about the formation of the Galaxy, constraining the time interval elapsed between globular cluster formation and the onset of the disc formation.

The VLT will probably contribute to the observational side of this effort with detailed abundance analyses for cluster giants, and by establishing the coherence between giant and dwarf metallicity scales. Measurements of the white dwarf cooling sequence may also provide an independent distance estimate, and even – under optimal circumstances – a white dwarf cooling age.

References

1. Bergeron P., Ruiz M.T., Leggett S.K. 1997, ApJS 108, 339
2. Carney B.W., Latham D.W., Laird J.B., Aguilar L.A. 1994, AJ 107, 2240
3. Cassisi S., Castellani V., Degl'Innocenti S. et al. 1998, to be published in A&A (astro-ph/9811329)
4. Chabrier G. 1999, ApJ 513, L103
5. Clementini G., Gratton R., Carretta E., Sneden C. 1999, MNRAS 302, 22
6. Cool A.M., Piotto G., King I.R 1996, ApJ 468, 655
7. De Marchi G., Paresce F. 1995, A&A 304, 202
8. Dubath P., Meylan G., Mayor M. 1992, ApJ 400, 510
9. Dubath P. Mayor M., Meylan G. 1993, ASP Conf. Ser. 48, 557
10. Elson R.A.W., Gilmore G.F., Santiago B.X., Casertano S. 1995, AJ 110, 682
11. Gratton R., Fusi Pecci F., Carretta E. et al. 1997, ApJ 491, 749
12. Hansen B. 1998, *Nature* 394, 860
13. King I.R., Stephens A., Boesgaard A.M., Deliyannis C.P. 1998, AJ 115, 666
14. King I.R., Anderson J., Cool A.M., Piotto G. 1998b, ApJ 492, L37
15. Paresce F., De Marchi G., Jedrzejewski R. 1995, ApJ 442, L57
16. Perlmutter S., Aldering G., Goldhaber G. et al. 1998, to be published in ApJ (astro-ph/9812133)
17. Pont F. 1997, PhD Thesis, Geneva Observatory
18. Pont F., Mayor M., Turon C., VandenBerg D.A. 1998, A&A 329, 87
19. Reid I.N. 1997, AJ 114, 161
20. Renzini A., Bragaglia A., Ferraro F.R. et al. 1996, ApJ 465, L23
21. Richer H.B., Fahlman G.G., Ibata R. et al. 1997. ApJ 484, 741
22. Richer H.B., Fahlman G.G., Rosvick J., Ibata R. 1998, ApJ 504, L91
23. Saumon D., Jacobson S.B. 1998, preprint (astro-ph/9812107)
24. Schmidt, B.P., Suntzeff N.B., Phillips M.M. et al. 1998, ApJ 507, 46
25. VandenBerg D.A., Bolte M., Stetson P.B. 1996, ARA&A 34, 461
26. Zinn R., West M.J. 1984, ApJS 55, 45

The Galactic Bulge: Prospects for the VLT

R. Michael Rich

Department of Physics and Astronomy, UCLA, Los Angeles. CA 90095-1562, USA

1 Introduction

The central bulge of the Galaxy is 100 times closer than the nearest extragalactic bulge, that of M31. Until the era of 100m telescopes arrives, the Milky Way bulge will remain the best venue for the detailed study of bulge populations. The VLT will have two significant advantages in this venture. First, the Southern Hemisphere location is ideal for observations of bulge fields, which are at $\delta \approx -30°$. The second advantage is the VLT itself, which will have the UVES echelle that will ultimately be fiber fed. The combination of the wide field imaging of the VLT survey telescope with multi-object high resolution spectroscopy on UT2 will produce the largest and best data sets yet seen in this subject area.

The challenge is for research using VLT to make an incremental advance in this area. And if the quality of the data set on the Galactic bulge approaches that which we have on disk stars today (cf. Edvardsson et al. 1997), what are the most pressing questions to address? How can we use the VLT to change the way we think about chemical enrichment and galaxy formation?

2 One Person's Historical View

The bulge as a stellar population was really discovered by Baade. Whitford's early infrared scans indicated a strong concentration of flux toward Sagittarius, but he unfortunately did not discover the Galactic Center. Baade's (1951) identification of a density maximum in the numbers of RR Lyrae stars, and correctly linking these with the old globular clusters, was a breakthrough. Shapley showed that globular clusters were concentrated toward a distant Galactic center, but Baade found the stellar population and also inferred its age. He went on to note the strong absorption lines in the bulge of M31, and correctly concluded that the population was very old, yet was able to enrich in metals in a short time (Baade, 1963). The insight is so brilliant and quotable that I give it now. It is fair to say that ESO and VLT *continue* this great tradition in stellar populations.

"We must conclude, then, then in the central region of the Andromeda Nebula we have a metal-poor Population II, which reaches -3^m for the brightest stars, and that underlying it there is a very much denser sheet of old stars, probably something like those in M67 or NGC 6752. We can be certain that these are enriched stars, because the cyanogen bands are strong, and so the metal/hydrogen ration is very much closer to what we observe in the Sun and in the present interstellar medium than to what is observed for Population II. And the process of enrichment probably has taken very little time. After the first generation of stars has formed, we can hardly speak of a 'generation', because the enrichment takes place so soon, and there is probably very little time difference. So the CN giants that contribute most of the light in the nuclear regions of the Nebula must also be called old stars; they are not young."

– Walter Baade in The Evolution of Galaxies and Stellar Populations, 1963, Harvard University Press: Cambridge, p. 256.

Although we have taken a rather complicated path to get there, our best current view is that Baade was essentially correct. However, the path of how we arrive to this view is interesting. And there remain some important unanswered questions for the next generation of stars to address.

Baade's insights were justly founded on Morgan's spectroscopy and the identification of large numbers of M giants toward the nucleus in Blanco's early surveys. The possibility that the bulge may be more metal rich than the Sun was seriously discussed in the early 60's. However, the controversy of super metal rich (SMR) stars cast a cloud on the whole issue. By the 1970's, the bulge was acknowledged to be old, but metal poor. At the same time, study of the Galactic center was increasing on two fronts. The nonthermal activity (and a possible central black hole) was recognized, along with the growing evidence for current starforming activity.

2.1 A Younger Bulge?

It was the latter that raised the possibility that the bulge might have a substantial intermediate age population. This was supported strong on two fronts. First, Habing and collaborators discovered infrared-luminous and OH/IR stars toward the Galactic Center (Winnberg et al. 1985; Habing et al. 1985. At least some of the most luminous members of the class may be intermediate age (Bloemmaert et al. 1998). One could still escape by arguing that the OH/IR stars are rare and unusual, and trace little of the bulge's mass. Catchpole et al. (1990) surveyed the inner 2° with a single channel infrared photometer, and found evidence that the most luminous IR AGB stars are concentrated to the inner 100 pc: the intermediate age stars could be widespread and account for a great deal of mass. Further evidence for intermediate age stars came from the period distribution of the Mira variables (e.g. Glass et al. 1995). In the bulge, there is an excess of long period Miras not found in globular clusters. In the Solar vicinity, such stars have the kinematics of young stellar populations, and this evidence further pointed toward a widespread intermediate age population.

From the theoretical quarter, it was realized that stellar bars might thicken to form bulges (Combes et al. 1990). Since disks are unstable to bar formation, a merger that produces a dissipatively formed disk, which becomes bar unstable, can eventually produce a central bulge. Not only was the Galactic bulge possibly intermediate age, perhaps most if not all bulges could potentially have been formed in the last few Gyr.

Because of the photometric crowding, high extinction, and spatial depth of the bulge, even HST imaging of Baade's Window, with turnoff photometry, could show that the bulge was mostly of order 8-15 Gyr old, but could not really fix the age more precisely than that (Holtzman et al. 1998).

3 An Ancient Bulge

Even the abundance ratios of McWilliam & Rich (1994) were not conclusive in arguing that the bulge must have an alpha-element enhanced composition that would have been expected from a starburst history. Mg and Ti were found to be enhanced, but Si and Ca followed normal Solar Neighborhood abundance trends. Abundance ratios have the potential to constrain the timescale for chemical enrichment (and thus, formation) of the bulge. But the 1994 data at $R = 16,000$ were simply too poor to do the job.

The key to constraining the age of the bulge is to use the age-sensitive ΔV_{TO}^{HB} indicator, the magnitude difference between the red clump and the turnoff point. As metal rich populations age, the luminosity of the He burning stars changes little while the turnoff fades. Ortolani et al. (1995) show that the field population in Baade's Window has the same turnoff-to HB magnitude difference as the old, metal rich globular clusters NGC 6553 and 6528. And this difference (3.6 mag) is also shared by the rest of the old halo globular clusters. Although calibration of this quantity to an actual age in Gyr is some time away (and undoubtedly depends on metallicity and other factors), this method overcomes the problems of reddening and distance modulus that have plagued previous efforts to establish the age of the bulge. The close agreement between the bulge and the globular clusters is significant.

Although Ortolani et al. (1995) strongly argues for an ancient bulge outside of the central 100 pc, it still is widely believed that the central 100 pc must have a substantial contribution from intermediate age stars. Now, this last hiding place for an intermediate age bulge is endangered. Surveys of the giant branch in the infrared suggest that any intermediate age population is largely absent outside of 100 pc (Frogel, Tiede, & Kuchinski 1999). Of course, this was somewhat known from the Catchpole et al. survey, but the modern data are far more conclusive. What one really wants is a measurement of the main sequence turnoff point in the center, to pursue the same approach as Ortolani et al. (1995) did in Baade's Window.

We have done such a measurement, using NICMOS onboard HST. We have imaged fields between 8 and 50 pc from the center, and to our surprise, every bulge field is dominated by old stars (Figure 1). The turnoff to HB difference is 3.6 mag. Surely, there is evidence for star formation at the very center: WR stars and massive O stars occur in abundance. The massive

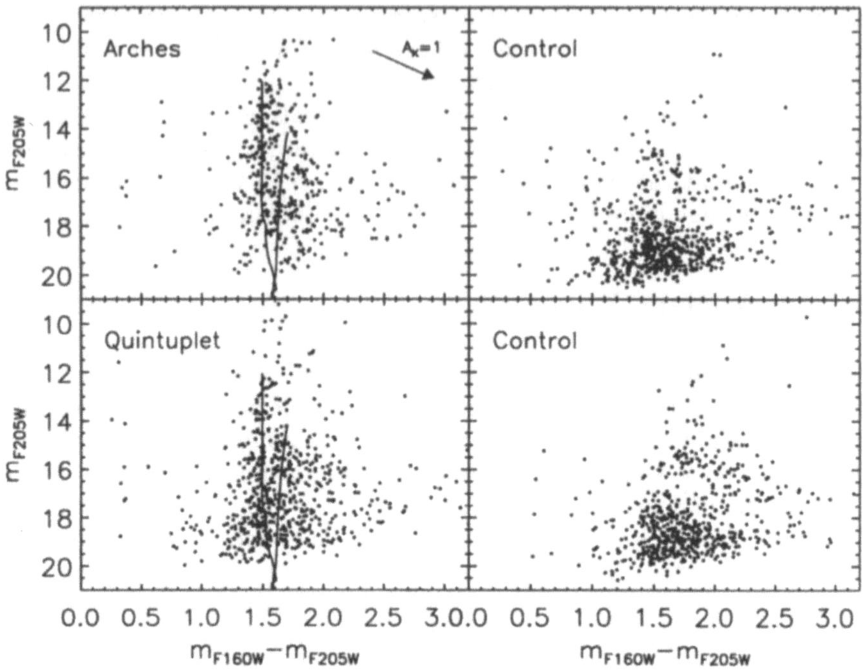

Fig. 1. This figure illustrates NICMOS/HST photometry of 2 populous young star clusters in the Galactic Center (Arches and Quintuplet) and control fields near the clusters, in the bulge (Figer et al. 1999). These clusters are within 20 pc of the nucleus. The zero-age main sequence and a 10 Gyr isochrone are illustrated. Notice the extended young main sequence in the Arches cluster, which is about 1 Myr old. In contrast, notice that the field has a clear red clump population and relatively few young stars. The clump population is even more evident in other fields. In both of these fields, the clump population is extended along the reddening line. The red clump, and old main sequence turnoff at m(205) = 18, are evidence that the field population near the nucleus is dominated by old ($> 5 Gyr$) stars.

Arches cluster has a dramatically top-heavy IMF (Figer et al. 1999). But the underlying stellar population is old, of order 10 Gyr and probably as old as the oldest globular clusters. The evidence is a clearly detected red clump population and main sequence rise. Only in the field 8 pc (!) from the nucleus is the intermediate age main sequence noted... and even in that field, the red clump is very much in evidence. This dominance of old stars poses some difficulty. If the star formation rate were to be merely 0.1 $M_{\odot} yr^{-1}$ on average it would be sufficient to account for most of the mass of the central bulge. Perhaps conditions in the nucleus favor only massive star formation; the long lived stars are not produced in these starbursts. The VLT may contribute to solving this mystery; adaptive optics and interferometry might be able to probe to the lower main sequence of the Arches cluster. HST cannot. We have tried, and the cluster is much too crowded. In any event, while the Sgr dwarf is getting shredded in the halo, the infall of such things has had little impact on the bulge as a whole, as judged by the lack of intermediate age stars.

It is noteworthy that there is growing evidence from the high redshift arena that spheroids (including at least some bulges) are ancient (cf. Stanford et al. 1998, Renzini 1998). One must be careful not to find the desired answer in the data, but the case for at least ellipticals in rich clusters being ancient is quite compelling.

4 Frontiers in Abundance Measurements

The area where VLT is poised to make the most dramatic contribution is in the area of high resolution spectroscopy. The present data have been painstakingly acquired, at low S/N and low resolution. And there are problems that even Keck data do not appear to resolve with certainty. The abundance of NGC 6553 is presently disputed, and the issue is vitally important because the implications are great. NGC 6553 has a strongly descending giant branch. TiO blanketing in the coolest red giants is so great that the tip of the RGB is at the level of the horizontal branch. The bulge color-magnitude diagram is similarly afflicted (Rich et al. 1998) and so is that of the M31 bulge (Jablonka et al. 1999) and halo (Rich et al. 1996). Barbuy et al. find that this cluster has [Fe/H]=–0.55, only 0.15 dex more metal rich than 47 Tuc. They find the alpha elements strongly enhanced, by +0.5 dex, and argue that Z (the overall metal abundance) is nearly Solar. On the other hand, Cohen et al. (1999) find [Fe/H]=–0.2 and an enhanced, but not dramatic, alpha abundance.

The difference is important because if old, metal rich populations are strongly enhanced in alpha elements, it is likely because the enrichment happened rapidly, before type I SNe emerged on the scene to produce significant amounts of iron. If elliptical/bulge populations formed *very* early, the composition of NGC 6553 might be representative of everything from the Galactic bulge, to the M31 bulge, on to the giant ellipticals in say, the Virgo cluster. Therefore it is vitally important to settle this issue, and to set the metal rich abundance scale. Very high resolution, high S/N data are needed, and UVES has the potential to provide those. With multi fiber feeds, one may contemplate spending two nights to get spectra of eight 19th mag bulge main sequence stars. Flexible scheduling may permit VLT to take advantage of the amplification of the odd main sequence star caused by microlensing.

If it turns out the alpha element enhancement is going on in NGC 6553, then it becomes interesting to map [α/Fe] as a function of Galactic latitude in the bulge. This will require of order 100 stars at 4 latitudes, probably the innermost Sgr I window, Baade's Window, and the Plaut field (these span about *300*, *500*, *and 1000* pc from the nucleus). Interior to about 800 pc, the descending giant branch is strongly evident, while outside of 1000 pc, it appears to be absent. Is this because the density of the most metal rich stars falls so dramatically off the nucleus? Or is there a global composition gradient, perhaps caused by the production of alpha elements in a very short-lived, violent central starburst? Although the examination of high redshift galaxy samples tell us much about the general process of galaxy evolution, they cannot constrain the history of the Milky Way. Nor can they easily constrain how massive galaxies such as our own were assembled. Stellar archaeology can tell us a great deal about the earliest formation history of our own galaxy. Low-mass stars preserve precious samples from the earliest phases of the Galaxy's formation. The VLT can be the tool to uncover this fossil record.

References

1. Barbuy, B., Renzini, A., Ortolani, S., Bica, E., & Guarnieri, M.D. 1999, A&A, 341, 539
2. Blommaert, J.A.D.L., Van der Veen, W.E.C.J., van Langevelde, H.J., Habing, H.J., & Sjouwerman, L.O. 1998 A&A 329, 991
3. Cohen, J., Gratton, R., et al. 1999 Preprint.
4. Combes, F., Debbasch, F., Friedli, D., & Pfenniger, D. 1990, A&A 233, 82
5. Edvardsson, B., Andersen, J., Gustafsson, B., Lambert, D.L, Nissen, P.E., & Tomkin, J. 1993, A&AS 102, 603
6. Figer, D.F., Kim, S., Morris, M., Serabyn, E., Rich, R.M, & Mclean, I.S. 1999, ApJ, in press.
7. Frogel, J.A., Tiede., G.P., & Kuchinski, L.E. 1999, AJ, 117, 2296
8. Glass, I.S., Whitelock, P.A., Catchpole, R.M., & Feast, M.W. 1995, MNRAS, 273, 383
9. Habing, H.J., Olnon, F.M., Chester, T., Gillett, F., & Rowan-Robinson, M. 1985, A& A, 152, L1
10. Holtzman, J.A., Watson, A.M., Baum, W.A. et al. 1998, AJ, 115, 1946
11. Jablonka, P., et al. 1999 ApJL, in press
12. McWilliam, A., & Rich, R.M. 1994, ApJS, 91, 749
13. Ortolani, S., Renzini, A., Gilmozzi, R., Marconi, G., Barbuy, B., Bica, E, &Rich, R.M. 1995, Nature, 377, 701
14. Stanford, S.A., Eisenhardt, P.R., & Dickinson, M. 1998, ApJ, 492, 461
15. Winnberg, A. Baud, B., Matthews, H.E., Habing, H.L., & Olnon, F.M. 1985, ApJ, L45

Globular Clusters in the Galactic Bulge

Sergio Ortolani

Department of Astronomy, University of Padova,
Vicolo dell'Osservatorio 5, Padova, I-35122 Italy

Abstract. We present the characteristics and the distribution of the clusters projected in the direction of the Galactic Bulge. The clusters are compared to the bulge stellar field population. It appears that these clusters are good tracers of the galactic bulge. The derived scenario is an old bulge stellar population with a peculiar chemical composition.

1 Overview

The galactic bulge is characterized by a high concentration of stars as well as of globular clusters in a rather small volume.

In a radius of about 4 Kpc from the galactic center there are about 50 known globular clusters (Barbuy et al., 1999a). Considering that the recent catalogues of the galactic globular clusters include about 140 objects (Harris, 1996), it is evident that the globular clusters are strongly concentrated toward the galactic center, and more than one third of them are actually in the region of the galactic bulge.

The galactic bulge contains also a large number of variable stars, mainly RR Lyrae and Mira type. In a note published in 1946, Baade presented his results on the first photographic survey in the galactic bulge, namely in the region which will be called later the Baade Window, where he identified for the first time the variable stars. He found 152 variables in an area of 0.38 square degrees. Most of them were RR Lyrae. Comparing this result with the Milky Way fields previously investigated, he argued that the stellar population of the Sagittarius region is different from the disk population and that it could be instead of type II. He concluded then that the halo of our galaxy *represents the continuation of the nuclear bulge.*

Until recently the situation didn't change much and there is still an open debate in the literature whether the bulge is the inner extension of the halo, having formed in the same process, or whether it formed much later, perhaps as a consequence of the dynamical evolution of the disk (Larson 1990, Renzini 1999, Raha et al., 1992, Wyse et al., 1997).

Thanks to microlenses surveys, the number of known RR Lyrae stars recently increased by more than one order of magnitude, but it is not yet completely clear if these stars are just the metal poor halo component extending inside the bulge or if they belong to the bulge (Alcock et al., 1998). From their pulsational properties only a small fraction of them seems to be metal rich. Still it is not easy to explain their presence. In fact the metal rich globular clusters in the halo contain very few, if any, RR Lyrae. Lee (1992) pointed out that the presence of numerous RR Lyrae stars with a nearly solar abundance can be explained only if they are older than the galactic halo.

The bulge includes also a large number of bright Miras and long period variables which have been interpreted in the past as evidence of a young population. This is still matter of controversy because of the degeneracy age-metallicity. Guarnieri et al. (1997) presented evidence for the case of very bright long period variables in a population with high metallicity and an old age.

New, interesting results are coming mainly from the the spectroscopy (see the review by Barbuy in these proceedings). There is today evidence that the bulk of the bulge stellar population is old and metal rich. The spectroscopic analysis recently obtained by Rich (1988), McWilliam and Rich (1994) and confirmed Sadler et al. (1996) and by Barbuy et al. (1999b) revealed (a) a broad metallicity distribution in the range $-1.5 < [Fe/H] < +0.5$, and (b) a strong [α-elements/Fe] enhancement. This signature indicates a fast enrichment by SNs II at early times in the formation of the bulge. A scenario with an old bulge formed during a rapid enrichment process can be derived (see the models by Matteucci and Brocato, 1990).

The direct study of the bulge age via color-magnitude diagrams (CMDs) is an obvious task to constrain the formation process of our galaxy. However the measurements are affected not only by crowding but also by differential reddening across the field of view and by contamination due to the disk stellar population seen projected on the galactic bulge. These effects produce an apparent selective upward shift (with an increasing factor at fainter magnitudes) of the turnoff, relatively to the HB, giving the impression of a younger age. Furthermore the color magnitude diagrams obtained in the field appear broadened because they contain stars with different metallicity and spread in depth.

Early data from HST WFPC1 observations in the Baade Window have been interpreted (Holtzman et al., 1993) as evidence for an intermediate age (around 10 Gyr) of the field population, but more recent results from Ortolani et al. (1995), which combine WFPC2 and ground observations, are in favour of an older age.

2 Observations of Bulge Globular Clusters

The globular clusters are simple stellar populations, so they are not affected by metallicity, age and distance spread. For this reason in the last 15 years we concentrated our studies on the globular clusters projected in the direction of the galactic bulge.

Most of the low galactic latitude clusters have been discovered by Terzan and collaborators more than 20 years ago (Terzan 1968 and references therein) from wide field infrared plates. So far no other systematic surveys, aimed to discover globular clusters, have been undertaken. The available sample is still far to be complete. Barbuy et al. (1998) showed that there still is an empty wide strip extending about 1 degree above and 2 degrees below the galactic plane. A concentration of clusters, at least 20-30, hidden by heavy extinction, is expected there. These objects could be discovered in a wide field near infrared survey covering about 20 square degrees.

Until recently very few data were available in the literature for most of the clusters in the direction of the galactic bulge (Racine and Harris, 1989, Armandroff, 1989). In our systematic study we revised previous classifications and derived new metallicities, distances for about 50 clusters in the area included between $l = \pm 20°$ and $b = \pm 15°$. This sample includes all the known globular clusters within 5 ° from the galactic center have been checked (Barbuy et al. 1998).

The method we used is based on the comparison of color-magnitude diagrams (CMDs) with templates. The best template is chosen according to the morphological comparison (which gives a first guess for the metallicity). The horizontal shift, derived from the best fit with the template, gives the reddening, while the vertical shift gives the apparent distance modulus. Integrated spectra have also been used as an additional guideline for the reddening and the metallicity. It is now possible to study the properties of this sample on a statistical base.

With a few exceptions the studied clusters are actually located nearby the galactic center.

Most of them are metal-rich clusters, characterized by a red and very short HB and a very extended and tilted giant branch.

Five objects, (10 % of our sample) turned out to be different objects (galaxies, HII regions, supernova remnants or open clusters, or simply non-existing, see e.g. Bica et al., 1995).

An important fraction of clusters classified in the previous literature as metal rich clusters resulted to be blue HB with very steep giant branches, resembling those of the most metal poor halo clusters such as M 30. The correct classification of a blue HB globular cluster in the bulge is more difficult than a red HB because the blue tail is blended with the disk main sequence stars. Furthermore integrated photometry or spectroscopy are often contaminated by the bright red giant stars of the bulge field and then they give high values of metallicity. This can explain why very few blue, metal poor HB clusters have been previously identified in the studied region.

Very few examples of intermediate type have been found.

The total number of the red HB clusters is comparable to the blue HB ones. The red HB clusters are much more concentrated toward the galactic plane, with a ratio of red HB clusters over blue HB of two in the inner ± 5 °, while it decreases below one at a distance greater than 10 ° from the galactic plane. This trend can be explained in a scenario of interloping metal rich bulge clusters and more metal poor halo clusters (Minniti, 1995). In spite of this different latitude trend, the two samples are well spatially mixed and their kinematics do not seem sharply different. This can be certainly due also to the low numbers, but we cannot exclude that at least some of the blue HB clusters formed in the bulge itself at early times, still maintaining orbits characteristic of the bulge.

These metal poor inner bulge clusters are of great interest because, in an inside-out formation model, they could be the oldest in the Galaxy. Combined deep VLT optical-near infrared observations below the turnoff and high resolution spectroscopy of the giant stars to get their chemical composition and kinematics, are needed to clarify the nature of these clusters.

3 Space Distribution, Kinematics and Ages

The method used to find the distances is a key aspect in any space distribution analysis of globular clusters. In the case of low latitude clusters, affected by heavy absorption, the problem is complicated because of the possible dependence of the ratio R of the selective to absolute extinction (reddening to extinction) on several parameters such as temperature, spectral type, chemical composition of the observed stars and on the absolute value of the absorption itself (Grebel and Roberts, 1995). Following Cardelli et al. (1989), instead, the ratio R should be nearly constant, with relevant exceptions only in very special regions of active stellar formation. Since we derive the absorption from the reddening (obtained from the horizontal shift of the CMDs) the uncertainty on this coefficient is dominating all the other distance uncertainties when the reddening is higher than E(B-V) = 0.3-0.4 mag. Current estimates of the ratio R are from 2.7 to 3.6. Tests made on globular clusters observed in optical and in the near infrared (where the reddening law is more stable) such as Terzan 2 (Ortolani et al., 1997 and Kuchinski et al., 1995) and NGC 6553 (Guarnieri et al. 1998) show that the current adopted value R=3.1 is too low, and a value around 3.3 could be a more acceptable compromise. However higher statistics are needed to reveal any trend.

With this caution we found that all but two clusters projected within 5 degrees from the galactic center are located within 4 Kpc from the galactic center. This is the radius of the Galactic Molecular Ring (Burton 1976) and the inner region coincides with the "hole" in the galactic disk, characterized by a pronounced lack of gas and population I tracers (Georgelin and Georgelin, 1976; Taylor and Cordes, 1993). This is a clear indication that the globular clusters projected in the direction of the inner galactic bulge belong to the bulge itself and are not connected to the disk.

The vertical distribution of the genuine inner bulge clusters within 5 degrees is strongly concentrated toward the galactic plane, following Kent's (1992) bulge exponential distribution, but it is quite different from that discussed by Armandroff (1989).

Our results from the kinematics confirm previous findings from Minniti (1995): the globular clusters in the galactic bulge show a slow rotation if compared to the disk, with a slope similar to that of the bulge field stars. There is no apparent ranking of blue and red HB clusters.

Combining the spatial distribution and the results from kinematics, one can conclude that the clusters projected in the direction of the galactic bulge share the same properties of the bulge stellar population, while they are completely decoupled from the galactic disk (see also Ibata and Gilmore, 1995).

As concerning the age, we detected the main sequence turnoff in three bulge clusters, NGC 6553, NGC 6528 (Ortolani et al., 1995) and NGC 6652 (Ortolani et al., 1994). Their HB - main sequence luminosity difference is comparable to the halo cluster 47 Tuc. In a first approximation this is an indication of a comparable age. NGC 6553 and 6528, in particular, have a chemical composition very similar to the Baade Window field giants (Barbuy et al., 1999b) and almost identical CMDs. Their luminosity functions, from the horizontal branch to the main sequence, are also very similar to that of the Baade Window (Ortolani et al., 1995, Fig. 3). This implies that the age of the bulge (including field stars and the metal rich globular clusters) is comparable to the age of the relatively metal rich clusters in the halo. From this point of view it appears that the original intuition by Baade was not too far from the results obtained with modern techniques.

The results of our analysis are based only on a few clusters. While a near infrared survey would allow the discovery of new, expected inner bulge clusters, deep optical and near infrared observations carried out with a big telescope such as VLT are needed to get the main sequence turnoff in a wider sample. Combining optical and infrared photometry it would be also possible to resolve the problem of the reddening and then of the distances. At the same time further spectroscopic analysis are needed to check the metallicities that we derived mainly from the morphology of the color-magnitude diagrams.

References

1. Alcock, C. et al. 1998, *ApJ* **492**, 190
2. Armandroff, T. 1989 *AJ* **97**, 375
3. Baade, W. 1946, *PASP* **58**, 249
4. Barbuy, B., Ortolani, S., Bica, E., 1999a, in preparation
5. Barbuy, B., Renzini, A., Ortolani, S., Bica, E., Guarnieri, M.D. 1999b, *A&A* **341**, 539
6. Barbuy, B., Bica, E., Ortolani, S. 1998, *A&A* **333**, 117
7. Bica, E., Clariá, J.J., Bonatto, C., Piatti, A.E., Ortolani, S., Barbuy, B. 1995, *A&A* **303**, 747
8. Burton, J.B. 1976, *Ann. Rev. A&A* **14**, 275
9. Cardelli, J.A., Clayton, G.C., Mathis, J.S. 1989, *ApJ* **345**, 245
10. Geogelin, Y.M., Georgelin, Y.P. 1976, *A&A* **49**, 57

11. Grebel, E.K., Roberts W.J. 1995*A&A Suppl.* **109**, 293
12. Guarnieri, M.D., Ortolani, S., Montegriffo, P., Renzini, A., Barbuy, B., Bica, E., Moneti, A. 1998, *A&A* **331**, 70
13. Guarnieri, M.D., Renzini, A., Ortolani, S. 1997, *ApJ* **477**, L21
14. Harris, W.E. 1996 *AJ* **112**, 1487
15. Holtzman, J.A. et al, 1993 *AJ* **106**, 1826
16. Ibata, R.A., Gilmore, G.F., 1995 *MNRAS* **275**, 605
17. Kent, S.M. 1992, *ApJ* **387**, 181
18. Kuchinski, L.E., Frogel J.A., Terndrup, D.M., Persson, S.E. 1995, *AJ* **109**, 1131
19. Larson, R.B. 1990, *PASP* **102**, 709
20. Lee, Y.W. 1992, *AJ* **104**, 1780
21. Matteucci, F., Brocato, E. 1990, *ApJ* **365**, 539
22. McWilliam, A., Rich, R.M. 1994, *ApJS* **91**, 749
23. Minniti, D. 1995, *AJ* **109**, 1663
24. Ortolani, S., Bica, E., Barbuy, B., 1994 *A&A* **286**, 444
25. Ortolani, S., Renzini, A., Gilmozzi, R., Marconi, G., Barbuy, B., Bica, E., Rich, R.M.: 1995, *Nature* **377**, 701
26. Ortolani, S., Bica, E., Barbuy, B., 1997 *A&A* **326**, 614
27. Racine, R., Harris, W.E. 1989, *AJ* **98**, 1609
28. Raha, N., Sellwood, J.A., James, R.A., Kahn, F. D. *Nature* **352**, 411
29. Renzini, A. 1999, astro-ph/9902108
30. Rich, R.M. 1988, *AJ* **95**, 828
31. Sadler, E.M., Terndrup, D.M., Rich, R.M. 1996,
32. Taylor, J.H., Cordes, J.M. 1993, *ApJ* **411**, 674
33. Terzan, A. 1968, *Comptes Rendus Acad. Sci. Paris* **267**, 1245
34. Wyse, R., Gilmore, G., Franx, M. 1997, *ARA&A* **35**, 637

Studying the Stellar Populations of the Local Group with VLT

Eline Tolstoy

ESO, Karl-Schwarzschild Str. 2, D-85748 Garching bei München, Germany

1 Introduction

The best chance we have to understand star formation and how it proceeds in the Universe is going to come from detailed studies of the numerous different environments found within the Local Group (LG). Present day star formation in our Galaxy occurs exclusively in metal rich environments (Z $\sim Z_{\odot}$), so if we want to study how low metallicity stars form (and thus understand observations of galaxies at high-redshift) we have to look beyond our Galaxy, to the smallest star forming dwarf galaxies, which can have extremely low metallicities (Z $\sim 0.02 - 0.05 Z_{\odot}$). Of course in its entirety a stellar population always contains the complete details of the star formation history of a galaxy, however this information is often hard to disentangle retroactively. We also have much to learn from the Magellanic Clouds (Z $\sim 0.1 - 0.3 Z_{\odot}$), although because they are undergoing interactions with our Galaxy and each other their evolutionary picture and its general applicability less obvious. In our LG there are also a number of "remnants", or galaxies which which currently do not form stars (*e.g.* the dSph, such as Carina, Leo I, Ursa Minor, etc..). It is not straight forward to draw parallels between galaxies which are forming stars and those which aren't. This is of course because star formation has such a dramatic impact upon a galaxy, and alternative methods have to be used to make the most basic of comparisons of properties (*e.g.* metallicity, mass, luminosity evolution). It is necessary to put all the dwarf galaxies into a global picture if we are to draw meaningful conclusions about their star formation properties (*e.g.* Ferrara & Tolstoy 1999). Many of the small LG galaxies contain direct evidence of complicated star formation histories (*e.g.* Smecker-Hane *et al.* 1994; Tolstoy *et al.* 1998; Gallart *et al.* 1999), which suggests that star formation patterns can change dramatically over long time scales. This kind of evolutionary behaviour can have a dramatic impact upon the accurate interpretation of galaxy redshift surveys (cf. Tolstoy 1998).

As in all scientific endeavors the most useful approach to solving a complex question is to apply a number of different techniques, compare the results and then, with the help of some theoretical insight, form a consistent picture. The most important advances are likely to come not only from the ability to push fainter and deeper into interesting regions of the Colour-Magnitude diagram (CMD), but also from the ability to carry out detailed spectroscopic analyses of individual *stars in the same* CMDs. The first images and spectra to come out of the Paranal Observatory from FORS1, (*e.g.* see Kudritski, this volume; Appenzeller, plenary talk), have offered the exciting two-fold promise: the ability to go very faint and exquisite image quality. This new era of the ground based 8-10m telescope promises to push forward our understanding of galaxy evolution via the study of nearby systems.

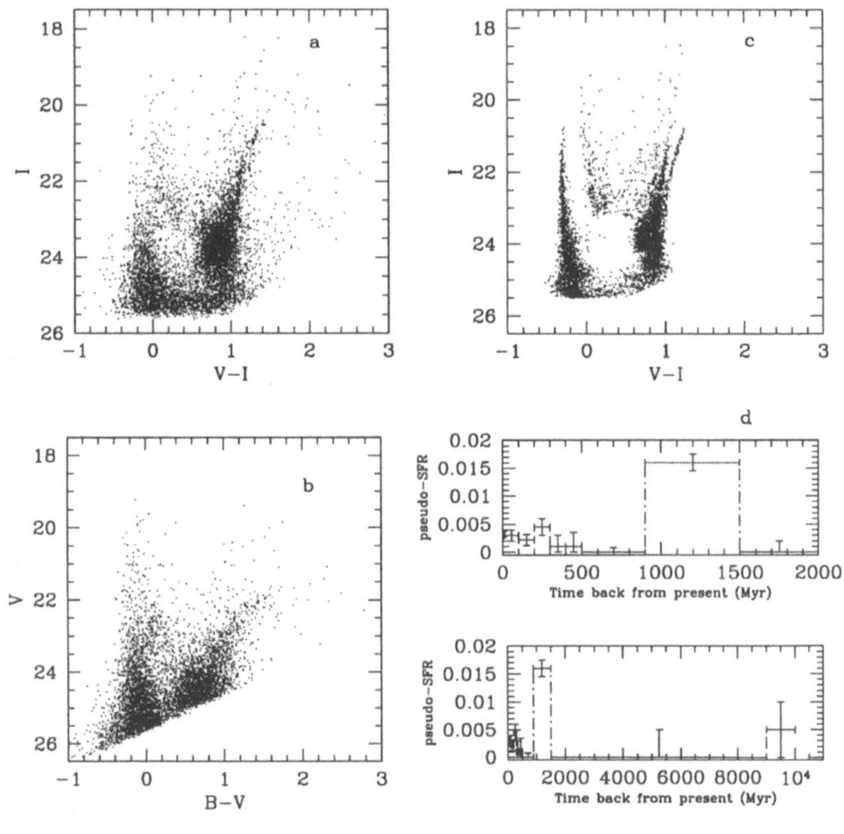

Fig. 1. Here are the results of the analysis of HST/WFPC2 data of nearby dwarf irregular galaxy Leo A (Tolstoy *et al.* 1998). In a. is the V−I, I CMD, 1 orbit exposure time per filter. In b. is the B−V, V CMD, 2 orbits in B. In c. is the best match Monte-Carlo simulation model (in V−I, I) found for these data convolved with the theoretical measurement error distribution, and in d. is the SFH that created the model CMD which best matches these data. See Tolstoy *et al.* (1998) for more details..

2 Colour-Magnitude Diagrams

The study of resolved stellar populations in CMDs as means of understanding the properties of nearby stars and galaxies has a distinguished historical tradition, going back to Leavitt almost a century ago, and including the careers of Hubble and Baade (*e.g.* Baade 1963). It was during the 1950's that the CMD provided the impetus to understand the, then, holy grail of stellar evolution, using the newly operating Palomar 200inch telescope, to observe Galactic star clusters. In the 1990s we have come so far that stellar evolution theory is understood in sufficient detail that we can make numerous predictions related to structures in a CMD (*e.g.* Bertelli *et al.* 1994; Tolstoy & Saha 1996; Girardi *et al.* 1996; Aparicio *et al.* 1996; Tolstoy 1999), based upon variations of age and metallicity in star forming regions many Gyr ago. Thus we have the ability to disentangle complex star formation histories of all nearby galaxies. with sufficiently deep and accurate CMDs.

Improved image quality is arguably of equal importance to an increase in aperture size. Digging deep in to a CMD is only a worthwhile exercise if the errors can be minimised. Much of the difficulty in interpreting CMDs comes from distinguishing effects which come from photometric errors due to image crowding and those caused by the star formation properties of a galaxy. The *Hubble Space Telescope* (HST) has been leading the way in providing high quality CMDs (*e.g.* Dohm-Palmer *et al.* 1997, 1998; Gallagher *et al.* 1998; Tolstoy *et al.* 1998; Cole *et al.* 1999; Gallart *et al.* 1999; see Figure 1), but the collecting area and instruments available on VLT will be important for further progress. In the case of low luminosity nearby galaxies HST images are completely uncrowded, and cover only a fraction of a galaxy, so using VLT can be an equally good option, given an average seeing of $\lesssim$ 0.6arcsec (*e.g.* see Fig 2). The resulting CMDs (Fig. 3) show a wealth of detail, and a much narrower red giant branch than previous observations. We also unequivocally define the presence blue main sequence stars in Antlia, with an age in the range 1-2 Gyr.

3 Metallicity Evolution

The global effects of stellar evolution upon the interstellar medium of a galaxy are surprisingly poorly understood, especially in metal poor environments. It is critical to our understanding of galaxy evolution to determine how the fraction of metals in a galaxy build up over time by the star formation process. This information is preserved in stellar abundance patterns of a galaxy, however distinguishing the effects of age and metallicity is often a complex task. Purely photometric determinations from CMD analysis cannot provide unique solutions because of age-metallicity degeneracy on the red giant branch. It is necessary to directly measure the abundance of a large sample stars to understand how the metallicity has Changed with time (e.g.

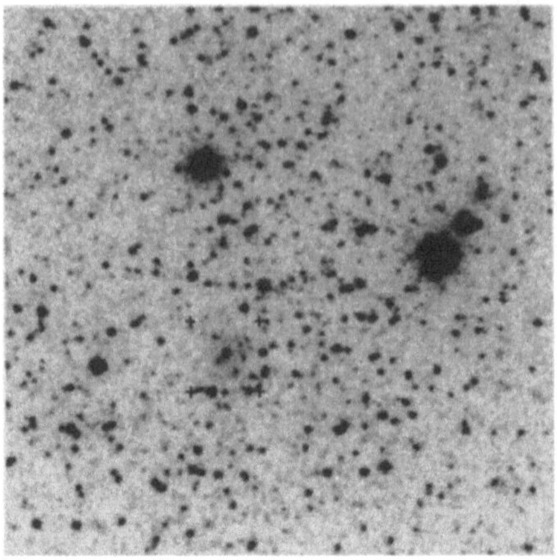

Fig. 2. This is the central section (50arcsec squared) of the combined I filter FORS1 images (total, 5400sec) of the Antlia dwarf galaxy taken under excellent seeing conditions on 30th and 31st January 1999. The average seeing was below 0.5 arcsec for the entire integration period. It can be seen that in this image crowding is not a problem, and the integrations could clearly be even longer. One point of note, is that an unusual source (in between the crosses here), remarked upon by Sarajedini *et al.* (1997), it is visible in the I band (I∼22.5) and not in the V (V<25). It is clearly a background galaxy in the FORS1 image.

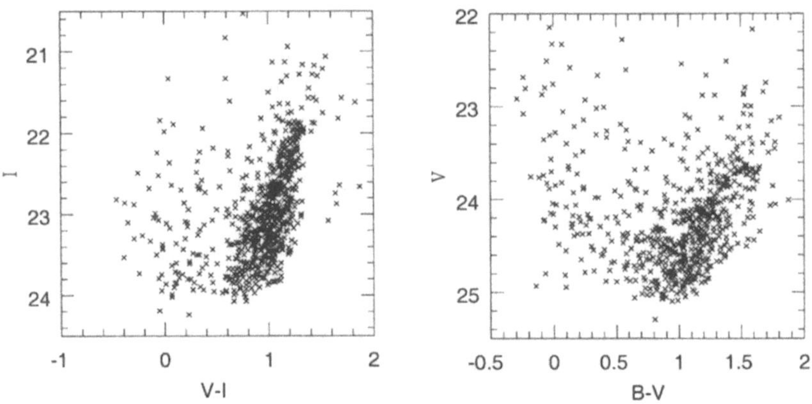

Fig. 3. Here we show the first CMDs from the FORS1 science verification data of the Antlia dwarf galaxy. They include stars within a circle of 3.3arcmin diameter centred on Antlia. The larger this area the more extensive will be the foreground contamination, but stars from Antlia are seen out to the edges of the FORS1 frame (7 arcmin across). The total integration times are 2400sec in B and V and 5400sec in I. There is a clear "blue plume" of stars, which is real. Inspection of the images shows that they are neither background galaxy confusion nor a cosmetic problem on the CCD. They could represent a comparatively young population of stars in Antlia (between 1 and 2 Gyr old). There is no evidence for extensive reddening internal to Antlia, but the calibrations used here are still preliminary. The (preliminary) distance from the tip of the red giant branch is 1.4 Mpc, slightly further away than previous estimates.

Edvardsson *et al.* 1993, in our Galaxy). If we can add independent metallicity information about individual stars to CMD analysis, then we can determine much more accurate star formation histories.

There are several observational approaches to this problem, all of which benefit from intermediate and high resolution spectrographs on large telescopes (*e.g.* HIRES, FORS1/2, UVES). The large aperture of VLT or Keck is crucial to overcome the limitation of the readout noise of the detectors, which is the limiting factor on smaller telescopes.

- Metallicity evolution can be predicted from HII region emission line diagnostics with some success (*e.g.* Matteucci & Tosi 1985; Kobulnicky & Skillman 1996; Pagel & Tautvaišienė 1998), although this is not independent of uncertain model assumptions. There are also a number of LG galaxies with very faint HII regions about which we know very little about the basic abundance measures.
- The direct relation between the Ca II triplet absorption index and stellar metallicities in metal poor systems, as defined by Da Costa & Armandroff (1995), affords a convenient and independent way to measure the relative metallicities of evolved red giant stars, which cover the age range 1–10 Gyr. This has been successfully applied to stellar populations in the Magellanic Clouds (*e.g.* Olszewski *et al.* 1991; Da Costa & Hatzidimitriou 1998).
- In the case of the brighter giants in nearest by galaxies, higher resolution spectroscopy is possible for an accurate analysis of abundance patterns [*e.g.* McCarthy *et al.* 1995 (M31); Shetrone, Bolte & Stetson 1998 (Draco); Venn 1999 (SMC)]. These detailed abundance analyses provide a more accurate picture of how chemical enrichment occurs in different environments. For example, the pattern of element enrichment (*e.g.* C, N, O, Na, Mg and Al to Fe) in globular cluster giants is found to differ from field Population II giants in the Galaxy halo (Pilachowski, Sneden & Kraft 1996; Shetrone 1996).

4 Summary

The power of high quality images and large collecting area of the VLT is impressive, and there are clearly many exciting discoveries waiting to be made. We hope for further steps forward from the study of resolved stellar populations with VLT, as from Palomar before it, because now we will be able to probe a much large range of star forming environment within the LG, and specifically very low metallicity environments. The tremendous gains in image quality and collecting area now available with the VLT on Paranal make it important to survey the resolved stellar populations of all the accessible galaxies in our LG. This will provide a uniform picture of the properties of stellar populations of galaxies with a wide variety of mass, metallicity, gas content etc. This has begun with FORS1 science verification observations in January 1999, with BVI images of the LG galaxy Antlia (see Fig. 3). The study of individual stars and star-formation regions in the nearby universe is the only way we will understand the observations of galaxy populations at high redshift.

Acknowledgments

The Science Verification data used in this paper are public, and were not presented at the Antofagasta meeting (only as press release images by Prof. I. Appenzeller in his plenary talk), the analysis here was made after the data were made public in May 1999. For more information, look at the ESO web page: `http://www.eso.org/science/ut1sv/`

References

1. Aparicio, A., Gallart, C., Chiosi, C., Bertelli, G. (1996) Ap.J.Lett, **469**, 97–100
2. Baade, W. (1963) "Evolution of Stars and Galaxies", ed. C. Payne-Gaposchkin
3. Bertelli, G., Bressan, A., Chiosi, C. *et al.* (1994) A.& A.Supp., **106**, 275–302
4. Cole, A.A., Tolstoy, E., Gallagher, J.S., Hoessel, J.G., Mould, J.R., Holtzman, J.A., Saha, A. & the WFPC2 IDT team (1999) A.J., in press (astro-ph/9905350)
5. Da Costa, G.S. & Armandroff, T.E. (1995) A.J., **109**, 2533–2552
6. Da Costa, G.S. & Hatzidimitriou, D. (1998) A.J., **115**, 1934–1945
7. Dohm-Palmer, R.C., Skillman, E.D., Saha, A., Tolstoy, E., Gallagher, J.S., Hoessel, J.G., Mateo, M., Chiosi, C. (1997) A.J., **114**, 2527–2544
8. Dohm-Palmer, R.C., Skillman E.D., Saha A., Tolstoy E., Mateo M., Gallagher J.S., Hoessel J.G., Chiosi C., Dufour R.J. (1998) A.J., **116**, 1227–1243
9. Edvardsson, B., Andersen, J., Gustafsson, B. *et al.* (1993) A.&A., **275**, 101
10. Ferrara, A. & Tolstoy E. (1999), submitted to MNRAS, (astro-ph/9905280)
11. Gallagher, J.S., Tolstoy, E., Dohm-Palmer, R.C., Skillman, E.D., Cole, A.A., Hoessel, J.G., Saha, A., Mateo, M. (1998) A.J. **115**,, 1869–1887
12. Gallart, C., Freedman, W. *et al.* (1999) A.J., in press (astro-ph/9906121)
13. Girardi, L., Bressan, A., Chiosi, C. *et al.* (1996) A.& A.Supp., **117**, 113–125
14. Kobulnicky, H.A. & Skillman, E.D. (1996) Ap.J. **471**, 211–236
15. Matteucci, F. & Tosi, M. (1985) MNRAS, **217**, 391
16. McCarthy, J.K., Lennon, D.J. *et al.* (1995) Ap.J.Lett., **455**, 135–138
17. Olszewski, E.W., Schommer, R.A. *et al.* (1991) A.J., **101**, 515–537
18. Pagel, B.E.J. & Tautvaišienė, G. (1998) MNRAS, **299**, 535–544
19. Pilachowski, C.A., Sneden, C., & Kraft, R.P. (1996) A.J., **111**, 1689–1704
20. Sarajedini, A., Claver, C.F. & Ostheimer J.C. (1997) A.J., **114**, 2505–2513
21. Shetrone, M.D. (1996) A.J., **112**, 1517–1535
22. Shetrone, M.D., Bolte, M. & Stetson, P.B. (1998) A.J., **115**, 1888–1893
23. Smecker-Hane, T.A. *et al.* (1994) A.J., **108**, 507–513
24. Tolstoy, E. & Saha A. (1996) Ap.J., **462**, 672–683
25. Tolstoy, E. (1998) "Dwarf Galaxies & Cosmology", eds. Thuan T.X. *et al.*, in press (astro-ph/9807154)

26. Tolstoy, E. (1999) "The Stellar Content of Local Group Galaxies", eds. Whitelock, P. & Cannon R., p. 218–230
27. Tolstoy, E., Gallagher, J.S., Cole, A.A., Hoessel, J., Saha, A., Dohm-Palmer, R.C., Skillman, E.D., Mateo, M. & Hurley-Keller, D. (1998) A.J., **116**, 1244–1262
28. Venn, K.A. (1999) Ap.J., in press, (astro-ph/9901306)

Fine Structure of the Red Clump in Local Group Galaxies

Léo Girardi[1,2]

[1] Dipartimento di Astronomia, Vicolo dell'Osservatorio 5, I-35122 Padova, Italy
[2] Max-Planck-Institut für Astrophysik, Karl-Schwarzschild-Str. 1, D-85740 Garching bei München, Deutschland

Abstract. Some fine structures can nowadays be identified in the high-quality colour-magnitude diagrams (CMD) of Local Group galaxies. The clump of red giants, for instance, may present a significant colour spread, and extensions to both brighter and fainter luminosities. Such features are predicted by population synthesis models which consider stars in the complete relevant ranges of ages and metallicities, and are potentially useful for constraining the star formation histories of the parent galaxies over scales of gigayears. We briefly comment the cases of fields in the Magellanic Clouds, M31, and the local CMD from *Hipparcos*.

1 Introduction

The red clump is a striking feature in the CMDs of intermediate-age and old open clusters, and in those of the nearest galaxies. It is composed mostly by low-mass stars in the stage of core helium burning (CHeB). In these stars, the onset of electron degeneracy after the central H-exhaustion postpones the He-ignition until the core mass grows to about $M_c \simeq 0.45\,M_\odot$. It causes all low-mass CHeB stars to have similar luminosities and hence position in the CMD.

In open clusters, the red clump typically presents a very small dispersion in both colour and magnitude. For instance, in M 67 the red clump is defined by 6 stars with $\sigma(B-V) \lesssim 0.05$, $\sigma(V) \lesssim 0.1$ mag (see e.g. Montogomery et al. 1993). This results from the small spread of masses and metallicities among these stars. On the contrary, red clump stars in galaxy fields present a significant spread in both mass and metallicity; as a consequence, these clumps are expected to present a fine structure in the CMD (Girardi 1999).

This fine structure is not evident in the case of most ground-based observations of nearby galaxies, for which the clump spread in the CMD reflects, to a large extent, the presence of large photometric errors. The *Hipparcos* satellite, however, has provided an impressive CMD of the nearby stars (Perryman et al. 1997), in which the red clump clearly presents an intrinsic structure (Girardi et al. 1998). HST has also provided very nice CMDs for the Magellanic Clouds (MC) and other nearby dwarf galaxies, in which sub-structures of the clump seem to be present. In the case of the MCs, however, too few clump stars can be sampled in a typical WFPC2 frame, so that the clump features are not so clear. This situation is expected to improve with the FORS

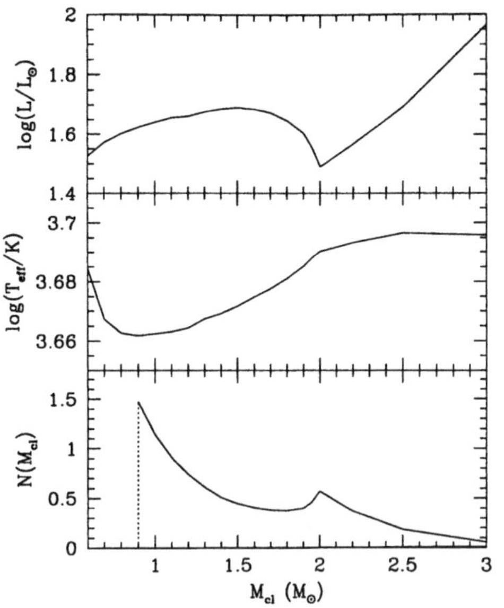

Fig. 1. For $Z = 0.019$, we present the mean luminosity (upper panel) and effective temperature (middle) of CHeB stars as a function of mass. The lower panel shows their mass distribution for a galaxy model with constant star formation rate up to 10 Gyr ago.

camera at VLT since it can provide CMDs with quality comparable to the HST ones (at least for the Magellanic Clouds), but sampling much larger fields.

2 Theoretical Expectations

In order to simulate the clump structure in different galaxy models, we make use of the extensive set of evolutionary tracks and isochrones of Girardi et al. (1999). This database covers a large interval of stellar initial masses (from 0.15 to 7 $M_{\odot}$) and metallicities (from $Z = 0.001$ to 0.03), with a very good mass resolution. Synthetic CMDs are then generated by means of a population synthesis tool, for any arbitrary history of star formation and chemical enrichment.

Fig. 1 illustrates the location of solar metallicity ($Z = 0.019$) CHeB stars in the HR diagram, as a function of mass, and their mass distribution in a model galaxy which formed stars at a constant rate from 0.1 to 10 Gyr ago, with a Salpeter IMF and with Reimers' mass-loss rates along the RGB (see Girardi 1999 for details). The following aspects are evident in this figure:

- stars with $M \lesssim 2\ M_{\odot}$ (i.e. low-mass stars) constitute most of the clump, distribute over a non-negligible range in $T_{\rm eff}$, and with an almost constant luminosity. They form the *main red clump* feature we are used to.
- stars with $M \simeq 2\ M_{\odot}$ not only occupy a particular region of the HR diagram (they are about 0.4 mag fainter, and slightly bluer than most clump stars), but represent a second peak in the mass distribution. Therefore, these stars define a fainter *secondary red clump*.

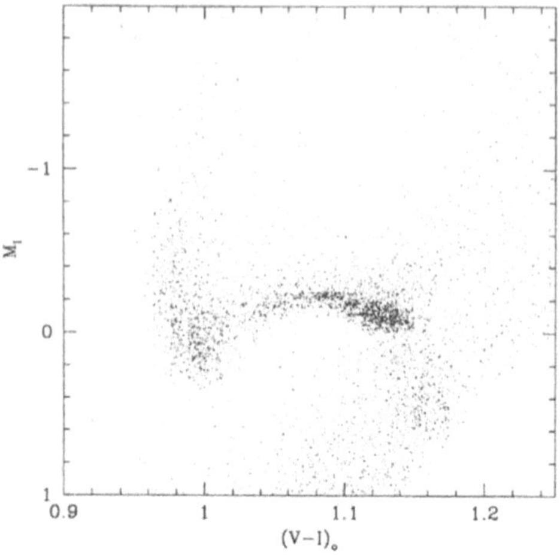

Fig. 2. Synthetic M_I vs. $V - I$ CMD for a galaxy model which assumes $Z = 0.019$ (with a small metallicity dispersion), and constant star formation rate. Only a small region of the CMD, centered on the red clump, is shown.

- stars with $M \gtrsim 2\ M_{\odot}$ (i.e. intermediate-mass stars) appear at higher luminosities and in non-negligible quantities. They may originate a plume of bright clump stars (or the so-called *vertical red clump*).

These structures are evident in the synthetic CMD of Fig. 2. We remark that essentially the same features appear at different metallicities, provided that $Z \gtrsim 0.004$. For lower metallicities, the colour spread of the clump gets very low, unless for the tail of lowest-mass clump stars, which may distribute over a blue horizontal branch if ages are high enough. On the other hand, models which assume an intrinsic age-metallicity relation for clump stars, tend to present the main clump distributed over a large colour interval, but still at an almost constant luminosity (Girardi 1999).

3 Observations and Practical Implications

From the above-mentioned work, it results that the clump structure in the CMD contains potentially useful information about the history of star formation and chemical enrichment of the parent galaxy. The most evident example is given by the secondary clump stars in a field: their number should be simply proportional to the SFR at an age of $\sim$ 1 Gyr ago (corresponding to an initial mass of 2 $M_{\odot}$ in the present models). The colour distribution of the main clump, instead, should reflect both the age and metallicity distribution of clump stars, over timescales of gigayears. Importantly, this information appears at a luminosity level which is at least 3 mag brighter than the oldest turn-offs of most galaxies.

Also, it is important to emphasise that the model predictions are supported by a number of interesting observations. For instance, a secondary clump has been noticed in the *Hipparcos* CMD of local stars (Girardi et al. 1998) and in some LMC fields observed by Bica et al. (1998). Also, Corsi et al. (1994) observed a minimum in the clump luminosities for LMC clusters of ages $\sim$ 1 Gyr, which corresponds to the secondary clump we mention here.

Other important observation which can be accounted for by present models is the nearly constancy of the clump I-band luminosity, M_I, with the $V-I$ colour, noticed by Paczyński & Stanek (1998), Stanek & Garnavich (1998), Udalski et al. (1998) and Stanek et al. (1998). This was interpreted as indication that the clump luminosity does not depend on the stellar population. If this were the case, the red clump would be an excellent standard candle, with a zero point precisely determined by means of *Hipparcos* data. Theoretical models instead indicate a non-negligible dependence of M_I on age and metallicity, amounting to a maximum of 0.6 mag (see also Cole 1998). At first sight, it seems in contradiction with the observations. However, constant M_I as a function of $V-I$ can easily be obtained in population synthesis models, if we assume either a nearly constant Z with varying age, or a Z decreasing with age according to a normal age–metallicity relation (see Girardi et al. 1998 and Girardi 1999, for details). The first case seems to describe reasonably what observed in the *Hipparcos* CMD, whereas the second case would correspond better to what observed in M 31.

It follows that the red clump may be a useful distance indicator only if we know the distribution of ages and metallicities inside the galaxy we are observing. Otherwise, systematic errors as large as $\sim$ 0.4 mag may be present in the distance modulus derived by means of this method.

Acknowledgements

Part of this work has been carried out at the Max-Planck-Institut für Astrophysik during a stay funded by the Alexander von Humboldt-Stiftung. The collaboration with A. Weiss, M.A.T. Groenewegen and M. Salaris was decisive during this period. I thank ESO for a travel grant to attend the VLT opening symposium.

References

1. Bica E., et al., 1998, AJ 116, 723
2. Cole A.A., 1998, ApJ 500, L137
3. Corsi C.E., et al., 1994, MNRAS 271, 385
4. Girardi L., Groenewegen M.A.T., Weiss A., Salaris M., 1998, MNRAS 301, 149
5. Girardi L., 1999, MNRAS, in press.
6. Girardi L., Bressan A., Bertelli G., Chiosi C., 1999, A&AS, submitted.
7. Montgomery K.A., Marschall L.A., Janes K.A., 1993, AJ 106, 181
8. Paczyński B., Stanek K.Z., 1998, ApJ 494, L219
9. Perryman M.A.C., et al., 1997, A&A 323, L49
10. Stanek K.Z., Garnavich P.M., 1998, ApJ 503, L131
11. Stanek K.Z., Zaritsky D., Harris J., 1998, ApJ 500, L141
12. Udalski A., et al., 1998, Acta Astr. 48, 1

Old Stars in Dwarf Irregular Galaxies: WLM and NGC3109

D. Minniti[1], M. Rejkuba[1], A. A. Zijlstra[2], M. V. Alonso[3], and M. Gregg[4]

[1] Depto. de Astronomía, P. Universidad Católica de Chile
[2] Department of Physics, UMIST, Manchester, UK
[3] Observatorio Astronómico Córdoba, Argentina
[4] Lawrence Livermore Natl. Laboratory and Univ. of California at Davis, USA

Abstract. We present evidence for the existence of old stellar halos in the dwarf irregular galaxies WLM and NGC3109, located in the outskirts of the Local Group. These halos consist of metal-poor ($[Fe/H] < -1$) and old ($t > 10^{10}$ yr) stars, as inferred from deep optical color-magnitude diagrams obtained with the ESO NTT and with STIS on-board the HST. Future VLT spectroscopy will allow us to measure the kinematics of stars in these halos and also to estimate the dark matter content of these galaxies.

1 dIrr Galaxies in the Outskirts of the Local Group

The current knowledge of galaxy properties along the Hubble sequence is summarized by Kennicutt (1998). It is interesting that the integrated photometric or spectroscopic properties can now be tested directly for nearby galaxies with observations of their constituent stars: classical extragalactic astronomy is being transformed into stellar astronomy. The best known galaxies are the one contained in the Local Group, where we can see down to the faint end of the galaxy luminosity function (in recent years a few more faint members have been added: Mateo 1998, Grebel 1998).

Recently, simulations of star formation histories of nearby galaxies have been used to predict their appearance in the past (Tolstoy 1998). This allows a comparison between the Local Group and observations of galaxies at cosmological distances. Most studied Local Group galaxies contain old stars, formed at least 10^{10} yr ago, at a present redshift of 4 or larger. The appearance of Local Group galaxies thus has become an important constraint in models of galaxy formation and evolution.

The most important are the dwarf galaxies. dIrr's are thought to be largely primordial, related to the original merger building blocks of the larger spirals. dSph's are gas-poor and located almost exclusively close to major galaxies indicating they have undergone significant evolution through interactions (Bingelli et al. 1987). They are the smallest stellar systems known to contain dark matter (Aaronson 1983, Olszewski et al. 1986), but obviously the original size of the system is not known. dIrr's also contain dark matter, as shown by their rising HI rotation curves (Carignan & Freeman 1988, Navarro et al. 1997). This dark matter may be associated with the disk or may form a spheroid.

The VLT will be able to study the kinematics and metallicity/age of individual stars of Local Group dIrr galaxies, and to derive information on the distribution of the dark matter. This is of particular interest for dIrr galaxies located at the outer fringes of the Local Group which were born and remained in isolation. Thus, their structure was little affected by interaction and can be extrapolated to their formation at high redshift. Below we discuss two specific cases.

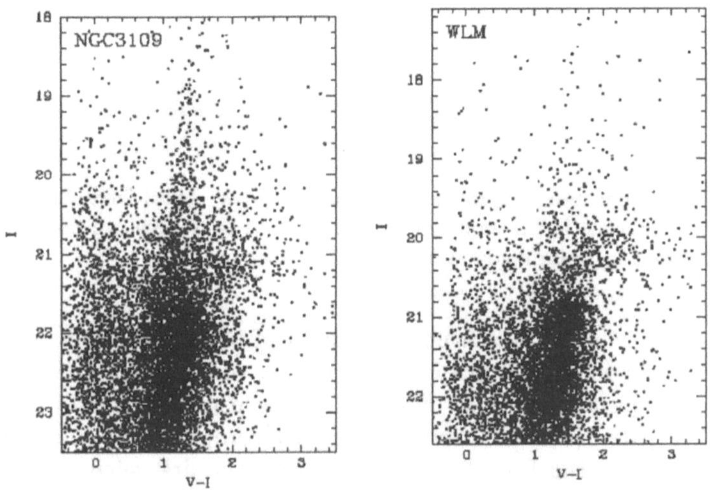

Fig. 1. Color-magnitude diagrams of the central regions (disks) of the dwarf irregulars NGC3109 and WLM obtained with EMMI at the ESO NTT.

2 The dIrr Galaxies WLM and NGC3109

WLM (also DDO221) and NGC3109 are dwarf irregular galaxies located at $D = 0.9$ Mpc and $D = 1.3$ Mpc, respectively. Note that WLM is fainter than the SMC, while NGC3109 is almost as bright as the LMC. Lying far away from the dominant massive spirals, both galaxies are considered to be members of the Local Group (van den Bergh 1994). From least action principle considerations (Peebles 1995), it is likely that these galaxies have remained in a low density environment, at the edge of the Local Group.

Figure 1 shows the color-magnitude diagrams of the disks of these galaxies. Multiple populations are present, with a wide range of ages going from very young ($t < 10^7$ yr) to very old ($t > 10^{10}$ yr). These diagrams are complicated by reddening and crowding, but can be simulated using theoretical isochrones in order to infer star formation histories (e.g. Aparicio 1998).

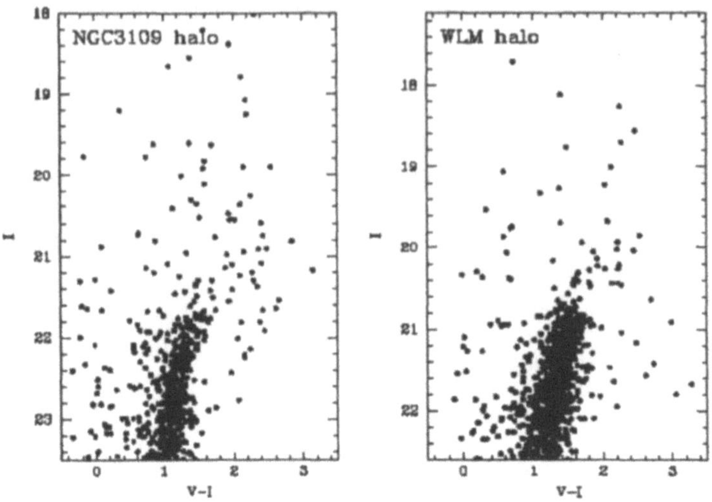

Fig. 2. Color-magnitude diagrams of the halos of the dwarf irregulars NGC3109 and WLM. The scale is the same as Figure 1. Note that only old and metal-poor stars are present in these fields away from the disks.

However, the color-magnitude diagrams of these disks do not tell the whole story. Aside from multiple populations, these galaxies have resolved structures. In particular, these normal looking dwarf irregulars have halos. Minniti & Zijlstra (1996) found an extended component of old and metal-poor stars around the Im IV–V galaxy WLM, the first such discovery in a faint dIrr galaxy. A similar halo was subsequently discovered in the brighter Im galaxy NGC3109 (Minniti et al. 1999). Figure 2 illustrates how the color-magnitude diagram changes in the halos of these galaxies, away from the disk. These halos are old ($t > 10^{10}$ yr) and metal-poor ($[Fe/H] < -1$), as revealed by the location and extent of the red giant branch.

Figure 3 shows a deep color-magnitude diagram obtained with STIS on board the Hubble Space Telescope. This diagram reaches more than 3 magnitudes fainter than the NTT data of Figures 1 and 2, and clearly shows a red clump – horizontal branch, which is somewhat extended to the blue. Blue horizontal stars, like RR Lyrae stars, are unambiguous tracers of old populations, confirming that these galaxies formed stars $> 10^{10}$ yr ago.

3 Progress to be Made with the VLT

The importance of galactic halos in terms of galaxy formation theories was recognized very early on (Eggen et al. 1962), because they are the oldest measurable tracers from primordial times. The finding of a halo in dwarf irregular galaxies argues for a generic mode of galaxy formation that requires a halo in the presence of a disk, regardless of galaxy size. This implies that formation

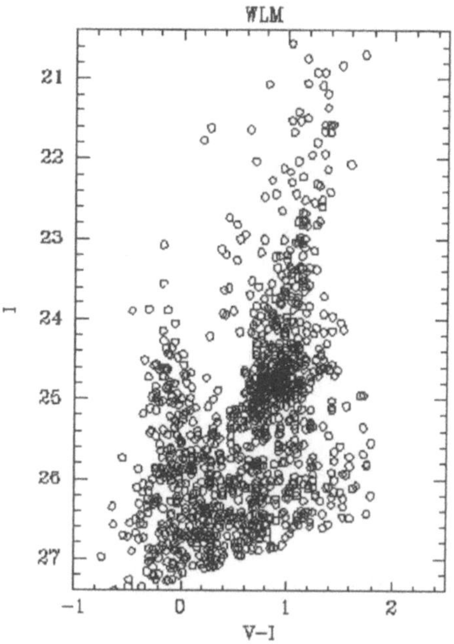

Fig. 3. Deeper color-magnitude diagram of a 28″ × 50″ field located at the interface between the WLM disk and halo obtained with STIS at the HST. The *CL* and *LP*-band magnitudes of STIS have been roughly transformed into the *V* and *I*-band magnitudes (see Gregg & Minniti 1997).

mechanisms are similar along the spiral Hubble sequence, regardless of disk type or luminosity.

The properties of the extended stellar component of WLM and NGC3109 resemble the halos of other larger galaxies. For example, the properties of the Milky Way stellar halo are: 1) extended, 2) old, 3) metal-poor, 4) no gradients, and 5) kinematically hot. The WLM and NGC3109 halos fulfill these properties, except that their kinematics are not known because the stars lie beyond reach of spectrographs at 4m class telescopes. This can now be attacked observationally with 8m telescopes like the VLT. If, as expected, the old stars form a spheroidal halo, with the VLT it will be possible for the first time to measure the halo velocity dispersion in an irregular galaxy. This would give the dark matter content, which can be directly compared to the results based on HI measurements in the disk. Indeed, the study of stellar halos of dwarf galaxies with VLT can now be used to infer their dark matter content and distribution.

Acknowledgements: This work was performed in part by the Chilean Fondecyt Project N. 1990440, and by the U. S. Department of Energy by Lawrence Livermore National Laboratory under Contract W-7405-Eng-48.

References

1. Aaronson 1983, ApJ, **266**, L11
2. Aparicio, A. 1998, in IAU Symp. 192 on "The Stellar Content of Local Group Galaxies", eds. P. Whitelock & R. Cannon, (ASP: San Francisco), p. 20
3. Bingelli, Tamman, & Sandage 1987, AJ, **94**, 251
4. Carignan & Freeman 1988, ApJ, **332**, L33
5. Eggen, O. J., Lynden-Bell, D., & Sandage, A. 1962, ApJ, **136**, 748
6. Grebel, E. 1998, in IAU Symp. 192 on "The Stellar Content of Local Group Galaxies", eds. P. Whitelock & R. Cannon, (ASP: San Francisco), p. 1
7. Gregg, M. D., & Minniti, D. 1997, PASP, **109**, 1062
8. Kennicutt, R. 1998, ARA&A, **36**, 189
9. Mateo, M. 1998, ARA&A, **36**, 435
10. Minniti, D., & Zijlstra, A. A. 1996, ApJ, **467**, L13
11. Minniti, D., Zijlstra, A. A., & Alonso, M. V. 1999, AJ, 117, 881
12. Navarro, Frenk, & White 1997, ApJ, **462**, 563
13. Olszewski et al. 1996, AJ, **111**, 750
14. Peebles, J. 1995, ApJ, **449**, 52
15. Tolstoy, E. 1999, astro-ph/9901245
16. van den Bergh, S. 1994, AJ, **107**, 1328

High Mass Class I Sources in M 17

Rolf Chini, Markus Nielbock, and Marcus Jütte

Astronomisches Institut, Ruhr-Universität Bochum, Universitätsstr. 150/NA 7, D-44780 Bochum, Germany

Abstract. The evolutionary stage of newly discovered MIR sources within M 17 has been investigated by 2.2, 10 and 20 µm imaging. Combining our N and Q data with NIR observations for 23 objects we obtain spectral energy distributions that qualify all sources to be of Class I. The observed luminosities range between 35 and 2300 $L_\odot$ indicating that the objects represent the youngest generation of massive early type stars in M 17 still surrounded by relics of their protostellar clouds. The morphology of the 10 and 20 µm emission reveals flattened structures in some cases suggesting for the first time the presence of circumstellar disks around massive stars.

1 Introduction

The age determination of the youngest stellar objects (YSOs) is still in its infancy as it is the discrimination between protostars and pre–mainsequence objects. Most YSOs are heavily embedded in the interior of dust clouds and – apart from heavy interstellar reddening – their energy output is dominated by reprocessed stellar light in the form of thermal dust emission. This latter fact, however, offers an opportunity to introduce an age scale based on the amount of circumstellar dust: Both quantities – age and dust – should be inversely proportional during the earliest evolutionary stages. Consequently, Lada [10] proposed to measure the amount of dust by the spectral slope $\alpha_{\rm IR}$ between 2 µm and 10 or 20 µm according to

$$\alpha_{\rm IR} = \frac{{\rm d}\log(\lambda F_\lambda)}{{\rm d}\log(\lambda)} \tag{1}$$

The earliest stages of evolution, where most of the observed luminosity is derived from accretion, are characterized by values of $\alpha_{\rm IR} > 0$; the corresponding objects are regarded as true protostars and referred to as Class I. Chini & Wargau [6] (hereafter CW) have shown that values of $\alpha_{\rm IR}$ between 0 and +3 correspond either to dust radiation of $300\,{\rm K} < T_{\rm d} < 600\,{\rm K}$. They cautioned, however, that extinction by interstellar (not circumstellar) dust may simulate positive α–values: in the extreme case, all stars suffering $A_V > 80$ mag, will appear as Class I objects. It should also be noted that this empirical classification scheme by Lada was based on observations of only five low–mass sources located in the Ophiuchus cloud.

CW introduced another youth criterion in the absence of data longward of 4.8 µm. They found empirically that the location of Class I sources is characterized by $(K-L) > 0.26(J-K)+0.25$ and $(K-M) > 0.37(J-K)+0.80$.

These relations are almost independent of the amount of extinction and measure directly the emission of circumstellar dust. As a consequence, they are equivalent to the Class I criterion ($\alpha_{IR} > 0$) and may be used as indicators for the evolutionary stage of a young object.

We are investigating the stellar content of the Omega Nebula (M 17, NGC 6618) which is one of the most luminous H II regions in our galaxy and a site of ongoing star formation. The major ionizing sources have been identified by Chini, Elsässer & Neckel [3] (hereafter CEN) by means of *UBVRI* photometry. Concerning the IMF of M 17 so far only massive stars between O and A have been found, whereas objects of lower mass have either escaped detection due to high extinction and/or confusion or have not been formed at all. Nevertheless, the process of star formation is still going on in that area as witnessed by several cocoon stars (Chini [4], Chini & Krügel [5], CW) and an ultra-compact HII region (Felli et al. [7]). In the present paper we report on the detection of the first high-mass Class I objects, some of which may be surrounded by circumstellar disks.

2 Observations

The region across M 17 observed in the present paper has a size of $\approx 400'' \times 430''$ and covers the field of the exciting O,B star cluster investigated by CEN at *UBVRI* and by CW at *JHKLM*. The *J* and *K* images were obtained at 2.2 m telescope at Calar Alto in June 1998 with the IR-camera MAGIC. We estimate the NIR fluxes to be accurate to about 0.05 mag. The 10 and 20 μm mosaics were obtained at the 2.2 m telescope on La Silla in June/July 1997, using the array camera MANIA (Böker et al. [1]). The field of view was $44.2'' \times 44.2''$ with a pixel resolution of $0.345'' \times 0.345''$. We produced a mosaic of 133 images at each waveband by offsetting the telescope after each frame by the chopper throw of 26″ along the chop direction of 137°; the overlap between adjacent scan lines was 32.5″. For reconstructing the maps we modified the program package MOPSI (Zylka 1995). While the integration time per frame was 30 seconds in the mapping mode, we observed 18 particular fields of interest with integration times between 120 and 180 seconds. The calibration uncertainties are 25% in *N* and 10% in *Q*.

3 The Evolutionary Stage

Our MIR maps reveal 23 compact sources, two of which are the previously detected Kleinmann–Wright–Object (Kleinmann & Wright [9]) and the ultra-compact HII region M 17-UC1 (Felli et al. [7]). Six sources have optical counterparts and coincide with stars observed by CEN while 11 sources coincide with previously detected NIR objects (CW, Bunn et al. [2]). The remaining 5 anonymous objects are only visible on our 10 and 20 μm images.

3.1 Spectral Indices

In order to investigate their evolutionary stage we calculate both the infrared spectral indices $\alpha_{\rm IR}$ as proposed by Lada [10] and the NIR excess relations as proposed by CW. Within the observational errors we find that all sources have positive $\alpha_{\rm IR}$ and thus are to be classified as Class I objects. Those sources with optical counterparts have an average $\alpha_{KN} = 0.75 \pm 1.21$ and correspond to the *"cocoon stars"* discovered by Chini & Krügel [5]. Those sources with NIR counterparts have $\alpha_{KN} = 1.78 \pm 0.59$ and thus are even clearer Class I objects. Both groups also fulfil the NIR colour criteria $(K - L) > 0.26(J - K) + 0.25$ and $(K - M) > 0.37(J - K) + 0.80$ proposed by CW. Their visual extinction ranges between $10 < A_V < 40$ mag with no systematic difference between optical and NIR objects. Due to the absence of NIR data the remaining anonymous objects have only lower limits of $\alpha_{\rm IR}$ indicating values above 3. The separation of $\alpha_{\rm IR}$ between the three groups suggests an evolutionary sequence with increasing influence of circumstellar dust from the visible to the pure MIR objects.

3.2 Luminosity

The total luminosity of the YSOs cannot be determined uniquely because of both interstellar and circumstellar dust. It is very likely that most of the luminosity is emitted longward of 20 µm, where observational data do not exist. The results on the *"cocoon stars"* by Chini & Krügel [5] show that only about 10 % of the total luminosity is emitted below 20 µm. Because the *"cocoon stars"* are contained in the present sample we may treat them as template objects and use their ratio of $L_{\rm obs}/L_{\rm tot}$ as a guess (or at least as a lower limit) for the remaining YSOs in M 17. The observed luminosities between 1.2 and 20 µm range from 35 to 2300 $L_\odot$, adopting a distance of 2.2 kpc. This transfers into total luminosities of 350 to 23000 $L_\odot$ suggesting that the YSOs represent the youngest generation of early type stars in M 17 and are still enshrouded by the remnants of their protostellar cloud. The fact that some of these objects can be observed even at wavelengths below 2.2 μm is due to the fact that the ionizing radiation and the stellar winds from the nearby O type stars have partly cleared the region. We suggest, that these M 17 objects are the first high mass counterparts of classical Class I sources.

3.3 Morphology

The 10 and 20 µm emission for 5 of the sources is clearly extended with axes ratios of ≈ 1.5. We have checked these cases on our K−band images in order to find double stars. Although we cannot completely rule out the existence of unresolved multiple MIR sources with no NIR counterparts it is very likely that we are dealing with flattened dust cocoons and/or circumstellar disks. Their linear dimensions correspond roughly to 15000 AU (FWHM), adopting a distance of 2.2 kpc.

3.4 M17 UC1

The ultra compact HII region M 17-UC1 has been investigated by Felli & Stanga [8] at NIR and mid-infrared wavelengths. Our 10 μm image in Fig. 1 shows a double source where M 17-UC1 is the object to the NE. The second source (to the SW) is displaced by 5.10″ relative to a similar object visible in the contour map by Felli & Stanga, which, however, is located SE of M 17-UC1. If this displacement would be due to a proper motion of the southern source between epochs 1980 and 1997, we derive a projected linear distance of $1.7 \cdot 10^{17}$ cm and a velocity of 3130 km/s. These values seem rather implausible. If we are dealing with the same object we have to suppose an error in the contour map by Felli & Stanga [8]. Our configuration is corroborated by an independent N–band image taken 1995 with TIMMI at the ESO 3.6 m telescope.

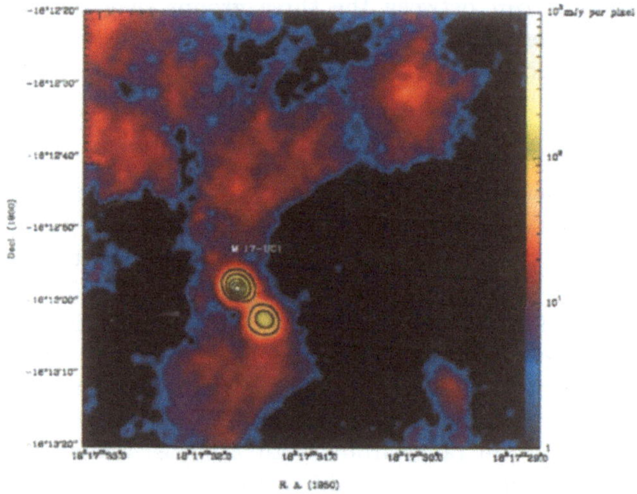

Fig. 1. M 17-UC1 at 10 μm obtained with MANIAC. Contour levels represent flux values of 50, 100, 200, 300, 400, 500, 600 and 700 mJy/pixel.

References

1. Böker, T., Storey, J. W. V., Krabbe, A., Lehmann, T. 1997, PASP **109**, 827
2. Bunn, J. C., Hoare, M. G., Drew, J. E. 1995, MNRAS **272**, 346
3. Chini, R., Elsässer, H., Neckel, Th. 1980, A&A **91**, 186 (CEN)
4. Chini, R. 1982, A&A **110**, 332
5. Chini, R., Krügel, E. 1985, A&A **164**, 175
6. Chini, R., Wargau, W. F. 1998, A&A **329**, 161 (CW)
7. Felli, M., Johnston, K. J., Churchwell, E. 1980, ApJ **242**, L157
8. Felli, M., Stanga, R. 1987, A&A **175**, 193
9. Kleinmann, D. E., Wright, E.L. 1973, ApJ **185**, L131
10. Lada, C. J. 1987, Star Formation: From OB Associations to Protostars. In: Peimbert, M., Jugaku, J. (eds.), Star Forming Regions. IAU Symp. **115**, 1

WR Stars in Star Forming Regions

Werner Schmutz[1,2] and Wiliam D. Vacca[3]

[1] Institute of Astronomy, ETH Zentrum, CH-8092 Zürich, Switzerland
[2] Physikalisch-Meteorologisches Observatorium Davos, CH-7260 Davos Dorf, Switzerland
[3] Institute for Astronomy, University of Hawaii, 2680 Woodlawn Dr., Honolulu, HI 96822 USA

Abstract. WR galaxies are characterized by broad emission features at 4686 A and occasionally at 4640 A and 5805 A. The features are attributed to stellar emission from a large population of WR stars. The presence of WR stars allows a determination of the age of the starburst. In a few WR galaxies the 4640 emission line has been measured to be as large or larger than the He II 4686 line. We demonstrate that a 4640/4686 ratio larger than unity cannot be produced by any known population of WR stars and discuss the implications for the interpretation of the star formation histories in WR galaxies.

1 WR Stars as Age Indicators

Wolf-Rayet (WR) galaxies are characterized by a broad emission feature at 4686 Å, occasionally accompanied by an additional broad feature at 4640 Å, in their integrated optical spectra (Conti 1991). The 4686 Å feature is attributed to stellar He II emission from a large population of Wolf-Rayet stars. The broad emission feature at $\sim$ 4640 Å seen in moderate and low-resolution spectra of a subset of WR galaxies (e.g., Allen, Wright & Goss 1976; Kunth & Schild 1986; Vacca & Conti 1992) is usually interpreted as the result of a blend of stellar N V $\lambda\lambda$4604, 4620, N III $\lambda\lambda$4634, 4641, and C III/IV $\lambda\lambda$4650, 4658 emission, with the nitrogen lines due to WR stars of the nitrogen sequence (WN stars) and the carbon lines due to WR stars of the carbon sequence (WC stars). Often, the entire broad emission blend between $\sim$ 4600 Å and $\sim$ 4700 Å – "the 4650 bump", or the "WR bump" (Arnault, Kunth, & Schild 1991) – which includes the He II λ4686 feature, is used to identify WR galaxies; the blend is often not observed at a resolution high enough to distinguish the various line contributions.

The presence of WR features in starburst galaxies provides a unique tool for age dating the burst episode. Spectral synthesis models indicate that strong WR features, such as He II 4686, can be observed in the integrated spectra of starbursts only during a brief interval ($\sim 2 - 7$ Myr) after a burst and only if the burst lasted a short time (less than a few Myr) (Meynet 1995; Schaerer et al. 1999). More precise age dating can be achieved by measuring the relative strengths of lines produced by WC stars and those produced by

WN stars and determining the ratio of the number of WC stars to the number of WN stars present in the starburst region (e.g., Schaerer & Vacca 1998).

Despite the enormous potential for using WR features to investigate young stellar populations, there are still reasons for caution in using the features to infer the properties and histories of starburst galaxies. These arise primarily from the fact that the various contributions to the stellar emission blends have never been observed with high enough spectral resolution to confirm their identities and measure their individual strengths. Schmutz & Vacca (1999) have recently addressed one of the problems resulting from this lack of high resolution data, the anomalous 4640/4686 line ratio seen in some WR galaxies. When present, the λ4640 feature is usually weaker than the broad He II λ4686 emission line. However, there are a few WR galaxies with a λ4640 feature as strong as the He II λ4686 emission and in at least one recently reported case, NGC 3049 (Schaerer, Contini, & Kunth 1999, hereafter SCK99), the λ4640 feature is even stronger than He II λ4686. In this paper we summarize the findings of Schmutz & Vacca (1999), who investigated the origin of the λ4640 feature and demonstrated that it is indeed very difficult to produce a λ4640 emission feature stronger than that of He II λ4686 with any mixture of known WR spectra. We then offer some tentative explanations for enhanced emission in the λ4640 feature and discuss the consequences on the interpretation of the physical properties and stellar populations of those starbursts with strong λ4640 emission.

2 Interpretation of the Broad Emission Feature at 4640 Å

The commonly-accepted interpretation of the strong λ4640 emission feature is that it is due to N III $\lambda\lambda$4634, 4641, with a potentially large contribution from C III $\lambda\lambda$4647, 4650. For example, in their study of five WR galaxies, SCK99 note: "The relative strength of (N III λ4640+C III λ4650)/He II λ4686 ranges from 0.9 to 2.5 with an average of 1.5, which is much larger than the average of 0.24$\pm$0.15 obtained for late-type LMC WN (WN L) stars (Schaerer & Vacca 1998). The excess flux can easily be attributed to the emission from the WC stars." In the following subsections we investigate whether the λ4640 feature and the enhanced 4650/4686 flux ratio are due to a combination of WC and WN stars (§2.2), or only WN stars (§2.3).

2.1 WC Stars

We fit the WR features at 5805 Å and 5696 Å of NGC 3049 with a scaled spectrum of the WC 6 star WR 50. This star has been selected because it has a 5696/5805 line ratio similar to that seen in NGC 3049. Alternatively, we can also reproduce the $\lambda\lambda$5805, 5696 carbon lines of the WR galaxy with a combination of the WC 4 star BR 10 and the WC 7 type WR 57. Both fits

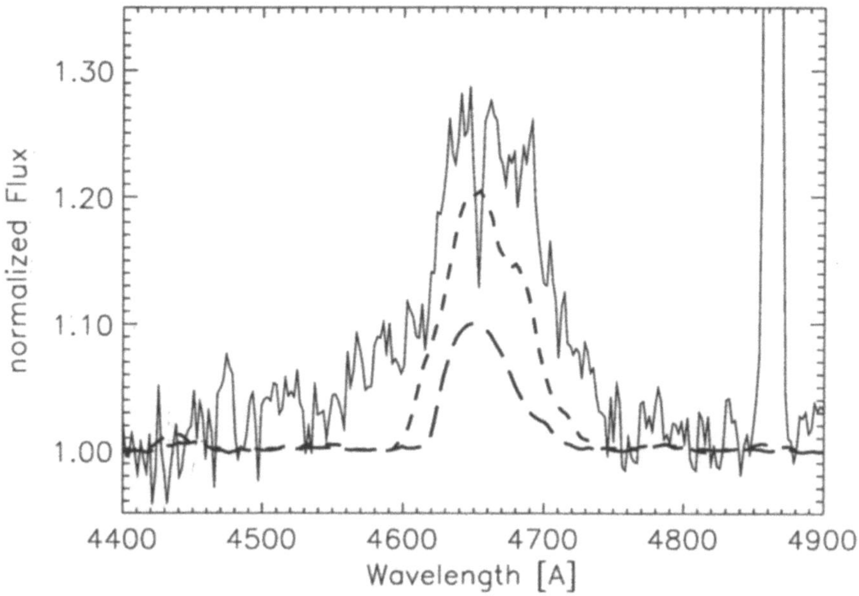

Fig. 1. The broad λ4650 complex in the spectrum of NGC 3049 (solid line, spectrum from SCK99) compared with the C/He features as predicted from the two fits to the broad feature at 5805 Å. The short dashed line denotes a combination of a WC 4 and a WC 7 star; the long dashed line denotes a WC 6 spectrum.

to the yellow WR bump are satisfactory but they yield somewhat different strengths for the 4650–4686 feature. In Fig. 1 we show the line strengths in the blue region of the stellar spectra with the same intensity scaling used to reproduce the yellow WR bump. The WC 6 star has a smaller ratio of blue to yellow emission lines than the WC 4+WC 7 combination. The cases shown in Fig. 1 are extremes and therefore the true value for NGC 3049 is most likely between the two. If we now subtract the two predictions for the WC star contribution from the observed blue WR bump we obtain the spectra shown in Fig. 2. The two resulting spectra differ in the absolute strengths of the emission features but in both cases the 4640/4686 ratio has not changed substantially from its original value. Even after subtraction of a WC emission component, we are left with a 4640/4686 ratio larger than unity. The reason is that the WC feature is not centered at the wavelength needed to produce the observed λ4640 bump. Therefore, an underlying WC star contribution cannot be responsible for a 4640/4686 ratio larger than unity. We stress that we are not claiming that WC stars do not contribute at all to the WR bump; we are merely stating that WC stars cannot be responsible for the enhanced 4640/4686 ratio.

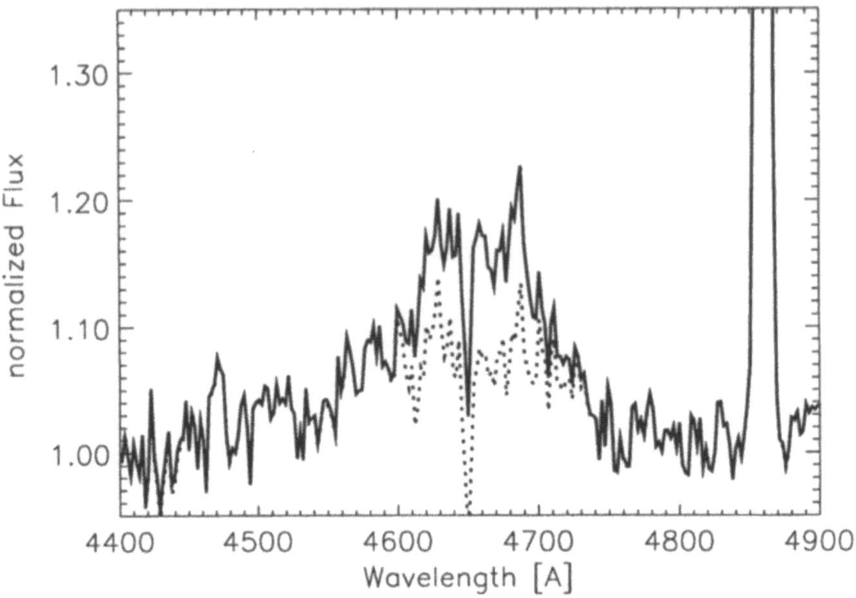

Fig. 2. The broad λ4650 complex in the spectrum of NGC 3049 after subtraction of the WC-star contribution indicated in Fig. 1. The solid line is the result after subtracting a WC 6 stellar spectrum shown in Fig. 1 and the dotted line is obtained after subtracting the WC 4-WC 7 combination.

2.2 WN Stars

As we have just shown, the λ4640 feature in the optical spectrum of NGC 3049 is as strong as the λ4686 emission even after subtraction of a possible contribution by WC stars. Is it possible that the 4640 Å emission arises primarily from WN L stars? Clearly, any mixture of WN L stars invoked to explain a 4640/4686 ratio larger than unity must be rather unusual since the average WN L star exhibits a 4640/4686 ratio much smaller than one (Schaerer & Vacca 1998). Here we ignore this constraint and try to find *any* WR star that will yield the ratio observed in NGC 3049. Of the nearly 200 WR stars for which we have access to optical spectra, only two have 4640/4686 ratios which approach the value observed in NGC 3049: WR 108 (a WN 9ha star) and WR 123 (a WN 8o star). (See Smith, Shara, & Moffat 1996 for the spectroscopic classifications of these stars.) High resolution optical spectra reveal that even in these most extreme WNL stars the peak intensities of the nitrogen lines are still smaller than that of the helium line (Crowther, Hillier, & Smith 1995). However, at low resolution – e.g., 10 Å which is the resolution of both the spectra of the WR galaxies (SCK99) and the stellar spectra we are using here (Torres-Dodgen & Massey 1988) – the two N III $\lambda\lambda$4634, 4641 com-

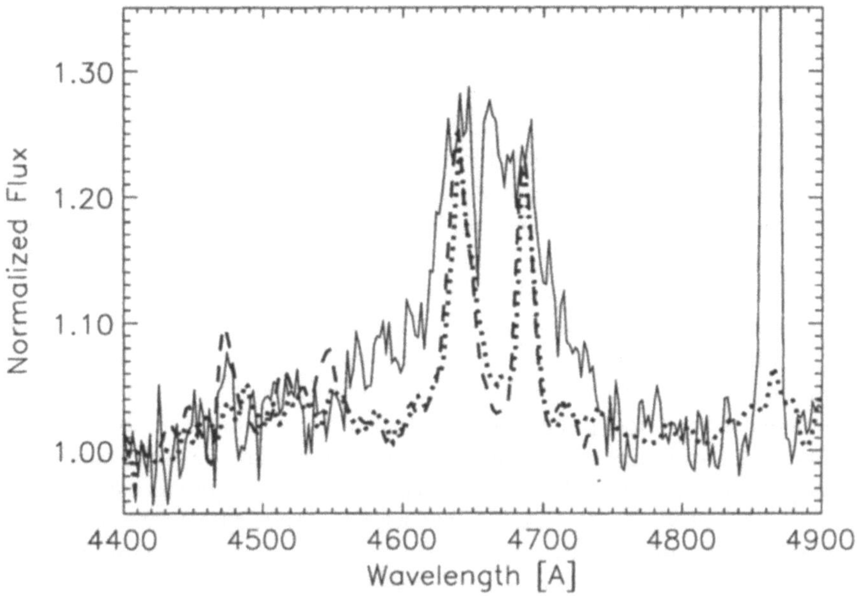

Fig. 3. The blue WR bump of NGC 3049 compared with the spectra of WR 108 (dotted line) and WR 123 (dashed line). These WN L stars are those with the most extreme known N III λ4640 to He II λ4686 line ratio.

ponents merge to form a λ4640 feature somewhat stronger than the helium line.

In Fig. 3 we directly compare the WR bump of NGC 3049 with the low-resolution spectra of WR 108 and WR 123. (We emphasize again that the spectral resolution is the same for all the objects.) We see that these WR spectra reproduce the observed 4640/4686 ratio but the stellar line widths are much narrower than that of the WR galaxy. There are of course WR stars with broader lines that would match the observed width. However, none of these has a N III λ4640 line strength that would even remotely approach the strength of He II λ4686. Thus, in order to reproduce the observed WR bump strength, not only are we required to select the WR star with the most intense nitrogen N III lines relative to He II among the ~ 200 known LMC and galactic WR stars, but we must also postulate a new, hitherto unknown, spectral type with terminal velocities faster than that of WR 108 and WR 123. Although the explanation of the WR bump with a population of WR stars with as-yet-unknown spectral types cannot be ruled out, it seems to us that a more satisfactory interpretation should be based on a mixture of the *known* WR types. We therefore conclude that the λ4640 feature in WR galaxies with large 4640/4686 ratios is most probably not due (solely) to N III λ4640 from a WR star population.

3 Conclusions

There are a few WR galaxies with a λ4640 feature as strong as the He II λ4686 emission and in at least one recently reported case, NGC 3049 (SCK99), the λ4640 feature is even stronger than He II λ4686. This result is surprising because there is no known population of WR stars which can produce a 4640/4686 line ratio greater than unity. Schmutz & Vacca (1999) suggested two possible explanations for the anomalous line ratio: contamination of the stellar emission blend by nebular lines due to continuum fluorescence and/or the presence of a large population of Of stars. Both possibilities would imply that the numbers of WR stars in some starburst galaxies may have been overestimated. This in turn would have a profound impact on the interpretation of the ages of the burst populations and the star formation histories in these objects. If Of stars are responsible for the anomalous line ratio, Schmutz & Vacca (1999) found that the instantaneous burst scenario cannot be correct for any high metallicity ($Z \geq Z_{\odot}$) region whose optical spectrum exhibits both an enhanced 4640/4686 flux ratio and C IV 5808 emission from WC stars. These regions must have experienced a "multiple-burst" star-formation event, composed of several instantaneous bursts separated by short time intervals (a few Myr). For low metallicity regions, the simultaneous presence of Of and WC emission lines can occur only during a 1 Myr time interval 3 Myr after a burst. In all cases, the support for an instantaneous burst scenario is considerably weakened by the smaller number of WR stars. Such uncertainties seriously hamper attempts to use the WR features as diagnostics for metallicity, star formation histories, or stellar population characteristics.

References

1. Allen, D.A., Wright, A.E., & Goss, W.M., 1976, MNRAS 177, 91,
2. Arnault, Ph., Kunth, D., & Schild, H., 1991, A&A 224, 73.
3. Conti, P.S. (1991) ApJ 377, 115.
4. Crowther, P.A., Hillier, D.J., & Smith, L.J., 1995, A&A 293, 172.
5. Kunth, D., & Schild, H., 1986, A&A 169, 71.
6. Meynet, G. 1995, A&A, 298, 767.
7. Schaerer, D., & Vacca, W.D., 1998, ApJ 497, 618.
8. Schaerer, D., Contini, T., & Kunth, D., 1999, A&A 341, 399 (SCK99).
9. Schmutz, W., & Vacca, W.D., 1999, NewAstr, in press.
10. Smith, L.F., Shara, M.M., & Moffat, A.F.J., 1996, MNRAS 281, 163.
11. Torres-Dodgen, A.V., & Massey, P., 1988, AJ 96, 1076.
12. Vacca, W.D., & Conti, P.S., 1992, ApJ, 401, 543.

Supernova Rates

Enrico Cappellaro[1] and Ferdinando Patat[2]

[1] Osservatorio Astronomico di Padova, vicolo dell'Osservatorio 5, I-35122 Padova, Italy
[2] European Southern Observatory, Alonso de Cordova 3107, Vitacura, Casilla 19001, Santiago 19, Chile

Abstract. We present our latest estimates of the SN rates in galaxies of the local Universe, based on the combined sample of five SN searches. We exploit the updated statistics to probe the SN rates against two different indicators of the SFR in a galaxy, namely the integrated color and the FIR properties. Finally, we describe a new project aimed to derive the rate of all types of SNe up to redshift $z < 0.5$.

1 Introduction

The evolution of the rate of supernovae (SNe) with redshift can be used to probe the history of the star formation rate (SFR) in galaxies, constrain the scenarios for galaxy formation and evolution and, at the same time, give useful indications on the SN progenitors. Indeed, several authors have published predictions of the SN rates as a function of redshift which are still waiting for verification (Jørgensen et al. 1997; Kobayashi et al. 1998, Porciani & Madau 1998, Ruiz-Lapuente et al. 1997; Sadat et al. 1998; Youngelson et al. 1998).

So far detailed estimates of the SN rates are available only for the local Universe (Cappellaro et al. 1997) and, only a very preliminary estimate of the rate of SN Ia at redshift $z \simeq 0.4$ have been published (Pain et al. 1996). In this contribution we present our latest estimates of the SN rates in galaxies of the local Universe and describe a project aimed to derive the rate of all types of SNe up to redshift $z < 0.5$.

2 SN Rates in the Local Universe

New estimates of the SN rates have been obtained by combining the logs of four photographic SN searches with that of the Evans' visual search (Cappellaro et al. 1999). In this way we have obtained a sample of 137 SNe in a reference sample of about 10^4 galaxies, accumulating the largest statistics ever assembled for this purpose.

For the calculation of the SN rates we used the control time method (Zwicky 1938). The total control time for our combined searches is ~ 25000 years for SN Ia. The computational recipe includes corrections for the search bias in the galaxy nuclear regions and in inclined spirals, and allows reliable estimates of *the total* errors.

Table 1. Rates of SNe from the combined logs of five SN searches.

galaxy	rate [SNu]*			
type	Ia	Ib/c	II	All
E-S0	0.18 ± 0.06	< 0.01	< 0.02	0.18 ± 0.06
S0a-Sb	0.18 ± 0.07	0.11 ± 0.06	0.42 ± 0.18	0.72 ± 0.14
Scd-Sd	0.21 ± 0.08	0.14 ± 0.07	0.86 ± 0.36	1.21 ± 0.46
Other	0.40 ± 0.19	0.22 ± 0.17	0.65 ± 0.37	1.26 ± 0.40
All	0.20 ± 0.06	0.08 ± 0.04	0.40 ± 0.15	0.68 ± 0.11

* 1SNu = $1\mathrm{SN}(100\mathrm{yr}){-}1(10^{10}\mathrm{L}_{\odot}^{\mathrm{B}})^{-1}$. SN rates in these units scale with the Hubble constant as $(H_0/75)^2$

The rate of core collapse SNe (Ib/c and II) can be used to estimate the average star formation rate in galaxies. Assuming that the mass of the progenitors of core collapse SNe is in the range $10 - 40$ $\mathrm{M}_{\odot}$ and a Salpeter mass function with index 1.35 we find the following average SFR($\mathrm{M} > 6\mathrm{M}_{\odot}$): in E-S0 < 0.01, in Sa-Sb 0.2 ± 0.1, and in Sbc-Sd 0.4 ± 0.1 $\mathrm{M}_{\odot}\mathrm{yr}^{-1}(10^{10}\mathrm{L}_{\odot}^{\mathrm{B}})^{-1}$.

Actually, it is expected that even in galaxies of the same morphological type there is a significant dispersion of the SFR depending on the recent star formation episodes. The different diagnostics which are used to probe SFR in galaxies have been reviewed in a recent paper by Kennicutt (1998). Because we are limited by the SN statistics we selected tracers that are available for large samples of galaxies, namely the integrated colors and far infrared (FIR) luminosities.

To test the dependence of SN rates on galaxy colors for each morphological type we grouped the galaxies into two subsamples, containing objects bluer and redder than the global average. We then computed separately the SN rates for each of these subsamples. We found that the rate of core collapse SNe (II+Ib/c) is significantly higher (by a factor 2) in spirals with bluer U-V color. Instead, the rate of SN Ia is, within the uncertainties, independent on galaxy colors. This is as expected, since galaxies with a bluer color are expected to host younger and more massive stars, that is to say more core-collapse SN progenitors, than redder galaxies. By using evolutionary synthesis models, it is possible to estimate the SFR corresponding to an observed galaxy color. If these models and our understanding of the SN progenitor scenarios are correct we would expect the same dependence on galaxy color of the SFR and the core collapse SN rate. As we show in Fig. 1, our results confirm this expectation.

Instead we did not found a dependence of the core-collapse SN rate on the integrated FIR emission of galaxies. This implies that, at least in normal galaxies, FIR emission is not an universal measurement of SFR. Actually,

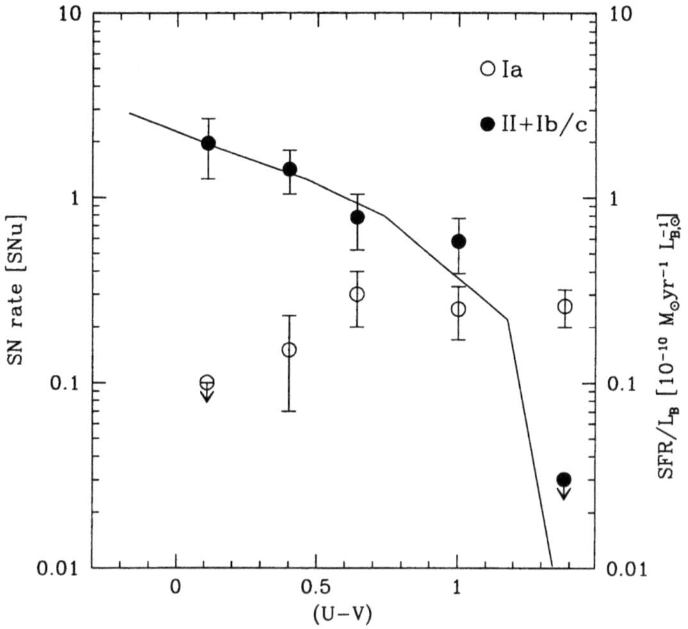

Fig. 1. SN rates in SNu (left-hand scale) in spirals with different U-V color. Filled symbols are core-collapse SN II+Ib/c and open circle are SN Ia. The line gives the SFR rate per unit B luminosity (right-hand scale) for galaxies with different colors as predicted by the evolutionary synthesis models of Kennicutt (1998). The surprising coincidence of the two scales can be partially understood by units conversion (cf. Cappellaro et al. 1999)

it is now clear that there are different contributing factors to FIR luminosities of normal galaxies. In particular, early spiral galaxies often exhibit low temperatures and relatively high FIR luminosities due to dust heating from the general stellar radiation field, which are not directly related to SFR (Kennicutt 1998).

Finally, we compared the rate of SNe in AGN host galaxies with that in normal galaxies, finding similar values and therefore there are no evidences that the nuclear engine is stimulating the SF in the host galaxy.

3 A New SN Search at Intermediate Redshifts

Though the rate of SNe in the local Universe is a necessary benchmark, for constraining stellar and galaxy models, the evolution of the SN rate with redshift would be really needed.

In principle we would like to probe the SN rates back to the epoch of galaxy formation, but this would require a very large telescope and an efficient infrared camera which are not yet available. On the other side, whereas even a single estimate of the rate of SNe at $z > 1$ could be of paramount importance for the discrimination between current models of galaxy formation, it is necessary to reconstruct the whole history of the SFR.

We therefore decided to start a long term project specifically conceived to measure SN rates in galaxies at increasing redshifts. As a first step, we are starting a new SN search in clusters at intermediate redshifts, using the wide field CCD camera which has just been installed at the ESO2.2m telescope (WFI@2.2) in these days.

Considering the performance of the WFI@2.2 we expect to be able to detect normal SNII up to $z = 0.3-0.4$. One of the problem for deriving the SN rates at intermediate redshift is that usually one does not know the distances and the morphological types of the surveyed galaxies. Again a conservative approach would be to concentrate our effort on a few selected fields where this information are or will become available. Another possibility is to use the same multicolor images used for the SN search and confirmation (see below) to derive colors, hence photometric redshifts, for all galaxies in the field.

It is clear that, if spectroscopic classification of all candidates is required, the bottle neck for building up of a sufficient statistics is the allocation of a 8-m class telescope (the candidates will be in the range 22-24 mag). For this reason we are currently exploring an alternative strategy in which the candidate confirmation is based on a multicolor light curve analysis, with an approach similar to that of photometric redshifts. The basic idea is that the different types of SNe have in general very characteristic light and color evolutions. This means that with a proper sampling of the light and color curves it will be possible to confirm the candidate SN and assign it a type though only with a certain confidence level. The fact that for each single event there will be a chance of misclassification will be compensated by the possibility to built a much larger statistics, which is mandatory for the aim discussed above. Nine nights at the WFI@2.2 and 4 half nights at the NTT and ESO3.6 have been already allocated to this program.

References

1. Cappellaro, E., et al. (1997) A&A 322, 431
2. Cappellaro, E., Evans, R., Turatto, M. (1999) astro-ph 9904225
3. Jørgensen, H.E., Lipunov, V.M., Panchenko, I.E., Postnov, K.A., Prokhorov, M.E. (1997) ApJ 486, 110
4. Kennicutt, R.C., (1998) astro-ph 9807187
5. Kobayashi, C., Tsujimoto, T., Nomoto, K., Hachisu, I., Kato, M. (1998) ApJ 503, L155
6. Pain, R., et al. (1996) ApJ 473, 356
7. Porciani, C., Madau, P. (1998) astro-ph 9810403
8. Ruiz-Lapuente, P., Canal, R. (1997) ApJ 497, L57
9. Sadat, R., Blanchard, A., Guiderdoni, B., Silk, J. (1998) A&A 331, L69
10. Yungelson, L., Livio, M (1998) ApJ 497, 168
11. Zwicky, F., (1938) ApJ 88,529

Massive Stars and Their Role in Spectra and Abundances of High Redshift Galaxies

André Maeder

Geneva Observatory, 1290 Sauverny, Switzerland

Abstract. There are great changes in our understanding of massive star evolution. Although it is a known fact that the populations of massive stars (O–stars, WR stars, blue and red supergiants) are very different in galaxies of different metallicities, the major role of axial rotation was not yet fully realised. Rotation dramatically influences all model outputs. Moreover, recent observations suggest faster rotation at lower metallicities, a result which may thus be particularly critical for the interpretation of the spectra and chemical abundances of high redshift galaxies.

1 Introduction

Massive stars play an essential role in the spectral and chemical evolution of galaxies, particularly in phases where the star formation rate is high. Spectral signatures of massive stars are observable in the integrated spectrum of galaxies up to redshifts as high as z=3, now observable with the VLT and Keck telescopes. These signatures are in particular:

O–stars	Nebular lines
	UV continuum
	UV resonance lines
	OI in the IR (63 μ)
WR stars	HeII 4686
WC stars	CIV 5808
Red supergiants	Ca triplet
	CO in the IR (2.3 μ)
	Brγ in the IR (2.16 μ)

The calibration of these signatures in terms of the stellar content provides detailed information on the populations of massive stars in galaxies (cf. Maeder and Conti 1994) and on their star formation rate. These signatures have to be well analysed so that we may disentangle the starburst and AGN contributions, as for example in Seyfert galaxies (cf. Heckman et al. 1997). Here we examine some properties of massive stars, responsible for the spectral signatures in the above table, as they are known from nearby galaxies.

2 Star Number Ratios in Nearby Galaxies

Table 1 shows some important number ratios for O–stars, red supergiants, WN and WC stars in nearby galaxies. These data come from Maeder and Conti (1994 and ref. therein); the ratios of red supergiants (RSG) to WR stars are from Massey (1998).

Table 1. Number ratios of massive stars in nearby galaxies

Galaxy	Z	WR/O	WC/WR	WC/WN	RSG/WR
M31	0.035	0.24	0.44	0.79	1.7
Milky Way					
ring 6–7.5 kpc	0.029	0.205	0.55	1.22	–
ring 7.5–9 kpc	0.020	0.104	0.48	0.92	∼ 1
ring 9.5–11 kpc	0.013	0.033	0.33	0.49	–
M33	0.013	0.06	0.52	1.08	∼ 4
LMC	0.006	0.04	0.20	0.26	–
NGC 6822	0.005	0.02	–	–	8.3
SMC	0.002	0.017	0.11	0.13	–
IC 1613	0.002	0.02	–	–	–

Table 1 shows that with decreasing metallicity Z the number ratios WR/O decrease, the same occurs for the ratio WC/WR, while the fraction RSG/WR is increasing. The basic physical reason for these behaviours is that a higher Z implies more ions, more spectral lines and more momentum transferred to the wind, and thus higher mass loss rates $\dot{M}$ (cf. Kudritzki et al. 1989). Higher $\dot{M}$–rates imply, in turn, that the minimum mass for forming WR stars (i.e. bare cores) is lower and also that the lifetimes spent as WR stars are increased (cf. Maeder and Meynet 1994).

Basically the He–burning phase is shared mainly between the blue supergiant (BSG), the red supergiant (RSG) and the WR stages:

$$t_{\mathrm{He}} \simeq t_{\mathrm{BSG}} + t_{\mathrm{RSG}} + t_{\mathrm{WR}} \tag{1}$$

Since higher Z implies larger t_{WR}, the remaining part of the He–burning phase ($t_{\mathrm{BSG}} + t_{\mathrm{RSG}}$) is therefore reduced. However, the exact sharing between t_{BSG} and t_{RSG} depends on the internal H–profile outside the He–core, and this is greatly influenced by any mixing process in stars (cf. Section 4). We note that, although t_{He} is basically fixed by nuclear physics, strong mass loss reduces the mass and luminosity of the WR stars and therefore makes t_{He} larger. Numerical models providing quantitative data for some of the signatures discussed in Section 1 were made for the case of constant star formation rate by Maeder and Meynet (1994) and for the case of starbursts by Meynet (1995) and Schaerer (1997).

3 The Need to Take Axial Rotation into Account for Massive Stars at Low Metallicities

Axial rotation has generally been considered as a secondary effect in stellar evolution. For massive stars this point of view is wrong for both theoretical and observational reasons. Theoretically the main mixing process, i.e. shear diffusion, is proportional to the thermal diffusivity, which imposes a very short mixing time scale in massive stars (cf. Maeder 1997). Observationally, there are several discrepancies with current models, in particular helium and nitrogen excesses observed in fast rotating O–type stars as well as in almost all supergiants (cf. Herrero et al. 1998; Venn et al. 1998). Also, a boron depletion accompanies the N–excesses in B-type stars, which was taken as an indication for deep mixing due to rotation (cf. Fliegner et al. 1996).

Recent results by Venn et al. (1998) show that, in A–type supergiants of the SMC, the N/H ratio may be as high as 10–20 times the average N/H ratio of the SMC. In the A–supergiants of the Milky Way, the N/H excesses reach at most a factor of 3–4. These results indicate a much stronger mixing in the A–supergiants of the SMC than in the galactic ones. Does that mean more fast rotators in the SMC than in the Galaxy?

Another fact tends to support this last view. It has been known for long that some clusters of the SMC and LMC contain very large fractions of Be stars, i.e. of fast rotators (cf. Feast 1972; Grebel et al. 1994). A systematic study of all young clusters in the SMC, LMC and the Galaxy for which reliable analyses of Be stars are available, has been made recently (Maeder et al. 1999). Fig. 1 shows the main results.

Fig. 1 clearly supports the above suggestion of a larger number of fast rotators at lower metallicity Z. This result seems general and not only limited to the SMC, since the same trend is observed within the Galaxy and also for the comparison of the LMC and the Galaxy. The origin of this very important relation is likely to be found in star formation. Several effects may intervene. A lower Z implies less dust and ions in star forming regions. The coupling of the magnetic field to the matter is then weaker, the ambipolar diffusion of the magnetic field should proceed faster, thus leading to less angular momentum losses by the central contracting body. During pre–main sequence evolution the weaker opacities at lower Z are leading to an earlier disappearance of the external convective zones and thus also to less magnetic coupling between the forming star and its surroundings. Finally, the importance and the survival lifetime of the accretion disk may increase with metallicity, thus favouring the dissipation of angular momentum at higher Z.

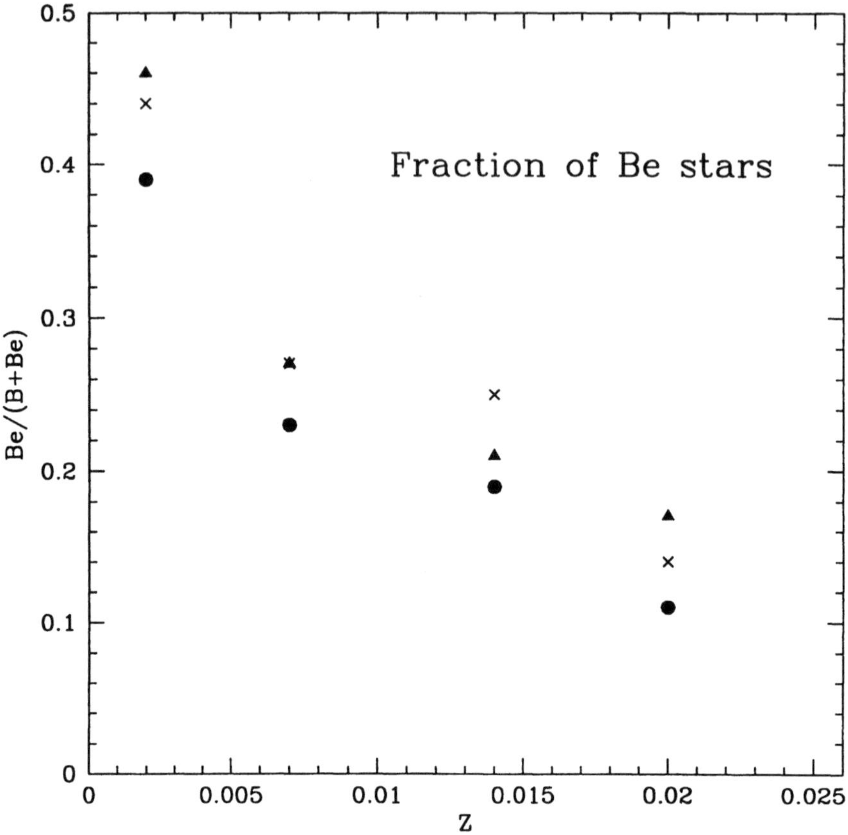

Fig. 1. The relative fraction of Be stars to B + Be stars in 21 clusters of the SMC, LMC and the Galaxy (inner and outer ring) as a function of the average metallicity Z. The triangles, crosses and dots refer to number counts made in different magnitude intervals, $M_V = -4, -2; -5, -2;$ and $-5, -1.4$ respectively.

4 Massive Stars with Rotation: Main Model Outputs

Models of rotating stars with highly developed internal physics have been constructed (cf. Meynet and Maeder 1999), taking into account: –the effects of the centrifugal force on the hydrostatic equilibrium and stellar shape, –shear mixing (Maeder 1997), –the meridional circulation (Maeder and Zahn 1998), the evolution of the internal Ω–profile, the changes of the $\dot{M}$–rates due to rotation, etc. The main results are the following ones:

– **MS tracks, lifetimes and abundances:**
  All model outputs depend on mass M, metallicity Z and rotation expressed here by the initial value of Ω/Ω_c, the fraction of the break–up velocity on the ZAMS. For a given M, there is a family of tracks as shown by Fig.

2, with increasing L for higher rotation. The MS lifetimes are increased, for example at 20 $M_\odot$, by 40% for an initial $\Omega/\Omega_c = 0.4$. Simultaneously, there are excesses of helium and of N/C: $Y_s = 0.47$ (instead of 0.252) and N/C = 7.8 (instead of 0.2) at the end of the MS in the above example.

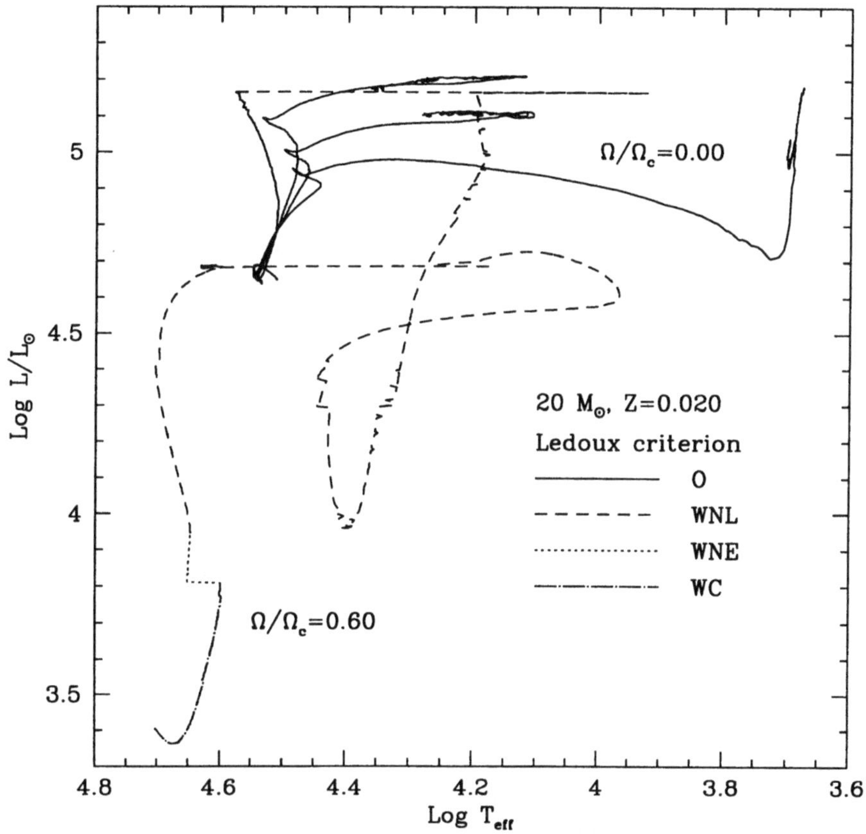

Fig. 2. Evolution of a 20 $M_\odot$ star with Z = 0.02 for various ratios Ω/Ω_c = 0, 0.3, 0.4 and 0.6. The fractions of the He lifetime spent in the RSG stage are respectively 0%, 18%, 33% and 34%. The corresponding N/H excesses are 0, 0.8, 1.0 and 1.2 dex in the A–supergiant phase.

- **Blue and red supergiant lifetimes:**
  At low Z, stars in the He–burning phase stay in the blue and do not reach the RSG stage. This is a major problem, since the observations show many RSG at low Z (cf. Langer and Maeder 1995). Interestingly enough, rotation brings the star to the red: for $\Omega/\Omega_c = 0.3$, a third of the He lifetime is

spent in the red, which allows us to solve the difficulty mentioned above, concerning the usually too small number of RSG predicted at low Z.

- **WR tracks, lifetimes and abundances:**
  For an initial $M \geq 40$ $M_{\odot}$, rotation forces the stars to enter the WR stage (as can be judged from surface abundances) before the end of the H–burning phase, thus avoiding the LBV and RSG phases. For fast rotation mass loss is very high, luminosity decreases while the WR lifetime increases. The increases in the WR lifetimes from $\Omega/\Omega_c = 0$ to 0.3 is about the same as that resulting from doubling the $\dot{M}$–rates (cf. Maeder and Meynet 1994), which is just what is required by the observed WR star numbers.
  An increase of rotation also changes the balance between the WN and WC stages: rotation strongly increases the WN phase, because there is a large zone with a mixture of H and He. Also, more transition WN/WC stars are produced in a way consistent with observations.

As a conclusion we emphasize that the models with rotation are unavoidable for correctly interpreting massive star populations, particularly at low metallicity according to the results of Section 3. Finally, we also point out that the mass of the C/O core left at the end of the He-burning phase is increased in case of moderate rotation, and this considerably influences the yields from supernovae.

References

Feast M.W. 1972, MNRAS **159**, 113
Fliegner J., Langer N., Venn K.A. 1996, A&A **308**, L13
Grebel E.K., Roberts W.J., Will J.M., de Boer K.S. 1994, Space Sci. Rev. **66**, 65
Heckmann T.M., Gonzalez–Delgado R., Leitherer C., Meurer G.R., Krolik J., Wilson A.S., Koratkar A., Kinney A. 1997, ApJ **482**, 114
Herrero A., Kudritzki R.P., Vilchez J.M., Konze D., Butler K., Haser S. 1992, A&A **261**, 209
Kudritzki R.P., Pauldrach A., Puls J., Abbott D.C. 1989, A&A **219**, 205
Langer N., Maeder A. 1995, A&A **295**, 685
Maeder A. 1997, A&A **321**, 134
Maeder A., Conti P.S. 1999, ARAA **32**, 227
Maeder A., Grebel E.K., Mermilliod J.C. 1999, A&A, in press
Maeder A., Meynet G. 1994, A&A **287**, 803
Maeder A., Zahn J.-P. 1999, A&A **334**, 1000
Massey P. 1998, ApJ **501**, 153
Meynet G. 1995, A&A **298**, 767
Meynet G., Maeder A. 1999, A&A, in press
Schaerer D. 1998, ApJ **467**, L17
Venn K.A., McCarthy J.K., Lennon D.J., Kudritzki R.P. 1998, PASP Conf. Ser. **131**, 177

Low and Intermediate Mass Stars: Learning Theory from Observations

Cesare Chiosi

Astronomy Department, Vicolo Osservatorio 5, 35122 Padova, Italy

Abstract. We summarize the results of stellar evolution theory obtained over the last two decades and highlight the points of disagreement with the present-day observational information. Arguments are given to favor stellar models in which the classical treatment of convection and mixing is abandoned and more complex schemes are adopted.

1 Introduction

Over the past decades the theory of stellar structure and evolution has reached such an unprecedented degree of complexity and sophistication to make it feasible not only the interpretation of even small details of the HR Diagram (HRD) of stellar populations in different astrophysical environments but also to enable us to make use of stars as cosmological clocks to date the epoch of the oldest (the first indeed) star forming episodes. Despite these achievements, there are aspects of the stellar physics, that are less firmly established than others thus hampering the potential of current models in interpreting the observational data. This is particularly true for convection and mixing.

Equally important progress has been made in our observing capability. Modern instrumentation provide a wealthy of data for stellar populations that enable us to tightly constrain the theoretical scenario.

The question immediately arises whether a cross comparison of theory and observation can help us understand better the basic physic of stellar models, and whether in particular we may infer from it the kind of mixing that is likely to occur in real stars. There are indeed several observational hints that in stars the convective regions ought to extend beyond the classical limit. Since correct stellar models bear very much on many astrophysical applications, those points of disagreement need to be urgently settled down.

2 Basic Stellar Evolution

Low Mass Stars (LMS). They undergo core He-flash in a strongly degenerate core, but soon after removing degeneracy burn helium quiescently, and later evolve through the AGB phase. The initial mass of such stars is lower than $M_{HeF} = 1.8 \div 2.2 M_{\odot}$. In this range there is another important limit mass to remember, i.e. the mass M_L below which stars burn central hydrogen in radiative conditions, whereas do it in a convective core above. This mass limit is about $1.2 \div 1.3 M_{\odot}$. The most noticeable features of LMS are: the extended RGB, along which stars climb at a suitable rate; the penetration of the envelope convection down to deep layers (1st dredge-up); the slowing down of the RGB evolution when the H-burning shell encounters the discontinuity in the chemical profile set by the external convection in previous stages (origin of the bump in the luminosity function); the loss of a significant fraction of the envelope mass near the RGB tip (most important in stars of about 0.7 $M_{\odot}$); the core He-burning in convective conditions along the HB, whose extension depends on the amount of mass lost at the RGB tip, metallicity, spread in the fractionary mass of the H-exhausted core, and other parameters; the early AGB phase with the 2nd dredge-up; the thermally pulsing AGB phase with the 3rd dredge-up; finally the complete removal of the envelope and the formation of a White-Dwarf.

Intermediate Mass Stars (IMS). They ignite helium non degenerately, but following central He-exhaustion develop an electron degenerate C-O core, and as AGB stars experience He-shell flashes or thermal pulses. The initial mass falls in the range $M_{HeF} < M < M_{up} = 8 \div 9 M_{\odot}$. Noticeable features of IMS are; the much less extended RGB; the 1st dredge-up episode bringing He and N (CNO products) to the surface; the absence of degeneracy in the core prior to He-exhaustion; the slow core He-burning taking place in two distinct regions of the HRD: near the Hayashi line and at higher T_{eff}s, or equivalently the occurrence of a loop whose extension is very sensitive (among other factors) to the size of the convective core and envelope; finally the intersection of the blue band of core He-burning models with the instability strip of the Cepheid stars. The AGB phase is terminated either by the envelope ejection and the formation of a C-O White Dwarf if $M_{HeF} < M < M_W$, or C-ignition with subsequent deflagration (or detonation) once the C-O core has grown to the Chandrasekhar limit of $1.4 M_{\odot}$ if $M_W < M < M_{up}$. The exact value of M_W depends on the efficiency of mass loss by stellar wind during the AGB phase (little mass is lost during the previous RGB stages).

Which Kind of Mixing? The local Schwarzschild criterion ($\nabla_R = \nabla_A$) provides the simplest evaluation of the extension of convective regions (cores, shells and envelopes) and the Mixing Length Theory (MLT) simplifies the complicated pattern of motions therein by saying that full, and instantaneous mixing takes place. *The reality is, however, more complicated than this simple picture.* First of all, inconsistencies are known to develop at the border of convective regions which to be cured give rise to the so-called *H-semi-convection* in core H-burning massive stars, *He-semi-convection* and *breathing convection* in core He-burning low and intermediate mass stars (see Chiosi et al. 1992 for a recent review). Secondly, convection and mixing are non-local phenomena that in principle can also penetrate (*overshoot*) into regions that are stable according to the Schwarzschild criterion. Finally, mixing requires a certain amount of time to occur, and may not be complete.

Unfortunately as soon as the classical scheme is abandoned, things lose simplicity and a unique solution does not exist, thus making uncertain the effects of convection and mixing. Nevertheless, it is worth recalling that

(i) the size of the central convective core affects the luminosity, effective temperature, and lifetime of the corresponding evolutionary phase;

(ii) the inward extension of the external convection may affect the surface abundances;

(iii) the size of the intermediate convective shells has less straightforward effects: in massive stars it may affect the extension of the blue loops, in intermediate and low mass stars it may alter the surface abundances during the latest evolutionary stages;

(iv) finally, the extension of the convective core during central H- and He-burnings affects the limit masses M_{HeF} and M_{up}, defining low and intermediate mass stars.

3 Do the Above Models Agree with Observations?

Hipparcos Data: Field HRD. Star counts in the Hipparcos HRD (Perryman et al. 1995) of the solar vicinity ($d \leq 100$ pc) make it evident that the region of the so-called Hertzsprung gap is very sensitive to overshooting on the main sequence for stars with mass $M \simeq 1.6 M_{\odot}$, and that the onset of this phenomenon should occur either abruptly or gradually at about $1.6 \div 1.7 M_{\odot}$ (Schröder 1998). Analyzing the same data to infer the past history of star formation in the galactic disk Bertelli et al. (1999) found that full compatibility between the number of MS and Post-MS stars, require some extra-mixing during the core H-burning phase of stars in mass range $1.5 \div 2 M_{\odot}$.

Hipparcos Data: Red Clump. As already recalled, the red clump is populated by stars developing degenerate He cores. The mean M_I of the clump stars is almost insensitive to the V-I color for a given population (composition) however it varies from site to site (Girardi et al. 1998). In addition to that, stars slightly heavier than M_{HeF}, should define a secondary, clumpy structure about 0.3 mag below the blue extremity of the red clump. Both features are well separated in the M_I vs. $V - I$ diagram of metal-rich stellar populations. The secondary clump is indeed detectable in the Hipparcos HRD for the solar vicinity. Since the stars in the secondary clump should represent a narrow range of masses their mass determinations (binary systems) can provide information about M_{HeF}. Preliminary studies hint for $M_{HeF} \simeq 1.7 \div 1.8 M_{\odot}$ thus suggesting some overshooting.

Intermediate Age Clusters. The young rich clusters of the Magellanic Clouds (MCs) are classical templates to which the results of stellar evolution theory are compared. Looking at the case of NGC 1866, (i) for the observed luminosity of the giants, there are too many stars above the predicted main sequence turn-off, whose number is a significant fraction of the number of giant stars; (ii) the predicted ratio of post main sequence stars to the main sequence stars was about four times the observed one. To cast light on this problem, Chiosi et al. (1989) introduced the concept of the integrated luminosity function of main sequence stars normalized to the number of giants (NILF), which simply reflects the ratio of core He- to H-burning lifetimes, and concluded that only models with substantial core overshooting could match the observed NILF. The analysis of other MC's clusters, like NGC 1831 and NGC 2164 (Vallenari et al. 1991, 1992), yielded similar results. This conclusion has been recently strengthened by Keller et al. (1998), who presented new HST photometry for several rich clusters of the MCs.

Old Open Clusters. The old open clusters (ages from 1 to 7÷8 $\times 10^9$ yr) having turn-off masses between 1 $M_\odot$ and 2 $M_\odot$, probe the stellar structure in the mass range, in which on the main sequence the transition occurs from radiative to convective cores, from pp chain to CNO cycle, and from very bright to much less evident RGBs. Long ago Barbaro & Pigatto (1984) pointed out that the CMD of these clusters (e.g. NGC 2420, NGC 3680, IC 4651, M67, and others) does not agree with the predictions from classical models. The main signatures of disagreement were the detailed shape of the turn-off and RGB, the clump of red stars (most likely core He-burners), and the number of stars brighter than the main sequence at the beginning of the sub-giant branch with respect to the main sequence stars. All this suggested that a certain amount of convective overshooting during the core the H-burning phase ought to occur. Aparicio et al. (1990) first tried to establish the amount of required overshoot and how its onset should depend on the star mass. They suggested that in this mass interval convective overshooting should gradually increase with the star mass. Their recipe has been adopted in all stellar models of the Padua library (see Bertelli et al. 1994). Strong support to this idea came from the study of the CMD of IC 4651 by Bertelli et al. (1992) and the systematic analysis of several old open clusters by Carraro & Chiosi (1994). Recently, Rosvick & VandenBerg (1998) and Gim et al. (1998) have re-addressed this topic analyzing the CMD of NGC 6819 and NGC 7789, two rich open clusters whose age is 2.5 and $\geq 1.6 - 1.7$ Gyr, and whose turn-off mass is about 1.6 and $1.8 M_\odot$ respectively. In both cases the morphology at the turn-of is such that *"only models with substantial core overshooting can explain"*. Another type of evidence comes from eclipsing binaries, for which good determinations of mass, radius, luminosity, and abundances are available for stars near the turnoff (Andersen et al. 1990; Napiwotzki et al. 1991; Nordstrom et al. 1996; Pols et al. 1997). Since many stars fall beyond the limit for the core H-burning phase of classical models, a cooler turnoff seems to be required, which was attributed to substantial overshooting.

Relative Frequency of AGB Stars. The ratio of star counts $R_2 = N_{AGB}/N_{HB} = t_{AGB}/t_{HB}$ is known to depend on the kind of mixing once the mass pertinent to the HB stars is determined (initial He content). The comparison of theoretical predictions for the lifetime ratio t_{AGB}/t_{HB} as derived from different mixing schemes with the ratio of star counts $R_2 = N_{AGB}/N_{HB}$ leads to the following results. In models calculated with the Schwarzschild criterion, the amount of mixed material is too low and in turn R_2 is too high. In models with semi-convection, the extension of the mixed region is too large and in turn R_2 is too low. Breathing convection makes things even worse because only half of the AGB stars can be accounted for. Finally, models with non local mixing are able to reconcile theory and observations if the overshooting region extends about $1 \times H_P$ across the Schwarzschild border (see e.g. Chiosi et al. 1987).

AGB & Carbon Stars. The TP-AGB phase of LMS and IMS is at base of our current understanding of the LF of AGB stars in general, the relative frequency of C-stars, the surface abundances of chemical elements (s-process included), the initial-final mass relationship, the stellar yields of several important elements, etc... This phase is quite complicate as it requires solid understanding of several phenomena such as

(i) inter-shell nucleosynthesis,
(ii) convective dredge-up,
(iii) envelope burning in most massive stars ($M \geq 3 \div 4 M_{\odot}$),
(iv) and mass loss by stellar winds.

Items (ii) and (iv) determine the overall uncertainty. The most recent TP-AGB models are by Marigo et al. (1996, 1998, 1999), Marigo (1998), and Wagenhuber & Groenevegen (1998) to whom the reader should refer for all details and exhaustive referencing. Perhaps the most important consequence of envelope burning is the breakdown at high luminosities of the core mass -luminosity ($M_c - L$) relation. While the luminosity ascent of low-mass AGB stars ($M \leq 3 M_{\odot}$) in which the H-burning shell is the dominant energy source is governed by this relation, massive AGB stars can rapidly reach much higher luminosities. This anticipates the ejection of the envelope. Current models can reasonably explain the LF of carbon stars, the distribution of these in the M_{bol} vs log(age) plane, and the surface abundances. Nevertheless there are a few debated points:

Faint C-Stars. With the standard efficiency of the 3rd dredge-up mechanism too few if none at all *low luminosity* C-stars are formed in contrast with the observations. This hints that more efficient external mixing is required (see Herwig et al. 1997; Marigo et al 1996, 1998, 1999).

Bright C-Stars. Why there are so few bright C-stars? Three explanations are possible: (1) Envelope burning (fresh C is converted into N via CNO cycle); (2) Very efficient stellar wind that removes the envelope before the C-phase is started; (3) Convective overshooting that lowers M_{up} down to $5 \div 6 M_{\odot}$ *instead* of the $8 \div 9 M_{\odot}$ of the classical models.

Initial-Final Mass Relationship: IFMR. Establishing the relation between the final mass M_F, left at the end of the AGB phase, and the initial mass M_I of the progenitor is of paramount importance for various aspects related to the previous evolution, mixing being first in the list. Unfortunately, at present both the observational and theoretical determinations of the IFMR still suffer from great uncertainty because of its semi-empirical nature. M_F requires some theoretical support (mass-radius relation) which, however, is quite well established (see Weidemann 1990; D'Antona & Mazzitelli 1990). M_I is more difficult to assess as it depends on the kind of mixing at work. The theoretical predictions show a clear disagreement with Weidemann (1990) for the solar vicinity, concerning both the shape of the relation and the value of M_{up}. A better accordance is found with the IFMR by Herwig et al. (1997) and the data by Jeffries (1997). These data seem to indicate the occurrence of convective overshooting during the previous phases (lower M_{up} and bigger M_{CO} cores and M_{WD} masses in turn at given M_I).

Anomalies in Chemical Abundances. Surface abundances of RGB stars are particularly useful probes of mixing taking place during the 1st dredge-up, because mass loss by stellar wind is less of a problem. Canonical models predict precisely the changes in abundances that should occur during the 1st dredge-up (see Iben & Renzini 1983): (i) substantial reduction of Li, Be, and B; (ii) decrease of C abundance by 30%; (iii) decrease in the $^{12}C/^{13}C$ ratio from 90 to about 20-30; (iv) increase of N abundance by a factor 0.44 $(C/N)_o$, where $(C/N)_o$ is the initial abundance ratio by number (net increase of about 80%); (v) modest increase in He, and decrease in H, and very little variation in O-abundance. As far as Li and Be are concerned, although there appears to be substantial agreement between classical predictions, there are indications that some extra-mixing may have occurred during the main sequence phase. In the case of $^{12}C/^{13}C$, there is only partial agreement with the observations. While the majority of stars appear to have $^{12}C/^{13}C$ ratios consistent with the theory, a sizable minority (30%) appear to have $^{12}C/^{13}C$ considerably lower. In the recent study by Charbonnel et al. (1998, and references therein) accurate data for Li, C, N and $^{12}C/^{13}C$ for five field giants with $[Fe/H] = -0.6$ and two stars in 47 Tuc are presented together with precise absolute magnitudes M_{bol} based on Hipparcos parallaxes. This allows the authors to establish an evolutionary sequence: $^{12}C/^{13}C$ drops from 20 to 7 for M_{bol} passing from 1 to 0.5, while Li disappears. The ratio $^{12}C/^{14}N$ decreases by 0.2 to 0.4 dex. The two stars in 47 Tuc show even lower $^{12}C/^{14}N$. These observations strongly indicate that extra-mixing occurs only when the stars reach the bump stage of the LF. The nature of this extra-mixing is still obscure (likely rotation according to the authors). Star to star abundance variations of C, N, O Na, and Al together with low $[Mg/Fe]$ and large $[Al/Fe]$ in globular cluster red giants have been reported (Shetrone 1996). Among other explanations, Denissenkov et al. (1998) suggest deep extra-mixing may occur in RGB stars even if problems remain for the $[Mg/Fe]$ vs

$[Al/Fe]$ anti-correlations. In addition to this, there are difficulties with the surface abundances of AGB stars and their descendents and the blue supergiant stars in the so-called Hertzsprung gap. Both topics are not addressed here.

4 Non-Local Mixing Schemes

H-semi-convection, He-semi-convection, breathing convection, and local overshooting stem from the simple consideration and inconsistency that convective elements reaching the layer at which the buoyancy acceleration vanishes (the Schwarzschild condition) still possess kinetic energy and therefore may penetrate (overshoot) into the surrounding stable regions.

If the physical ground of convective overshoot is simple, its formulation and efficiency are much more uncertain giving rise to a variety of solutions and companion evolutionary models. See for instance Maeder (1975), Bressan et al. (1981), Matraka et al. (1985), Bertelli et al. (1985) Xiong (1990), Canuto (1998a,b), Canuto & Dubovikov (1998), Canuto & Mazzitelli (1991, 1992), Grossman et al. (1993), Grossman (1996), Grossman & Taam (1996), Deng et al. (1996) Ventura et al. (1998) and references therein, In a way or another the various attempts seek to address the following questions:

(1) What drives the penetration of convection into surrounding stable regions that are thus affected by mixing?

(2) What is the thermodynamic structure of the unstable and potentially unstable regions?

(3) What is the time scale of mixing? Instantaneous or over a finite (long) period of time? What is the mechanism securing either full or partial homogenization of the unstable regions?

(4) Are the gradients in molecular weight inhibiting the penetration of convective elements into radiative regions?

Attempts to answer the above questions have generated more complex schemes, in which overshooting is described as non local phenomenon and mixing is formulated as a diffusive process taking place over suitable time scales. The properties of stellar models based on non-local diffusive overshoot have been recently summarized by Deng et al. (1996), Chiosi (1998) and Salasnich et al. (1998) to whom the reader should refer for all details.

References

1. Aparicio A., Bertelli G., Chiosi C., Garcia-Pelayo J. M., 1990, A&A 240, 262
2. Andersen J., Nordstrom B., Clausen J. V., 1990, ApJ 363, L33
3. Barbaro G., Pigatto L., 1984, A&A 136, 355
4. Bertelli G., Bressan A., Chiosi C., 1985, A&A 150, 33
5. Bertelli G., Bressan A., Chiosi C., 1992, ApJ 392, 522
6. Bertelli G., Bressan A., Chiosi C., Fagotto F., Nasi E., 1994, A&AS 106, 275

7. Bertelli G., Nasi E., Bressan A., Chiosi C., 1999, A&A submitted
8. Bressan A., Bertelli G., Chiosi C., 1981, A&A 102, 25
9. Canuto V., 1998a, ApJ 505, L47
10. Canuto V., 1998b, ApJ 508, L103
11. Canuto V., Dubovikov M., 1998, ApJ 493, 834
12. Canuto V., Mazzitelli I., 1991, ApJ 370, 295
13. Canuto V., Mazzitelli I., 1992, ApJ 389, 724
14. Carraro G., Chiosi C., 1994, A&A 287, 761
15. Charbonnel C., Brown J. A., Wallerstein G., 1998, A&A 332, 204
16. Chiosi C., 1998, In *Stellar Astrophysics for the Local Group*, eds. A. Aparicio, A. Herrero, F. Sanchez, (Cambridge: Cambridge Univ), p. 1
17. Chiosi C., Bertelli G., Bressan A., 1987, In *late stages of Stellar Evolution*, eds. S. Kwok & S.R. Pottasch, (Dordrecht: Reidel Publishing Company), p. 213
18. Chiosi C., Bertelli G., Meylan G., Ortolani S., 1989, A&A 219, 167
19. D'Antona F., Mazzitelli I., 1990, ARA&A 28, 139
20. Deng L., Bressan A., Chiosi C., 1996a, A&A 313, 145
21. Denissenkov P. A., Da Costa G. S., 1998, A&A 333, 926
22. Gim M., VandenBerg D. A., Stetson P. B., Hesser J. E., Zurek D. R., 1998, PASP 110, 1318
23. Girardi L. A., Groenevegen M. A. T., Weiss A., Salaris M., 1998, MNRAS 301, 149
24. Grossman S. A., Narayan R., Arnett D., 1993, ApJ 407, 284
25. Grossman S. A., 1996, MNRAS 279, 305
26. Grossman S. A. Taam R.E., 1996, MNRAS 283, 1165
27. Herwig F. Blöcker T, Schönberner D., El Eid M., 1997, A&A 324, L81
28. Iben I. Jr., Renzini A., 1983 ARA&A 21, 271
29. Jeffries R. D., 1997, MNRAS 292, 177
30. Keller S., Bessell M., Da Costa G., 1998, In *New Views of the Magellanic Clouds*, IAU Symp. 190, in press
31. Maeder A., 1975, A&A 40, 303
32. Marigo P., 1998, A&A 340, 463
33. Marigo P., Bressan A., Chiosi C., 1996, A&A 313, 545
34. Marigo P., Bressan A., Chiosi C., 1998, A&A 331, 564
35. Marigo P, Girardi L., Bressan A., 1999 A&A 344, 123
36. Matraka B., Wassermann C., Weigert A., 1985, A&A 107, 283
37. Napiwotzki R., Schönberner D., Weidemann V., 1991, A&A 243, L5
38. Nordstrom B., Andersen J., Andersen M. I., 1996, A&AS 118, 407
39. Perryman M.A.C., Lindegren L., Kovalevsky J., Turon C., Hog E., et al., 1995, A&A 304, 69
40. Pols O. R., Tout C. A., Schröder K. P., Eggleton P. P., Manners J., 1997, MNRAS 289, 869
41. Rosvick J. M., VandenBerg D. A., 1998, AJ 115, 1516
42. Salasnich, B., Bressan A., Chiosi, C., 1999 A&A 342, 131
43. Shetrone M.D., 1996, AJ 112, 2639
44. Schröder K. P., 1998, A&A 334, 901
45. Vallenari A., Chiosi C., Bertelli G., Meylan G., Ortolani S., 1991, A&AS 87, 517
46. Vallenari A., Chiosi C., Bertelli G., Meylan G., Ortolani S., 1992, AJ 105, 1100
47. Ventura P., Zeppier A., Mazzitelli I., D'Antona F., 1998, A&A 334, 953
48. Wagenhuber J., Groenevegen M. A. T., 1998, 340, 183
49. Weidemann V., 1990, ARA&A 28, 103
50. Xiong D. R., 1990, A&A 323, 31

Abundance Determinations in Massive Blue Stars Within the Local Group

Artemio Herrero[1,2], Rolf–Peter Kudritzki[3,4], Stephen J. Smartt[5], Kim Venn[6], Norbert Przybilla[4], Daniel J. Lennon[5], James K. McCarthy[7], and M. Ilusión Monteverde[1]

[1] Instituto de Astrofísica de Canarias, E-38200 La Laguna, Spain
[2] Departamento de Astrofísica, Universidad de La Laguna, Avda. Francisco Sánchez s/n, E- 38071 La Laguna, Spain
[3] Universitäts–Sternwarte München, Scheinerstr.1, D-81679 München 80, RFG
[4] Max–Planck Institut für Astrophysik, Karl–Schwarzschild–Str. 1, D–85740, Garching, Germany
[5] Isaac Newton Group, Apdo. 368, E-38700 Santa Cruz de La Palma, Spain
[6] Department of Physics and Astronomy, Macalester College, St. Paul, MN 55101, USA
[7] Palomar Observatory, California Institute of Technology, Pasadena, CA 91125, USA

Abstract. We present the current status of our knowledge on the photospheric abundances of massive blue stars (with spectral types O-A) in the Milky Way and Local Group galaxies, and the implications of these for stellar and galactic structure and evolution. We also consider possible extensions of this project to observe individual stars in galaxies beyond the Local Group. Finally, we indicate how the VLT and a number of its instruments (mainly FORS, UVES and FLAMES/GIRAFFE) will help advance these studies.

1 Introduction

Atmospheric abundances play a key role in our understanding of the different physical processes that occur in the interior of massive blue stars (MBS, from O and B-type dwarfs to O, B and A-type supergiants). Their strong winds (driven by metal lines) and short lifetimes means that they play a critical role in galactic structure and chemical evolution through strong interaction with the surrounding ISM. Further, the so called Wind–momentum Luminosity Relation has recently been proposed as a method for determining extragalactic distances. This appears to be critically dependent on stellar metallicity, but the effect can be calibrated which both enhances the method and also provides fundamental knowledge of the system's ISM abundances (see the contribution by R.P. Kudritzki in these proceedings).

To derive abundances in MBS we first obtain stellar parameters from model atmosphere simulations and then perform line formation calculations for the chemical elements we are interested in. The theoretical techniques can combine a number of astrophysical methods : LTE or NLTE; plane–parallel or spherical geometry; hydrostatic or hydrodynamic calculations; inclusion or not of line–blanketing, micro- and macro-turbulence, rotation, broadening mechanisms, etc. Further, different combinations of approaches are usually selected by different authors to analyse different (and sometimes the same!) spectral types, as the relative importance of these effects varies with stellar parameters. The influence of these effects on derived results can vary for different chemical elements in a given star, for different ionization stages of a given element, or even for different lines of a given ionization stage of an element.

This complexity suggests that a consistent analysis technique should be a *differential* approach, were we compare the same line in stars of the same (or similar) spectral type using a consistent set of theoretical calculations. Hence we can derive quite accurate abundance *differences* between stars which is only weakly dependent on the model atmosphere physics used - providing of course the latter does have some physical plausibility. We believe that using such consistent methods carefully that we can derive meaningful abundance determinations in both massive main-sequence stars, and their evolved descendents.

2 The Milky Way

Galactic B-dwarfs due to their relatively young age (10 Myrs), represent the *current* local chemical composition of the ISM, and hence are directly comparable to H II regions. Their optical and UV spectra are well reproduced by present model atmosphere calculations, and photospheric abundances can be reliably derived (see [1], [2], [3], [4], [5]). A long running discrepancy existed for many years concerning the radial oxygen abundance gradient in the Milky Way. The value derived from B-dwarfs was significantly different to that from H II region studies. However Smartt & Rolleston [6], have now obtained a result from B-type stars consistent with the H II region gradient by re-analysing an extensive data-set using consistent careful differential technique. Their result has been recently confirmed in [7]. Closer to the galactic centre, Smartt et al. [8] using B-dwarfs and A-type main-sequence stars found that the abundance of O does not keep increasing below 5 kpc, while Si and Mg do. Why the α–elements behave in a different manner is difficult to explain in the framework of standard nucleosynthetic theories. In the anti-centre direction it appears that the abundance pattern in B-dwarfs supports a common origin for C and N, distinct from that of α–elements, [9],[10].

Fewer results on photospheric abundances of O-type main-sequence stars and supergiants have been produced. Herrero et al. [11] found in their analysis of 24 O-stars that nearly all O-type supergiants, and some dwarfs with large rotational velocities showed enhanced He abundances, in contradiction with predictions from standard evolutionary theories (see f.e. [12]). However, whether this so-called He-discrepancy is due to deficiencies in the model atmospheres or in the evolutionary calculations, is still unclear. Recent work ([13], [14]) indicates that introducing microturbulence into the calculations can reduce the derived He abundances for supergiants later than O9. Similar results have been found by Villamariz & Herrero (in prep.), but they conclude that the effect cannot explain the whole discrepancy, especially for earlier types. Much effort has been made to explain these and other observed stellar properties as a consequence of rotationally induced mixing (see the contribution by A. Maeder in these proceedings, or [15]). In support of this work, an analysis of 11 O-types with moderate (projected!) rotational velocities in Cyg OB2 (a very young OB association) indicates that no He enhancements exist ([16]). This result suggests a lack of mixing in the Cyg OB2 stars due to the low rotational velocities or to the youth of the association (or both).

However, the strongest evidence for such mixing may not come purely from He, which is probably only moderately enhanced and is also difficult to determine (because of line saturation effects). The abundance ratios of C, N and O, may be affected to a much greater degree by CNO burning and subsequent mixing to the stellar surface. In Table 1 we list the evidence favouring the hypothesis that the atmospheres of evolved MBS have been contaminated by CNO mixing. We should point out that in Table 1 we are comparing different spectral types, techniques and model calculations.

Table 1. Evidence for CNO mixing in OBA supergiants

Stars	References	Comments
A- supergiants	[17]	
B- supergiants	[1], [18], [19]	
O9.7 supergiants	[20]	Microturbulence included
OBN stars	[21]	Rapid rotators
early O supergiants	[22],[23]	Fit to the UV spectrum

3 The Magellanic Clouds

The metal deficiency of the Magellanic Clouds is reflected in the strength of metal lines in massive supergiants (see Fig 1). Large metallicity differences like those present between the Milky Way and the SMC forces us to redefine the line diagnostics used for spectral typing (see [24]).

Abundance analyses of OB-types consistently reproduce the metal deficiencies of both the LMC and the SMC, similar to what is found with other

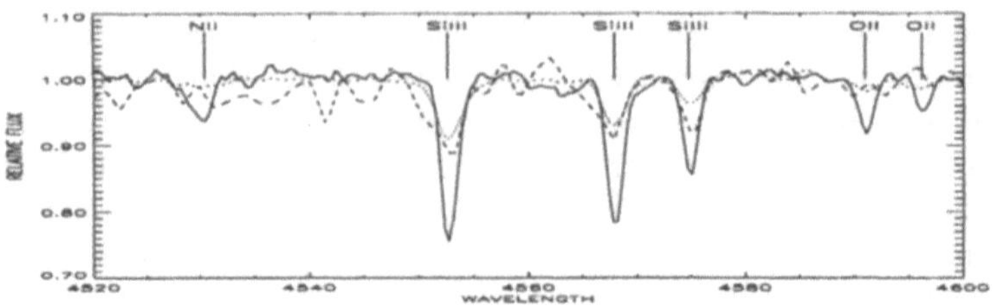

Fig. 1. A comparison of B- supergiants of similar spectral type in the Milky Way (HD 14956, B1.5 Ia; solid), the SMC (SK 191, B1.5 I; dotted) and M33 (110-A, B1.5 Ia$^+$; dashed)

methods. In [25], [26] and [27] values of −0.7, −1.1 and −0.8 dex respectively are obtained for the mean SMC metallicity as compared to the Sun; while [26], [27] and [28] obtain -0.8, -0.3 and -0.3 dex for the LMC. This agrees well with H II regions studies. However, local abundance scatter can be large, as has been shown in [29] for NGC 330.

Evidence for CNO core-surface mixing in OB supergiants is not conclusive, due to small sample numbers and possible inconsistencies in analysis techniques. Haser et al. ([27]), for example, in their analysis of four Magellanic Cloud stars, only find consistent evidence of CNO mixing in one of them. Contrary to this, [30] implies that the abundances of 10 A-supergiants in the SMC are not only consistent with a low metal content, but also with photospheric contamination by core CNO-burning products and even by elements from the first dredge–up. The current available evidence suggests that mixing from the stellar interior probably does occur in Magellanic Clouds massive stars; however its dependence on the initial stellar metallicity requires further quantitative study.

4 M31 and M33

Telescopes of 4m aperture can be pushed to their current limits to obtain high-quality spectra of B and A-type supergiants in M31 and M33, as in [31] and [32]. In Fig. 1 we see the comparison of MW, SMC and M33 B-supergiants of similar spectral types, showing the feasibility of abundance determinations for stars in M33.

Using spectra of this quality, we have derived the radial O gradient in M33 using B-type supergiants, and obtained a result in agreement with H II region analysis [33]. A further, more in-depth study of the stellar sample used indicates that results are reasonably consistent with the nebular ones, but that a simple radial gradient for O may not fully describe the situation. Silicon abundances in the stars correlate well with those for oxygen. The spectra are of 2Å resolution, and the errors estimated in the analysis may explain the slight inconsistencies found between the stellar and nebular results.

This sort of work will be improved using telescopes of the 8m and 10m generation. Even before the publication of the oxygen abundance gradient result in [33], McCarthy et al. ([34]), using spectra obtained at the Keck I telescope with HIRES, published iron abundances for two M33 A-type supergiants, one close to the centre of M33 and the other one much further out. The small numbers involved preclude any quantitative comparisons, but the goal in this work is to determine, [α/Fe] ratios at different galactocentric distances in M33, which will allow constraints to be placed on models of galaxy formation and evolution ([35]). In addition, determining spectral parameters together with metallicity will allows us to calibrate the Wind momentum – Luminosity Relation as a function of this metallicity.

Similar studies have been made in M31 ([36], [37]), allowing a preliminary comparison with H II region metallicity gradients. The advantage of using stellar spectra is that we can determine and compare abundances of a larger number of elements formed in different nucleosynthetic sites (see Fig. 2).

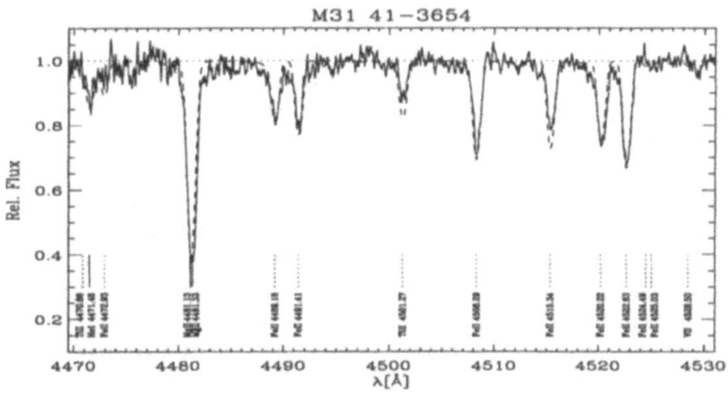

Fig. 2. Spectral line fits to the M31 star 41–3654. Only a small portion of the spectrum is shown, but the quality of the fit in the full range 3800 to 7800 Å is similar, and comparable to galactic standards (from Przybilla, PhD thesis, in prep.)

5 The VLT Role

Even at the resolution of FORS the VLT can make a fundamental contribution to our knowledge of abundances in very distant MBS, as has been repeatedly shown in the present conference (see for example the contributions by I. Appenzeller and R.P. Kudritzki). However, the full capacities of the VLT will be best exploited (for our work on stellar abundances) by using FLAMES+GIRAFFE on the VLT-2, as Table 2 illustrates. There we summarise the VLT instrumental possibilities when observing MBS in the Local Group, and even beyond, to the distance of the Virgo cluster.

Table 2. Use of the VLT and its instrumentation to observe MBS in Local Group galaxies and beyond. Exposure times are in seconds. N is the number of objects that can be observed in a single exposure in the blue

Instrument	R	S/N	t_{exp}	m_v	N	Comments
FORS	1000	120	3600	20.0	19	bright MBS in LG
		70	28 800	22.5	19	brightest MS in Virgo
UVES	10000	100	3600	17.5	1	brightest MS in LG
GIRAFFE	15000	100	3600	18.5	130	bright MS in LG

References

1. Gies D.R., Lambert D.L. (1992) Ap. J. **387**, 673–700
2. Kilian J. (1992) A&A **262**, 171–187
3. Cunha K., Lambert D.L. (1994) Ap. J. **426** 170–191
4. Smartt S.J. (1996) PhD Thesis, Queen's University of Belfast
5. Vrancken M. (1998) PhD Thesis, Vrije Universiteit Brussel
6. Smartt S.J., Rolleston, W.R.J. (1997) Ap. J. **481**, L47–L50
7. Gummersbach C.A., Kaufer A. et al. (1998) A&A **338**, 881–896
8. Smartt S.J., Venn, D.A., Dufton P.L., Lennon D.J., Rolleston, W.R.J. (1999) A&A *submitted*
9. Smartt S.J., Dufton P.L., Rolleston W.R.J. (1996) A&A **310**, 123–131
10. Hibbins R.E., Dufton P.L. et al. (1998) A&A **332**, 681–685
11. Herrero A., Kudritzki R.P. et al. (1992) A&A. **261**, 209–234
12. Schaller G., Schaerer D. et al. (1992) A&ASS **96**, 269–331
13. McErlean N.D., Lennon D.L., Dufton P.L. (1998) A&A **329**, 613–623
14. Smith K.C., Howarth I.D. (1998) MNRAS **299**, 1146–1158
15. Heger A. (1998) PhD Thesis, Technische Universität München
16. Herrero A., Corral L.J. et al. (1999) A&A, in press
17. Venn K.A. (1995) Ap.J. **449**, 839–862
18. Fliegner J., Langer N., Venn K.A. (1996) A&A **308**, L13–L16
19. McErlean N.D., Lennon D.L., Dufton P.L. (1999) A&A, in press
20. Smith K.C., Howarth I.D., Siebert K.W. (1998) Boulder–Munich II: Properties of Hot, Luminous Stars, ASP Conf. Series Vol. 131, ed. I.D. Howarth, p. 153–158
21. Schönberner D., Herrero A. et al. (1988) A&A **197**, 209–222
22. Pauldrach A.W.A., Kudritzki R.P. (1994) A&A **283**, 525–560
23. Taresch G., Kudritzki R.P. et al. (1997) A&A **321**, 531–548
24. Lennon D.J. (1996) A&A 317, 871–882
25. Rolleston W.R.J., Dufton P.L. et al. (1993) A&A **277**, 10–20
26. Reitermann A., Baschek B. et al. (1990) A&A **234**, 109–118
27. Haser S.M., Pauldrach A.W.A. et al. (1998) A&A **330**, 285–305
28. Rolleston W.R.J., Brwon P.J.F. et al. (1995) A&A **315**, 95–104
29. Lennon D.J., Dufton P.L. et al. (1996) A&A **314**, 243–250
30. Venn K.A. (1999) Ap.J., in press
31. Herrero A., Lennon D.J. et al. (1994) A&A **287**, 885–892

32. Monteverde M.I., Herrero A. et al. (1996) A&A **312**, 24–32
33. Monteverde M.I., Herrero A. et al. (1997) Ap.J. **474**, L107–L110
34. McCarthy J.K., Kudritzki R.P. et al. (1996) Ap.J. **455**, L135–L138
35. Matteucci F., Chiappini C. 1999, in Proc. ESO Conf.on Chemical Evolution from Zero to High Redshift, Springer Conference Series.
36. McCarthy J.K., Kudritzki R.P. et al. (1997) Ap.J. **482**, 757–764
37. Venn K.A., McCarthy J.K. et al. (1998) Boulder–Munich II: Properties of Hot, Luminous Stars, ASP Conf. Series Vol. 131, ed. I.D. Howarth, p. 177–187

Abundances in Sagittarius: Present State and Perspectives for the Use of VLT

Piercarlo Bonifacio

Osservatorio Astronomico di Trieste, Via G.B. Tiepolo 11, 34131, Trieste, Italia

Abstract. Sagittarius, the nearest external galaxy, will be amenable to detailed abundance studies with VLT and other 8m class telescopes. Such data, in conjunction with the similar data of our own galaxy, will allow a deeper understanding of chemical evolution. Our study conducted with NTT shows the presence of metal-rich stars with radial velocities compatible with Sagittarius membership and of stars as metal-poor as [Fe/H] $= -1.5$. In this talk I shall address the way in which VLT instruments will allow to clarify this intricate situation.

1 Introduction

The Sagittarius dwarf galaxy discovered [4,5] as a comoving group ($V_{hel} \approx 140 \mathrm{kms}^{-1}$) with a velocity dispersion as small as $2\mathrm{kms}^{-1}$, is the nearest external galaxy and will be amenable to detailed abundance studies with VLT and other 8m class telescopes. These shall be important for two distinct reasons: on the one hand, we shall be able to study chemical evolution in an environment different from our Galaxy; on the other hand, if Sagittarius shows any chemical signature, this should allow us to identify Sagittarius debris, currently populating the Galactic Halo.

The colour-magnitude diagram of Sgr shows a wide Red Giant Branch (RGB), which is interpreted as evidence of a spread in metallicity of Sgr. Marconi et al [9] conclude that the RGB of Sgr lies between that of 47 Tuc ([Fe/H]$= -0.71$) and M2 ([Fe/H]$= -1.58$). In the following I shall describe the work done with the ESO NTT to determine spectroscopic metallicities of giants in Sgr and shall address the potentiality of VLT on this issue.

2 NTT Observations

We selected, from the Marconi et al [9] sample, stars on the RGB which ought to display the spread in metallicity present in Sgr. We used the ESO NTT telescope with the EMMI instrument in Multi Object Spectroscopy (MOS) mode. The dispersing element was grism # 5 providing a resolution of about 1500. We have obtained spectra for a total of 57 objects in field Sgr1 of [9] on June 19th 1996. In addition, we obtained long slit spectra of one of the stars observed with the MOS and of star HD 190287 on September 18th 1998.

Of the 57 stars observed 23 matched the criterion $100\,\mathrm{kms}^{-1} \leq v_{hel} \leq 180\,\mathrm{kms}^{-1}$ [6], which we adopted, to ascribe membership to Sagittarius. For these

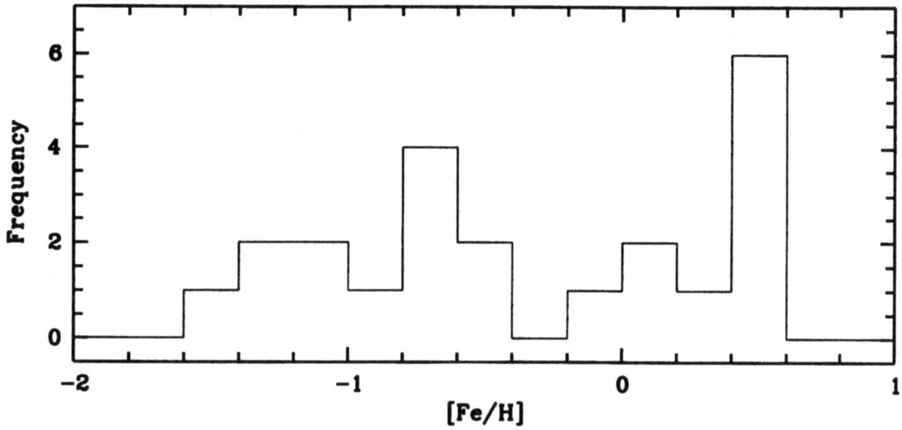

Fig. 1. Metallicity distribution for 22 radial velocity members of Sgr

23 stars the average radial velocity is 136 $kms^{-1} \pm 18\ kms^{-1}$, in good agreement with previous determinations; the r.m.s. is dominated by the error in the measure of the radial velocity and not by the velocity dispersion of Sgr.

3 Abundance Estimates from Low Resolution Spectra

Our approach has been to define spectral indices which measure some prominent spectral features which may be used for abundance estimates. We defined six spectral indices, two of which measure the Mg I b triplet and the remaining four measure essentially iron and iron-peak elements. The location of the indices may be found in Fig.1 of reference[10] . The indices are all measured with respect to a common pseudo-continuum, defined by six quasi-continuum windows. The aim is to be able to determine both [Fe/H] and [Mg/Fe].

We computed the values of the indices for a small grid of synthetic spectra (T=4750 K, 5000 K, 5250 K; log g = 2.50 [Fe/H]= $-1.5, -1.0, -0.5, 0.0, +0.5$; $\xi = 2kms^{-1}$) computed with the SYNTHE code [7]. The spectra where broadened with a gaussian profile of 210 kms^{-1} to match the resolution of the observed spectra. The input model-atmospheres were computed with version 9 of the ATLAS code[7] switching off the overshooting option. Models with [Fe/H]= -1.5 and -1.0 were computed using $\alpha-$enhanced opacity distribution functions. For each star the temperature was determined from the $(V-I)_0$ colour[9] through the calibration of[1] . Strictly speaking this calibration is valid only for dwarf stars, however the $V-I$ colour has only a weak dependence on gravity, especially in the colour range we are interested in. We excluded from analysis SgrM 172 ($V-I=1.81$), because it is much cooler than all the other stars. The metallicity was determined from the four iron-sensitive indices by interpolating in the table of computed indices.

3.1 Zero-Points

We made the same analysis for the Sun using the solar atlas[8], degraded at the same resolution as our grism spectra, as an observed spectrum and a small grid of synthetic spectra appropriate for the Sun (T=5777, log g =4.44, [Fe/H]= $-0.5, 0.0, +0.5$). The difference between these derived metallicities and the meteoritic iron abundance, is small (around 0.1–0.2 dex). On the assumption that these differences reflect inadequacies of our atomic data and model atmosphere and that the systematics is the same for the Sun and for our program stars, we treated these as zero-point shifts.

3.2 Cross-Checks

We determined the metallicity of the Sun from a twilight spectrum taken with the MOS. The deduced [Fe/H] was $-0.32, -0.04, -0.01, +0.10$ for the four indices respectively. These are not 0.00 because of the noise in the observed spectrum. One of the indices (F1) yields an abundance considerably different from the others. To obtain the final metallicity we decided to use a central estimate, defined as the mean of the two central values, which is akin to the median and more robust than the mean. In this case the central estimate is [Fe/H]= -0.02 , in good agreement with the expected value of 0.00.

Our program stars are considerably cooler and more luminous than the Sun. We analysed the spectrum of the cool metal-poor giant HD 190287 ([Fe/H]= -1.34, $T_{\rm eff} = 5300$), taken with EMMI, grism # 5 and a long-slit. We obtain [Fe/H] = $-1.52, -1.22, -1.23, -1.37$, for the four indices, which yields a central estimate [Fe/H]= -1.30.

We compared a spectrum of one of our program stars (SgrM 139) through the EMMI long-slit with its MOS spectrum. The derived metallicities were the same ($[\rm Fe/H]_{MOS} = -0.05$; $[\rm Fe/H]_{LS} = +0.02$) and also the radial velocities were consistent within errors ($v_{MOS} = +147\rm kms^{-1}$; $v_{LS} = +124\rm kms^{-1}$).

3.3 Error Estimates

In order to estimate the random errors in our metallicity estimates we used a Monte Carlo simulation. We took a synthetic spectrum ($T_{\rm eff}$ = 5000K [Fe/H]= -0.5) to which we added Poisson noise so that S/N = 10 (i.e. the S/N of our spectra) and estimated the metallicity from this noisy spectrum. The mean of 10000 realizations was $<$ [Fe/H] $>= -0.56$ and the standard deviation was 0.25 dex, we take this as our estimate of the random error.

In addition one should consider the errors arising from uncertainties in the atmospheric parameters. The largest error comes from the microturbulent velocity. At this resolution we are forced to rely only on strong lines which are very sensitive to microturbulence. The value of ξ must be assumed since there is no way to fix it from the spectrum. The associated error is larger for the more metal-rich stars, if the microturbulence is 1 $\rm kms^{-1}$ the metallicities increase by $\Delta([Fe/H]) = 0.24 \times [Fe/H] + 0.38$. A change in effective temperature by 150 K brings about a change of 0.2 dex in metallicity as does a change of surface gravity by 0.5 dex.

3.4 Previous Results and Future Work

Preliminary results of this work have been already presented[10,2]. There are two main differences with the results presented here. The first is that here we use pre-tabulated synthetic indices rather than iteratively performing spectrum synthesis for each star and each index. The second is that in the present work the indices are measured with respect to the pseudo-continuum, rather than with respect to an estimate of the true continuum, obtained by iteratively re-normalizing the observed spectrum using the current estimate of the best-matching synthetic spectrum, as done previously. The re-normalization procedure was abandoned because we realized that it is strongly dependent on the initial metallicity estimate. In fact it tends to relax the observed spectrum onto the first guess. Differences on the order of 0.2 dex are obtained from the same observed spectrum but initial metallicity guesses differing by 0.5 dex. Our present results are in agreement with our previous estimates at the level of 0.1 dex except for the two most metal-poor stars (SgrM 124 and SgrM 115) for which we find a metallicity which is 0.32 and 0.41 dex lower.

We have so far not determined [Mg/Fe] because the indices M1 and M2 depend on both [Mg/H] and [Fe/H]. Our grid of synthetic spectra includes only "standard" values of [Mg/Fe] (0.0 down to [Fe/H]= −0.5 and +0.4 below). On the other hand, our preliminary results suggest different [Mg/Fe] ratios, especially for $-0.5 < [Fe/H] < 0.0$. On the contrary the F indices are essentially independent of the abundance of α elements. Determining [Mg/Fe] is still possible, once [Fe/H] is fixed through the F indices, but requires a grid of synthetic spectra computed for a range of [Mg/Fe] values or direct spectrum synthesis on a star by star basis. This shall be the object of our future work.

4 Results on Sagittarius

The metallicity distribution of our 22 candidate Sagittarius members is shown in Figure 1. The peak at metallicity +0.5 reflects our choice to avoid extrapolation at metallicities higher than our most metal-rich grid point; when a star showed indices indicative of higher metallicity we assigned to it the value +0.5.

The distribution appears to be distinctly tri-modal: there is a metal-rich population whose mean metallicity is yet undefined and there are two metal-poor populations which peak around −1.2 and −0.5 respectively. The existence of the two metal-poor populations is in substantial agreement with previous findings[9]. The presence of a metal-rich population is something new and unexpected. It is not yet clear whether the metal-rich stars belong to Sagittarius or are Bulge interlopers, although in the latter case their number is uncomfortably large and may require a revision of our understanding of Bulge kinematics. An independent study based on Keck High-Res spectra[11], finds that 2 out of 7 stars, with radial velocity compatible with Sagittarius membership, are in fact metal-rich. Yet another photometric survey[3] suggests the possible existence of a RGB sequence considerably more metal-rich than 47 Tuc.

Although we are at an early stage to draw definitive conclusions, circumstantial evidence is arising for a metal-rich population in Sagittarius. This does not fit with our present understanding of chemical evolution of dwarf spheroidals, however we should keep an open mind and stick to the observations.

5 Perspectives for the Use of VLT

VLT will quite likely play a fundamental role in the study of Local Group Galaxies and Sagittarius in particular. With UVES it shall be possible to do a detailed line by line analysis for stars along the RGB to below the Horizontal Branch ($V_{HB} \approx 18.2$). The Turn-off $21 \leq V \leq 21.5$ will still be out of reach of UVES though.

New perspectives are opened by FORS. The resolution shall be at most 1/2 of that used by us at NTT, however the indices we have defined measure such strong features that they will be usable even at this lower resolution, as shown by simulations we carried out. Moreover the MOS slitlets of FORS have sharp edges and it will be possible to flux-calibrate the spectra, thus allowing to measure spectral indices in absolute flux. This will recover all the information contained the continuum. The spectral coverage (350 - 590 nm) will allow to use the Ca II H and K lines and probably other iron-related indices in the range 400 - 500 nm . Finally with VLT+FORS a S/N of about 100 can be reached in one hour for stars of $V \approx 18.5$, with this S/N our Monte Carlo simulations predict the random error in metallicity determinations to be around 0.02 dex.

The use of a 3m class telescope, such as NTT, for preliminary observations aimed at the determination of radial velocities has proved to be too time-consuming to be worth the effort. FORS will be able to provide good quality spectra from which radial velocities, metallicities and key abundance ratios, such as [Mg/Fe] and [Ca/Fe], can be determined.

The real killer in the studies of Local Group galaxies shall be Giraffe fed by 130 fibres in MEDUSA mode. The highest resolution achieved by Giraffe (15000) is high enough to perform a classical line by line analysis. However the number of spectra we shall get (several hundreds per night) will preclude the use of this approach. We will have to find new ways to estimate abundances from these spectra which will be able to cope with the data flow provided by Flames+Giraffe.

One possibility is suggested by our low-resolution experience: define spectral indices that can be related to abundances of particular elements and tabulate their value from synthetic spectra. The estimation of an abundance requires then only an interpolation in an appropriate table.

Different indices will be appropriate for different temperature and luminosity regimes. Now is the time to investigate both theoretically and observationally which features will provide useful indices. For example some indices may be defined to measure essentially strong iron lines others essentially weak iron lines, the balance of the two will allow to determine microturbulence.

The index approach is not the only possibility. Other very promising techniques such as autocorrelation and cross-correlation should be investigated. A whole range of methods could complement each other to extract information from the spectra, however if any particular star seems deserving of special attention, all the spectra will be nicely archived and one can go back and do the good old line by line analysis.

6 Acknowledgements

I wish to thank G. Marconi, P. Molaro and L. Pasquini for allowing me to present results of common work in advance of publication. This work is based on observations collected at the European Southern Observatory, Chile, ESO N° 57.E-0586.

References

1. Alonso A., Arribas S., Martinez-Roger C. (1996) The empirical scale of temperatures of the low main sequence (F0V-K5V). A& A 313:873
2. Bonifacio P., Pasquini L., Molaro P., Marconi G. (1998) Spectroscopy of giants of the Sagittarius dwarf galaxy, In:Galaxy evolution:Connecting the distant Universe with the local fossil record Paris, Meudon 21-25 September 1998
3. Bellazini M., Ferraro F.R., Buonanno R. (1999) The Sagittarius Dwarf Galaxy Survey (SDGS) I MNRAS, in press astro-ph/9812344
4. Ibata R.A., Gilmore G., Irwin M.J. (1994) A dwarf satellite galaxy in Sagittarius. Nature 370:194
5. Ibata R.A., Gilmore G., Irwin M.J. (1995) Sagittarius: the nearest dwarf galaxy. MNRAS 277:781
6. Ibata R.A., Wyse R.F.G., Gilmore G., Irwin M.J., Suntzeff N.B. (1997) The Kinematics, Orbit, and Survival of the Sagittarius Dwarf Spheroidal Galaxy. AJ 113:634
7. Kurucz R.L. (1993) SAO CD-ROM No. 13,18
8. Kurucz R.L., Furenlid I., Brault J., Testerman L. (1984) Solar Flux Atlas from *296 to 1300 nm*. *Nat. Sol. Obs.* Atlas No.1

9. Marconi G., Buonanno R., Castellani M., Iannicola G., Molaro P., Pasquini L., Pulone L. (1998a) The stellar content of the Sagittarius Dwarf Galaxy. A& A 330:453
10. Marconi G., Bonifacio P., Pasquini P., Molaro P. Spectroscopy of red giants of the Sagittarius dwarf galaxy (1998b) In: The Stellar Content of Local Group Galaxies, conference held in Cape Town, South Africa, 7-11 September 1998, to be published by ASP, edited by Patricia Whitelock and Russell Cannon, p. 8.
11. Smecker-Hane T., McWilliam A., Ibata R.A. (1998) Chemical Abundances in the Sagittarius Dwarf Galaxy. AAS 192:6613

The Old Population of the Magellanic Clouds

Monique Spite[1] and Vanessa Hill[2]

[1] Observatoire de Paris-Meudon
[2] ESO (Garching)

Abstract. The brightest old stars in the LMC are as faint as V=16 (V=16.5 in the SMC). The VLT with UVES or GIRAFFE will provide, for the first time, high resolution spectra of old stars in the Magellanic Clouds, which will enable to measure abundance ratios of different elements. We show that, for example, observing the variation of the [O/Fe] ratio as a function of [Fe/H], strongly constrains the SFR and the IMF in the Magellanic Clouds.
As another example, we show that, thanks to the high multiplex capabilities of GIRAFFE (and even FORS), it will be possible to determine the metallicity distribution and the velocity dispersion of large samples of the very old RR Lyrae stars, and thus to check the SFR, and discuss the existence of a pressure supported halo around the LMC.

1 Introduction

The VLT at Paranal will be the first 8m class telescope able to observe the Magellanic Clouds, and thus, it will be an opportunity for ESO to add a major contribution to our knowledge of these two close-by galaxies. In this talk I will insist on the main advances that are expected to come from the study of the old stellar populations of the Magellanic Clouds with the spectrographs UVES and GIRAFFE (on KUEYEN) and also with FORS (on ANTU).

Among the young stars, many bright massive stars still survive (with $age < 10^8 yrs$ $M > 10 M_\odot$ and $V < 14$). High resolution spectra of these stars (hot dwarfs and giants or cool supergiants) could be readily obtained with 4m class telescopes and thus the chemical composition of the young stars in the Magellanic Clouds is rather well known. In particular, it has been found that the metallicity of these young stars of the LMC is about twice lower than the metallicity of the young stars of our Galaxy. The factor is about 4 for the young stars of the SMC.

But even the brightest star of the old population of the Magellanic Clouds ($ages > 8 - 15.10^9 yrs$) is about 4 magnitudes fainter that the young population: only stars with masses lower than $1 M_\odot$ remain (all the brighter, more massive stars of these ages have long died). However, with UVES or GIRAFFE, it will become possible to obtain high resolution high S/N spectra of these faint stars.

The study of the old population of the Magellanic Clouds will strongly constrain the Initial Mass Function and the Star Formation Rate, quantities which are critical for building models of galaxy evolution.

In the following discussion, we will mainly discuss the case of the Large Magellanic Cloud.

2 The Evolution of the Magellanic Clouds

The chemical evolution of the Magellanic Clouds depend on two fundamental parameters : the initial mass function (IMF) and the Star Formation Rate (SFR).

2.1 The Initial Mass Function

Adopting the nomenclature of Scalo (1986), with $n(M)\ dM \propto M^{\gamma-1}$, it has been found in different systems of our Galaxy that $\Gamma \approx -1.35$ (cf. in particular Elmegreen (1997). The most direct way to determine the IMF is to estimate the masses of stars in a cluster (or association) : since all the stars in a given cluster have the same age, they differ only by their masses. From a direct comparison of the color magnitude diagram to theoretical evolutionary tracks it is possible to deduce the masses of the stars. The most important sources of uncertainty are the errors on the evolutionary tracks and on the conversion "color/effective temperature".

Before 1994, most of the measurement pointed toward a slope of the IMF steeper in the LMC than in the Galaxy (fewer massive stars formed compared to lower mass stars). For example Hill et al.(1994) found from UBV photometry of stars in associations, $\Gamma \approx -2$. But a little later, Massey et al. (1995) improving the determination of the temperature of the hottest stars of the associations, found $\Gamma \approx -1.3 \pm 0.3$. This value of the IMF in the LMC has been confirmed by HST photometry of clusters in the LMC: in R136 Hunter et al. (1995) measured $\Gamma \approx -1.22 \pm 0.06$ and in NGC 1818 Hunter et al. (1997) measured $\Gamma \approx -1.23 \pm 0.08$

2.2 The Star Formation Rate

For a long time, it was assumed that in the LMC, the cluster formation rate reflected the star formation rate (Baade, 1963). On the other hand, it was believed that the probability for a LMC cluster to be disrupted was very low, because of the small number of molecular clouds in the LMC. As a consequence, it was thought that a direct counting of the clusters as a function of time would give a good representation of the evolution of the star formation rate. But it is now well established that in the LMC, there are two groups of clusters, one born about 15 Gyrs ago and the other one spanning over the past 3 Gyrs (Fig. 1, following Geisler et al. 1997). There are practically no clusters with ages between 5 and 12 Gyrs. This lack of clusters corresponds to a lack in metallicity: no clusters have metallicity in the range $-1.8 < [Fe/H] < -1.1$dex (Fig. 1).

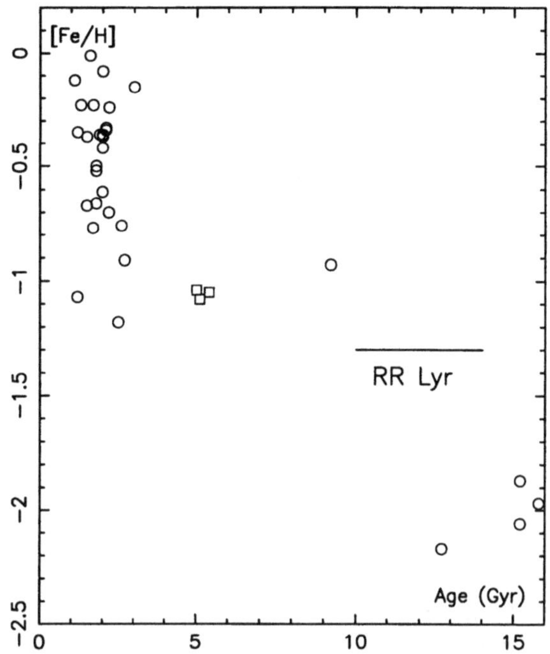

Fig. 1. The metallicity of the clusters of the LMC (following Geisler et al 1997 [open circles], and Sarajedini (1998) [open squares] is compared to the metallicity of the RR Lyrae stars following Hill & Beaulieu 1997.

However Hill and Beaulieu (1997) who studied the RR Lyrae stars discovered in the LMC as part of the EROS experiment, could show (by an indirect method) that these old stars (their age is larger than 10 Gyrs) have a mean metallicity of about [Fe/H] $= -1.3$dex, just corresponding to the gap in the clusters metallicity distribution (Fig. 1). In a recent paper van den Bergh (1998) remarks that there are no clusters between 12 and 5 Gyrs but during this period the metallicity has increased by a factor of about 4. This factor is difficult to explain if no stars have been formed.

It is also possible to deduce the star formation rate from the color magnitude diagrams in different regions of the LMC. With the HST, accurate diagrams can be obtained several magnitudes below the main sequence turnoff, these observations provide a significant improvement over ground-based studies (Holtzman et al. 1997, Geha et al. 1998). With a given IMF, different color magnitude diagrams are simulated with different star formation history and compared to the observations. Following Geha et al., the best model is obtained with a star formation rate which remains roughly constant for most of the LMC's history and then increases by a factor of 3 about 2 Gyr ago (the slope of the IMF has been taken constant and equal to −1.35). This model predicts that a significant metal poor population exists in the LMC, about 20% of the stars should have a metallicity lower than [Fe/H]=−1.3dex.

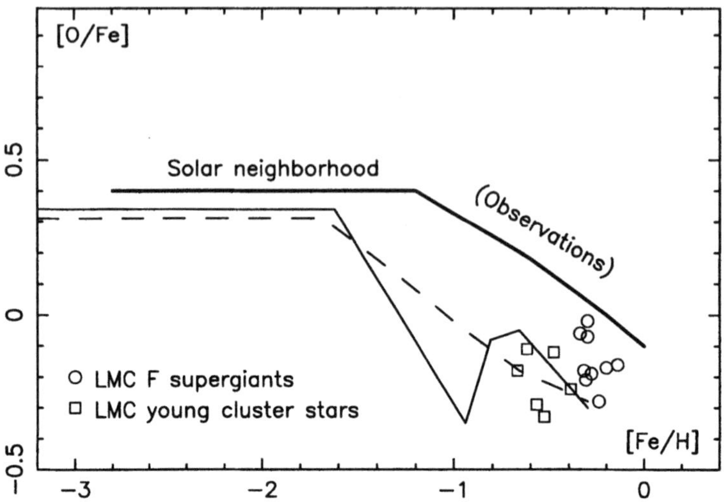

Fig. 2. The predictions of the models of Tsujimoto et al. computed with slightly different IMF and two different SFR are compared to the observations of the ratio [O/Fe] in the LMC. In the first model (full line) it has been assumed that an initial burst occurs from the epoch of galaxy formation, with a duration of 1 Gyr, and that a quiescent phase of negligible star formation continues until the last 3 Gyr . In the second model (dashed line), the star formation is assumed to be continuous. To constrain the parameters of the models it is necessary to observe the ratio [O/Fe] in more metal deficient stars of the LMC as it has been done in our Galaxy.

3 The Abundance Ratios in the Magellanic Clouds

The chemical composition of the stellar atmospheres reflects the chemical composition of the interstellar medium from which the stars have been formed. As a consequence, from the chemical composition of stars of various ages, it is possible to deduce the chemical composition of the interstellar medium at different times. The ISM is enriched little by little, by material expelled by supernovae explosions and stellar winds, and the chemical composition of the material expelled depends on the mass of the progenitor star. For example, the massive type II supernovae will expel oxygen rich material, whereas the less massive type I supernovae will expel iron rich material and thus the ratio [O/Fe] will depend on the initial mass function and on the star formation rate at different epochs.

It is now well known that *young* stars of the Magellanic Clouds are metal deficient relative to the young Galactic stars. In the LMC [Fe/H]≈ -0.3 and in the SMC [Fe/H]≈ -0.7dex. However, the star formation started at about the same time in the LMC and the Milky Way: the ages of the oldest globular clusters have been found very similar, (about 15 Gyr), in the two galaxies (Olsen et al. 1998). The star formation started perhaps a little later in the

SMC: following Mighell et al. (1998) the age of the oldest globular cluster in the SMC, NGC 121, is about 12 Gyr.

From a careful study of the abundance ratios in F and K supergiants of the SMC and the LMC, Hill et al. (1995, 1999) could show that the ratio [O/Fe] was the same in the young supergiants of the SMC, the LMC and the Milky Way. In (Fig. 2) these observational results for the LMC are compared to some models of Tsujimoto (1995) computed with different IMF and SFR. It is clear that the observations of the young stars alone (which give a snapshot of the chemical composition *at present time*), cannot constrain the models. It is necessary to determine element ratios such as [O/Fe] also in older, more metal-poor stars of the LMC. The value of [O/Fe] at early times in the LMC ([Fe/H] $< -2.$dex) is very sensitive to the IMF, whereas the shape of the relation depends mainly on the SFR.

With UVES and later GIRAFFE, it will become possible to obtain high resolution spectra of old K giants in the Magellanic Clouds. It will then become possible to study the ratios of the different elements in the Clouds and thus to constrain the IMF and the SFR.

4 The RR Lyrae Stars and the History of the SFR in the LMC

The RR Lyrae are very old low mass stars ($M \approx 0.8 M_{\odot}$) which cross the instability strip on the horizontal branch, and thus pulsate. Their period is less than 1 day, their age $>$ 10Gyr and their mean absolute magnitude $< Mv > \approx 0.5$. They are used as distance indicators and also as tracers of chemical and dynamical properties of old stellar populations.

It has been shown that, in a cluster, the mean period of the RR Lyrae is a function of the metallicity of the cluster (cf. for example Sandage 1993). From an analysis of the period distribution of the RR Lyrae stars detected in the LMC during the EROS and MACHO experiment (search for microlensing events), it has been shown that the mean period of the RR Lyrae in the LMC is about 0.54d, a period which corresponds to a metallicity of about −1.3dex (see for example Hill & Beaulieu 1997). This metallicity corresponds almost exactly to the gap in metallicity distribution of the LMC clusters (Fig. 1). The simplest interpretation is to suppose that the bulk of the RR Lyrae stars (and thus of the old population of the LMC) has been formed later than the oldest clusters of the LMC and thus, would confirm a more continuous SFR in the LMC.

It would be very important to check directly the metallicity distribution of statistically significant samples of these variable stars, this will become possible with the VLT, thanks to the high multiplex capabilities of GIRAFFE (and even FORS, although is has a lower spectral resolution). The RR Lyrae in the LMC are very faint objects (between 19 and 20) but their metallicity can be deduced from low resolution spectra by the the ΔS method (comparison of the intensity of the Hydrogen lines and the Calcium K line).

5 Conclusion

With the VLT, it will become possible to check the influence of the IMF and of the SFR on the chemical evolution on the Magellanic Clouds. If it is confirmed that the star formation has increased much more monotonically than the cluster formation, it will remain to be explained why no cluster have been formed during more than 6 Gyr (or why they have been destroyed). Following van den Bergh (1998), strong shocks due to cloud-cloud collisions might promote the formation of clusters. These bursts of cluster formation could be triggered by dynamical interactions between the LMC the SMC and the Galaxy.

References

1. Baade, W. (1963) Evolution of stars and galaxies (Cambridge, Harvard University Press
2. Elmegreen, B. G. (1997) ApJ 486, 944
3. Geisler, D., Bica, E., Dottori, H., Claria, J. J.,Piatti, A. E., Santos, J. F. C. Jr (1997) AJ 114, 1920
4. Geha, M. C., Holtzman, J. A., Mould, J.R., Gallagher, J. S. III, Watson, A. M. et al. (1998) AJ 115, 1045
5. Hill, R. J., Madore, B. F., Freedman, W. L. (1994) ApJ 429, 204
6. Hill, V., Andrievsky, S., Spite, M. (1995) A&A 293, 347
7. Hill, V., Beaulieu, J.-P. (1997) in "Variable stars and the astrophysical returns of the microlensing surveys", R.Ferlet , J.-P.Maillard & B.Raban eds, Editions Frontières p. 267
8. Hill , V., Spite, M. (1999) in "Galaxy evolution: Connecting the distant Universe with the Local Fossil Record", M.Spite ed., Kluwer, in press
9. Holtzman, J. A., Mould, J.R., Gallagher, J. S. III, Watson, A. M., Grillmair, C. J. et al. (1997) AJ 113, 656
10. Hunter D.A., Shaya, E. J., Holtzman, J. A., Light, R. M., O'Neil, E. J., Lynds, R. (1995) ApJ 448, 179
11. Hunter, D. A., Light, R. M., Holtzman, J. A., Lynds, R., O'Neil, E. J., Grillmair, C. J. (1997) ApJ 478, 124
12. Massey, P., Lang, C. C., Degoia-Eastwood, K., Garmany, C. D. (1995) ApJ 438, 188
13. Mighell, K. J., Sarajedini, A., French, R. S. (1998) AJ 116, 2395
14. Olsen, K. A. G., Hodge, P. W., Mateo, M., Olszewski, E. W., Schommer, R. A., Suntzeff, N. B., Walker, N. R. (1998) MNRAS 300, 665
15. Sandage, A. (1993) AJ 106, 687 and 703
16. Sarajedini, A. (1998) AJ 116, 738
17. Scalo, J. M. (1986) Fund Cosmic Phys. 11. 1
18. Tsujimoto, T., Nomoto, K., Yoshii, Y., Hashimoto, M., Yanagida, S., Thielemann, F.-K. (1995) Mon. Not. R. Astron. Soc. 277, 945
19. van den Bergh, S. (1998), ApJ 507, L39

The Deep Thick Disk as Seen by UVES

Klaus Fuhrmann

Universitäts-Sternwarte München

Abstract. On the base of recent results from high-resolution spectroscopy and solid space-borne astrometry for a sample of nearby F- and G-type stars, some prospects are discussed for an *in situ* study of the Galactic thick-disk population with UVES.

1 A New Horizon

Stellar population studies are one of the most promising means in galactic astrophysics. The potentially provide the detailed record of our Milky Way Galaxy, its formation and evolution, and can give important clues to our understanding of extragalactic sources as well. The stumbling-block is, of course, first to identify and separate the various entities on stage. This is in fact an enormous task and involves questions such as

- *Is the Galactic disk made up of separate "thin" and "thick" components, and if so, what is their relationship?*
- *Is there only a single halo population, or do we have to approve the existence of at least a "low" and "high" halo, prograde and/or retrograde rotating populations?*
- *Which is the role played by Galactic globular clusters? How are they associated with the disk and halo population(s)?*
- *What can be learned from old open clusters with respect to the formation of the thin disk?*
- *Is the thin disk itself a monolithic entity, or does it disintegrate into a "young" and "old" thin disk?*
- *What is, or has been, the role of mergers, such as Sagittarius, or other dwarf galaxies?*
- *What is the nature of the Galactic bulge, and how is it related to the other populations?*

the list is easily extensible, but we shall mainly focus here on the Galaxy's thick-disk population. Since it became popular in 1983 by Gilmore & Reid's star count analyses at the south Galactic pole, it progressively turned out to be the population that tips the scales for an improved understanding of the ancient epoch of the Milky Way Galaxy. Until recent, photometric surveys were the only means to study the spatial properties of unevolved stars of this population *in situ.* The VLT's UVES spectrograph now offers a new horizon for a deep spectroscopic study of faint, long-lived F and G dwarfs at high resolution *and* high signal-to-noise in the southern hemisphere.

2 Is the Milky Way's Thick Disk a Distinct Population?

Many photometric and spectroscopic surveys have shown, and presumably everybody agrees upon, that the Galactic disk stars are embedded in a "pressure"-supported halo, which dominates the stellar content on large scales and is clearly a separate entity. We refer in this context to the diagrams that have been published by Schuster, Parrao & Contreras Martínez (1993, their Fig. 5) and Carney et al. (1996, their Fig. 1), in which large samples of stars ($N > 10^3$) are compared for their Galactic rotational velocities versus metallicity and which convincingly demonstrate the discreteness of the disk and halo.

The question however arises whether the disk itself is separable into a thin and thick component. Although we know that a single ~350 pc vertical scale height population cannot account for the observations at the Galactic poles (e.g. Majewski 1992), the fact that thick- and thin-disk stars have somewhat overlapping properties in kinematics and chemistry repeatedly caused doubts as to whether the thick disk has a disjoint nature, or is instead the mere extension of the thin disk. This degeneracy is however overcome by the introduction of *abundance ratios* from spectroscopy and, in particular, by careful stellar *age determinations*, which have recently become possible on the base of accurate space-borne astrometry in combination with improved stellar interior calculations (cf. Bernkopf 1998).

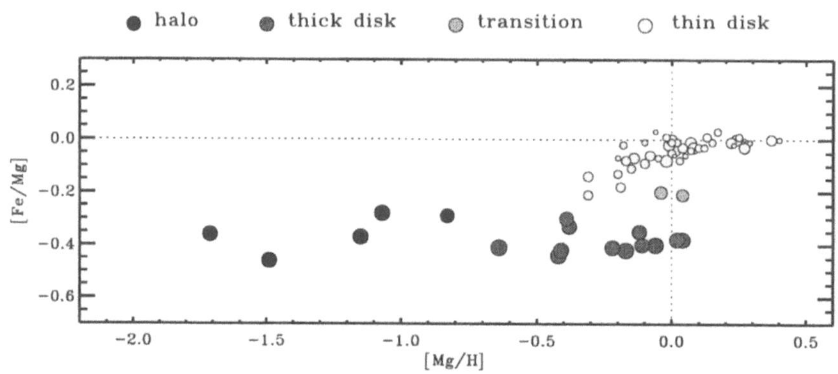

Fig. 1. Abundance ratios [Fe/Mg] vs [Mg/H] from a high-resolution spectroscopic study of nearby F and G stars. Circle diameters are in proportion to the stellar ages. Different *stellar populations* are given with various grayscale symbols as indicated in the legend on top, and are based on chemistry, kinematics and age (see text for details)

Thus we suggest the separability of thick- and thin-disk stars in the [Mg/H]-[Fe/Mg] abundance plane of Fig. 1 for a recent high-resolution spectroscopic study of ~60 F and G stars, most of which are long-lived objects (cf.

Fuhrmann 1998). The stellar ages are introduced to this diagram by small (large) circle diameters for young (old) objects. The data of Fig. 1 immediately imply a stellar metal content to be anything but a clock (the most metal-poor halo star is as old as the most metal-rich thick-disk star). Instead, we rather identify the *relative* iron deficiency, i.e. low [Fe/Mg] values, to be much better correlated with age. The basic information that allows for the suggested separation of the disk populations is in fact age, since all those stars that are labeled as thick-disk objects are older than ~12 Gyr, whereas the thin-disk stars do not exceed ~9 Gyr. The resulting star formation gap (there are only two "transition" stars with intermediate ages in Fig. 1) thereby explains the good chemical separability of thick- and thin-disk stars in the [Mg/H]-[Fe/Mg] abundance plane, since the production of iron via type Ia supernovae must have been an ongoing process in that phase.

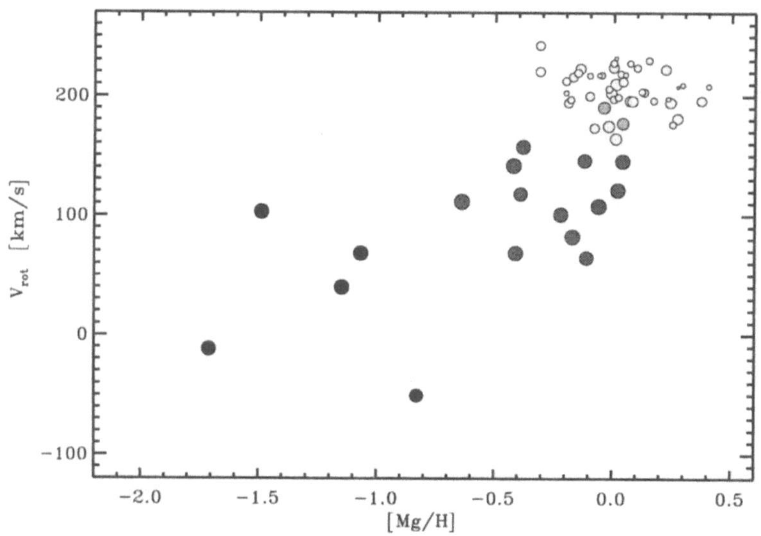

Fig. 2. Galactic rotational velocities versus metallicity. Grayscaling and symbol sizes have the same meaning as in Fig. 1

In Fig. 2 we now proceed with the discussion of the stars' Galactic rotational velocities $V_{rot} = V_{lsr} + 220$ km s^{-1} in the [Mg/H]-V_{rot} diagram. A striking feature herein is that both disk populations are much better discernible than might have been expected on the base of the above mentioned large surveys. On closer inspection of the tabulated space velocities in Carney et al. (1994), it turns out that thick-disk stars in common with our analysis have a mean offset of $\langle \Delta V_{rot} \rangle \sim -20$ km s^{-1}. Thus there is direct evidence for an *asymmetric drift of no* less than −50 km s^{-1}, instead of −30 km s^{-1}, as favoured by Carney, Latham & Laird (1989). In fact, the analysis of all identified disk stars of our sample even results in $\langle V_{rot} \rangle = 112$ km s^{-1} for the thick disk, compared to $\langle V_{rot} \rangle = 205$ km s^{-1} for the thin disk.

At present the most vulnerable point of the above given results is however the small sample size: most of the investigated stars are nearby and it is therefore only the thin-disk population which is reasonably represented. A more detailed study of the properties of the thick-disk stars, independent of any biases in metallicity or kinematics, inevitably requires much deeper surveys on scales of kiloparsecs.

3 Faint Stars in UVES' Net

Whereas spectroscopic studies of F and G dwarfs at high resolution and high signal-to-noise were often restricted to the solar neighborhood (<100 pc), a detailed analysis of a representative sample of thick-disk stars should preferably commence no less than three vertical scale heights above the thin-disk, i.e. at least 1 kpc above the Galactic plane. Fig. 3 gives an illustration of the relative fractions of disk and halo stars that might be expected in response to the gravitational sieve (note however that the actual distribution of stars has a strong dependence on the rather ill-known local normalization of the thick disk). Thus at a distance modulus $(m - M) = 10$ the typical apparent magnitudes of long-lived turnoff stars will be close to $m_V \sim 14$, that is, easily within reach for the UT2+UVES (cf. ESO 1998) at R >50000 and S/N >100.

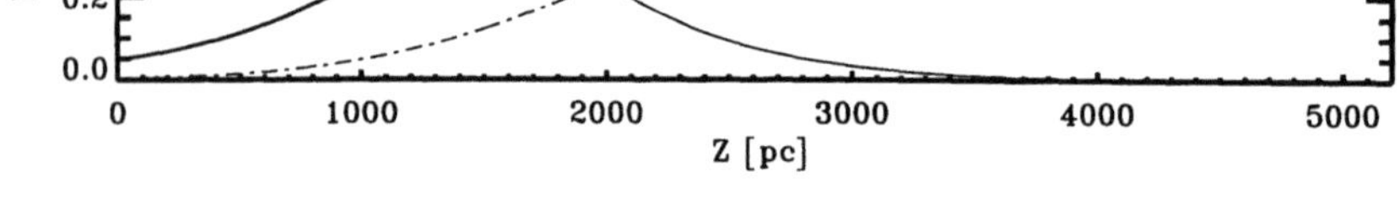

Fig. 3. Relative fractions of disk and halo stars as a function of distance Z from the Galactic plane. Local normalizations and density laws are indicated on top

Consequently *the derived stellar parameters can be as accurate as for very nearby stars.* But what does this mean quantitatively? How trustworthy are, for instance, the distances derived from spectroscopy, which are a necessary datum to obtain the stars' space velocities? And what is the uncertainty of the deduced stellar luminosities and effective temperatures for the stellar age-datings? What means do we have to identify spectroscopic binaries?

An overall framework to tackle these questions for F and G stars has recently become available; its name is: Hipparcos. Though rather restricted to nearby stars, the wealth of data from this astrometric satellite is indeed overwhelming. It not only revolutionizes the accuracy in stellar age-datings from "*qualitative to quantitative*" for many stars, but also helps to scrutinize the uncertainties in surface gravity, effective temperature and mass, and thereby, radius, metallicity and distance. The latter information is particularly valuable for the stars' kinematics, and to detect and discard spectroscopic binaries or other "enigmatic" objects from a comparison of spectroscopic and astrometric distances.

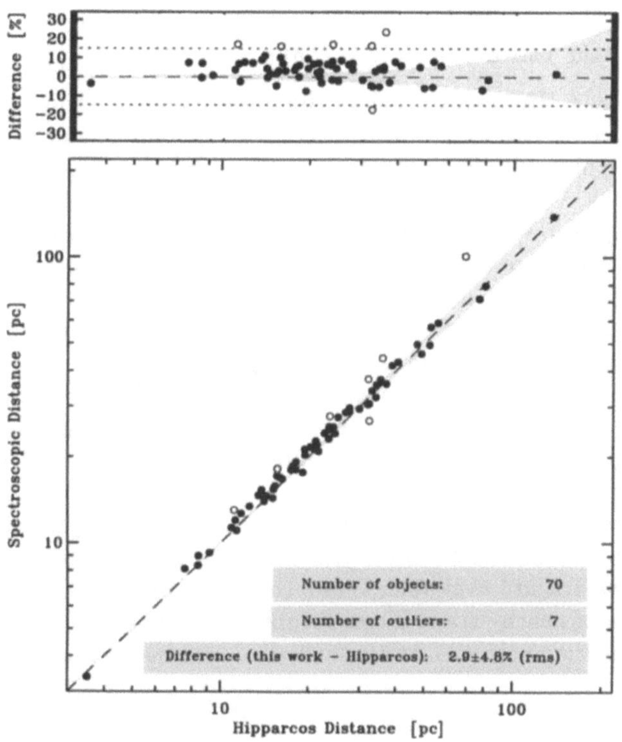

Fig. 4. Comparison of the spectroscopic versus astrometric distance scale. The Hipparcos data have typical parallax errors of 1 mas, as illustrated by the shaded wedge. Individual uncertainties of the spectroscopic data should not exceed 15% (dotted curve in the upper panel), with no dependence on distance. The spectroscopic distances result in ***higher*** values by 2.9% on average, and a statistical ***rms*** error of 4.8%. Outliers (e.g. spectroscopic binaries) are marked by open circles

This comparison is shown for our sample of stars in Fig. 4. As is immediately evident, there is a good correlation that also allows for an identification of outliers (open circle stars). Hence, what can be learned from Fig. 4 is that (for a more detailed discussion, see Fuhrmann 1998), in principle,

- the spectroscopically derived surface gravities are indeed very trustworthy (i.e. $\Delta \log g \leq 0.1$ dex), for this stellar parameter dominates the overall accuracy of the displayed correlation
- the effective temperature scale cannot be systematically affected for those stars, where log g is obtained from the LTE iron ionization equilibrium, since there is a close coupling of log g and T_{eff}. This applies to all stars as hot as, or cooler, than the Sun (about 50% in Fig. 4)
- the well-determined surface gravities give rise to trustworthy stellar metallicities
- the spectroscopic distance scale is systematically affected by less than 3% (which corresponds to less than 0.03 dex in $\Delta \log g$), and, most important, the *rms* uncertainty is confined to 5%

In other words, Hipparcos provides the robust reference frame that places much confidence in the forthcoming studies of *distant* F and G dwarfs. In particular, we point out that accurately determined spectroscopic distances, characterized by a 5% *rms* uncertainty, are indeed realizable on Galactic scales, as long as UVES produces the high resolution, high S/N spectra, and provided that there is no, or only weak, interstellar absorption, which is however the case at the Galactic poles.

Although the question of spectroscopic binaries is of course a matter of concern, high resolution + high S/N is again a good means to detect many of these objects: our study of the nearby thick-disk (and halo) stars provides evidence that most (if not all) of these objects have negligible rotational velocities ($v\sin i < 2$ km s^{-1}) and chromospheric activities, unless binarity gives rise to the contrary.

In summary, the above results suggest that accurate analyses of long-lived F- and G-type dwarfs can be done with standard analyses techniques, once high quality spectra are available. While this assessment is largely based on Hipparcos data of nearby stars, a comparably accurate study of more distant objects, such as the ancient Galactic thick disk, is *now* certainly within reach for UVES.

References

1. Bernkopf, J. (1998) A&A 332, 127
2. Carney, B.W., Latham, D.W., Laird, J.B. (1989) AJ 97, 423
3. Carney, B.W., et al. (1994) AJ 107, 2240
4. Carney, B.W., et al. (1996) AJ 112, 668
5. ESO (1998) Doc. No. VLT-MAN-ESO-13200-0000
6. Fuhrmann, K. (1998) A&A 338, 161
7. Gilmore, G., Reid, N. (1983) MNRAS 202, 1025
8. Majewski, S.R. (1992) ApJS 78, 87
9. Schuster, W.J., Parrao, L., Contreras Martínez, M.E. (1993) A&AS 97, 951

Abundances of Intermediate/Low Mass Stars in the Galactic Bulge

Beatriz Barbuy

University of São Paulo/Fapesp, C.P. 3386, São Paulo 01060-970, Brazil

Abstract. Detailed abundance analyses available for stars in the Galactic bulge are reviewed. These consist of 5 intermediate-mass stars in the Galactic center, 12 field stars in Baade's Window and a few stars in the bulge globular clusters NGC 6528 and NGC 6553. Clearly this is a field which will greatly benefit from the advent of high resolution spectroscopy with 8m class telescopes.

1 Introduction

The bulk of the bulge stellar population is old, showing metallicities in the range $-1.5 < [Fe/H] < +0.5$. At the same time there is a large number of Miras, which seem to be a mixture of young/intermediate age Miras, together with luminous Miras belonging to the old population.

It seems to us that the bulge stellar population is essentially old as predicted in models of galaxy formation "inside-out" (Larson 1990; Renzini 1999), but there seems to be as well a young, disk-like population, resulting from the dynamical evolution of a bar (Raha et al. 1992; Spergel et al. 1996; Pfenniger 1993).

If the bulge is formed rapidly in the beginning of the Galaxy life, a fast enrichment by SNs II at early times would result in a signature through [α-elements/Fe] enhanced. If the bulge is disk-like, then one would expect [α-elements/Fe] = 0.0.

In order to try to solve this question we are carrying out a program of spectroscopic determination of abundances in bulge stars.

2 Intermediate Mass Stars

The result of a wide range of metallicities is the presence of different kinds of evolved stars, as illustrated in Fig. 3 by Rich (1992).

The Miras in particular can be as well massive and of intermediate/young age, but also low-mass and old. Wood & Bessell (1983) proposed that the bulge contained a substantial fraction of 1 Gyr Miras, and Harmon & Gilmore (1988) suggested an important fraction of intermediate age Miras, based on IRAS surveys. Miras are a suitable tracer of the bulge structure, since they are bright in the infrared. According to Catchpole et al. (1999), Miras form a bar. This would be therefore probably a young, disk-like, intermediate mass class of Miras.

However, let us point out that Miras belonging to a metal-rich and old stellar population are also luminous (Guarnieri et al. 1997). Consequently, the two kinds of populations should be disentangled somehow before drawing conclusions about the bulge shape and age.

Very young massive cool stars present at the Galactic center were analysed by Sellgren et al. (1998) and Ramirez et al. (1998). They obtained IR spectra for a sample of the brightest stars right at the Galactic center: IRS 7, 11, 19, 22, 24, having obtained [Fe/H] = 0.0, typical of young disk stars.

3 Old Low-Mass Stellar Population

The metallicity distribution of low-mass field stars in Baade's Window (Rich 1988; McWilliam & Rich 1994) and that for bulge globular clusters within 5° of the Galactic center (Barbuy et al. 1998) are very similar.

Detailed determinations of metallicity and abundance ratios for bulge stars are not many, due to the difficulties to observe such faint and reddened stellar populations. Low resolution spectra obtained by Rich (1988) indicated a mean abundance for bulge stars of [Fe/H] = +0.25. Sadler et al. (1996) derived [Mg/Fe] ≈ +0.3 from measurements of the Mg_2 index in about 400 stars. Idiart et al. (1996) derived [Mg/Fe] = +0.45 from an integrated spectrum of Baade's Window.

McWilliam & Rich (1994, hereafter MR94) presented a high-resolution spectroscopic study of 12 bulge K giants of Baade's Window. Their main results are: [Fe/H] = -0.25, [Mg/Fe] = [Ti/Fe] = +0.3, [Ca/Fe] = [Si/Fe] = 0.0, [O/Fe] = 0.0 and [Eu/Fe] = slightly enhanced.

The interesting technique of observing during a microlensing event has been used by Lennon et al. (1996) to obtain high resolution spectra of a G0 IV-V dwarf. A metallicity around [Fe/H] ≈ +0.3 to +0.6 was derived.

4 The Bulge Globular Clusters NGC 6553 and NGC 6528

Ortolani et al. (1995) have shown that NGC 6528 and NGC 6553 should be nearly coeval with the halo clusters, and that the luminosity function of these clusters is very similar to that of field stars in Baade's Window, confirming that the clusters and bulge field form a same stellar population, making the clusters ideal laboratories for bulge studies.

Since 1990 we have been carrying out studies of globular clusters in the bulge, through imaging and spectroscopy. In Barbuy et al. (1998) we gathered the results obtained for all clusters within 5° of the Galactic center. In the present paper we present the first results on detailed abundances of these clusters.

5 Detailed Analysis and Abundance Ratios

A detailed analysis was carried out for two stars in NGC 6553 (Barbuy et al. 1999a) and one star in NGC 6528, using high resolution échelle spectra obtained at the ESO 3.6m telescope using CASPEC.

The effective temperatures were derived from HST VI and ESO 2.2m IRAC2 JK photometry. The reddening was corrected according to Guarnieri et al. (1998), where it was derived by comparing the colour of the subgiant branch of NGC 6553 to that of 47 Tuc. It is important to point out a correction to Table 5 of Barbuy et al. (1999a): the temperatures derived from (V-I) and (J-K) using relations by McWilliam (1990) are not correct in that table, and they should be around 4000 K, therefore in agreement with the final temperatures adopted. The effective temperatures so derived were checked for excitation equilibrium using FeI lines of different excitation potential.

An important matter is also the model atmosphere employed: we used a model atmosphere grid by B. Plez (private communication), which is an extension of the grid by Plez et al. (1992).

In order to derive the metallicity [Fe/H]: (a) we only used FeI and FeII lines for which laboratory oscillator strengths gfs by Fuhr et al. (1988) were available; (b) the measurement of equivalent widths was carried out by tracing a continuum close to the observed continuum, and to compensate for a possible slightly higher continuum, the curve-of-growth was traced along the upper envelope of equivalent widths - see Barbuy et al. (1999a).

Elemental abundances were derived by computing synthetic spectra line-by-line. In Table 1 are given the resulting abundance ratios for a mean of two stars in NGC 6553 (see more details in Barbuy et al. 1999a), one star in NGC 6528 (Barbuy et al. 1999b) and a mean of results for 12 bulge K giants by MR94. The results for NGC 6528 are now complete (see Barbuy 1999). Chemical evolution models for the bulge were presented by Matteucci & Brocato (1990), Matteucci et al. (1999), where a inside-out model for the formation of the Galaxy is adopted. An increased star formation efficiency, of a factor 10 relative to the solar neighbourhood is assumed, and a timescale for the bulge formation of about 10^7 yr is obtained. The main result is an enhancement of α elements, with the exception of calcium: [Ca/Fe] $\approx$ 0.0 as given in Matteucci et al. (1998), in very good agreement with observations.

6 Comments on Results by Cohen et al. 1999

Reasons for a difference in metallicity by 0.35 dex relative to Cohen et al. (1999) for NGC 6553 are partly mentioned in Sect. 7.1 of Cohen et al.'s paper, but some incorrect comparisons are indicated there:

(i) Model atmosphere: we employed Plez et al. (1992, 1997) grid of atmospheric models, whereas Cohen et al. used Kurucz model atmosphere. A difference of about *0.1dex could* come from the models.

(ii) Cohen et al. state that Barbuy et al. (1999a) used oscillator strengths gfs fitted to the solar spectrum with the Holweger-Müller model atmosphere. This is incorrect: for the derivation of metallicity [Fe/H] and abundance ratios, we only used laboratory gfs by Fuhr et al. (1988), Martin et al. (1988), Wiese et al. (1969) and the compilation of laboratory gfs by MR94.

(iii) We believe that the major reason for the difference in metallicity between Barbuy et al. (1999a) and Cohen et al. (1999) comes from the temperature scale adopted in Cohen et al.: it is clear that the new metallicity scale by Gratton et al. (1996) and Carretta & Gratton (1997) is a consequence of their hotter temperature scale and in their scale essentially all clusters are more metal-rich by 0.2 dex relative to other analyses.

In any case, there are so few observations and analyses of individual stars in these clusters, that clearly new high S/N and high resolution spectra for stars of a wide range of temperatures will be of great interest. This is only achievable with 8m class telescopes.

7 Super-Metal-Rich Stars in the Solar Neighbourhood

Grenon (1998, 1999) presented a sample selected from the NLTT proper motion catalogue, for which Hipparcos proper motions, Geneva photometry and radial velocities with Coravel are available.

The selected sample consists of metal-rich stars with orbits typical of bulge stars presently in their apocenter, close to the solar neighbourhood. Hipparcos data allowed to derive an age of 10-11 Gyrs for this sample (Grenon 1998).

Small subsamples of these stars were analysed in Barbuy & Grenon (1990) and Castro et al. (1997), where [Fe/H] > 0, [s-elements/Fe] < 0 (Castro et al. 1997), [α-elements/Fe] ≥ 0 were found. These subsamples were too small and further observations are necessary.

8 Conclusions

Concerning intermediate mass stars, essentially no abundance determinations are available, except for the analysis of young intermediate-mass stars IRS 7, 11, 19, 22, 24, located right at the Galactic center by Ramirez et al. (1998) and Sellgren et al. (1998). Many papers are devoted to Miras in the bulge, dealing with IRAS and ISO observations, which enable to use Miras as tracers of the bulge shape (e.g. Catchpole et al. 1999). The Miras are supposed to belong to an intermediate-mass disk population, but luminous Miras can also belong to a metal-rich old population (Guarnieri et al. 1997). Probably the two contributions are present, and probably they are dominated by the disk component, as concerns Miras.

As regards low-mass bulge stars, a first detailed analysis of low-mass bulge stars was carried out by McWilliam & Rich (1994). In Barbuy et al. (1999a,b) we presented a high resolution detailed abundance analysis of individual stars in the bulge globular clusters NGC 6528 and NGC 6553, which show [Fe/H] ≈ -0.6 combined to a strong enhancement of α-elements. For both clusters we obtain $[Z/Z_{\odot}] \approx -0.1$. Note that 47 Tuc with $[Z/Z_{\odot}] = -0.32$, would show a difference relative to these metal-rich clusters in the enhancement of α elements. Note that also ellipticals and bulges of spiral galaxies show enhancements of magnesium (Worthey et al. 1992).

A comparison of these results with those for field bulge stars by McWilliam & Rich (1994) show strong similarities, as for example, low [Ca/Fe] ≈ 0. Note that Cohen et al. (1999) recent paper finds conflicting results, i.e., solar metallicity and solar abundance ratios for NGC 6553.

Our results show that the bulk of stellar populations in the Galactic bulge seem to be old, and their chemical enrichment proceeded in a fast timescale at early times (see discussions in Renzini 1998, 1999).

Table 1. Abundance ratios in NGC 6553, NGC 6528, and bulge field K giants (MR94)

[X/Fe]	NGC6553	NGC6528	MR94
[Fe/H]	-0.55	-0.6	–
Na	+0.65	+0.4	+0.20
O	–	+0.35	+0.03
Mg	+0.33	+0.40	+0.35
Al	+0.50	+0.4	+0.58
Si	+0.35	+0.6	+0.14
Ca	+0.32	0.0	+0.14
Ti	+0.60	+0.60	+0.37
Y	-0.20	–	+0.05
Zr	-0.40	-0.40	-0.53
Ba	-0.10	0.0	+0.20
La	+0.13	–	+0.09
Eu	+0.0	0.0	+0.26

References

1. Barbuy, B. 1999, in *Galaxy evolution: connecting the distant universe with the local fossil record*, eds. M. Spite, F. Crifo, Springer-Verlag, in press
2. Barbuy, B., Bica, E., Ortolani, S. 1998, *A&A* **333**, 117
3. Barbuy, B., Grenon, M. 1990, in *Bulges of Galaxies*, ESO Conf. and Workshop Proceedings 35, eds. B.J. Jarvis & D.M. Terndrup, p. 83
4. Barbuy, B., Renzini, A., Ortolani, S., Bica, E., Guarnieri, M.D. 1999a, *A&A* **341**, 539
5. Barbuy, B., Renzini, A., Ortolani, S., Bica, E. 1999b, in preparation
6. Carretta, E., *Gratton*, R.G. 1997, *A&AS* **121**, 95

7. Castro, S., Rich, R.M., Grenon, M. Barbuy, B., McCarthy, J. 1997, *AJ* **114**, 376
8. Catchpole, R.M., Whitelock, P., Feast, M.W., Hughes, S.M.G., Alard, C., Olivier, E. 1999, in IAU Symp. 192, eds. P. Whitelock & R. Cannon, ASP, in press
9. Cohen, J.G., Gratton, R.G., Behr, B.B., Carretta, E. 1999, ApJ, submitted
10. Fuhr, J.R., Martin, G.A., Wiese, W.L. 1988, *Atomic Transition Probabilities: Iron through Nickel*, Journal of Physical and Chemical Reference Data, vol. 17, suppl. no. 4
11. Gratton, R.G., Carretta, E., Castelli, F. 1996, *A&A* **314**, 191
12. Grenon, M. 1998, in Highlights of Astronomy, in press
13. Grenon, M. 1999, in *Galaxy evolution: connecting the distant universe with the local fossil record*, eds. M. Spite, F. Crifo, Springer-Verlag, in press
14. Guarnieri, M.D., Renzini, A., Ortolani, S. 1997, *ApJ* **477**, L21
15. Guarnieri, M.D., Ortolani, S., Montegriffo, P., Renzini, A., Barbuy, B., Bica, E., Moneti, A.: 1998, *A&A* **331**, 70
16. Harmon, R., Gilmore, G. 1988, *MNRAS* **235**, 1025
17. Idiart, T., Freitas Pacheco, J.A., Costa, R.D.D. 1996, *AJ* **111**, 1169
18. Larson, R.B. 1990, *PASP* **102**, 709
19. Lennon, D.J., Mao S., Fuhrmann, K., Gehren, T. 1996, *ApJ* **471**, L23
20. Martin, G.A., Fuhr, J.R., Wiese, W.L. 1988, *Atomic Transition Probabilities: Scandium through Manganese*, Journal of Physical and Chemical Reference Data, vol. 17, suppl. no. 3
21. Matteucci, F., Brocato, E. 1990, *ApJ* **365**, 539
22. Matteucci, F., Romano, D., Molaro, P. 1999, *A&A* **341** , 458
23. McWilliam, A. 1990, *ApJS* **74**, 1075
24. McWilliam, A., Rich, R.M. 1994, *ApJS* **91**, 749 (MR94)
25. Ortolani, S., Renzini, A., Gilmozzi, R., Marconi, G., Barbuy, B., Bica, E., Rich, R.M.: 1995, *Nature* **377**, 701
26. Pfenniger, D. 1993, IAU Symp. 153, p. 387
27. Plez, B., Brett, J.M., Nordlund, A. 1992, A&A, 256, 551
28. Plez, B. 1997, private communication
29. Raha, N., Sellwood, J.A., James, R.A., Kahn, F. D. *Nature* **352**, 411
30. Ramirez, S.V., Sellgren, K., Terndrup, D.M., Carr, J.S., Balachandran, S., Blum, R.D. 1998, in IAU Symp. 184, eds. Y. Sofue, p. 67
31. Renzini, A. 1998, astro-ph/9801209, in The Young Universe, eds. S. D'Odorico, A. Fontana, E. Giallongo, *ASP Conf. Ser.* **146**, 298
32. Renzini, A. 1999, astro-ph/9902108
33. Rich, R.M. 1988, *AJ* **95**, 828
34. Rich, R.M. 1992, in *The Center, Bulge, and Disk of the Milky Way*, ed. L. Blitz, Kluwer Acad. Pub., p. 47
35. Sadler, E.M., Terndrup, D.M., Rich, R.M. 1996, *AJ* **112**, 117
36. Sellgren, K., Carr, J.S., Balachandran, S. 1998, in IAU Symp. 184, eds. Y. Sofue, p. 21
37. Spergel, D.N., Malhotra, S., Blitz, L. 1996, in *Spiral Galaxies in the Near-IR*, eds. D. Minniti, H.-W. Rix, Springer-Verlag, p. 128
38. Wiese, W.L., Martin, G.A., Fuhr, J.R. 1969, *Atomic Transition Probabilities: Sodium through Calcium*, NSRDS-NBS 22
39. Wood, P.R., Bessell, M.S. 1983, *ApJ* **265**, 748
40. Worthey, G., Faber, S.M., Gonzalez, J.J. 1992, *ApJ* **398**, 69

Abundances and Kinematics of Halo Stars

Poul E. Nissen

Institute of Physics and Astronomy, University of Aarhus, Denmark

Abstract. Recent determinations of abundances of the elements in nearby stars with halo kinematics are reviewed and discussed in terms of chemical evolution models for the Galaxy. Particular attention is given to trends and scatter of various abundance ratios for the most metal-poor *and* the most metal-rich halo stars including new puzzling results for the oxygen-to-iron ratio based on near-UV OH lines. The need for high spectral resolution studies of element abundances in halo stars with different kinematical properties is emphasized, and a program for determining oxygen abundances in the most metal-poor stars using UT2 and UVES is suggested.

1 Introduction

In the following a short review of recent studies of element abundances in halo field stars is given. Particular attention is paid to trends and scatter of various abundance ratios for the most metal-rich and the most metal-poor halo stars including new puzzling results for the oxygen-to-iron ratio. The need for high spectral resolution studies of abundances in halo stars with different kinematical properties is emphasized, and a program for determining oxygen abundances in the most metal-poor stars using UT2 and UVES is discussed.

2 The Metal-Rich Halo Stars

From recent surveys of high velocity stars it has become clear that stars with halo kinematics are not limited to metallicities $[Fe/H] < -1.0$. A small fraction of stars with metallicities in the range $-1.0 < [Fe/H] < -0.5$ have a galactic rotational velocity component $V_{rot} < 50\ km\,s^{-1}$, and are well separated from the thick disk stars in the V_{rot}–[Fe/H] diagram (Schuster et al. [1], Carney et al. [2]).

In a recent paper, Nissen & Schuster [3] have determined abundances for two groups of main-sequence stars in the metallicity range $-1.3 < [Fe/H] < -0.5$, one group with thick-disk kinematics ($V_{rot} > 150\ km\,s^{-1}$) and the other one with halo kinematics ($V_{rot} < 50\ km\,s^{-1}$). The stars were selected from Strömgren photometry to have about the same narrow ranges in effective temperature and surface gravity. From high resolution spectra it was then possible to derive very precise differential abundance ratios. The results for oxygen are shown in Fig. 1. Whereas all thick-disk stars follow the trend of [O/Fe] for the galactic disk derived by e.g. Nissen & Edvardsson [4] from the

[OI] λ6300 line, the majority of the halo stars (9 out of 14) have O/Fe ratios about 0.2 dex below the disk stars. As shown by Nissen & Schuster the same 9 halo stars also have low ratios of Na, Mg, Si, Ca, and Ti with respect to Fe. Furthermore, these 'low-α' stars tend to have larger galactic orbits than the other halo stars suggesting that they have been accreted from dwarf galaxies with an early burst of star formation (Pagel & Tautvaišienė [5]) or formed in outer regions of the Galaxy with a star formation rate slow enough to incorporate the ejecta from SNIa at a low value of [Fe/H] (Gilmore & Wyse [6]).

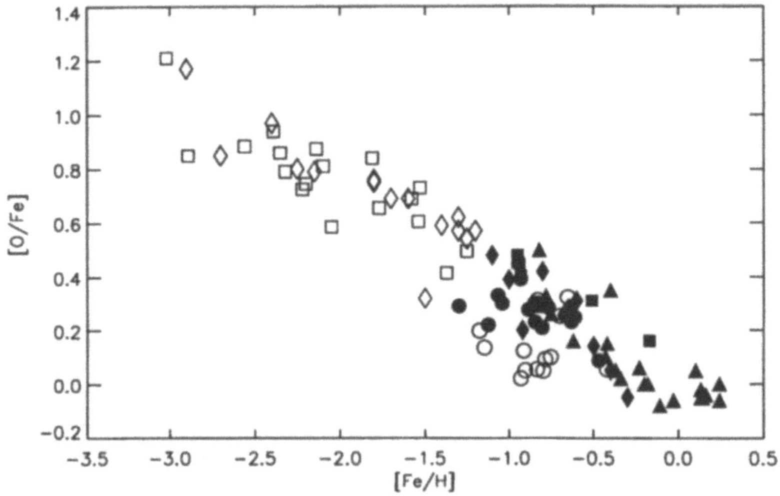

Fig. 1. [O/Fe] vs. [Fe/H]. Open symbols refer to stars with halo kinematics and filled symbols to disk stars. (*triangles*) Nissen & Edvardsson [4]; (*circles*) Nissen & Schuster [3]; (*diamonds*) Israelian et al. [16]; (*squares*) Boesgaard et al. [17].

A few additional halo stars with 'low-α' abundances and large galactocentric orbital distances have been discovered by King [7] and Carney et al. [8]. Furthermore, Jehin et al. [9] have found very significant variations and correlations for the abundances of α-, s-, and r-elements relative to iron for a group of 21 turnoff stars with $-1.3 < [Fe/H] < -0.7$. Most of the stars have thick-disk kinematics, but two of the 'low-α' stars have halo kinematics.

We conclude that abundance ratios of metal-rich halo stars are not universal functions of [Fe/H]. The degree of the scatter and possible correlations with kinematics are, however, open questions. Clearly, a large survey aiming at determining very precise abundance ratios for a sample of say 200 halo stars with $-1.5 < [Fe/H] < -0.5$ would be very interesting. Probably, most of the high resolution spectra needed could be obtained with 3-4 meter telescopes, but in order to get sufficient precision of equivalent widths of weak lines in the fainter stars 8-10 meter telescopes may be required.

3 The Very Metal-Poor Halo Stars

Two important papers on abundance ratios in halo stars with [Fe/H] < -2.5 have been published by McWilliam et al. [10] and Ryan et al. [11]. The main results are: (i) The abundance of the α-elements relative to Fe seems to be nearly constant at $[\alpha/\mathrm{Fe}] \simeq 0.4$, but slight trends in e.g. Mg/Fe and a cosmic scatter at a level of ± 0.1 dex cannot be excluded. Aluminum and the neutron-capture elements show large underabundances relative to iron with a very significant scatter of up to 2 dex. Most surprising are the results for the iron-peak elements. Sc/Fe and Co/Fe increase with decreasing [Fe/H] below [Fe/H] < -2.5, whereas Cr/Fe and Mn/Fe decrease rather steeply.

Several possible interpretations of these results have been discussed. The large scatter in the neutron-capture element abundances suggests that we are seeing effects of single supernovae events, whereas the rather well-defined trends of the ratios between the iron-peak elements can be explained if the supernovae yields are tightly correlated with explosion energy and hence with the amount of pure hydrogen that the ejecta is mixed with (Ryan et al. [11], Shigeyama & Tsujimoto [12], and Tsujimoto & Shigeyama [13]). Alternatively, the unusual abundance ratios at [Fe/H] $\simeq -3.5$ could be due to Pop. III stars formed before reionization at redshift $z \simeq 5$ (Ostriker & Gnedin [14]); subsequent Pop. II supernovae have then diluted the Pop. III ejecta to solar ratios for the iron-peak elements.

It would be very interesting to observe a large sample of these very metal-poor stars to refine our knowledge about trends and scatter especially for stars below [Fe/H] $\simeq -3.5$ of which only a handful has been observed so far. Thanks to the HK survey of Beers and collaborators (Beers [15]) about 20 stars with [Fe/H] < -3.5 are known. An obvious task for UVES would be to observe these stars in the blue spectral region with a resolution of about 40 000. In addition, a survey of several hundred stars with [Fe/H] < -2.0 would probably be very rewarding in particular because the radial velocity distribution of the HK survey stars shows evidence of two sub-populations, one with a dispersion of $\simeq 60\ \mathrm{km\,s^{-1}}$ and one with $\simeq 130\ \mathrm{km\,s^{-1}}$ (Beers [15]).

4 Oxygen Abundances

Oxygen is the most abundant element next to hydrogen and helium, and the abundance of oxygen is of high interest in connection with nucleosynthesis in massive stars, chemical evolution models, ages of stars, and production of light elements by cosmic ray spallation processes.

The abundance of oxygen in halo stars may be determined from the forbidden OI line at 6300Å, the OI triplet at 7774Å and OH lines in the near-UV or the IR. Unfortunately, results from these features do not agree; oxygen abundances of dwarfs derived from the triplet and the OH lines in the UV are much higher than abundances of giants derived from the forbidden line.

This disagreement has been sharpened by two recent works on near-UV OH lines in unevolved stars by Israelian et al. [16] and Boesgaard et al. [17]. As shown in Fig. 1 the results of the two works agree very well although different sets of OH lines were applied. An independent study by Reetz [18] also shows a continuously increasing trend of [O/Fe] towards more metal-poor stars.

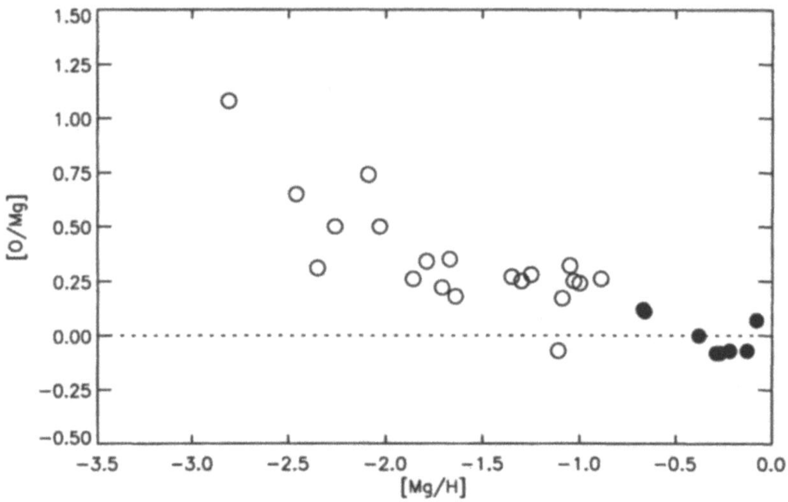

Fig. 2. [O/Mg] vs. [Mg/H]. Open symbols refer to stars with halo kinematics and filled symbols to disk stars. Sources for the abundances are given in the text

The [O/Fe] data shown in Fig. 1 is in strong disagreement with oxygen abundances of halo giant stars derived from the [OI] λ 6300 line by e.g. Barbuy [19], who found a nearly constant level [O/Fe]$\simeq$ 0.4 from [Fe/H] $= -1.0$ to -3.0. Either oxygen has been depleted in giants, which is not supported by standard stellar models, or some of the methods of determining oxygen abundances are affected by severe systematic errors as a function of [Fe/H]. Systematic differences in the iron abundances between giants and dwarfs may also contribute to the large difference in O/Fe.

It is often stated that chemical evolution models predict a constant O/Fe in the halo and hence are in agreement with the [OI] giant data. This is, however, not correct. As shown by Chiappini et al. [20] a realistic model taking into account that the more massive stars produce higher O/Fe ratios and have shorter lifetimes than the less massive stars predicts a rise in [O/Fe] from about 0.3 dex at [Fe/H] $= -1.0$ to a level of 0.6 to 0.8 dex at [Fe/H] $= -3.0$ depending on which SNII yields are used.

The need for accurate oxygen abundances for many more metal-poor stars becomes particular clear when [O/Mg] is plotted vs. [Mg/H] as in Fig. 2, where the oxygen abundances come from Israelian et al. [16] and Boesgaard et al. [17] and the Mg abundances from Magain [21] [22], Ryan et al. [23], Fuhrmann et al. [24], Fuhrmann [25], Edvardsson et al. [26] and Nissen et al. [27]. In general the Mg abundances in these papers agree well and therefore average values were adopted. An exception is BD+03 740 for which Magain [22] found [Mg/Fe] = 0.54, whereas Fuhrmann et al. [24] derived an unusual low value of [Mg/Fe] = −0.28. For this star we adopted [Mg/Fe] = 0.54.

As seen from Fig. 2, [O/Mg] seems to increase strongly for [Mg/H] below −2.0. It is not possible to explain such a trend and the high O/Mg ratios by canonical Pop. II supernovae yields. Instead, one has to invoke Pop. III stars with very high oxygen yields. The number of stars below [Mg/H] $\simeq$ −2.0 is, however, very few, the scatter in O/Mg is large, and as discussed above the oxygen abundances derived from the near-UV OH lines may suffer from systematic errors. Clearly, we need a large set of reliable oxygen and magnesium abundances below [Mg/H]$\simeq$ −2.0.

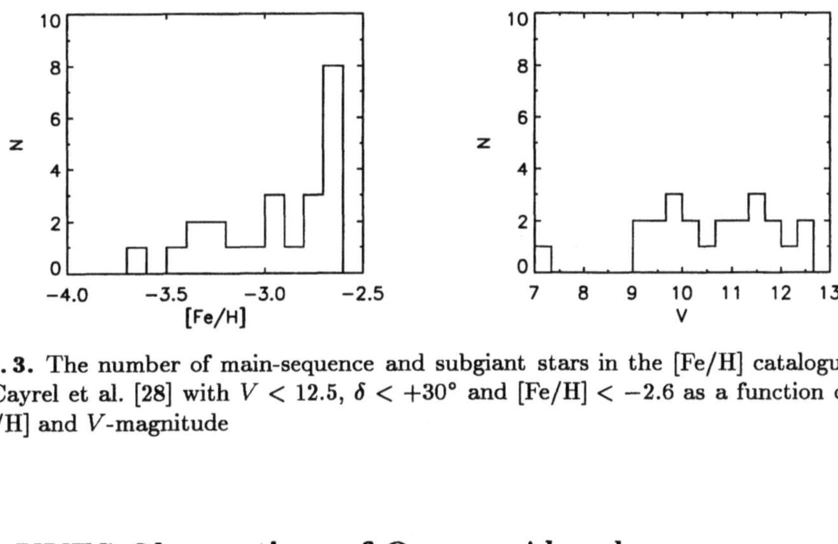

Fig. 3. The number of main-sequence and subgiant stars in the [Fe/H] catalogue of Cayrel et al. [28] with $V < 12.5$, $\delta < +30°$ and [Fe/H] < −2.6 as a function of [Fe/H] and V-magnitude

5 UVES Observations of Oxygen Abundances

In the case of very metal-poor dwarf stars only the near-UV OH lines can be used to derive oxygen abundances. Other spectral features are too faint. The observations are, however, difficult due to the high atmospheric extinction, 1.5 to 2 mag, around 3200Å and the rather low flux of F and G stars at this wavelength. Even with the large collecting area of UT2 and the good efficiency of UVES in the UV it is estimated that about 3 hours of integration time will be needed to observe a $V = 12.5$ mag star with a resolution of $R \simeq 50\,000$ and $S/N \simeq 100$. This means that only a few stars from the Beers et al. HK survey can be reached. A search in the [Fe/H]-catalogue of Cayrel de Strobel [28] for main-sequence and subgiant stars with $V < 12.5$, declination $\delta < +30°$

and [Fe/H] < −2.6 resulted, however, in a list of 23 stars. Fig. 3 shows the distribution of these stars as a function of [Fe/H] and magnitude. As seen a very interesting sample could be reached with UVES on UT2. Furthermore, it is noted that the distribution is rather flat in V suggesting that additional very metal-poor stars can probably be found between $V = 11$ and 12.5.

We conclude that UVES could contribute with oxygen abundances from near-UV spectra for a much larger sample of halo stars with [Fe/H] < −2.5 than has been observed up to now. From the same spectra one could also derive beryllium abundances from the BeII resonance lines at 3130Å and nitrogen abundances from the NH band at 3360Å. Such a program should, however, be supplemented by a careful study of stars at higher metallicities to see if consistent oxygen abundances can be derived from all available spectral features.

References

1. Schuster W.J., Parrao L., Contreras Martínez M.E. 1993, A&AS 97, 951
2. Carney B.W, Laird J.B., Latham D.W., Aguilar L.A. 1996, AJ 112, 668
3. Nissen P.E., Schuster W.J. 1997, A&A 326, 751
4. Nissen P.E., Edvardsson B. 1992, A&A 261, 255
5. Pagel B.E.J., Tautvaišienė G. 1998, MNRAS 299, 535
6. Gilmore G., Wyse R.F.G. 1998, AJ 116, 748
7. King J.R. 1997, AJ 113, 2302
8. Carney B.W., Wright J.S., Sneden C., Laird J.B., Aguilar L.A., Latham D.W. 1997, AJ 114, 363
9. Jehin E., Magain P., Neuforge C., Noels A., Parmentier G., Thoul A.A. 1998, A&A 341, 241
10. McWilliam A., Preston G.W., Sneden C., Searle L. 1995, AJ 109, 2757
11. Ryan S.G., Norris J.E., Beers T.C. 1996, ApJ 471, 254
12. Shigeyama T., Tsujimoto T. 1998, ApJ 507, L135
13. Tsujimoto T., Shigeyama T. 1998, ApJ 508, L151
14. Ostriker J.P., Gnedin N.Y. 1996, ApJ, L63
15. Beers T.C. 1999, in *Connecting the distant Universe with the local fossil record*, eds. M. Spite and F. Crifo, Kluwer (in press)
16. Israelian G., García Lopez R., Rebolo R. 1998, ApJ 507, 805
17. Boesgaard A.M., King, J.R., Deliyannis C.P., Vogt S.S. 1999, ApJ 117, 492
18. Reetz J. 1998, Dissertation, Ludwig-Maximilians-Universität, Munich
19. Barbuy B., 1988, A&A 191, 121
20. Chiappini C., Matteucci F., Beers T.C., Nomoto K. 1999, ApJ (in press)
21. Magain P. 1987, A&A 179, 176
22. Magain P. 1989, A&A 209, 211
23. Ryan S.G., Norris J.E., Bessell M.S. 1991, AJ 102, 303
24. Fuhrmann K., Axer M., Gehren T. 1995, A&A 301, 492
25. Fuhrmann K. 1998, A&A 338, 161
26. Edvardsson B., Andersen J., Gustafsson B., Lambert D.L., Nissen P.E., Tomkin J. 1993, A&A 275, 101
27. Nissen P.E., Gustafsson B., Edvardsson B., Gilmore G. 1994, A&A 285, 440
28. Cayrel de Strobel G., Soubiran C., Friel E.D., Ralite N., Francois P. 1997, A&AS 124, 299

Tidal Tails Around Galactic Globular Clusters

Georges Meylan[1], Stéphane Leon[2,3], and Françoise Combes[3]

[1] ESO, Karl-Schwarzschild-Strasse 2, D-85748 Garching, Germany
[2] ASIAA, Institute of Astronomy and Astrophysics, Academia Sinica, P.O. Box 1-87, Nankang, Taipei 115, Taiwan
[3] DEMIRM, Observatoire de Paris, 61, Avenue de l'Observatoire, F-75015 Paris, France

Abstract. We present hereafter our study of the 2-D structures of the tidal tails associated with 20 galactic globular clusters, structures obtained by using the wavelet transform to detect weak structures at large scale and filter the strong background noise for the low galactic latitude clusters (Leon, Meylan & Combes 1999). We also present N-body simulations of globular clusters, in orbits around the Galaxy, in order to study quantitatively and geometrically the tidal effects they encounter (Combes, Leon & Meylan 1999).

1 Introduction

In addition to the effects of their internal dynamical evolution, globular clusters suffer strong dynamical evolution from the potential well of their host galaxy (Gnedin & Ostriker 1997, Murali & Weinberg 1997). These external forces speed up the internal dynamical evolution of these stellar systems, accelerating their destruction. Shocks are caused by the tidal field of the galaxy: interactions with the disk, the bulge and, somehow, with the giant molecular clouds, heat up the outer regions of each star cluster. The stars in the halo are stripped by the tidal field. All globular clusters are expected to have already lost an important fraction of their mass, deposited in the form of individual stars in the halo of the Galaxy (see Meylan & Heggie 1997 for a review).

Recent N-body simulations of globular clusters embedded in a realistic galactic potential (Oh & Lin 1992; Johnston et al. 1999) were performed in order to study the amount of mass loss for different kinds of orbits and different kinds of clusters, along with the dynamics and the mass segregation in tidal tails. Grillmair et al. (1995) in an observational analysis of star counts in the outer parts of a few galactic globular clusters found extra-cluster overdensities that they associated partly with stars stripped into the Galaxy field.

We present hereafter our study of the 2-D structures of the tidal tails associated with 20 galactic globular clusters, obtained by using the wavelet transform to detect weak structures at large scale and filter the strong background noise for the low galactic latitude clusters (Leon, Meylan & Combes 1999). We also present N-body simulations of globular clusters in orbits around the Galaxy, in order to study quantitatively and geometrically the tidal effects they encounter (Combes, Leon & Meylan 1999).

2 Observations and Data Reduction

Our sample clusters share different properties or locations in the Galaxy, with various masses and structural parameters. It is of course necessary to have very wide field imaging observations. Consequently, we obtained, during the years 1996 and 1997, photographic films with the ESO Schmidt telescope. The field of view is $5.5° \times 5.5°$ with a scale of $67.5''$/mm. The filters used, viz. BG12 and RG630, correspond to B and R, respectively. All these photographic films were digitalized using the MAMA scanning machine of the Observatoire de Paris, which provides a pixel size of 10 μm . The astrometric performances of the machine are described in Berger et al. (1991).

The next step – identification of all point sources in these frames – was performed using SExtractor (Bertin & Arnouts 1996), a software dedicated to the automatic analysis of astronomical images using a multi-threshold algorithm allowing good object deblending. The detection of the stars was done at a 3-σ level above the background. This software, which can deal with huge amounts of data (up to 60,000 $\times$ 60,000 pixels) is not suited for very crowded fields like the centers of the globular clusters, which were simply ignored. We performed a star/galaxy separation by using the method of star/galaxy magnitude vs. log(star/galaxy area).

For each field, we constructed a B vs. $(B - V)$ color-magnitude diagram, on which we performed a field/cluster star selection, following the method of Grillmair et al. (1995), since cluster stars and field stars exhibit different colors. In this way we can identify present and past cluster members from the fore- and background field stars by identifying in the CMD the area occupied by cluster stars. The envelope of this area is empirically chosen so as to optimize the ratio of cluster stars to field stars in the relatively sparsely populated outer regions of each cluster.

3 Wavelet Analysis

With the assumption that the data can be viewed as a sum of details with different typical scale lengths, the next step consists of disentangling these details using the space-scale analysis provided by the Wavelet Transform (WT, cf. Slezak et al. 1994; Resnikoff & Wells 1998). Any observational signal includes also some noise, which has a short scale length. Consequently the noise is higher for the small scale wavelet coefficients. We performed Monte-Carlo simulations to estimate the noise at each scale and apply a 3-σ threshold on the wavelet coefficients to keep only the reliable structures. In this way it is possible to subtract the short-wavelength noise without removing details from the signal which has longer wavelengths.

The remaining overdensities of the cluster-like stars, remaining after the application of the wavelength transform analysis to the star counts, are associated with the stars evaporated from the clusters because of dynamical relaxation and/or tidal stripping by the galactic gravitational field.

It is worth emphasizing that we have taken into account the following strong observational biases: (i) bias due to the clustering of galactic field stars; (ii) bias due to the clustering of background galaxies; (iii) bias due to the fluctuations of the dust extinction, as observed in the IRAS 100-μm map.

4 Observational Results

NGC 5139 $\equiv \omega$ Centauri, the most massive galactic globular cluster (Meylan et al. 1995), currently crossing the disk plane, is a nearby globular cluster located at a distance of 5.0 kpc from the sun. Its relative proximity allows to reach the main sequence for the star count selection. We estimate, taking into account the possible presence of mass segregation in its outer parts, that about 0.6 to 1 % of its mass has been lost during the current disk shocking event. Although this cluster has one of our best tail/background S/N ratios, it is by far not the only one exhibiting tidal tails.

Considering all 20 clusters of our sample, we reach the following conclusions (see Leon, Meylan & Combes 1999 for a complete description of this work):

- All the clusters observed, which do not suffer from strong observational biases, present tidal tails, tracing their dynamical evolution in the Galaxy (evaporation, tidal shocking, tidal torquing, and bulge shocking).
- The clusters in the following sub-sample (viz. NGC 104, NGC 288, NGC 2298, NGC 5139, NGC 5904, NGC 6535, and NGC 6809) exhibit tidal extensions resulting from a recent shock, i.e. tails aligned with the tidal field gradient.
- The clusters in another sub-sample (viz. NGC 1261, NGC 1851, NGC 1904, NGC 5694, NGC 5824, NGC 6205, NGC 7492, Pal 5, and Pal 12) present extensions which are only tracing the orbital path of the cluster with various degrees of mass loss.
- NGC 7492 is a striking case because of its very small extension and its high destruction rate driven by the galaxy as computed by Gnedin & Ostriker (1997). Its dynamical "twin" for such an evolution, namely Pal 12, exhibits, on the contrary, a large extension tracing its orbital path, with a possible shock which happened more than 350 Myr ago.
- The presence of a break in the outer surface density profile is a reliable indicator of some recent gravitational shocks.

Our recent CCD observations with the Wide Field Imager at the ESO/MPI 2.2-m telescope will soon provide improved results, because of the more accurate CCD photometry. These observations will allow more precise observational estimates of the mass loss rates for different regimes of galaxy-driven cluster evolution.

5 Numerical Simulations

We tried with extensive numerical simulations to reproduce the above observations. We *performed* N-body simulations of globular clusters, in orbits

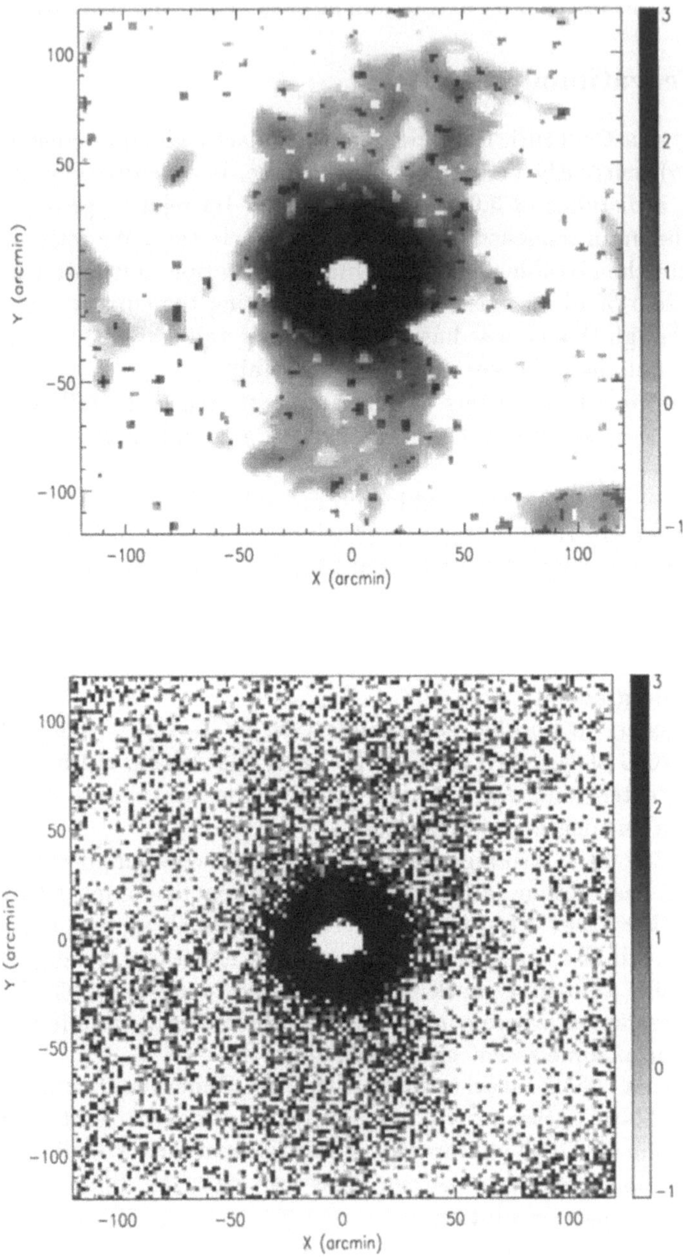

Fig. 1. NGC 5139 $\equiv \omega$ Centauri. In the upper panel, filtered image of color-selected star-count overdensities using the Wavelet Transform to be compared with the raw star counts in the lower panel. The upper panel displays the full resolution using the whole set of wavelet planes.

around the Galaxy, in order to study quantitatively and geometrically the tidal effects they encounter. The N-body code used is an FFT algorithm, using the method of James (1977) to avoid the periodic images. With N = 150,000 particles, it required 2.7s of CPU per time step on a Cray-C94.

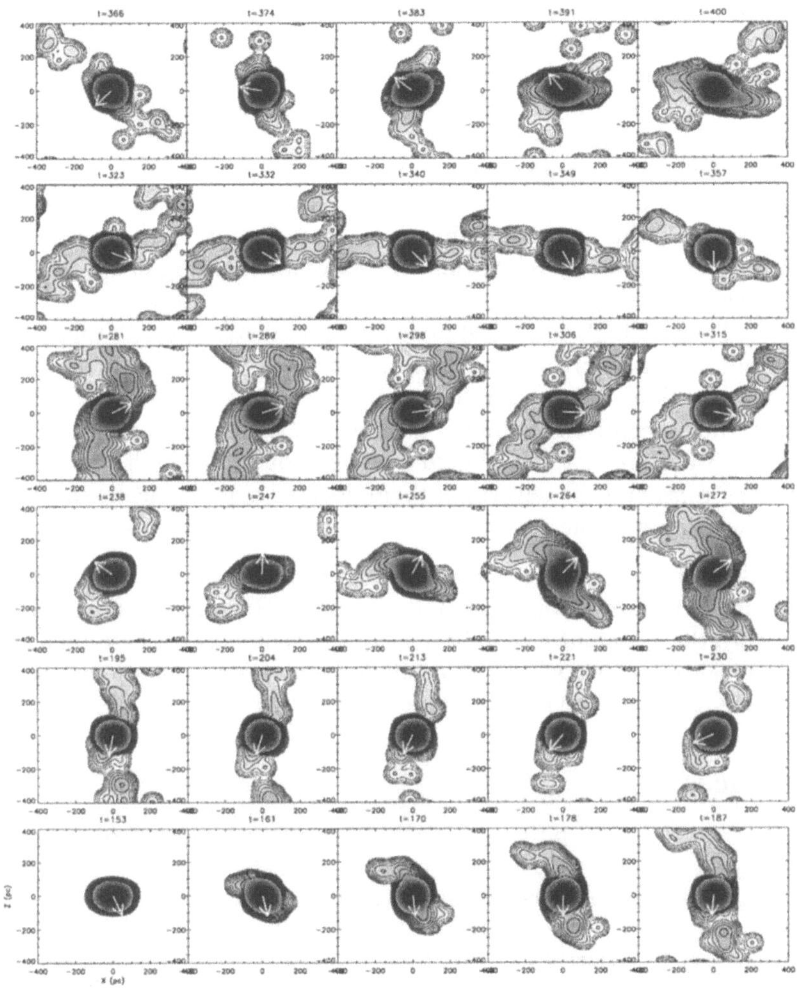

Fig. 2. Tidal tails mapped at different epochs with the wavelet algorithm applied to one of our simulations. The direction perpendicular to the galactic plane is indicated by the arrow. The time sequence starts with the lower-left panel and ends with the upper right one. The third panel exhibits tails which are quite reminiscent of what is observed in NGC 5139 ≡ ω Centauri (see Fig. 1).

The globular clusters are represented by multi-mass King-Michie models, including mass *segregation* at initial conditions. The Galaxy is modelled as realistically as possible, with three components, bulge, disk and dark halo: the bulge is a spherical Plummer law, the disk is a Miyamoto-Nagai model, and the dark matter halo is added to obtain a flat Galactic rotation curve.

The main conclusions of our simulations can be summarized as follows (see Combes, Leon & Meylan 1999 for a complete description of this work):

- All runs show that the clusters are always surrounded by tidal tails and debris. This is also true for those that suffered only a very slight mass loss. These unbound particles distribute in volumic density like a power-law as a function of radius, with a slope around −4. This slope is much steeper than in the observations where the background-foreground contamination dominates at very large scale.
- These tails are preferentially composed of low mass stars, since they are coming from the external radii of the cluster; due to mass segregation built up by two-body relaxation, the external radii preferentially gather the low mass stars.
- For sufficiently high and rapid mass loss, the cluster takes a prolate shape, whose major axis precesses around the z-axis.
- When the tidal tail is very long (high mass loss) it follows the cluster orbit: the observation of the tail geometry is thus a way to deduce cluster orbits. Stars are not distributed homogeneously through the tails, but form clumps, and the densest of them, located symmetrically in the tails, are the tracers of the strongest gravitational shocks.

Finally, these N-body experiments help to understand the recent observations of extended tidal tails around globular clusters (Grillmair et al. 1995, Leon et al. 1999): the systematic observations of the geometry of these tails should provide much information on the orbit, dynamics, and mass loss history of the clusters, and on the Galactic structure as well.

References

1. Berger J., Cordoni J.P., Fringant A.M., Guibert J., Moreau, O., Reboul H., Vanderriest C., 1991, A&AS, 87, 389
2. Bertin E., Arnouts S., 1996, A&AS, 117, 393
3. Combes F., Leon S., Meylan G., 1999, submitted
4. Gnedin O.Y., Ostriker J.P., 1997, ApJ, 474, 223
5. Grillmair C.J., Freeman K.C., Irwin M., Quinn P.J., 1995, AJ, 109, 2553
6. James R.A., 1977, J. Comput. Phys. 25, 71
7. Johnston K.V., Sigurdsson S., Hernquist L., 1999, MNRAS, 302, 771
8. Leon S., Meylan G., Combes F., 1999, submitted
9. Meylan G., Heggie D.C., 1997, A&AR, 8, 1-143.
10. Meylan G., Mayor M., Duquennoy A., Dubath P., 1995, A&A, 303, 761
11. Murali C., Weinberg M. D., 1997, MNRAS, 291, 717
12. Oh K.S., Lin D.N.C., 1992, ApJ, 386, 519
13. Resnikoff H.L., Wells R.O., 1998, Wavelet Analysis, Springer, Berlin Heidelberg
14. Slezak E., Durret F., Gerbal D., 1994, AJ, 108, 1996

PARALLEL WORKSHOP 3:

FROM EXTRASOLAR PLANETS TO BROWN DWARFS

Edited by

Michel Mayor and Francesco Paresce

Diversity of Planetary Systems: Formation Scenarios and Unsolved Problems

Jack J. Lissauer[1] and D.N.C. Lin[2]

[1] Space Science Division, MS 245-3, NASA Ames Research Center, Moffett Field, CA 94035, USA
[2] Astronomy Department, University of California, Santa Cruz, CA 95064, USA

Abstract. Models of planet formation and of the orbital stability of planetary systems are described and used to discuss possible characteristics of undiscovered planetary systems. Modern theories of star and planet formation, which are based upon observations of the Solar System and of young stars and their environments, predict that rocky planets should form in orbit about most single stars. It is uncertain whether or not gas giant planet formation is common, because most protoplanetary disks may be depleted before solid planetary cores can grow large enough to gravitationally trap substantial quantities of gas. A potential hazard to planetary systems is radial decay of planetary orbits resulting from interactions with material within the disk. Planets more massive than Earth may decay rapidly and sweep up smaller planets in their path. The potential of the VLT to improve our understanding of planetary formation is discussed briefly.

1 Introduction

What are the characteristics of planetary systems around stars other than the Sun? How many planets are typical? What are their masses and compositions? What are the orbital parameters of individual planets, and how are the paths of planets orbiting the same star related to one another? These questions are difficult to answer because planets are so faint that none have yet been directly observed over interstellar distances. Radial velocity surveys have demonstrated that planets with masses and orbits quite different from those within our own Solar System are present around main sequence stars in our region of the galaxy (Mayor & Queloz 1995, Marcy & Butler 1998 and references therein; Hatzes, this volume). All of the extrasolar planets thus far discovered induce variations in stellar reflex motion much larger than would a planetary system like our own, and surveys accomplished to date are consequently strongly biased against detecting low mass and long period planets. Our own Solar System may represent a biased sample of a different kind, because it contains a planet with conditions suitable for life to evolve to the point of being able to ask questions about other planetary systems (Wetherill 1994).

Extrapolating from the small and biased sample of planets that have been detected to a model of the variety of planetary systems which may be present elsewhere in the galaxy is a daunting challenge surely fraught with pitfalls. Detailed predictions are almost certain to be erroneous. However, the substantial progress made over the past few decades towards understanding the origins and dynamical stability of planetary systems makes it possible to assess hypothesized common attributes and scaling relations of planetary systems in a quantitative manner (Lissauer 1995).

The nearly planar and almost circular orbits of the planets in our Solar System argue strongly for planetary formation within flattened circumstellar disks. Observations by Goodman *et al.* (1993) indicate that typical star-forming dense cores in dark molecular clouds have specific angular momentum $> 10^{21}$ cm^2 s^{-1}. When these clouds undergo gravitational collapse, an order of magnitude reduction from their present dimension (~ 0.1 pc) would lead to the formation of rotationally-supported disks analogous to the primordial solar nebula (cf. Terebey *et al.* 1984, Cassen *et al.* 1985). Observational evidence for the presence of disks of Solar System dimensions around pre-main sequence stars has increased substantially in recent years (O'Dell *et al.* 1993, Mundy *et al.* 1996, McCaughrean & O'Dell 1996; McCaughrean, this volume). The existence of disks on scales of a few tens of astronomical units is inferred from the power-law spectral energy distribution in the infrared over more than two orders of magnitude in wavelength. The power law index in these systems is consistent with that expected from either: (1) a steady-state viscous accretion disk with $\dot{M} \sim 10^{-7}$–10^{-8} M$_\odot$ yr^{-1}, or (2) the reprocessing of radiation emitted near the accreting central star by a geometrically-thin opaque disk (Adams *et al.* 1987, Chiang & Goldreich 1997). Observations of infrared excesses in the spectra of young stars suggest that the lifetimes of protoplanetary disks span the range of $10^6 - 3 \times 10^7$ years (Strom *et al.* 1993, Hartmann *et al.* 1998, Hillenbrand *et al.* 1998).

The orbital spacing of planets is an important factor in determining how many planets are likely to exist within habitable zones. Although modern theories of planetary growth do not yield deterministic "Bode's Law" formulae for the orbits of planets (Wetherill 1990), characteristic orbital spacings do arise. These scalings suggest that spacings between planets grow roughly in proportion to the distance from the central star, but that separations also depend on the masses of the star and planets in the system and on quasi-random stochastic factors.

Terrestrial planets are believed to grow by pairwise accretion of solid bodies until the spacing of planetary orbits becomes large enough that the configuration is stable for the lifetime of the system (Safronov 1969, Wetherill 1990, Lissauer 1995). The initial stages of the formation of gas giant planets probably also involved binary agglomeration of solid bodies, with accretion of significant quantities of gas initiated once the heavy-element core had acquired enough mass to gravitationally trap hydrogen and helium (Pollack *et al.* 1996; Wuchterl *et al.* 2000), the two cosmically most abundant elements, which together are believed to make up over 98% of the mass entering protoplanetary disks. Issues involving the ultimate sizes and spacings of gas giant planets are complex and poorly understood (Lissauer *et al.* 1995), and provide a major source of uncertainty for modeling the potential diversity of planetary systems.

Stability criteria for planetary systems are reviewed in Section 2. Theoretical models of planetary growth based upon observations of our Solar System and of star-forming regions are summarized in Section 3. (See Lissauer 1993 for a more detailed review.) Section 4 discusses the theory of disk-planet interactions and planetary migration. Section 5 reviews other models of the recently-discovered extrasolar planetary systems. Section 6 concludes this article with a brief discussion of observations which may be made by the VLT that could substantially improve our understanding of the process of planetary formation and the diversity of planetary systems.

2 Stability of Orbital Configurations

Dynamical stability criteria provide some of the best constraints on the plausible range of planetary system properties. A system containing only two planets, both of which are on initially circular orbits, will never experience close approaches (i.e., is Hill stable) provided the ratio of the separation in orbital radii, Δa, to the planets' semimajor axes, a, satisfies

$$\frac{|\Delta a|}{a} > 2.4\left(\frac{M_1 + M_2}{M_\star}\right)^{1/3} \tag{1}$$

where a is the average semimajor axis of the planets and M_1, M_2 and $M_\star$ are the masses each of the planets and that of the star, respectively (Gladman 1993). Deterministic chaos also plays an important role in the stability of planetary orbits. Trajectories tend to be chaotic whenever resonances overlap. Strong resonances are more closely spaced near a planet, and resonances overlap over a range in semimajor axis centered on the perturbing planet. In order to avoid the resonance overlap region of a planet of mass M_p, a test particle must be sufficiently distant that

$$\frac{|\Delta a|}{a} > 1.5\left(\frac{M_p}{M_\star}\right)^{2/7} \tag{2}$$

(Duncan *et al.* 1989). Note that for both Hill stability and stability against strong chaos, the minimum separation between two planets increases with planetary mass, but the dependence is much slower than linear. Thus, the merger of planets on adjacent orbits usually increases the stability (dynamical lifetime) of a planetary system.

Although separation in semimajor axis is very important in determining the stability of a planetary system, other factors are also involved. Resonances can either increase orbital stability, e.g., Neptune and Pluto, or lead to instability, such as at the 3:1 Kirkwood gap in the asteroid belt. Eccentricities and inclinations also affect stability. As both eccentricity and inclination require excess energy, they give the systems more freedom of motion, and thus allow for greater instability.

Perturbations from several planets add in a complicated, nonlinear manner. Nonetheless, formulae based on generalizations of the Hill-Jacobi exclusion zone (Eq. 1) and the resonance overlap criterion of the restricted 3-body problem (Eq. 2) appear to be quite useful for assessing the stability of our planetary system and the satellite systems of the giant planets (Lissauer 1995). However, as the root cause of most long-term orbital instabilities is resonances, these criteria provide at best rules of thumb for judging stability, not a precise formulation that always works. For planets of a given mass, the number of them which can lie within a given logarithmic distance interval increases slowly (roughly as the two-sevenths power) with stellar mass. Smaller planets may be more closely spaced. See Lissauer (1995) for a more detailed and quantitative discussion of proposed scaling rules for planetary spacings and masses and Levison *et al.* (1998) for the results of numerical experiments on the stability of various hypothetical systems of giant planets.

The planets in our Solar System orbit close enough to each other that the final phases of planetary growth could have been the merger or ejection of planets on unstable orbits. However, the low eccentricities of the orbits of the outer planets imply that some damping process, such as accretion/ejection of numerous small planetesimals or interactions with residual gas within the protoplanetary disk, must also have been involved (cf. Lissauer *et al.* 1995). Although t he eccentricities of the giant planets are presently quite small, mutual gravitational perturbation between the planets induces a gradual increase in their eccentricity until their orbits become unstable and begin to cross each other on a timescale τ_x. Analytic theories suggest that $\tau_x \sim 10^{18}$ yr for the outer planets in our Solar System (Murray & Holman 1999). However, these models neglect solar mass loss and the effects of passing stars, which should shorten the lifetime of the system by several orders of magnitude (Duncan & Lissauer 1998).

If sufficiently massive and dense planets exist far enough from the star that they are not too deep within its gravitational potential well, they can eject material into interstellar space. Oort Cloud comets are believed to be icy planetesimals which were sent outwards at very close to the Solar System escape velocity, and were perturbed into long-lived orbits by nearby stars, interstellar material or the galactic tidal field. The highly nonlinear nature of this process suggests that larger mass protoplanetary disks may produce only slightly more massive planetary systems; however, the complexities of gas accretion by planets and of disk dispersal time scales cast some doubt upon this conclusion.

Quantitative theories of celestial mechanics are developed primarily to study our Solar System. In this case, the major planets have well separated nearly circular orbits such that they have undergone very little orbital evolution. However, the common finding of eccentric extrasolar planets and the discoveries of three Jupiter-mass planets around the star Upsilon Andromedae suggest that the above dynamical processes may provide some general selection criteria which determine the eventual orbital evolution of planetary systems (see §5). The actual initial conditions are determined by the planet formation processes.

3 Planet Formation

Even a very slowly rotating molecular cloud core has far too much angular momentum to collapse down to an object of stellar dimensions, so a significant fraction of the material in a collapsing core falls onto a rotationally-supported disk orbiting the pressure-supported (proto)star. Such a disk has the same initial elemental composition as the growing star. At sufficient distances from the central star, it is cool enough for $\sim 1-2\%$ of this material to be in solid form, either remnant interstellar grains or condensates formed within the disk. During the infall stage, the disk is very active and probably highly turbulent, as a result of the mismatch of the specific angular momentum of the gas hitting the disk with that required to maintain Keplerian rotation (Cassen & Moosman 1981). Gravitational instabilities and viscous and magnetic forces may add to this activity (Adams & Lin 1993, Lin & Papaloizou 1996). When the infall slows substantially or stops, the disk becomes more quiescent. Interactions with the gaseous component of the disk affect the dynamics of small solid bodies, and the growth from micron-sized dust to kilometer-sized planetesimals remains poorly understood (Weidenschilling & Cuzzi 1993).

The dynamics of larger solid bodies within protoplanetary disks are better characterized. The primary perturbations on the Keplerian orbits of kilometer-sized and larger planetesimals in protoplanetary disks are mutual gravitational interactions and physical collisions. These interactions lead to coagulation (and in some cases erosion and fragmentation) of planetesimals. In principle, gravitational encounters stir planetesimal random velocities up to the escape speed from the largest common planetesimals in the swarm (Safronov 1969). But, the equilibrium magnitude of the velocity dispersion is determined by the dynamical interaction between different populations of planetesimals (Aarseth *et al.* 1993; Kokubo & Ida 1996, 1998). The most massive planets have the largest gravitationally-enhanced collision cross-sections, and accrete almost everything with which they collide. If the random velocities of most planetesimals remain much smaller than the escape speed from the largest bodies, then these "planetary embryos" grow extremely rapidly. The size distribution of solid bodies becomes quite skewed, with a few large bodies growing much faster than the rest of the swarm, in a process referred to as "runaway accretion" (Wetherill & Stewart 1989). Eventually, planetary embryos accrete most of the small bodies within their gravitational reach, and the runaway growth phase ends (Ida & Makino 1993).

Slower growth continues (at least for terrestrial-type planets) as the eccentricities of planetary embryos are pumped up by long-range mutual gravitational perturbations (Chambers *et al.* 1996, Kokubo & Ida 1996). As planetary masses increase, they become more efficient at stirring random velocities of neighboring bodies. An increase in the velocity dispersion would enlarge the "feeding zone" and reduce the capture cross-section of the planetary embryos. In this limit, several large bodies may coexist as the growth of the most massive embryos is suppressed by their own gravitational perturbation on the nearby planetesimals (Kokubo & Ida 1998). Eventually, the mutual gravitational interaction between these embryos would excite their eccentricity to values larger than that of the Solar System planets and prolong buildup (Wetherill 1990). The inferred growth timescale of a few Earth-mass embryos at the present location of Jupiter and Saturn is much longer than the persistence timescale for typical protostellar disks. However, according to the conventional theory of planetary origins, the formation of $\sim 10\ M_\oplus$ (Earth mass) cores is necessary prior to the accretion of gas onto giant planets such as Jupiter and Saturn (see below). Runaway accretion in a disk containing at least ~ 3 times the amount of solids that has been retained by the planets is one proposed solution (Lissauer 1987). Another possible avenue to bypath the growth bottle neck is the orbital evolution of protoplanetary cores (Bryden & Lin 1999). Most encounters between the field planetesimals and planetary embryos or protoplanetary cores lead to close scattering rather than direct collisions (cf. Greenzweig & Lissauer 1992). The resultant recoils induce a Brownian motion (Tanaka & Ida 1997) which continually moves the cores to unperturbed regions. The fresh supply of planetesimals with low random velocities enables the embryos and protoplanetary cores to maintain a high rate of growth and low eccentricity (Bryden *et al.* 1999b).

Protoplanetary cores with a fraction of an Earth mass have surface escape velocities larger than the gas sound speed in the protostellar disk. Such a growing core first attains a quasi-static atmosphere which undergoes Kelvin-Helmholtz contraction as the energy released by the planetesimal and gas accretion is radiated away at the photosphere. The contraction timescale is determined by the efficiency of radiation transfer, which is relatively low in some regions of the envelope. Spherically symmetric (1-D) models show that the minimum contraction timescale is a rapidly decreasing function of the cores' mass (Bodenheimer and Pollack 1986). The maximum gas accretion rate becomes comparable to the planetesimal bombardment rate only after the cores have already acquired $\sim 10 - 20\ M_\oplus$. In some models (Pollack *et al.* 1996), this growth bottleneck prolongs the timescale of gaseous planetary formation to be comparable to the observed persistence timescales of protostellar disks such that it remains unclear whether such a sequence of events could induce the efficient formation of giant planets in most protoplanetary disks (Wetherill 1994).

The growth of gaseous planets through envelope contraction would be more rapid if energy transport were more efficient. Multi-dimensional numerical simulations show that the stellar tidal torque induces a rapid rotation on the gas which accumulates near the cores (Lin *et al.* 2000). If such motion could lead to rotational instability (Goldreich & Schubert 1967) and induce meridional circulation, the rate of heat transfer may be enhanced. After a core has acquired $\sim 10 - 20$ $M_\oplus$, its growth is primarily due to gas accretion. Thereafter, the contraction rate of the envelope increases rapidly as it is replenished by the disk gas flowing across the planet's Roche lobe. The planet's tidal perturbation also excites density waves in the disk and induces the convergence of streamlines to form pronounced trailing wakes (Goldreich & Tremaine 1979, Lin & Papaloizou 1979, 1993). These waves carry a negative/positive angular momentum flux as they propagate away from the planet to the disk interior/exterior to its radius. When these waves are dissipated, angular momentum is effectively removed/supplied from/to the disk. A protoplanet significantly perturbs the structure and evolution of the disk when its tidal torque exceeds the viscous stress of the disk gas. Thereafter, a protoplanet induces the formation of a gap in the disk unless its Roche radius is less than the disk thickness (Papaloizou & Lin 1984). The depletion of gas in the gap reduces the gas supply to the contracting envelope and effectively terminates the planet's mass increase when the growth timescale become longer than the disk persistent timescale (Bryden *et al.* 1999a). Such a self-regulated growth limit provides a possible explanation to the observed mass distribution of extrasolar giant planets, most of which have $M \sin i < 4$ M_J (Marcy & Butler 1998).

Another uncertainty associated with the emergence of planets is their orbital migration as a consequence of angular momentum exchange between the disk and the planet while the later is still accreting from the disk (Goldreich & Tremaine 1980, Ward 1986). Based on a recent calculation which indicates that the torque exerted by the planet on the outer disk is stronger than that on the inner disk, Ward (1997) suggested that embedded planets migrate inwards on a timescale $\sim 10^5 (M_p/M_\oplus)^{-1}$ yr. If this process occurs in protostellar disks, protoplanets would migrate towards the stellar surface once their masses $M_p > 1$ $M_\oplus$ because their growth timescale (Lissauer & Stewart 1993) would then become longer than this migration timescale. Resolution of this rapid migration dilemma may require the complete and nonlinear analysis of the disk response to the protoplanet in the corotation regions which may be quite complex.

4 Migration of Giant Planets

Orbital migration of a planet becomes unavoidable after it has acquired a sufficient mass to open up a gap in the disk. After the termination of their growth through gap formation, protoplanets continue to remove angular momentum from the inner disk (Lin 1986). Observations of classical T Tauri stars indicate that gas in the inner region of protostellar disks is being continually depleted by accretion onto the young stellar objects. Without a mass supply, the surface density and the tidal angular momentum transfer rate interior to the planet's orbit decrease. In the outer disk, these quantities maintain their value as gas is prevented from viscously diffusing inwards by the protoplanet's tidal torque. The imbalance between the inner and outer disk leads to the inward orbital migration of the protoplanet. If the planet's mass is less than that of the disk, its orbital migration is coupled to the viscous evolution of the disk (Lin & Papaloizou 1986, Takeuchi *et al.* 1996), which occurs on the timescales of $\sim 10^6$ yr.

These considerations have led to the speculation that tidal evolution would cause some first-born protogiant planets to migrate towards and eventually merge with their host stars. This "infant mortality" would continue until the nebula mass is depleted to such an extent that the residual gas can no longer induce any significant evolution of the protogiant planets. According to this scenario, Jupiter and Saturn may be the last survivors (Lin 1995).

Migration speeds are expected to increase as planets moves closer to the stellar surface, so no one expected to find any direct evidence for such a process. But several short-period planets have been discovered in the past few years (Mayor & Queloz 1995, Marcy & Butler 1998). These planets, henceforth referred to as "vulcans", after the hypothetical planet once believed to travel about the Sun within the orbit of Mercury, have periods less than one week. It is possible that these vulcans may have formed substantially farther from the star and subsequently migrated inwards to their current short-period orbits (Lin *et al.* 1996). A planet would terminate its migration less than one-tenth of an AU from the star if (1) the disk does not extend to the stellar surface or (2) the host star induces angular momentum transfer to the planet's orbit.

For the first possibility, interaction between the disk and an intense ($>$ 1 kG) stellar magnetic field, has been suggested (Konigl 1991) as the cause for the modest observed rotation period ($P_* \sim 8$ days) of classical T Tauri stars (Bouvier et al. 1993; Choi & Herbst 1996) being much longer than that required to shed mass and angular momentum from the equator. Such a magnetic field strength may be consistent with recent measurements (Guenther 1997). In this scenario, the stellar field is assumed to induce a cavity out to the magnetospheric radius (a few $\times\ 10^{11}$ cm) where the local Keplerian period of the disk matches the stellar spin rate. When a protoplanet migrates interior to the 2:1 resonance of the gas at the magnetosphere radius, the protoplanet is decoupled from the disk and its migration is stalled. The main uncertainties here are in the observed distribution of stellar spin periods. The discovery of several fast rotating classical T Tauri stars casts doubts on the general effectiveness of magnetosphere-disk interaction and the common existence of magnetospheric cavity (Stassun *et al.* 1999).

The second possibility arises in the limit of a weak stellar magnetic field. In this case, the disk's inner boundary extends to the star's photosphere and the host stars are likely to be rapid rotators during the T Tauri phase. A tidal equilibrium may be established at a radius outside the corotation radius such that the planet's loss of angular momentum to the disk is compensated by its gain from the host star (Lin et al. 1996). The protoplanet migration is temporarily interrupted at the tidal equilibrium radius. However, on a timescale comparable to or shorter than the observed disk lifespan ($\sim 10^6$ yr), this transfer of angular momentum would induce the star to spin down and the planet to resume its inward migration. Nonetheless, after the closest planets are consumed by the star, it spins up and this process recurs for later protoplanetary arrivals until the disk is sufficiently depleted. Thus, the spin period of the protoplanet is expected to oscillate around a few days as observed in typical classical T Tauri stars. Note that this alternative explanation for the relatively slow rotation speed does not require the magnetosphere-disk interaction.

The main uncertainties associated with these scenarios are: 1) the common existence of sufficiently strong magnetic field in classical T Tauri stars and 2) the efficiency of tidally induced angular momentum exchange (Terquem *et al.* 1998). Furthermore, the discovery of planets with orbital periods ranging from 15 days to three years, which feel negligible tidal torque from their star and are unlikely to have entered a clear zone of the disk, cast doubt upon this model. More complex interaction between several planets and different regions of the protostellar disk may be required to account for the origin of these intermediate-period planets (Lin 1997). The discovery of the multiple-planet system around υ Andromedae may provide important clues and constraints (see below).

5 Other Models for the Formation of Planets Observed About Other Stars

Each of the planets identified in radial velocity surveys prior to the VLT Opening Symposium in March 1999 orbited a different star. This is in sharp contrast to our Solar System, where planets of comparable size have orbital periods within a factor of two or three of their neighbors. These systems thus appear to be more overstable than is our own (see Section 2), which may indicate different mechanisms important in the formation process. However, the recently-announced multiple-planet system around υ Andromedae (Butler *et al.* 1999) appears to be substantially closer to dynamical instability than does our Solar System (Lissauer 1999).

Some of the known extrasolar giant planets move on quite eccentric ($0.2 < e < 0.7$) orbits. These eccentric orbits may be the result of stochastic gravitational scatterings among massive planets (which have subsequently merged or been ejected to interstellar space, Rasio *et al.* 1996, Weidenschilling & Mazari 1996, Lin & Ida 1997), by perturbations of a binary companion (Holman *et al.* 1997), or by past stellar companions if the now single stars were once members of unstable multiple star systems (Laughlin & Adams 1999). However, as neither scattering nor migration offer a simple explanation for those planets with nearly circular orbits and periods from a few weeks to a few years, the possibility of giant planet formation quite close to stars should not be dismissed (Bodenheimer *et al.* 1999).

Prior to the discovery of extrasolar planets, models of planetary growth suggested that most single solar-type stars possess planetary systems which are grossly similar to our Solar System. It was realized that stochastic factors are important in planetary growth, so that the number of terrestrial planets (as well as the presence or lack of an asteroid belt) would vary from star to star, even if their protoplanetary disks were initially very similar (Wetherill 1990). The difficulty in accreting giant planet atmospheres prior to dispersal of circumstellar gas suggested that many systems might lack gas giants (Wetherill 1994, Lissauer 1995). The low eccentricities of the giant planets in our Solar System (especially Neptune) are difficult to account for (Lissauer *et al.* 1995), so systems with planets o n highly eccentric orbits were viewed as possibilities, although researchers did not hazard to estimate the detailed characteristics of such systems. A maximum planetary mass similar to that of Jupiter was suggested as a possibility if Jupiter's mass was determined by a balance between a planet's gap clearing ability and viscous inflows (Lin & Papaloizou 1979), although it was noted that the value of the viscosity could well vary from disk to disk. Orbital migration of some giant planets towards their parent star was also envisioned (Goldreich & Tremaine 1980), but since migration rates increased as the planet approached the star, such planets were expected to be accreted by their star (Lin 1986, Ward 1986, Lissauer 1995), and the existence of numerous giant planets with orbital periods ranging from a few days to several weeks was not predicted.

It must thus be admitted that theoretical models based upon observations within our Solar System failed to predict the types of planets thus far detected by radial velocity surveys. On the other hand, note that the surveys which prove that such planets exist also show that they are fairly rare, occurring in fewer than ten percent of the systems. The radial velocity surveys conducted thus far are quite biased in favor of detecting massive planets orbiting close to stars, and planets similar to those in our own Solar Systems would not yet have been detected. Thus, it is possible that the vast majority of single Sun-like stars possess planetary systems quite similar to our own. Alternatively, although theoretical considerations suggest that terrestrial planets are likely to grow around most Sun-like stars, they may typically be lost if most systems also contain giant planets which migrate into the central star.

6 Observing Extrasolar Planets and Protoplanetary Disks

Until recently, our ability to understand how planets form was constrained because we had observed only one planetary system, our own Solar System. Several extrasolar planets have been discovered within the past few years, and many more are likely to be found in the upcoming decades. The VLTI is likely to play a major role in detecting extrasolar planets during the first few decades of the 21st century. Astrometric measurements by the VLTI should detect planets indirectly. Direct imaging of giant planets may also be possible.

This work was supported by NSF through grant AST-9714275, NASA through its Planetary Geology & Geophysics Program under RTOP 344-30-50, its Origins program under NAG5-4277, 7515, and 8196, and an astrophysics theory program which supports a joint Center for Star Formation Studies at NASA-Ames Research Center, UC Berkeley, and UC Santa Cruz.

References

1. Aarseth, S., Lin, D. N. C., Palmer P. (1993) Evolution of planetesimals II. numerical simulations. Astrophys. J. **403**, 351–376.
2. Adams, F., Lada, C. J., Shu, F. H. (1987) Spectral evolution of young stellar objects. Astrophys. J. **312**, 788–806.
3. Adams, F., Lin, D. N. C. (1993) Transport processes and the evolution of disks. In Protostars and Planets III (E.H. Levy and J.I. Lunine, Eds.), pp. 721–748. University of Arizona Press, Tucson.
4. Bodenheimer, P., Hubickyj, O., Lissauer, J. J. (1999) Models of the in situ formation of detected extrasolar giant planets. Icarus, in press.
5. Bodenheimer, P., Pollack, J. B. (1986) Calculations of the accretion and evolution of giant planets. The effects of solid cores. Icarus **67**, 391–408.
6. Bouvier, J., Cabrit, S., Fernandez, M., Martin, E. L., Matthews, J. M. (1993) Coyotes - part one - multisite UBVRI photometry of 24 pre main sequence stars of the Taurus-Auriga cloud. Astron. Astrophys. Suppl. **101**, 485–498.
7. Butler, R. P., Marcy, G. W., Fischer, D. A., Brown, T. W., Contos, A. R., Korzennik, S. G., Nisenson, P., Noyce, R. W. (1999) Evidence for multiple companions to Upsilon Andromedae. Astrophys. J., submitted.
8. Byrden, G., Lin, D. N. C. (1999) On the formation of multiple planetary system. in preparation.
9. Byrden, G., Chen, X., Lin, D. N. C., Nelson, R. P. Papaloizou, J. C. B. (1999a) Tidally induced gap formation in protostellar disks: gap clearing and suppression of protoplanetary growth. Astrophys. J. **514**, 344–367.
10. Byrden, G., Lin, D. N. C., Ida, S. (1999b) Protoplanetary formation I: Neptune Astrophys. J., in press.
11. Cassen, P. M., Moosman, A. (1981) On the formation of protostellar disks. Icarus **48**, 353–376.
12. Cassen, P., Shu, F. H., Terebey, S. (1985) Protostellar disks and star formation: An overview. In Protostars and Planets II, (D.C. Black and M.S. Matthews, Eds.), pp. *448–483*. University of Arizona Press, Tucson.

13. Chambers, J. E., Wetherill, G. W., Boss, A. P. (1996) The stability of multi-planet systems. Icarus **119**, 261–268.
14. Choi, P. I., Herbst, W. (1996) Rotation periods of stars in the Orion nebula cluster: The bimodal distribution. Astron. J. **111**, 283–298.
15. Chiang, E., Goldreich, P. (1997) Spectral energy distributions of T Tauri stars with protostellar disks. Astrophys. J. **490**, 368–376.
16. Duncan, M., Quinn, T., Tremaine, S. (1989) The long-term evolution of orbits in the Solar System: A mapping approach. Icarus **82**, 402–418.
17. Duncan, M., Lissauer, J.J. (1998) The effect of post-main-sequence solar mass loss on the stability of our planetary system. Icarus. **134**, 303–310.
18. Gladman, B. (1993) Dynamics of systems of two close planets. Icarus **106**, 247–263.
19. Goldreich, P., Schubert, G. (1967) Differential rotation in stars. Astrophys. J. **150**, 571–587.
20. Goldreich, P., Tremaine, S. (1979) The excitation of density waves at the Lindblad and corotation resonances by an external potential. Astrophys. J. **233**, 857–871.
21. Goldreich, P., Tremaine, S. (1980) Disk-satellite interactions. Astrophys. J. **241**, 425–441.
22. Goodman, A. A., Benson, P. J., Fuller, G. A. Myers, P. C. (1993) Dense cores in dark clouds. VIII -Velocity gradients. Astrophys. J. **407**, 175–184.
23. Greenzweig, Y., Lissauer, J. J. (1992) Accretion rates of protoplanets. II- Gaussian distribution of planetesimal velocities. Icarus **100**, 440–463.
24. Guenther, E. W. (1997) Magnetic fields in T Tauri stars. In Herbig-Haro Flows and the birth of stars; IAU Symp. no. 182 (B. Reipurth and C. Bertout eds), pp. 465–474. Kluwer Dordrecht.
25. Hartmann, L., Calvet, N., Gullbring, E., D'Alessio, P. (1998) Accretion and the evolution of T Tauri disks. Astrophys. J. **495**, 385–400.
26. Hillenbrand, L., Strom, S. E., Calvet, N., Merrill, K. M., Gatley, I., Makidon, R. B., Meyer, M. R. Skrutskie, M. F. (1998) Circumstellar disks in the Orion Nebula cluster. Astron. J. **116**, 1816–1841.
27. Holman, M. J., Touma, J., Tremaine, S. (1997) Chaotic variations in the eccentricity of the planet orbiting 16 Cygni B. Nature **386**, 254–256.
28. Ida, S., Makino, J. (1993) Scattering of planetesimals by a protoplanet: Slowing down of runaway growth. Icarus **106**, 210–227.
29. Kokubo, E., Ida, S. (1996) On runaway growth of planetesimals. Icarus **123**, 180–191.
30. Kokubo, E., Ida, S. (1998) Oligarchic growth of protoplanets. Icarus **131**, 171–178.
31. Konigl, A. (1991) Disk accretion onto magnetized T Tauri stars. Astrophys. J. Lett. **370**, 39–43.
32. Laughlin, G., Adams, F. C. (1998) The modification of planetary orbits in dense stellar clusters. Astrophys. J. Lett. **508**, 171–174.
33. Levison, H. F., Lissauer, J. J., Duncan, M. J. (1998) Modeling the Diversity of Outer Planetary Systems. Astron. J. **116**, 1998–2014.
34. Lin, D. N. C. (1986) The nebular origin of the Solar System. In The Solar System: Observations and Interpretations. (M. Kivelson, Ed.) pp. 28–77. Prentice-Hall, Englewood Cliffs.

35. Lin, D. N. C. (1995) Protostellar disk and the solar nebula. In Molecular Clouds and Star Formation (C. Yuan and J.H. Yau, Eds.) pp. 261–292. World Scientific, Singapore.
36. Lin, D. N. C. (1997) Planetary formation in protostellar disks. In Accretion phenomena and related outflows, IAU Coll. 163 (D.T. Wickramasinghe, G.V. Bicknell, and L. Ferrario eds). pp. 321–330. Astron. Soc. Pac., San Francisco.
37. Lin, D. N. C., Bodenheimer, P., Richardson, D. C. (1996) Orbital migration of the planetary companion of 51 Pegasi to its present location. Nature **380**, 606–607.
38. Lin, D. N. C., Ida, S. (1997) On the origin of massive eccentric planets. Astrophys. J. **477**, 781–791.
39. Lin, D. N. C., Papaloizou, J. C. B. (1979) Tidal torques on accretion disks in binary systems with extreme mass ratios. Mon. Not. R. Astr. Soc. **186**, 799–812.
40. Lin, D. N. C. and Papaloizou, J. C. B. (1986) On the tidal interaction between protoplanets and the protoplanetary disk III - Orbital migration of protoplanets. Astrophys. J. **309**, 846–857.
41. Lin, D. N. C. Papaloizou, J. C. B. (1993) On the tidal interaction between protostellar disks and companions. In Protostars and Planets III (E.H. Levy and J.I. Lunine, Eds.), pp. 749–835. University of Arizona Press, Tucson.
42. Lin, D. N. C., Papaloizou, J. C. B. (1996) Theory of accretion disks II: application to observed systems. Annu. Rev. Astron. Astrophys. **34**, 703–748.
43. Lin, D. N. C., Papaloizou, J. C. B., Terquem, C., Bryden, G., Ida, S. (2000) Orbital evolution and planet-star tidal interaction. Protostars and Planets IV, V. Mannings, S. Russel and A.P. Boss, eds. (Tucson: University of Arizona Press), in press.
44. Lissauer, J. J. (1987) Timescales for planetary accretion and the structure of the protoplanetary disk. Icarus **69**, 249–265.
45. Lissauer, J. J. (1993) Planet formation. Annu. Rev. Astron. Astrophys. **31**, 129–174.
46. Lissauer, J. J. (1995) Urey Prize lecture: On the diversity of plausible planetary systems. Icarus **114**, 217–236.
47. Lissauer, J. J. (1999) Three planets of Upsilon Andromedae. Nature **398**, 659-660.
48. Lissauer, J. J., Pollack, J. B., Wetherill, G. W., Stevenson D. J. (1995) Formation of the Neptune system. In Neptune and Triton (D. Cruikshank, Ed.) University of Arizona Press, Tucson, pp. 37–108.
49. Lissauer, J. J., Stewart, G. (1993) Growth of planets from planetesimals. In Protostars and Planets III (E.H. Levy and J.I. Lunine, Eds.), pp. 1061–1089. University of Arizona Press, Tucson.
50. Marcy, G. W., Butler, R. P. (1998) Detection of extrasolar giant planets. Annu. Rev. Astron. Astrophys. **36**, 57–97.
51. Mayor, M., Queloz, D. (1995) A Jupiter-mass companion to a solar-type star. Nature **378**, 355–359.
52. McCaughrean, M. J., O'Dell, C. R. (1996) Direct imaging of circumstellar disks in the Orion Nebula. Astron. J. **111**, 1977–1986.
53. Mundy, L. G., Loonet, L. W., Erickson, W., Grossman, A., Welch, W. J., Forster, J. R., Wright, M. C. H., Plambeck, R. L., Lugten, J., Thornton, D. D. (1996) Imaging the HL Tauri Disk at lambda = 2.7 Millimeters with the BIMA Array. Astrophys. J. Lett. **464**, 169–173.

54. Murray, N., Holman, M. (1999) The origin of chaos in the outer Solar System. Science **283**, 1877–1881.
55. O'Dell, C. R., Wen, Z., Hu, X. (1993) Discovery of new objects in the Orion nebula on HST images - Shocks, compact sources, and protoplanetary disks. Astrophys. J. **410**, 696–700.
56. Papaloizou, J. C. B., Lin, D. N. C. (1984) On the tidal interaction between protoplanets and the primordial solar nebula. I - Linear calculation of the role of angular momentum exchange. Astrophys. J. **285**, 818–834.
57. Pollack, J. B., Hubickyj, O., Bodenheimer, P., Lissauer, J. J., Podolak, M., Greenzweig, Y. (1996) Formation of the giant planets by concurrent accretion of solids and gas. Icarus **124**, 62–85.
58. Rasio, F. A., Ford, E. B. (1996) Dynamical instabilities and the formation of extrasolar planetary systems. Science **274**, 954–956.
59. Safronov, V. S. (1969) Evolution of the Protoplanetary Cloud and Formation of the Earth and Planets. Moscow, Nauka Press (in Russian). English translation: NASA TTF-677, 1972.
60. Stassun, K. G., Mathieru, R. D., Mazeh, T., Brba, F. J. (1999) Examining disk-regulated rotation: The rotation period distribution of pre-main sequence stars in and around the Orion Nebula. Bull. Amer. Astron. Soc. **194**, 6806.
61. Strom, S. E., Edwards, S., Skrutskie, M. F. (1993) Evolutionary timescales for circumstellar disks associated with intermediate- and solar-type stars. In Protostars and Planets III (E.H. Levy and J.I. Lunine, Eds.), pp. 837–866. University of Arizona Press, Tucson.
62. Takeuchi, T., Miyama, S., Lin, D. N. C. (1996) Gap formation in protoplanetary disks. Astrophys. J. **460**, 832–847.
63. Tanaka, H., Ida, S. (1997) Distribution of planetesimals around a protoplanet in the nebula gas. Icarus. **125**, 302–316.
64. Terebey, S., Shu, F. H., Cassen, P. (1984) The collapse of the cores of slowly rotating isothermal clouds. Astrophys. J. **286**, 529–551.
65. Terquem, C., Papaloizou, J. C. B., Nelson, R. P., Lin, D. N. C. (1998) On the tidal interaction of a solar-type star with an orbiting companion: excitation of g-mode oscillation and orbital evolution. Astrophys. J. **502**, 788–801.
66. Ward, W. R. (1986) Density waves in the solar nebula: Differential Lindblad torque. Icarus **67**, 164–180.
67. Ward, W. R. (1997) Survival of planetary systems. Astrophys. J. Lett. **482**, 211–215.
68. Weidenschilling, S. J., Cuzzi, J. N. (1993) Formation of planetesimals in the Solar Nebula. In Protostars and Planets III (E. H. Levy and J. I. Lunine, Eds.) pp. 1031–1060. University of Arizona Press, Tucson.
69. Weidenschilling, S. J., Marzari, F. (1996) Gravitational scattering as a possible origin for giant planets at small stellar distances. Nature **384**, 619–621.
70. Wetherill, G. W. (1990) Formation of the Earth. Ann. Rev. Earth Planet Sci. **18**, 205–256.
71. Wetherill, G. W. (1994) Possible consequence of absence of "Jupiters" in planetary systems. Astrophys. Space Sci. **212**, 23–32.
72. Wetherill, G. W., Stewart, G. R. (1989) Accumulation of a swarm of small planetesimals. Icarus **77**, 330–357.
73. Wuchterl, G., Guillot, T., Lissauer, J. J. (2000) Giant Planet Formation. Protostars and Planets IV, V. Mannings, S. Russel and A.P. Boss, eds. (Tucson: University of Arizona Press), in press.

Solar Nebulae and Replenished Dust Disks with Planets

Pawel Artymowicz

Stockholm Observatory, Stockholm University,
SE-13336 Saltsjöbaden, Sweden

Abstract. Having studied the primordial, protoplanetary gas disks, and older, dusty "remnant" disks in infrared-emitting (Vega-type) circumstellar systems, we begin to observe and interpret the intermediate-stage or transition disks, with similar quantities of gas and dust. The disks may harbor planets leaving characteristic fingerprints on disk structure. However, dust can sometimes migrate rapidly within disks, mimicking planetary influences.

1 Introduction

Having studied for several decades the massive, primordial gas disks, and more recently the dusty "remnant" disks in infrared-emitting (Vega-type) planetary systems, we are now ready to observe and interpret theoretically the intermediate-age or transition disks. Such disks may contain comparable amounts of dynamically coupled gas and dust. Disks can be perturbed by embedded planet(s) providing opportunity for planet discovery, and valuable diagnostics for planetary properties. However, dust migration in disks sometimes mimics planetary influences.

2 How Planets Grow in Disks

For the forming planetary systems at the T Tau or HAeBe stage of stellar evolution, we do not have yet direct spatially-resolved observations to verify the existence of planets and their interaction with primordial disks. However, theory allows us to understand and predict the most obvious effects of such interaction: gap opening, flow of mass onto the growing planet, and the modification of planetary orbit by the disk gravity (back-reaction of disk on planet).

Two standard gap opening criteria, a viscous and a thermal one, have been formulated and widely used (e.g, Goldreich & Tremaine 1980, Lin & Papaloizou 1993). The gap does not open unless viscous torque in the disk is overpowered by gravitational torque of the planet (for which the mass ratio of the planet must exceed 40/Re, where Re is the Reynolds number in the disk). *Thermal criterion* requires, in addition, that the Roche lobe of the planet exceeds the disk scale height. High-resolution numerical hydrodynamical calculations are now possible to study gap opening, spiral wave production (with associated torques), and other processes.

Figure 1 shows the snapshot from our Piecewise Parabolic Method (PPM) simulation of an accretion disk resembling standard, low-mass, solar nebula, with embedded Neptune-class planet (25 Earth mass-body). This simulation shows a significant gap in the disk, characterized by a density ratio in/outside the gap of reaching 1:10. Unexpectedly, this happens despite the fact that the viscous criterion is violated by a factor of 2.5 and the thermal one by a factor of 5 (standard criteria require ~100 Earth masses for gap opening). More work on gap-opening criteria is clearly needed.

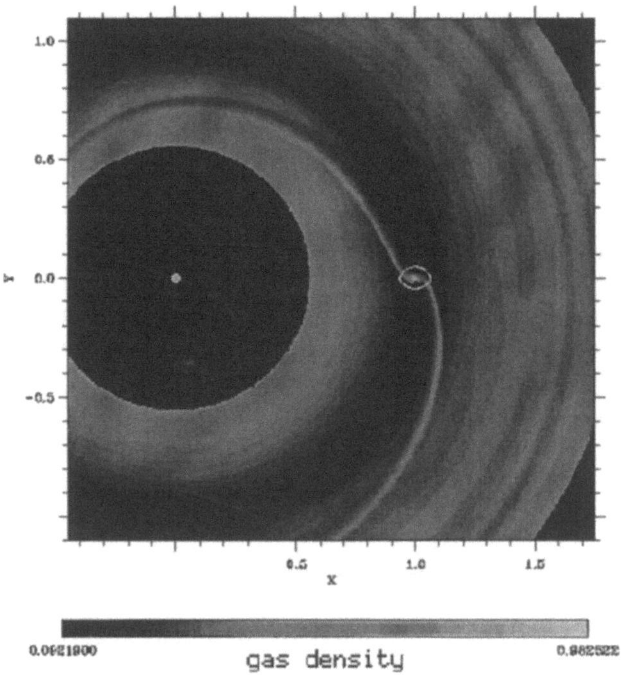

Fig. 1. A 25-Earth mass protoplanet begins to open a gap in the surrounding solar nebula (viscosity parameter $\alpha = 0.006$, thickness ratio $z/r = 0.05$). Gas density in arbitrary units is color-coded. The orbit of the protoplanet has eccentricity $e = 0.01$. The ovals centered at position (1,0) in this and the next figure are size markers showing the Roche lobe of a Jupiter-mass planet in a circular orbit

The standard theory also predicted that once the gap opens, proto-giant planets stop growing. Artymowicz & Lubow (1996) proposed that gas can be accreted through the gap onto the planet (from inside and outside disk parts), at a rate governed by disk viscosity and pressure (thickness). Several groups confirmed the gas flows through gaps, and obtained similar numerical results on the planet growth time scale under typical disk conditions (Artymowicz 1997a, Artymowicz et al 1997, Kley 1999, Bryden et al. 1999, Lubow et al. 1999). Nevertheless, differences in the interpretation of this phenomenon persist. Invoking very low viscosity coefficients, Bryden et al. stress that the planet's mass at which the growth *practically* terminates may not be not totally different than those predicted by standard gap opening criteria. Artymowicz et al. emphasize that Jupiter mass was *not* limited by gap opening, and that accretion disk properties may vary, leading to a wide spectrum of giant planet masses including occasional brown dwarf-type bodies. For example, solar nebula with $\alpha \sim 10^{-3}$ doubles the proto-Jupiter mass in less than 1 Myr, which makes such a system applicable to extrasolar but not the solar giant planets. The newly found "superplanets" an order of magnitude more massive than Jupiter (such as 70 Vir, and other) may be a natural product of sufficiently massive, hot, viscous, and/or atypically long-lived, disks.

Figure 2 illustrates the continued accretion onto a Jupiter-mass planet on circular orbit, located in a disk gap, in a PPM calculation. Only a portion of the disk is shown, although the whole disk was simulated. Similar high-resolution simulations with ZEUS code were performed by Lubow et al. (1999). They established that most mass enters the Roche lobe after first hitting the spiral shock wave near Lagrange equilibrium points.

How are the orbits of the growing giant planets evolving? We only know partial answers. The issue is greatly complicated by the presence of two, in principle comparable, sources of energy (E) and angular momentum (L) given to or removed from the planet: long-range gravitational forces and the E and L advected onto the planet, or planet+satellite disk system, with the accreting nebular gas. Much more work needs to be done in this area before we quantify all the important rates. Important new phenomena might result from advective terms. While it is well known that gravity pushes the gap-opening planets onto their central stars by robbing the planet of its L, the opposite may happen if the planet is already close to the star, the outer disk is much more massive than whatever gas remains between the planet and the star, and the accretion of high-L outer-disk gas cancels out or reverses the inward migration, perhaps saving the planet from plunging into the star.

The disk-planet interaction may be responsible for the apparent mass-eccentricity correlation among extrasolar planets (Mazeh et al. 1997), recently accentuated by the discovery of the first multiplanet system around normal star (υ Andromedae; Butler et al. 1999). In that system, two dominant planets exhibit eccentricities of about 0.2 and 0.4, for minimum masses of 2 and 4 Jupiters, correspondingly. Although eccentricity of orbits may well depend on the planet-planet perturbations both during the formative period and in the subsequent main-sequence period, we stress here that the typical result of any N-body interaction would be to circularize larger bodies at the cost of increasing elliptical deformation of smaller bodies. Often this effect can be described in terms of the tendency for energy equipartition, where energy

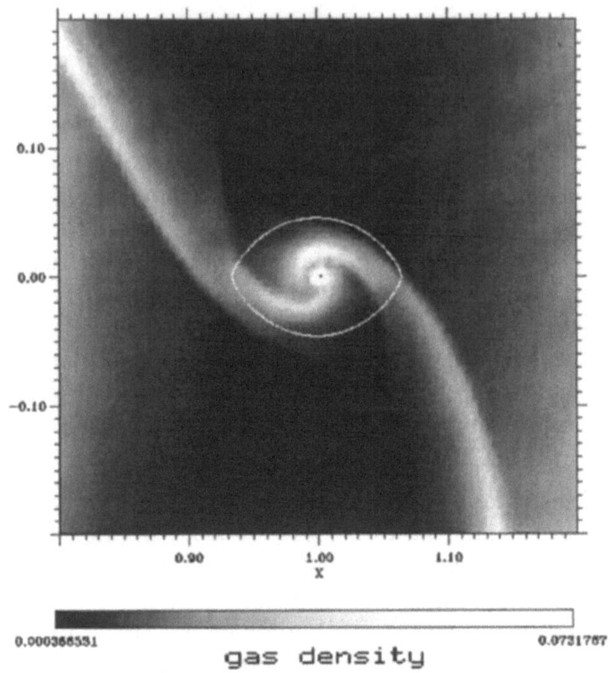

Fig. 2. As Fig. 1, except for a 1 Jupiter-mass protoplanet. Only the vicinity of the planet is shown, giving insight into the intricate pattern of shock waves and gas inflow into the Roche lobe

is taken to mean the epicyclic energy of non-circular motion. Even small-N systems like two planets and the star exhibit the above behavior, which is why the first calculations of the long-term evolution of υ And show that it tends to transform from the a direct to an inverse correlation between eccentricities and masses of planets (if it survives at all; see Lissauer 1999, Laughlin & Adams 1999). Since υ And apparently survived 2-3 Gyr in its current direct-correlation state despite the planet-planet interaction, we can suspect that such a situation was established by disk-planet interaction in the PMS phase of evolution.

3 Planetesimal Disks

The Vega-type or β Pic-type disks, surrounding $\sim$ 20% of all normal stars, are dust debris disks with some presence of gas, produced by disruption (by heat, cratering and mutual collisions) of swarms of planetesimals. β Pictoris and other prominent systems are similar to but 10^2 to $>10^4$ times more dusty than our own Zodiacal Light disk, replenished by comets and asteroids (Artymowicz 1997b, Vidal-Madjar 1998).

3.1 Beta Pictoris-Type Disks

Signs of bodies larger than planetesimals have been found in Vega-type systems. The ever-present central disk gaps are usually taken as evidence for the dynamical clearing by the gravity of planets (the gap sizes usually span the planetary region in our system). The best evidence to date comes from the β Pic disk. Seen almost edge-on, the disk has a central gap, and right outside of it (at a distance comparable with the Kuiper Belt of our own system) the disk is seen to be asymmetric, as if warped by an unseen planet orbiting inside the gap (Figure 3). Mouillet et al. (1997) showed that a variety of Jupiter/sun-like configurations could provide the warp, provided that the outer disk parts are inclined to the planet's orbit by about 3 degrees. At least one planets gravity seems to be required to cause the frequent infall events of planetesimals onto the star or its vicinity (Beust & Morbidelli 1996).

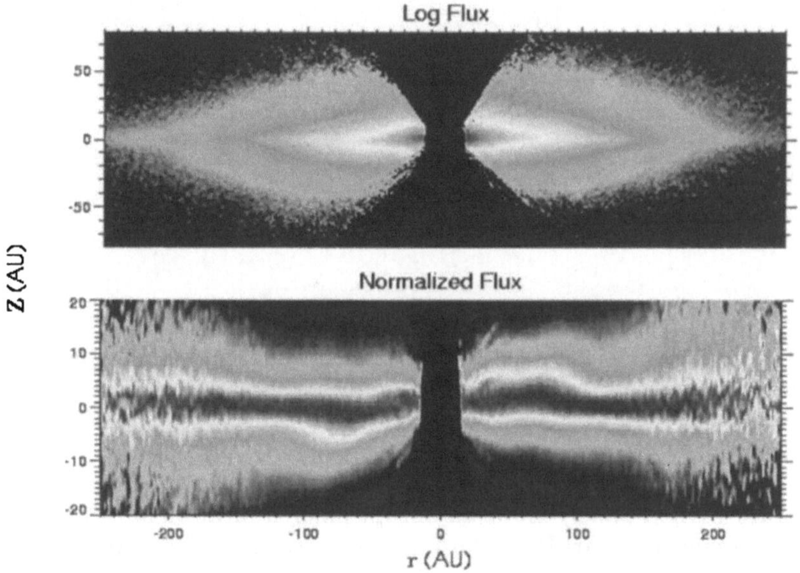

Fig. 3. One of the best scattered-light images of the β Pic disk (upper panel) was obtained with STIS/HST by Heap et al. (1999). A warp-like feature at projected distances 15-100 AU from the star is seen in the lower panel, where the image was renormalized to unit intensity in each pixel column, and stretched vertically by a factor of four.

4 Transitional Systems

Several transitional objects, of approximately 10 Myr of age, have recently been imaged. Situated in the evolutionary sequence between the massive, gas-dominated primordial solar nebulae and the much lighter, replenished dust disks, the disks of HR 4796A and HD 141569 show evidence of relatively sharp features.

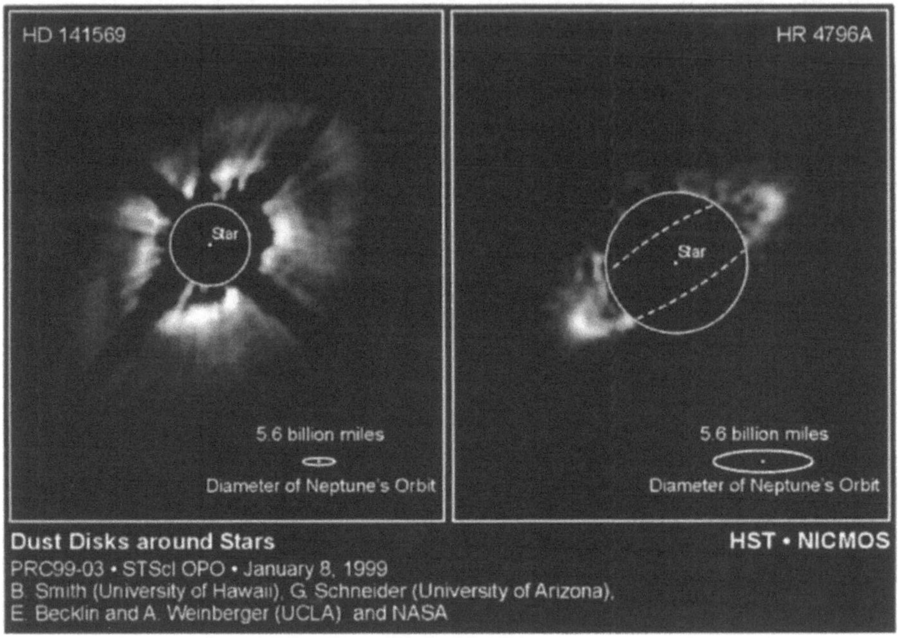

Fig. 4. 10-million yr old stars HR 4796A and HD 141569 with dust disk imaged by NICMOS. The dust disks show structure such as gaps and rings (edges), which may be due to the presence by planets.

Figure 4 (right panel) shows an abrupt inner truncation at ~ 70 AU from the main-sequence dwarf HR 4796A, suggestive of planetary tidal truncation radius (Koerner et al. 1999, Schneider et al. 1999). The planet may reside inside the gap, and one reason for the sharpness of the feature may be the presence of more gas (amount comparable with that of dust) than in β Pic, which has a smoother gap edge. We base this prediction on the gas drag, which restricts and organizes the motion of dust in the disk, allowing sharp gaps seen in many numerical simulations of gas disks. In addition, the presence of relatively larger amount of gas than in β Pic is also required by the necessity to prevent (again, by gas drag) the so-called dust avalanches (Artymowicz 1997b), which otherwise would destroy a full-blown planetary system disk together with its replenishing reservoir of planetesimals in time much shorter than the age of HR 4796A (8-10 Myr). Since, however, the amount of gas in the system is less than 7 Earth masses, the disk cannot be the primordial protoplanetary nebula. The disk must be in the final stage of clearing of the primordial gas. Detailed observations of its gap region may reveal the fingerprints of the planet(s).

A gap-like depletion has been observed in a disk of a young ($\sim$ 10 Myr) Herbig Ae star HD 141569 (see Figure 4, left panel; the radial spikes in this NICMOS image are instrumental artifacts.) Here we witness a feature superficially similar to the planet-induced gap in Figs. 1 and 2. In this interpretation, the planet is strangely far-away ($\sim$ 250 AU) from the star. It does not have to be very massive; we estimate by comparison with our numerical models (cf. Fig. 1) that a Neptune-class body on nearly circular orbit would be sufficient.

Even though these two transitional systems provide a happy playground for disk+planets dynamicists, we should be cautious before interpreting everything we see as a proof of the presence of a planet. There may be processes involving dust in disks that create large-scale non-uniformities, including non-axisymmetric ones. The somewhat forced example of the such process would be a collision of two planetesimals, injecting a large quantity of dust locally into the disk, then some of that dust being immediately picked up by stellar radiation pressure and forming a dust avalanche (multiplying exponentially on its way out by striking larger disk grains). Perhaps a a less contrived scenario for producing axisymmetric features like rings and enhanced edges is provided by dust migration through gas. Recall that the standard drag-induced migration is inward due to partial pressure support of gas disks vs. the Keplerian rotation of dust. However, in disks where the gas:dust ratio is not extremely high, and where dust grains are subject to substantial cancellation of gravitational attraction by radiation pressure from the star, the dust will move *outwards*.

For instance, consider a model gas disk around HD 141569, with a total gas mass of 0.7 Earth masses only. Trajectories of olivine+pyroxene dust particles released from a common point above the disk plane can be studied. Results show that grains smaller than 0.75 μm (radius) have radiation pressure/gravity ratio $\beta > 5$, and only those larger than 3.5 μm, have $\beta < 1$ and thus remain in disk on elliptic orbits, forming the visible disk. However, such orbits migrate outward in a matter of several to tens of orbital periods (!). Gas drag is caused by the azimuthal velocity difference of the disk (very slightly radiation-pressure supported) and dust grains (moderately to strongly radiation supported). As a result, dust migration and fractionation may occur. Most of the visible dust might migrate to the outer edge of the disk, where further migration would be prevented by the lack of gas. On the other hand, silicate sand grains (not affected by radiation pressure) would accumulate in the inner part of the disk by the familiar gas drag effect. This, and other complex, dust-related disk problems involving mutual collisions, need to be seriously addressed before we can say with confidence that HD 141569 or any other disk indeed houses the suspected planets.

References

1. Artymowicz, P. (1997a), On the formation of eccentric superplanets, in Brown Dwarfs and Extrasolar Planets, PASP conf. ser., Eds. R. Rebolo, E. L. Martin, and M. R. Zapatero-Osorio, 152-161
2. Artymowicz, P. (1997b), Beta Pictoris: An Early Solar System? Ann. Rev. Earth & Planet. Sci., **25**, 175-219
3. Artymowicz, P., Lubow, S. H. (1996), Mass Flow Through Gaps in Circumbinary Disks, Astroph. J. **476**, L77-L81
4. Artymowicz, P., Lubow, S.H., & Kley, W. (1997), Planetary systems and their changing theories. In Planetary Systems - The Long View, eds. L. Celnikier et al, Editions Frontieres, in print
5. Beust, H., Morbidelli, A. (1996), Mean-Motion Resonances as a Source for Infalling Comets toward beta Pictoris, Icarus **120**, 358-370
6. Bryden, G., Chen, X., Lin, D.C.N., Nelson, R.P., & Papaloizou, J.C.B. (1999), Tidally induced Gap Formation in Protostellar Disks: Gap Clearing and Suppression of Planetary Growth, Astroph. J., **514**, 344-367
7. Butler, P., Marcy, G., Fischer, D., et al (1999) Evidence for multiple companions to upsilon Andromedae. Astroph. J., in print
8. Goldreich, P., & Tremaine, S. (1980), Disk-satellite interactions, Astroph. J.,**241**, 425-441
9. Heap, S. R., Lindler, D. J., Lanz, T. M., et al. (1999), STIS Coronagraphic Observations of β Pictoris, Astroph. J., in print
10. Kley, W. (1999), Mass flow and accretion through gaps in accretion discs, MNRAS, **303**, 696-710
11. Koerner, D. W., Ressler, M. E., Werer, M. W., Backman, D. E. (1999), Mid-Infrared Imaging of a Circumstellar Disk around HR 4796: Mapping the Debris of Planetary Formation, Astroph. J. **503**, L83-L87
12. Lin, D.N.C., & Papaloizou, J.C.B. (1993), in Protostars and Planets III, ed. E. H. Levy & J. I. Lunine (Tucson: Univ. of Arizona Press), 749-835
13. Lissauer J. (1999), Three planets for Upsilon Andromedae, Nature, **398**, 659-661
14. Lubow, S. H., Seibert M., Artymowicz, P. (1999) Disk Accretion Onto High-Mass Planets. Astroph. J., in print
15. Mazeh, T., Mayor, M., and Latham, D.W. (1997), Eccentricity versus Mass for Low-Mass Secondaries and Planets, Astroph. J. **478**, 367-373
16. Mouillet, D., Larwood, J. D., Papaloizou, J. C. B., (1997), A planet on an inclined orbit as an explanation of the warp in the Beta Pictoris disc, MNRAS, **292**, 896-904
17. Schneider, G., Smith, B., Becklin, E. et al. (1999), NICMOS Imaging of the HR 4796A Circumstellar Disk, Astroph. J. **513**, L127-L130
18. Vidal-Madjar, A., des Etangs A. L., Ferlet, R. (1998), Beta Pictoris, a young planetary system? A review, Pl. Space Sci. **46**, 629-648

Evolution of Low-Mass Stars and Substellar Objects. Contribution to the Galactic Mass Budget

G. Chabrier[1], I. Baraffe[1], F. Allard[1], and P.H. Hauschildt[2]

[1] Centre de Recherche Astrophysique de Lyon (UMR CNRS 5574), Ecole Normale Supérieure de Lyon, 69364 Lyon Cedex 07, France
[2] Dpt. of Physics & Astronomy, University of Georgia, Athens, GA 30602, USA

Abstract. We briefly summarize our present knowledge of the theory of low-mass stars and substellar objects and their contribution to the Galactic population.

1 Introduction

The search for substellar objects (SSO) has bloomed over the past few years with the unambiguous identification of several free floating brown dwarfs (BDs) and with the discovery of numerous planets orbiting stars outside the solar system. Substellar objects therefore exist and ongoing and future observational projects are likely to reveal dozens more of such objects. Two groups, the Lyon group and the Tucson group, have aspired to derive a complete theory of the evolution and the spectral signature of low-mass, dense objects, from Sun-like to Saturn-like masses, covering 3 orders of magnitude in mass and 9 in luminosity, bridging the gap between stars and gaseous planets (see e.g. Burrows *et al.*, 1997 and Baraffe *et al.*, 1998 and references therein). These two groups have incorporated the best possible physics aimed at describing the mechanical and thermal properties of these objects - equation of state, synthetic spectra, non-grey atmosphere models. Both groups have "met in the middle" by successfully identifying the main spectral properties of the benchmark BD Gl229B, providing a determination of its mass and age (Marley *et al.*, 1996; Allard *et al.*, 1996). We refer to the afore-mentioned papers and the references therein for a complete description of the physics entering these models. Only a brief outline of these characteristics is given below.

2 The Physics of Sub-Stellar Objects

2.1 Interior

Central conditions for SSOs are typically $T_c \lesssim 10^5$ K and $\rho_c \sim 10^2$-10^3 g cm^{-3}, characterizing a strongly coupled electron-ion plasma. Moreover, in the envelope, the electron binding energy is of the order of the Fermi energy

$Ze^2/a_0 \sim E_F$ so that *pressure*-dissociation and ionization take place along the interior. Very recently laser-driven shock wave experiments at Livermore have achieved pressures on liquid D_2 of up to 5 Mbar at high temperature (Collins *et al.*, 1998), exploring for the first time the regime of pressure-dissociation and ionization, therefore probing the equation of state (EOS) under conditions characteristic of SSO interiors. As shown in Figure 3 of Collins *et al.*, the experimental hugoniot revealed the excellent behaviour of the EOS developed by Saumon and Chabrier (Saumon *et al.*, 1995 and references therein). In particular the pronounced compressibility observed in the dissociation domain ($\rho/\rho_0 = 5.88$) agrees remarkably well with the theoretically predicted value. These experiments open a new window in physics and astrophysics by constraining the physics of the interior of SSOs in laboratory experiments.

2.2 Atmosphere

In the atmosphere, the hydrostatic equilibrium condition $d\tau = \bar{\kappa} dP/g$, where $g = Gm/R^2 \approx 10^3$-10^5 cm s^{-2} is the characteristic surface gravity of SSOs, yields $P_{ph} \sim g/\bar{\kappa} \approx$0.1-10 bar and $\rho_{ph} \approx 10^{-6}$-$10^{-4}$ g cm^{-3} at the photosphere. Collisional effects are important under such conditions and introduce sources of absorption like e.g. the collision-induced absorption (CIA) between H_2-H_2 or H_2-He below $\sim$ 5000 K (see e.g. Borysow *et al.*, 1985). In the effective temperature range characteristic of low-mass stars (LMS) and SSOs ($T_{\rm eff} \lesssim 5000$ K), numerous molecules form, in particular metal oxides and hydrides (TiO, VO, FeH, CaH), the major absorbers in the optical, and CIA H_2, H_2O, CO which dominate in the infrared (see Allard *et al.*, 1997 for a review). The situation becomes even more complicated for SSOs, due to the changes in molecular chemistry which occur across the atmospheric temperature range from 2000 K (the coolest stars) to 170 K (Jovian conditions). At 2000 K, most of the carbon is locked into carbon monoxide CO, while the oxygen is found in water vapor H_2O, dominantly, and in titanium TiO and vanadium VO monoxides. Below $\sim$ 1800 K, the dominant equilibrium form of carbon is no longer CO but CH_4 (Fegley & Lodders, 1996). As confirmed by the observation of Gl229B (Oppenheimer *et al.*, 1995), methane features begin to appear in the infrared while titanium dioxide and silicate clouds form at the expense of TiO, modifying profoundly the opacity of the atmosphere. For Jovian-like atmospheres, the dominant equilibrium form of nitrogen is NH_3 ($T_{\rm eff} \lesssim 600$ K) and below $T_{\rm eff} \sim 200$ K water condenses to clouds at or above the photosphere. As shown by Tsuji *et al.*(1996) and Jones & Tsuji (1997), there is evidence for condensation of metals and silicates into grains (e.g. TiO into $CaTiO_3$, Mg, Si into $MgSiO_3$) for $T_{\rm eff} \lesssim 3000$ K, i.e. at the bottom of the main sequence.

A proper inclusion of this complex atmospheric chemistry, and of a proper calculation of unknown cross-sections, represent the main challenge for a correct description of the spectral signature and the evolution of substellar objects. This grain formation process has been included in the calculations of the Tucson group only in term of condensation. The species are precipitated according to their condensation curves. If a species has condensed, it is left at its saturated vapor pressure (Burrows *et al.*, 1997). There is no inclusion of the radiative transfer effect of the grains. These models can thus be considered as grain-free opacity models. Recently, the Lyon group has extended similar calculations by implicitly including all condensed species into the radiative transfer equations (see Allard & Hauschildt, these proceedings). The atmosphere profiles were matched with the interior profiles at an optical depth deep enough to lie on the internal adiabat (see Chabrier & Baraffe, 1997).

3 Color-Magnitude Diagram

The present evolutionary calculations have been conducted with three different atmosphere models: (i) for hot objects, i.e. massive or young enough, to preclude the formation of grains ($T_{\rm eff} \gtrsim 2800$ K), we have used the most recent grainless NGEN atmosphere models (Hauschildt *et al.*, 1999); (ii) for objects in the range 3000-1000 K, i.e. from the bottom of the MS down to Gliese229B-like objects, we have used complete atmosphere models which include the grain opacity sources in the transfer equations, the so-called DUSTY models (Allard & Hauschildt, these proceedings); (iii) for objects below 3000 K down to Jovian temperatures, we have also considered cases where the condensates settle rapidly below the photosphere and - although modifying the atmosphere EOS - do not participate to the opacity, the so-called COND models. This is similar to the Burrows *et al.*(1997) calculations and is motivated by the relative absence of grain features in the atmosphere of objects below $T_{\rm eff} \sim 1000$ K, i.e. Gliese229B-like and cooler objects. We found the effect of grain opacity (DUSTY) to affect only moderately the H-burning minimum mass. Models with grainless atmosphere yield $m = 0.072\, M_\odot$, $L = 5 \times 10^{-5}\, L_\odot$ and $T_{\rm eff} = 1700$ K at the H-burning limit, whereas models with grain opacity give $m \approx 0.07 M_\odot$, $L \approx 4 \times 10^{-5}\, L_\odot$ and $T_{\rm eff} \approx 1600$ K, for solar composition.

Figure 1 displays a K vs J-K color-magnitude diagram (CMD). In terms of colors there is a competing effect between grain and molecular opacity sources for objects at the bottom and just below the MS. As already identified for e.g. Gl229B (Allard et al., 1996; Marley et al., 1996) CH_4 and CIA of H_2 lead to bluer infrared colors for SSOs, as illustrated by the NEG and COND models, whereas grain opacity results in a severe reddening. Indeed, the present DUSTY models reproduce the spectra of the DENIS objects (Tinney *et al.*, 1998) and of the long puzzling object GD165B, which lies at the very edge of the H-burning limit (Kirkpatrick *et al.*, 1999). As $T_{\rm eff}$ decreases, grains settle below the photosphere and the DUSTY track will merge with the *COND* tracks. Observations of objects in the region between

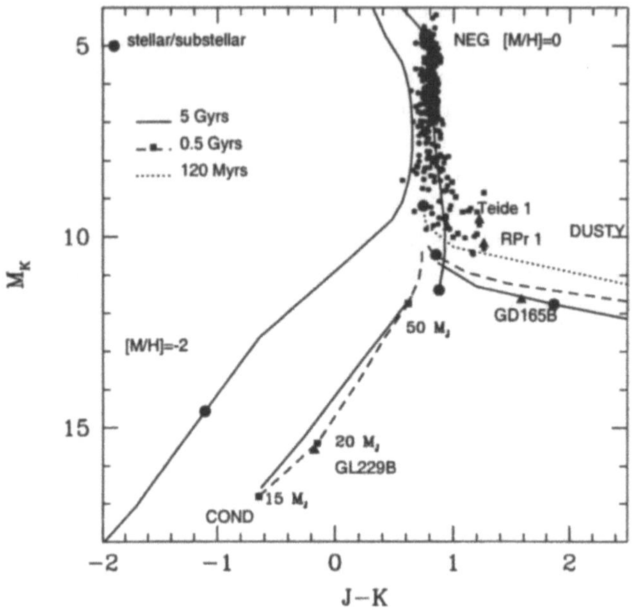

Fig. 1. M_K vs (J-K) diagram for 3 isochrones, namely 5×10^9 yrs, 5×10^8 yrs and 1.2×10^8 yrs (the age of the Pleiades). The small circles and squares correspond to the observation of main sequence stars (Leggett, 1996) and of Pleiades objects (Bouvier *et al.*, 1998), respectively. Some identified BDs are also indicated

these two extreme cases will be the next confirmation of the present theory. In spite of this strong absorption in the IR, however, BDs around 1500 K radiate nearly 90% (99% with dust) of their energy at wavelengths longward of 1 μm and infrared colors are still preferred to optical colors (at least for solar metal abundance), with M, L, K as the favored bands, and $M_M \sim M_L \sim 11$, $M_K \sim 12$ at the H-burning limit, at 5 Gyr. Direct observation of the characteristic fluxes of SSO's are now within reach with several observational projects, like SOFIA, SIRTF, ISAAC (Burrows *et al.*, 1997; Allard & Hauschildt, these proceedings).

Several BD surveys are presently conducted in young clusters and it is important to develop accurate (non-grey) models for pre-MS stars and young BD's. The more massive of such objects will be hot enough so that grain formation does not occur, but for $t \lesssim 5 \times 10^7$ yrs, proper evolutionary calculations must consider two effects: (i) first of all the influence of the initial conditions must be carefully examined because they will significantly affect the mass-magnitude relationship; (ii) for young objects, the gravity is small ($\log g \lesssim 3$) and sphericity effects in the resolution of the transfer equations might come into play. The derivation of such complete models is under work,

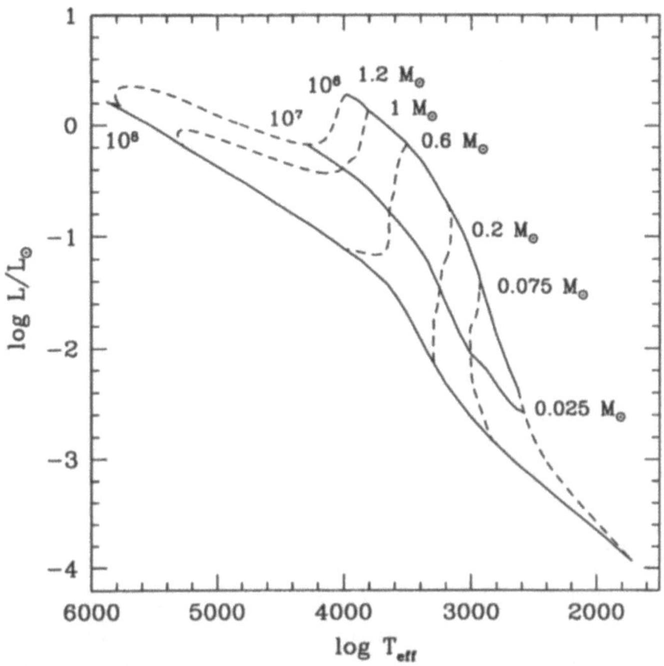

Fig. 2. Pre-main sequence isochrones (models are available from ftp site - see at the end of the paper)

but Figure 2 displays the present pre-MS models (Baraffe *et al.*, 1998) based on non-grey plane-parallel atmosphere models in a theoretical HR diagram.

4 Stellar and Brown Dwarf Mass Functions

4.1 Stellar Mass Function

The determination of the contribution of LMS and SSOs to the Galactic mass budget requires the correct determination (slope and normalization) of the mass-function (MF) down to the hydrogen burning limit, in order to have a solid foundation for extrapolation into the substellar domain. This issue, however, is presently not completely settled for the Galactic disk, and significant differences exist between the MF inferred from parallax-determined luminosity functions (LF) (Kroupa, 1995) and from the HST photometric LF (Gould, Flynn & Bahcall, 1997) (see Méra, Chabrier & Schaeffer, 1998, Fig. 1). It is relatively safe to say, however, that the MF keeps rising down to the H-burning limit, although with a slope shallower than a Salpeter MF. The LMS ($m \leq 0.6\,M_\odot$) mass densities inferred from the integration of the two afore-mentioned MFs (nearby and HST) yield, respectively,

$\rho_{LMS} \approx (1.9 \pm 0.1) \times 10^{-2}$ and $\rho_{LMS} \approx (2.6 \pm 0.2) \times 10^{-2}$ $M_\odot pc^{-3}$ (Méra *et al.*, 1998). Adding up the contribution from more massive stars and stellar remnants yields $\rho_\star \approx (4.0\text{-}4.6) \pm 0.3 \times 10^{-2}$ $M_\odot pc^{-3}$ in the Galactic disk, i.e. a surface density $\Sigma_\star \approx 31 \pm 2$ $M_\odot pc^{-2}$. This corresponds to a stellar *number*-density $n_\star \sim 0.06$ (HST) and ~ 0.35 pc^{-3} (nearby).

For the spheroid the question is more settled with a MF with $\alpha \lesssim 1$ (where $dN/dm \propto m^{-\alpha}$) and a stellar density $\rho_\star < 4.0 \times 10^{-5}$ $M_\odot pc^{-3}$, less than 1% of the required dynamical density (Graff & Freese, 1996; Chabrier & Méra, 1997; Gould *et al.*, 1998), i.e. a number density $n_\star < 10^{-3} pc^{-3}$. Extrapolating into the BD domain yields for the spheroid $n_{BD} < 10^{-4}$ pc^{-3}, $\rho_{BD} < \rho_\star/10$, i.e. a microlensing optical depth $\tau \sim 10^{-9}$, about 1% of the value measured toward the LMC. The puzzle remains unsolved for the ***dark halo*** MF, although both the HDF observations and the narrow-range of the observed time-distribution of the microlensing events towards the LMC strongly suggest an IMF different from a Salpeter one below $\sim 1\,M_\odot$ (see e.g. Chabrier, 1999).

4.2 Brown Dwarf Mass Function

A proper census of the number of brown dwarfs ($m < 0.07\text{-}0.09\,M_\odot$, depending on the metallicity) has significant implications for our understanding of how stars and planets form. The determination of the BD MF is a complicated task. By definition BDs never reach thermal equilibrium and most of the BD's formed at the early stages of the Galaxy will have fainted to very low-luminosities ($L \propto 1/t$, see e.g. Burrows & Liebert, 1993). Thus observations will likely be biased towards young and massive BD's. The age uncertainty is circumvented when looking for the BD MF in stellar clusters since objects in that case are likely to be coeval. The Pleiades cluster has been extensively surveyed and several BDs have been identified down to $\sim 25\,M_J$ (Martin *et al.*, 1998; Bouvier *et al.*, 1998). A single power-law function from ~ 0.4 to $0.04\,M_\odot$ seem to adequately reproduce the observations with some remaining uncertainty in the exponent $\alpha \sim 0.6 - 1.0$ (Bouvier *et al.*, 1998; Martin *et al.*, 1998). By examining the various observations of Doppler radial velocity surveys, Basri and Marcy (1997) reach the conclusion that all data are consistent with $0 \lesssim \alpha \lesssim 1$ (see Basri, these proceedings for an updated analysis of such data). On the other hand Mayor *et al.*(1997) find for the mass function of the companions of the G and K dwarf sample observed with CORAVEL a power-law function with $\alpha \sim 0.4$ from 0.4 down to $0.005\,M_\odot$. The determination of the MF in young clusters is potentially very interesting. Indeed young clusters did not have time to experience evaporation in the outer regions or mass segregation in the central regions and the present-day MF should reflect relatively closely the initial MF. Such a MF determination, however, is hampered by two observational and theoretical problems. From the observational point of view, young objects are immersed into dust and a proper determination of the LF requires a correct determination of the *differential* reddening in the cluster. From the theoretical point of view, none of the MFs determined up to now in young clusters include a careful examination of the effects mentioned in §3 (initial conditions, sphericity, non-grey effects) and thus are of dubious validity. It is thus certainly premature to claim a robust determination of the MF in young clusters.

An independent, powerful information on the stellar and substellar MF comes from microlensing observations. Indeed, the time distribution of the events provides a (although model-dependent) determination of the mass distribution and thus of the minimum mass of the dark objects: $dN_{ev}/dt_e = E \times \epsilon(t_e) \times d\Gamma/dt_e \propto P(m)/\sqrt{m}$, where E is the observed exposure, i.e. the number of star×years, ϵ is the experimental efficiency, Γ is the event rate and $P(m)$ is the mass probability distribution. The analysis of the published 40 MACHO + 9 OGLE events towards the bulge is consistent with a rising mass function at the bottom of the MS with a minimum mass $m_{inf} \sim 0.05\,M_\odot$, whereas a decreasing MF below 0.2 $M_\odot$ is excluded at the 95% confidence level (Han & Gould, 1995; Méra *et al.*, 1998). Although the time distribution might be affected by various biases (e.g. blending) and robust conclusions must await for larger statistics, the present results suggest that in order to explain both star counts and the microlensing experiments, a substantial amount of SSOs must be present in the galactic disk. Indeed, extrapolation of the stellar MF (§4.1) into the BD domain down to 0.05 $M_\odot$ yields for the Galactic disk $\rho_{BD} \sim 4.0 \times 10^{-3}\,\mathrm{M_\odot pc^{-3}}$, i.e. $\Sigma_{BD} \approx 3\,\mathrm{M_\odot pc^{-2}}$, i.e. a BD *number*-density comparable to the stellar one, $n_{BD} \sim 0.1\,pc^{-3} \sim n_\star$.

For the spheroid, extrapolation of the afore-mentioned stellar MF yields $\rho_{BD} \lesssim 10^{-5}\,\mathrm{M_\odot pc^{-3}}$, $n_{BD} \lesssim 10^{-4}\,pc^{-3}$, whereas for the dark halo the normalization is about 2 orders of magnitude smaller (Chabrier & Méra, 1997).

4.3 Planet Mass Function

It is obviously very premature to try to infer the mass distribution of exoplanets. This will first require a clear, both theoretical and observational, distinction between planets and brown dwarfs. An interesting preliminary result, however, comes from the observed mass distribution of the secondaries conducted by Mayor and collaborators. As shown by Mayor *et al.*(1997), there is a strong discontinuity in the mass distribution at $m_2/\sin i \approx 5\,M_J$, with a clear peak below this limit. If confirmed this would suggest that planet formation in a protoplanetary disk is a much more efficient mechanism than BD formation, which results from cloud collapse and fragmentation.

5 Conclusion and Perspectives

As mentioned in the introduction, accurate models for low-mass stars, brown dwarfs and giant planets are needed to shed light on the observable properties of these objects and to provide guidance to the ongoing and future surveys aimed at revealing their contribution to the Galactic population. From this point of view, tremendous progress has been made in the recent years with the derivation of consistent evolutionary model and synthetic spectra calculations which accurately reproduce the observed sequences of globular clusters, Pleiades, field objects and of the benchmark BD Gl229B in *various photometric passbands* (see references mentioned in the Introduction). Moreover the afore-mentioned LMS models have been shown to accurately reproduce the observational mass-magnitude relationship (Henry & McCarthy, 1993; Henry *et al.*, 1999) both in the optical and in the infrared (Baraffe *et al.*, 1998) and to yield a consistent, coeval sequence for the quadruple system *GG-Tau*, whose component masses extend from 1.2 $M_\odot$ down to $\sim 0.035\, M_\odot$ (White *et al.*, 1999), a formidable test for the theory. On the other hand, stringent constraints on the theory of dense/cool objects are now provided by laboratory high-pressure experiments, for the interior, and by various spectroscopic and photometric observations of LMS and SSO's. Any theory aimed at describing the mechanical and thermal properties of these objects *must* be confronted to these experimental/observational constraints in order to assess any degree of validity. Improvement in the theory will proceed along with the discovery of many more SSO's, hopefully bridging the gaps on either side of Gl229B from the bottom of the MS to Jupiter-like objects. Observation of a transit EGP would allow the determination of the radius and the mass of the object, providing a formidable constraint on the theory. For 51-Peg-like objects, the probability to observe such a transit $p = d_\star/D$, where $d_\star$ and D are the stellar and orbital diameters, respectively, is $p \sim 10\%$, by no means negligible.

The increasing number of observed LMS and SSOs, together with the derivation of accurate models, will allow eventually a robust determination of the stellar and substellar mass functions, and thus of the exact density of these objects in the Galaxy. As discussed in §4, present MF determinations in various Galactic regions point to a slowly rising MF near and below the H-burning limit, with a BD number density comparable to the stellar one. Whether this behaviour is universal (although we already know it is certainly not the case for the dark halo), whether it is consistent with a general log-normal form, must await confirmation from future observations. On the other hand, the amazingly rapid pace of exoplanet discoveries should yield the determination of the planetary MF. These combined informations will allow the determination of the BD minimum mass and planet maximum mass.

As seen, the physics of SSOs involves an amazingly large domain of physics and astrophysics, from the fundamental N-body problem to star formation and Galactic evolution, and will certainly remain a lively domain of the early 21st century astronomy.

The present grainless models are available from:

ftp ftp.ens-lyon.fr, username: anonymous
ftp > cd /pub/users/CRAL/ibaraffe
ftp > get BCAH98_models and BCAH98_models_BD

References

1. Allard, F., Hauschildt, P.H., Alexander, D., & Starrfield, S, 1997, ARA&A, **35**, 137
2. Allard, F., Hauschildt, P.H., Baraffe, I., & Chabrier, G., 1996, ApJ, **465**, L123
3. Baraffe, I., Chabrier, G., Allard, F. and Hauschildt, P.H., 1998, A&A, **337**, 403
4. Basri, G., & Marcy, G., 1997, AIP Conf. Ser., **393**, 228
5. Borysow, A., Trafton, L., Frommhold, L. & Birnbaum, G., 1985, ApJ, **296**, 644
6. Bouvier, J., Stauffer, J.R., Marin, E.L., Barrado Y Navascues, D., Wallace, B., & Bejar, V.J.S., 1998, A&A, **336**, 490
7. Burrows A., Marley, M., Hubbard, W.B., Lunine, J.I., Guillot, T., Saumon, D., Freedman, R., Sudarsky, D., Sharp, C. 1997, ApJ, **491**, 856
8. Burrows A., & Liebert, J., 1993, Rev. Mod. Phys., **65**, 301
9. Chabrier, G., 1999, ApJ, **513**, L103
10. Chabrier, G., and Méra, D., 1997, A&A, **328**, 83
11. Chabrier, G. and Baraffe, I., 1997, A&A, **327**, 1039
12. Collins, G.W., *et al.*, 1998, *Science*, **281**, 1178
13. Fegley B., & Lodders K., 1996, ApJ, **472**, L37
14. Gould, A., Flynn, C., and Bahcall, J., 1998, ApJ, **503**, 798
15. Graff, D. and Freese, K., 1996, ApJ, **456**, L49
16. Han, C. and Gould, A., 1995, ApJ, **447**, 53
17. Hauschildt, P.H., Allard, F., Baron, E., 1999, ApJ, **512**, 377
18. Henry, T., & McCarthy, D.W., 1993, AJ, **106**, 773
19. Henry, T., et al., 1999, ApJ, **512**, 864
20. Jones, H., & Tsuji, T., 1997, ApJ. **480**, L39
21. Kirkpatrick, J. D., Allard, F., Bida, T., Zuckerman, B., Becklin, E.E., Chabrier, G., and Baraffe, I., 1999, ApJ, in press
22. Kroupa, P., 1995, ApJ, **453**, 358
23. Leggett, S.K., Allard, F., Berriman, G., Dahn, C.C. & Hauschildt, P.H., 1996, ApJS, **104**, 117
24. Marley, M.S., Saumon, D., Guillot, T., Freedman, R., Hubbard, W.B., Burrows, A., Lunine, J.I. 1996, Science, **272**, 1919
25. Martin, E., Basri, G., Zapatero-Osorio, M.R., Rebolo, R., Lopez, R.J. Garcia, 1998, ApJ, **507**, L41
26. Mayor, M, Queloz, D., Udry, S., & Halbwachs, J-L., 1997, in *Rencontre de Blois*
27. Méra, D., Chabrier, G., and Schaeffer, R., 1998, A&A, **330**, 937
28. Oppenheimer, B.R., Kulkarni, S.R., Nakajima, T., Matthews, K. 1995, Science, **270**, 1478
29. Saumon, D., Chabrier, G., and VanHorn, H.M., 1995, ApJS, **99**, 713
30. Tinney, C., Delfosse, X., Forveille, T., & Allard, F., 1998, A&A, **338**, 1066
31. Tsuji, T., Ohnaka, K. and Aoki, W. 1996a, A&A, **305**, L1
32. Tsuji, T., Ohnaka, K., Aoki, W., Nakajima. T. 1996b, A&A, **308**, L29
33. White, R., Ghez, A., Reid, I., Schultz, G., 1999, astro-ph/9902318

Extrasolar Giant Planets: Masses and Luminosities from In-Situ Formation Theories

Günther Wuchterl

Institut für Theoretische Astrophysik der Universität Heidelberg, Tiergartenstraße 15, D-69126 Heidelberg, Germany

Abstract. Their expected luminosities make young giant planets favorable for the first direct detection of an extrasolar planet. The giant planet formation process is relatively slow with expected formation times ranging from comparable to the star formation timescale up to the nebula lifetime, depending on the formation theory. Therefore quantitative models of giant planet formation have to be considered when estimating young giant planet properties at pre-main sequence stellar ages. This is especially important for free-floating giant planet candidates where planetary nature is inferred from luminosities, without independent mass determinations. I will discuss observables of young- and proto-giant planets as they follow from the disk-instability and nucleated-instability hypothesis, respectively. The luminosity exceeds $10^{-4}\,L_{\odot}$, even for a Saturn-mass protoplanet, during the brief, $3\,10^{4}$ year period around maximum accretion. Luminosity maxima of giant planets occur at ages of 1–10 Myr, depending on the details of the formation process. To infer planetary nature in a young population, the properties of young planets have to be compared to those of proto brown dwarfs, as they follow from the respective formation theories.

1 Luminosities for Young Giant Planets

Young[1] giant planets are much more luminous than present solar system giant planets, with $\sim 10^{-9}\,L_{\odot}$ for Jupiter. Even a $20\,M_{\oplus}$ protoplanet can radiate at $\sim 10^{-6}\,L_{\odot}$, $= 1\,\mu L_{\odot}$, the typical luminosity for young giant planets (cf. [10]). They are therefore much easier to detect, especially when they are at large orbital distances. Unfortunately solar system dimensions appear on the sub-arcsecond scale even in the closest young, star-formation regions. But ejection of young planets after close encounters with a massive body might still provide planetary objects separated far enough from their parent systems for convenient observation. In such a situation, that facilitates direct detection, the mass and therefore the planetary nature cannot be determined

[1] This work has been supported by the Österreichischer Fonds zur Förderung der wissenschaftlichen Forschung (FWF) under project numbers S-7305–AST, S-7307–AST and the Deutsche Forschungsgemeinschaft (DFG), SFB 359 (key research project of the German national science foundation on 'Reactive flows, diffusion and transport').

from the gravitational interaction with the primary of the parent system. It has to be estimated by direct study of the candidate object. This requires knowledge of the properties to be expected for such young giant planets, most prominently for detection purposes, their luminosity. If, in a known range of ages for the region under investigation, there is nothing else that can be as faint as a young planet at its brightest, the candidate object may be considered to be a planet.

Usually luminosities of young giant planets are taken from planetary evolution models, that were originally designed to understand solar system giant planet luminosities at ages comparable to the present solar system age. These models give relatively high luminosities ($> 10^{-6}\,L_\odot$) for ages < 10 Myr, for a Jupiter mass (cf. [6]). They do not consider the formation process but assume an arbitrary hot initial state and argue that due to the short cooling times in the beginning, these initial conditions are rapidly forgotten and the results become independent of the initial assumptions. But the giant planet formation process itself occurs on the 10 Myr timescale. Therefore luminosities from such 'cooling' calculations cannot be used in the first 10^5–10^8 yr because they still show the memory of the artificial initial state [9].To obtain physically plausible luminosities for young giant planets they have to be calculated from formation theories. I present results obtained in the nucleated instability hypothesis, where a core grows first, and the disk instability hypothesis, that involves giant gaseous protoplanets. I compare those luminosities to the luminosities for stars and brown dwarfs, as they follow from the respective formation theories for the first Myrs.

2 Luminosities and Formation Theories

The relatively high luminosities of young giant planets occur for two main reasons: (1) the planets are still hot because the cooling just started and (2) the accretion process itself adds to the luminosity, e.g., when gas flowing onto the protoplanet is stopped at its 'surface'. The two reasons are interdependent because energy radiated away during the accretion process will not be stored into the interior of the planet for later release during the cooling phase. Early high luminosity phases due to the accretion process leave less energy for later, high luminosity cooling phases. Because of the unimportance of nuclear energy generation, the luminosity radiated over all time is limited to the binding energy of the cooled planet. Hence the accretion history is important for the early cooling history and I will give examples from two formation theories.

All calculations that I present use a new, time-dependent convection model [13] that is derived from the model of Kuhfuß (1987), [7]. It closely approximates standard mixing length theory in a local, static limit and has been successfully tested in calculations of the Sun and RR-Lyrae pulsations [13]. To determine the protoplanetary luminosities, the equations of radiation hydrodynamics with spherical symmetry, supplemented by the convection model, are solved with a 'protoplanetary' version of the VIP-code [4], where convection is added as outlined in [12]. The convection model is used with the standard parameters given in [13]. Equations of state (as in [10]) and opacities are for solar composition. Opacities include the contribution of dust particles with 'interstellar' properties and Alexander and Ferguson (1994) values, [1], in the molecular range (compilation by [5]). Deuterium burning and convective mixing are included. For the formulation of equations and boundary conditions see [10].

2.1 The Nucleated Instability Hypothesis

The Nucleated Instability Hypothesis is based on the 'planetesimal hypothesis' that assumes that planets grow within circumstellar disks via accretion of km-sized solid bodies, the so-called planetesimals. Sufficiently massive planetesimals embedded in a gravitationally stable preplanetary nebula can capture large amounts of gas and become the cores of giant planets. In this hypothesis one formation mechanism accounts for small solar system objects, terrestrial planets and giant planets. It is the only formation model for Uranus and Neptune type planets (see [14] for a review). In Figs. 1 and 2 results

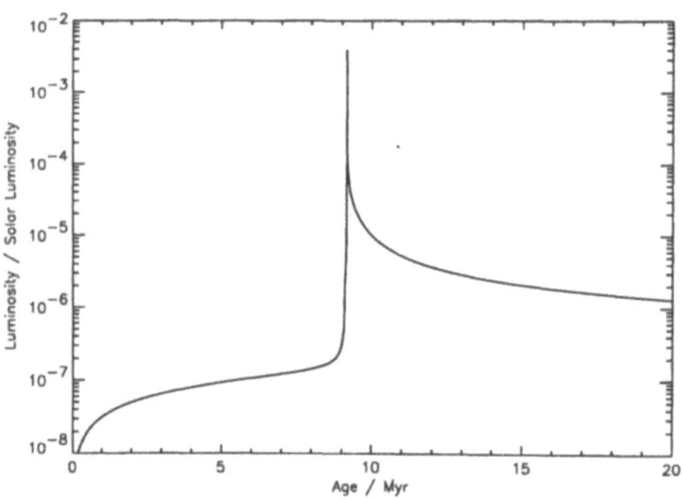

Fig. 1. Luminosity of a young giant planet that grows at an orbital distance of 5.2 AU from a solar mass star, according to the nucleated instability hypothesis. The final mass is 106 $M_\oplus$ with 9.2 $M_\oplus$ in the core, i.e. similar to Saturn. The peak luminosity is reached at the moment of maximum accretion, $\dot{M} \sim 0.14\,M_\oplus/yr$. That moment is reached after the core has grown to 9.2 $M_\oplus$ with a constant accretion rate of $\dot{M}_{core} = 10^{-6}\,M_\oplus/yr$. The width of the luminosity peak is 30000 yr at $10^{-4}\,L_\odot$, cf. Fig. 2

from a nucleated instability model for a proto giant planet growing at an orbital distance of 5.2 AU in a minimum mass nebula [8] and constant core accretion rate of $10^{-6} M_\oplus$/yr are shown (for a discussion see [14]). When the envelope mass reaches 97 $M_\oplus$ the protoplanet is shut off from the nebula and the following evolution is calculated at constant mass.

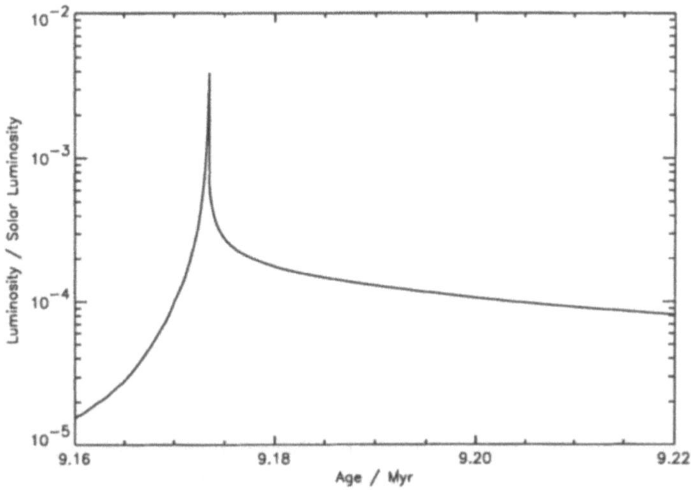

Fig. 2. Luminosity of a protogiant planet in the vicinity of maximum accretion. Detail of Fig. 1. The young giant planet is brighter than $10^{-4} L_\odot$ for about 30000 yr

2.2 Disk Instability and Giant Gaseous Protoplanets

Protoplanetary disks are considered to be gravitationally stable and self-stabilizing (see [14], for a discussion in the context of giant planet formation). But it has been claimed that a disk instability might be triggered by an external perturbation and giant planets might form via a tenuous intermediate stage as giant gaseous protoplanet (GGPP), [2]. The luminosity of a GGPP in the first Myr is estimated here by a radiation hydrodynamical calculation with the same assumptions as above, for the nucleated instability hypothesis, but without a core. The initial condition was a gravitationally unstable nebula condensation of a Jupiter-mass, M_J and a temperature of 10 K. Initially the GGPP had similar properties as the banana-shaped density enhancements found in [2] (mean density $8 \times 10^{-10}\,\mathrm{g\,cm^{-3}}$, central density $3.3 \times 10^{-8}\,\mathrm{g\,cm^{-3}}$). The results for this case and for $5\,M_J \sim 0.005\,M_\odot$ are displayed as the lowest two full lines in Fig. 3. Because a Jupiter mass is optically thick under nebula conditions, the $1\,M_J$ GGPP does not collapse immediately, but cools for ~ 1 Myr at a luminosity around a $\mu L_\odot$. The luminosity rise at the end of the depicted evolution indicates the onset of the final collapse to planetary densities.

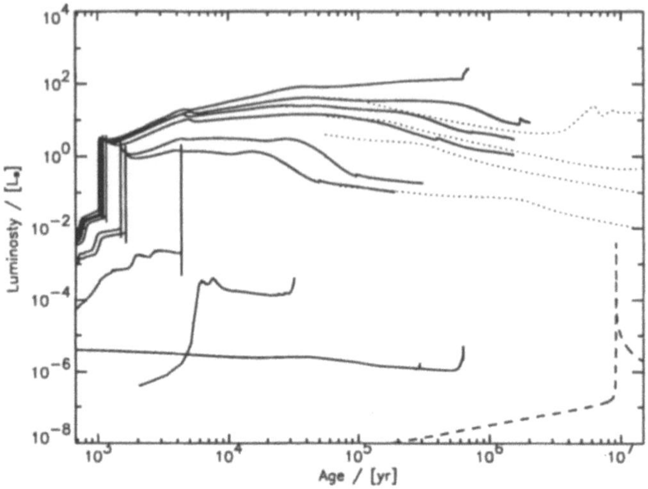

Fig. 3. Comparison of young stars, young brown dwarfs and young planets in the first 10 Myr as calculated from formation theories. The hydrodynamic evolution of gravitationally unstable cloud fragments of $10, 2, 1, 0.5, 0.1\,M_\odot$ (protostars), $0.05, 0.01\,M_\odot$ (proto brown dwarfs), and $0.005, 0.001\,M_\odot$ (giant gaseous protoplanets) are depicted as full lines, from top to bottom, respectively, together with the Saturn-mass proto giant planet (dashed) that has been calculated according to the nucleated instability hypothesis as discussed in Sect. 2.1 (cf. Figs. 1,2). Age is measured from the moment when the Rosseland-mean optical depth exceeds unity. Note the initial rises in luminosity, that are not present in evolutionary calculations that do not consider the formation process. For the planets, the formation of the core and the cooling in the case of the lowest mass GGPP, delay the luminosity peaks on the Myr timescale. Clearly the formation process has to be considered when discussing luminosities in a young population. The luminosity evolution for $2, 1, 0.5, 0.1\,M_\odot$ (dotted lines from top to bottom) as obtained from standard quasi-hydrostatic stellar pre-main sequence calculations ([3], 'MLT+Alexander') are plotted for comparison.

3 Zero Age for Planets, Brown Dwarfs and Stars

To compare the luminosities of objects of various masses at a given age a general definition of zero age is necessary. Because the thermally controlled Kelvin-Helmholtz contraction is the dominating evolutionary process in young gas spheres, whether they are giant planets, brown dwarfs or stars, I propose, to use the instant of time, when the interior of the gas-spheres is thermally enclosed for the first time, as age zero. The enclosure means that photons emitted as a consequence of an evolutionary change in the gas sphere cannot escape from the entire object directly and as soon as it changes — as is the case in the earliest isothermal cloud phases — but that they are radiated from a photosphere. Energy transfer to the photospheric bottleneck then delays the cooling and determines how the thermal reservoir in the interior is emptied. The cooling history can than be used to define an age since the heat reservoir was formed. As a practical definition for age zero I propose to use the instant of time when the Rosseland mean optical depth of a gaseous object exceeds unity. For a protostar this instant of time is practically at the end of the isothermal phase of the collapse. With that definition the luminosities of the protoplanets can be compared to stars and brown dwarfs of similar age.

4 A Comparison of the Luminosities of Young Planets, Brown Dwarfs and Stars

To finally address the question whether a young giant planet can be distinguished by its luminosity from a young brown dwarf or a young star of a similar age I present radiation hydrodynamical calculations of the formation of stars and brown dwarfs for the identical physical assumptions that were used to discuss planet formation above. The resulting luminosities in the first 10 Myr are compared to those of the giant planets in Fig. 3. I will discuss one important issue that follows from Fig. 3: the time delay between the 'initial' luminosity rise of stars and planets. Whereas protostars of all masses that are shown dramatically increase their luminosity within 2000 yr of cloud condensation and becoming optically thick, the luminosity increase for giant planets (with masses $< 0.05\,M_{\odot}$) is much slower. They need up to 10 Myr to reach their maximum accretion phase and, as a consequence, their maximum luminosity which is the starting point for their final contraction and cooling phase. This delay has an interesting consequence: A Saturn mass proto giant planet with the main accretion nucleated after 9 Myr almost outshines a $0.1\,M_{\odot}$ star of the same age, for a few 1000 yr.

5 Conclusion

The luminosity for a Saturn-mass protoplanet exceeds $10^{-4}\,L_{\odot}$, during the brief, $3\,10^{4}$ year period around maximum accretion. Luminosity maxima of giant planets occur at ages of 1–10 Myr, depending on the details of the formation process. To infer planetary nature in a young population, the properties of young planets have to be compared to those of proto brown dwarfs, as follows from the respective formation theories.

References

1. Alexander, D. R. and J. W. Ferguson (1994) Astrophys. Journ. **437**, 879
2. Boss, A. P. (1998) Evolution of the Solar Nebula IV – giant gaseous protoplanet formation. Astrophys. J. **503**, 923–937
3. D'Antona, F. and Mazzitelli, I. (1994) New Pre-Main-Sequence Tracks for $M \leq 2.5 M_\odot$ As Tests of Opacities and Convection Model. Astrophys. Journ. Suppl. Ser. bf 90, 457–500
4. Dorfi, E. A. and Feuchtinger, M. U. (1995) Adaptive Radiation Hydrodynamics of Pulsating Stars. Comp. Phys. Comm. **89**: 69–90
5. Götz, M. (1993) Die Entwicklung von Proto-Gasplaneten mit Drehimpuls — Strahlungshydrodynamische Rechnungen, Dissertation, Univ. Heidelberg
6. Burrows, A. and 8 other authors (1998) Advances in the Theory of Brown Dwarfs and Extrasolar Giant Planets. In: Rebolo R., Martin E. L. and Zapatero Osori M. R., (Eds.) Brown Dwarfs and Extrasolar Giant Planets. ASP Conf. Ser. **134**, 354–368
7. Kuhfuß, R. (1987) Ein Modell für zeitabhängige, nichtlokale Konvektion in Sternen. Dissertation, Tech. Univ. Munich.
8. Mizuno, H. (1980) Formation of the Giant Planets. Prog. Theor. Phys. **64**, 544–557
9. Stevenson, D. J. (1982) Formation of the giant planets. Planet. Space Sci. **30**, 755–764.
10. Wuchterl, G. (1991) Hydrodynamics of Giant Planet Formation II: Model Equations and Critical Mass. Icarus, **91**, 39–52
11. Wuchterl, G. (1993) The Critical Mass for Protoplanets Revisited: Massive Envelopes Through Convection. Icarus **106**, 323–334
12. Wuchterl, G. (1995b): Time dependent convection on self adaptive grids. Comp. Phys. Comm. **89**, 119–126
13. Wuchterl, G. and Feuchtinger M.U. (1998) A simple convection model for self-gravitating fluid dynamics. Astron. Astrophys. **340**, 419–430
14. Wuchterl, G., Guillot, T. and Lissauer J. J. (1999) Giant Planet Formation. In Protostars and Planets IV, Mannings, Boss and Russel, (eds.), Univ. Arizon Press **in press**

Imaging Exosolar Dust Disks

Paul Kalas

Space Telescope Science Institute, Baltimore, MD 21218, USA

Abstract. Replenished circumstellar dust disks have been resolved around seven main sequence stars at optical to submillimeter wavelengths. We review these observational results, and discuss the implications for exoplanetary systems.

1 Introduction

Captain Jacob (1859) described the orientation and morphology of Zodiacal light, the only circumstellar dust disk observable with the naked eye:

The light at its brightest was considerably fainter than the brighter portions of the milky way ... The outline generally appeared of a parabolic or probably elliptical form, and it would seem to be excentric as regards the sun, and also inclined, though but slightly, to the ecliptic.

Astronomical instruments that would permit astronomers to image exozodiacal dust disks appeared 80 years later with advances in solar astronomy. During the early 1900's, solar astronomers were faced with the challenge of inventing an instrument that could detect the solar corona in the absence of a total eclipse. Lyot (1939) finally invented the solar coronagraph, consisting of an opaque occulting spot at the focal plane of the telescope, and a "Lyot stop" at the pupil plane to mask diffracted light. Apparently, the Lyot coronagraph was not utilized for other purposes, such as searching for faint companions to nearby stars. The possibility that exozodiacal dust clouds might be more readily detectable than exoplanets due to their greater surface areas also did not attract attention.

Thirty years later, unmanned interplanetary vessels had replaced the sea captains and studied the Zodiacal dust cloud *in situ* (Leinert et al. 1976). Scientists developed a "planetary" coronagraph to observe Saturn's rings and moons in support of the Voyager spacecraft missions (Larson & Reitsema 1979, Jewitt et al. 1981). After the Infrared Astronomical Satellite (IRAS) revealed that Vega, Fomalhaut, ϵ Eri, and β Pic are surrounded by dust clouds, Smith and Terrile (1984) used their "stellar" coronagraph (Vilas & Smith 1987) to uncover a disk-like reflection nebulosity within 1000 AU of β Pic.

Because the lifetime of dust around these nearby main-sequence stars is orders of magnitude shorter than the ages of the central stars, the discoveries pointed to the existence of orbiting exoplanetesimals that replenish the detected grains via the same erosional process that comets and asteroids replenish the Zodiacal disk. The existence of pulsar planets was reported eight years later (Wolszczan & Frail 1992), followed by the discoveries of 17 planets around nearby main sequence stars (Marcy et al. 1999). Thus, by the end of the century, we had determined that nearby stars possess an inventory of objects comparable to that of our solar system (except for terrestrial planets in habitable zones).

2 Characteristics of Replenished Disks

Since IRAS observations of Vega provided the first evidence of replenished grains around a nearby star, the term "Vega Phenomenon" refers to the exosolar analog to our Zodiacal and trans-Neptunian dust system (Backman & Paresce 1993). Stars showing the Vega Phenomenon generate emission at 12, 25, 60 and 100 μm in excess of the expected stellar output. The basic experimental technique selects main sequence stars from stellar catalogs, and cross-correlates their listed positions to the positions of point sources detected by IRAS. From the spectral types and visible/NIR photometry one then estimates the photospheric spectral energy distribution (SED), and by comparison to the measured SED, determines the level of FIR excess (Fajardo-Acosta et al. 1998). Correcting for selection effects, 10% - 30% of A - K dwarfs display the Vega Phenomenon (this fraction is considered a lower limit; Backman & Gillett 1987, Aumann 1988).

A key result is that the SED's indicate a deficit of grains near the stars (Backman & Paresce 1993). High-resolution imaging in the optical, mid-IR, and sub-mm confirm the the existence and radii of central holes in the dust distribution. Table 1 summarizes a few properties of the spatially resolved replenished dust disks.

By "high-resolution" we mean that images now exist that are near diffraction limits. In the optical, HST (2.4 m) WFPC2 and STIS cameras have been used to map the β Pic disk at $\sim 0.1 - 0.2''$ resolution (FWHM of stellar images; Burrows et al. 1995, Heap et al. 1997). These images confirm a central depleted region and reveal a warp in the disk midplane. In the NIR, the ADONIS AO coronagraph on the ESO 3.6 m maps the inner warp with $\sim 0.12''$ resolution, and a theoretical simulation supports the existence of a planet inclined by $\sim 3^o$ to the disk midplane (Mouillet et al. 1997a,b). Also in the NIR, HST NICMOS resolves the disk around HR 4796 with $\sim 0.16''$ resolution (Smith et al. 1998).

In the mid-IR, cameras used on Keck (10 m) image the HR 4796 disk at 10 and 20 microns with $\sim 0.3''$ (20 AU) and $\sim 0.5''$ (34 AU) resolution, respectively (Jayawardhana et al. 1998, Koerner et al. 1998). The recent operation of adaptive optics at Keck II promises $0.05''$ resolution at H, corresponding to a mere 0.16 AU at the distance of ϵ Eri. In the sub-mm (850 μm), SCUBA at the James Clerk Maxwell Telescope (15 m) resolves dust emission around Vega, Fomalhaut, β Pic, and ϵ Eri with $14''$ resolution (Holland et al. 1998, Greaves et al. 1998).

Table 1. Properties of resolved dust disks

Central Star	D(pc)	Sp. Type	Age (yr)	Mass ($M_\odot$)	r_{disk} (AU)	r_{hole} (AU)
Sun	–	G2V	4.6x10^9	10^{-14}	50	30
Beta Pic	19.3	A5V	1-3x10^8	3x10^{-7}	>1300	20
Vega	7.8	A0V	3x10^8	3x10^{-8}	80	26
Fomalhaut*	7.7	A3V	1-3x10^8	5x10^{-8}	150	30
Epsilon Eri	3.2	K2V	5-10x10^8	3x10^{-8}	75	30
HR 4796A*	67.1	A5V	6-10x10^6	10^{-6}-10^{-7}	80	40
55 Cnc*	12.5	G8V	5x10^9	10^{-10}	60	35
BD +31° 643*	330	B5V	10^6-10^7	>10^{-6}	6,600	2300

In Table 1, an asterisk after the star name indicates a multiple stellar system. Because gravitational perturbations by a multiple system may retard the settling of gas and dust into a thin, dense disk plane, the naive prediction is that multiple systems are less likely to possess planetesimals and replenished dust disks. Observationally this is not the case - half the replenished disks in Table 1 exist in multiple stellar systems. If planetesimal growth is not hindered by stellar multiplicity, then planet growth may succeed around stellar systems of all types. In fact, about 1/3 of the radial-velocity-detected planetary systems are in multiple stellar systems.

Half the sample in Table 1 consists of early type stars, meaning that there is little overlap in the stellar samples utilized by disk researchers and radial-velocity searches for planets (55 Cnc is the exception). Early type stars are best at illuminating their surroundings, but rotation and a lack of features in the spectra make them non-ideal for radial velocity searches (the cutoff is about F5). One potential obstacle for planet growth around early type stars is the shorter timescale for clearing primordial dust and gas due to enhanced radiation fields - the luminosities of Vega, Fomalhaut and β Pic are 60, 13, and 8.7 $L_\odot$, respectively. The existence of replenished dust disks around these stars indicates that planetesimal and planet growth may be equally successful around early type stars.

The dust masses in Table 1 are estimated from the far-infrared excesses and sub-mm measurements. Given the age estimates, Table 1 indicates that dust mass decreases roughly as time^{-1} for the first 10^9 yr (Zuckerman & Becklin 1993; Koerner et al. 1998). For the solar system, cratering records indicate that the number of planetesimals raining down on the terrestrial planets decreased roughly as time^{-3} for the first 10^9 yr (and as time^{-1} for the next 3x10^9 yr; Chyba 1990). The discrepancy over the first 10^9 yr may arise from different disk evolution rates in and out of planetary growth regions. The dust masses in Table 1 refer to the dustiest part of each system, regions corresponding to our Kuiper Belt, not the terrestrial planet region of our solar system. Thus, a comparison of our early solar system's evolution to that of exosolar disks is *hindered* by a lack of information about our trans-Neptunian dust complex, and about dust mass in the depleted regions ("holes") for stars in Table 1.

Table 1 lists a range of ages that shows Fomalhaut, Vega and β Pic could be coeval (Barrado y Navascues et al. 1997). Why, then, does the β Pic disk contain one order of magnitude more dust than Fomalhaut and Vega? Whitmire et al. (1992) propose an exogenic process whereby a star passes through an interstellar cloud, temporarily enhancing the erosion of planetesimals. Artymowicz and Clampin (1997) present evidence refuting this idea, and champion the endogenic explanation that β Pic is simply younger. Kalas and Jewitt (1996) invoke both processes - 1000 AU scale asymmetries in the β Pic disk indicate an exogenic perturbation (e.g. a passing star or brown dwarf) that "stirs" the otherwise quiescent planetesimal system, which is still well-populated due to β Pic's young age. Another possibility is that disk evolution is dictated by the presence or absence of Jovian-mass planets - the timescale for planetesimal clearing will be significantly longer if a Jupiter-mass planet is missing (Wetherill 1994).

3 Links to Exoplanets

Spectra of the replenished dust disks reveal the composition of the planetesimal parent bodies, which in principle provide the material for building exoplanet cores and modifying their surfaces and atmospheres. A prominent 10 μm emission feature detected in the line of sight to β Pic (Telesco and Knacke 1991), and other Vega-like stars (e.g. Fajardo-Acosta et al. 1993, Sylvester et al. 1996), is attributed to Si-O stretching vibrations in micron-sized silicate particles. More detailed observations and models of β Pic's silicate feature (Knacke et al. 1994, Li & Greenberg 1998) robustly support the hypothesis of β Pic comets replenishing the short-lived, silicate-rich dust component.

Spectroscopic monitoring also reveals variable, transient, and almost always redshifted lines of neutral and ionized atomic species (reviewed by Lagrange et al. 1999). A model involving families of cometesimals rapidly sublimating in front of β Pic's photosphere is dynamically viable if the bodies are perturbed from a 4:1 mean motion resonance with a planet (Beust et al. 1996, Beust & Morbidelli 1996).

Images of dust disks constrain the otherwise elusive inclination, i, of the exosolar system to the line of sight (e.g. Fig. 4 of Kalas & Jewitt 1996). Thus, the sin i ambiguity for exoplanetary masses can be solved (e.g. Trilling and Brown 1998). Inclination also shows if a planetary system will eclipse the stellar photosphere. With eclipse data, one can infer the radius of a planet, and, given the mass from radial velocity searches, its density.

Model fits to β Pic give $i \geq 85^o$ (Kalas & Jewitt 1995). Lecavelier et al. (1997) and Lamers et al (1997) discuss a 0.06 magnitude variation of β Pic's light curve, which can be modeled either by an occulting planet, or a dust cloud. Thus, the favorable case of β Pic illustrates the difficulty of assigning unique explanations to any "events" in the light curve of a star, because a dust cloud can produce the same light signature as a planet.

HR 4796A has a more favorable viewing geometry because the disk images show a relatively dust free line of sight through the disk hole to the photosphere. A planetary transit requires $\tan i > a / R_*$, where a is the semi-major axis of the planet and R_* the stellar radius. If we take $R_* = 1.4 \times 10^9$ m, and $i \sim 75^o$ (Smith et al. 1998), then $a < 5 \times 10^9$ m (0.04 AU). Considering that four radial-velocity-detected exoplanets have $a \leq 0.05$ AU, HR 4796A presents a good opportunity to test for planetary transits.

Observations and modeling of the Zodiacal and Kuiper Belt grains (Dermott et al. 1994, Liou et al. 1996), and β Pic particles (Roques et al. 1994, Mouillet et al. 1997) demonstrate that planets gravitationally modify the structure of a collisionless dust disk. Possible features are central dust depletions, dust concentrations at the orbital radius of a planet, dust rings at mean motion resonances, and vertical warps created when a planet's orbital plane is inclined to that of the disk. Central dust depletions are observationally the most common features, and Table 1 lists their radii as well as the outer disk radii. The models indicate that a hypothetical planet must have ≥ 5 $M_\oplus$ to maintain a dust-free central region (by either ejecting particles outward, or injecting them toward destruction near the star).

If planet formation is a unique explanation for the central holes, then each instance of a central disk cavity would be indirect observational evidence for planet formation. If we estimate the fraction of stars displaying the Vega Phenomenon (F_V), and the fraction of Vega-like stars with central holes (F_h), then to first approximation the fraction of stars that have formed planets is $F_P \sim F_V \times F_h$. The observational evidence suggests $F_V \geq 0.1$ and $F_h \sim 1$, giving $F_P \geq 10\%$. This is twice the value indicated by radial velocity searches (Marcy et al. 1999). The difference may represent the relative frequencies of terrestrial planet to giant planet formation (e.g. Zuckerman et al. 1995).

Ice sublimation near a star could also contribute in generating a disk hole (e.g. Backman et al. 1992; Jura et al. 1998). However, the non-thermal UV photoevaporation of ice grains discussed by Artymowicz (1994, 1997) does not produce a sharp ice boundary. Also, the low luminosities of late-type stars such as ϵ Eri would not cause ice sublimation at the observed hole radii (Table 1). If no other depletion mechanism is proposed, then each instance of a central disk cavity is indirect evidence for planet formation. The hole radii in Table 1 therefore give us the approximate radius of each exosolar system.

4 Summary

Seven exosolar dust disks have been imaged at diffraction-limited resolutions. These observations confirm the existence of inner dust cavities previously inferred from SED's. The cavity radii are comparable to the semi-major axis of Neptune. *If the only explanation* for the cavities is planet formation, then the observed frequency of the Vega Phenomenon approximates the frequency of planet formation, $F_P \geq 10\%$, and the cavity radii, $r \sim 30$ AU, represent a typical solar system size.

Future work should focus on testing other physical mechanisms that could explain the disk cavities. We also argue for photometric monitoring of HR 4796A, given that its viewing geometry favors planetary transits with relatively little interference from the circumstellar disk.

Acknowledgements

We thank D. Backman, P. Artymowicz and D. Jewitt for valuable discussions. This paper is a shortened version of a review to appear in *Earth, Moon and Planets.*

References

1. Artymowicz, P. 1997, *Ann. Rev. Earth Planet. Sci.*, 25, 175
2. Artymowicz, P. 1994, in *Circumstellar Disks and Planet Formation*, ed. R. Ferlet and A. Vidal-Madjar (Editions Frontieres). p. 47
3. Artymowicz, P.,& Clampin, M. 1997, *ApJ*, 490, 863
4. Aumann, H.H. 1988, *AJ*, 96, 1415
5. Backman, D.E., Dasgupta, A. & Stencel, R.E. 1995, *ApJlett*, 450, L35
6. Backman, D. E. & Paresce, F. 1993, in *Protostars and Planets III*, ed. E. H. Levy and J. I. Lunine (University of Arizona Press: Tucson), p. 1253
7. Backman, D. E., Gillett, F. C. & Witteborn, F. C. 1992, *ApJ*, 385, 670
8. Backman, D. E. & Gillett, F. C. 1987, in *Cool Stars, Stellar Systems, and the Sun*, ed. J. L. Linsky and R. E. Stencel (Springer: Berlin), p. 340
9. Barrado y Navascues, D., Stauffer, J.R., Hartmann, L. & Balachandran, S.C. 1997, *ApJ*, 475, 313
10. Beust, H. & Morbidelli, A. 1996, *Icarus*, 120, 358
11. Burrows, C.J., Krist, J.E., & Stapelfeldt, K.R. 1995, *BAAS*, 187, 32
12. Chyba, C.F. 1990, *Nature*, 343, 129
13. Dermott, S.F., et al. 1994, *Nature*, 369, 719
14. Dominik, C., Laureijs, R. J., Jourdain de Muizon, M. & Habing, H.J. 1998, *A&A*, 329, L53
15. Fajardo-Acosta, S.B., Telesco, C.M., & Knacke, R.F. 1998, *AJ*, 115, 2101
16. Greaves, J.S. et al. 1998, *ApJ*, 506, L133
17. Heap, S.R., Lindler, D.J., & Woodgate, B. 1997, *BAAS*, 191, 4702
18. Holland, W.S., et al. 1998, *Nature*, 392,788
19. Jacob, W.S. 1859, *MNRAS*, 20, 4
20. Jayawardhana, R., et al. 1998, *ApJ*, 503, L79
21. Jewitt, D.C., Danielson, G.E., & Terrile, R.J. 1981, *Icarus*, 48, 536
22. Jura, M, et al. 1998, *ApJ*, 505, 897
23. Kalas, P., & Jewitt, D. 1996, *AJ*, 111, 1347
24. Kalas, P., & Jewitt, D. 1995, *AJ*, 110, 794
25. Knacke, R.F., et al. 1993, *ApJ*, 418, 440

26. Koerner, D. W., Ressler, M.E., Werner, M.W., & Backman, D.E. 1998, *ApJ*, 803, L83
27. Lagage, P.O. & Pantin E. 1994, *Nature*, 369, 628
28. Lagrange, A-.M., Backman, D., & Artymowicz, P. 1999, to appear in the proceedings of *Protostars and Protoplanets IV*, ed. V. Mannings, A. Boss, and S. Russel.
29. Lamers, H.J.G.L.M., Lecavelier des Etangs, A., & Vidal-Madjar, A. 1997, *A&A*, 328, 321
30. Larson , S.M. & Reitsema, H.J. 1979, *BAAS*, 11, 558
31. Lecavelier des Etangs, A., Vidal-Madjar, A., Burkl, G., Lamers, H.J.G.L.M., Ferlet, R., Nitschelm, C., & Sevre, F. 1997, *A&A*, 328, 311
32. Leinert, C., Link, H., Pitz, E., & Giese, R. H. 1976, *A&A*, 47, 221
33. Liou, J.-C., Zook, H.A. & Dermott, S.F. 1996, *Icarus*, 124, 429
34. Li, A. & Greenberg, J.M. 1998, *A&A*, 331, 291
35. Lyot, M.B. 1939, *MNRAS*, 168, 603
36. Marcy, G.W., Cochran, W.D. & Mayor, M. 1999, to appear in the proceedings of *Protostars and Protoplanets IV*, ed. V. Mannings, A. Boss, and S. Russell.
37. Mouillet, D., Larwood, J.D., Papaloizou, J.C.B. & Lagrange, A.M. 1997a *MNRAS*, 292, 896
38. Mouillet, D., Lagrange, A-.M., Beuzit, J.-L. & Renaud, N. 1997b, *A&A*, 324, 1083
39. Roques, F., Scholl, H., Sicardy, B., & Smith, B. 1994, *Icarus*, 108, 37
40. Smith, B.A., et al. 1998, *BAAS*, 193, 1382
41. Smith, B. A., & Terrile, R. J. 1984, *Science*, 226, 1421
42. Sylvester, R. J., Skinner, C. J., Barlow, M. J., & Mannings, V. 1996, *MNRAS*, 279, 915
43. Telesco, C.M. & Knacke, R.F. 1991, *ApJ*, 372, L29
44. Trilling, D.E. & Brown, R.H. 1998, *Nature*, 395, 775
45. Vilas, F., & Smith, B. A. 1987, *App. Optics*, 26, 664
46. Wetherill, G.W. 1994, *Ap.& Space Sci.*, 212, 23
47. Whitmire, D.P., Matese, J.J., & Tomley, L.J. 1988, *A&A*, 203, L13
48. Wolszczan, A. & Frail, D. 1992, *Nature*, 355, 145
49. Zuckerman, B., Forveille, T. & Kastner, J.H. 1995, *Nature*, 373, 494
50. Zuckerman, B. & Becklin, E.E. 1993, *ApJ*, 414, 793

The Stellar Mass Function

Francesco Paresce

European Southern Observatory

Abstract. Considerable progress has been made recently in more accurately determining the frequency distribution of stellar masses in the critical 0.1 to 1 $m_\odot$ range in both the field and in clusters and associations. I suggest that almost everywhere one chooses to look, be it in the halo or in globular clusters or in very young clusters, the MF can be reasonably well approximated within the current measurement error by a log-normal distribution function with a mean mass in the range 0.2–0.5 $m_\odot$ and width of order 1. Since this result applies equally well to 1 Myr and 10 Gyr old stellar systems, this form might be close if not coincident with the IMF after modifications by dynamical evolution. This simple hypothesis finds plausibility in the fact that a mass function of this type is a natural consequence of a wide variety of currently fashionable star formation scenarios but may be difficult to understand in conjunction with a rising mass function below the H burning limit.

1 Introduction

The IMF is a critical ingredient in our understanding of a large number of basic astronomical phenomena such as the formation of the first stars, the reionization and metal enrichment of the IGM, and galaxy formation and evolution of the star formation rate with z. It also plays a dominant role in star formation theories as the end result of molecular cloud contraction and fragmentation and its possible sensitivity to environmental conditions (gas pressure, density, temperature, sound speed distance, metallicity etc). Moreover, the IMF is one of the important factors determining the rate of cluster disruption via internal and external evolution (relaxation and tidal shocking) and, in consequence, of the possible dark matter content of galaxy halos. In this latter context, a power law IMF with a linear slope > 2 extending to very low masses is required to substantially affect the baryonic mass budget of the halo [2], [9].

The actual measurement of the IMF is a complex multi-step process whose ultimate precision and reliability rests heavily on a very careful quantitative analysis of all sources of possible random and systematic error. The first step is to obtain the luminosity function (LF) of the clearly specified sample in stars per volume per magnitude interval. The basic uncertainties here are due to sample contamination (cluster membership, orbit diffusion), incompleteness at the faint end, malmquist bias and calibration errors. Then, the LF has to be converted into a present day mass function (MF) in stars per volume per mass interval. This step is affected mainly by uncertainties due to the mass-luminosity and color-magnitude relations, the age, distance and extinction of the stars, their evolution, mass segregation and unresolved binaries. Finally, the IMF in stars ever formed per volume per mass interval depends crucially in most cases on age and the star formation rate and any subsequent effects such as external dynamical evolution of a cluster in a galactic tidal field.

Because of these difficulties, it has not been easy in the past to pin down the IMF of any stellar sample with truly satisfactory precision. Probably as a consequence, two extreme views on the stellar IMF seem to have emerged recently. At one extreme, there are claims that the IMF is by now understood well enough and that it is clearly universal [7]. At the other extreme, the claim is that there is no such thing as an observable universal IMF [21], [22] and that only recourse to theory can offer any hope to unravel the conundrum. In view of the fact that "the IMF is a theoretician's nightmare because of the dependence on the details of the physical processes" [22], it would seem rather disingenuous to abandon all hope on the observational determination of the IMF. Although it is certainly true that the observational situation is nowhere near as clear or as determined as the first school would like to admit, there are enough elements emerging from the chaos to allow perhaps a first tentative outline of the problem from this point of view. The key issue in my opinion is the proper selection of the most accurate measurements of and most fairly determined uncertainties on the IMF in as wide a range of stellar environments as possible. In the next few sections, I attempt such a critical selection.

2 Field Stars

The present situation regarding the MF for field stars is summarized in Figure 1 where I have drawn the deduced MF in several galactic field populations. For the halo, the LF of Dahn et al. [3] confirmed by Fuchs and Jahreiss [6] and Gizis and Reid [8] was converted to a MF by Chabrier and Mera [2] and corrected for binaries by Graff and Freese [9]. Holtzman et al. [12] is used for the bulge and Martini and Osmer [16] Reid and Gizis [19] and Kroupa [13] for the disk. The measurement peculiarities and sources of uncertainties in these measurements are exhaustively discussed and quantitatively evaluated by these authors so that they should be regarded as the most precise and up to date determinations of this function in the $\sim$ 3 to 0.1 $m_\odot$ range. I have distilled their measurements into their quoted best fitting power law in the appropriate mass range. Errors associated with the measurements of the slopes and masses of several dex in both axes should be considered typical. The data is shown in logarithmic mass units on the abscissae and the MF represented as the log dN/d log m in the ordinates. The slope of the power law is then $\Gamma = 1+\alpha$ where α is the slope in linear mass units ($\alpha = -2.35$ for the Salpeter IMF). The vertical position of the lines is arbitrary, of course, so I have shifted them up or down for enhanced visibility.

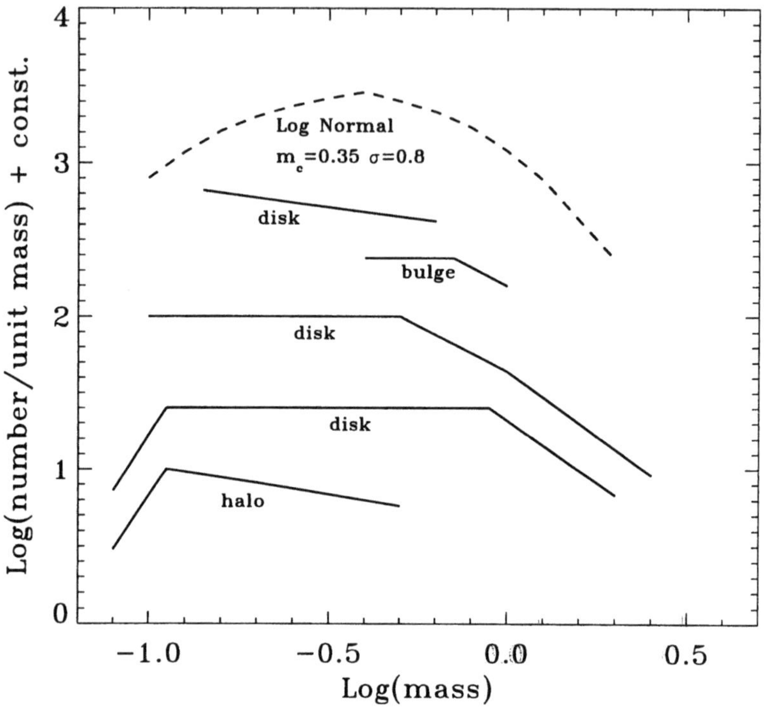

Fig. 1. The mass function of field stars in logarithmic units. The various curves represent from top to bottom: a log-normal distribution function with $m_c = 0.35$ $m_\odot$ and width $\sigma = 0.8$, disk [16], bulge [12], disk [13], disk [19] and halo [3].

A striking aspect of the results shown in this figure is the similarity between the various MFs despite the substantial differences in environment and physical characteristics such as metallicity and age. To be sure there are possible variations and inconsistencies in the details but overall the trend is pretty clear, namely a Salpeter-like increase in numbers with decreasing mass from ~ 3 to 1 $m_\odot$ followed by a definite break and flattening extending down to 0.1–0.2 $m_\odot$ with slopes in the $\Gamma = 0$ to -0.5 range. Moreover, all three measurements that reach down to the H burning limit with reasonable completeness clearly indicate a turnover below ~ 0.2 $m_\odot$. The only physically sensible function that can do justice to these measurements in the 0.1 to 1 $m_\odot$ range taken as a whole is a log-normal distribution of the type (also shown in Figure 1):

$$\log \frac{dN}{d\log m} = A - \frac{1}{2 < \sigma >^2} \left[\ln \frac{m}{m_c} \right]^2$$

with $m_c \sim 0.2$–$0.5\,\mathrm{m}_\odot$ and width $\sigma \sim 1$. In fact, reversing the process, namely convolving a log-normal distribution function of this type with the limited mass resolution and counting statistics presently available would easily generate the segmented power law mass functions shown in Figure 1.

3 Cluster Stars

A similar analysis can be carried out for stars in clusters. Here the data available is more extensive and varied. The use of clusters for determining the stellar MF has certain advantages over that of the field. These include the much greater homogeneity of the sample in terms of age, metallicity, distance etc but disadvantages include crowding and uncertain evolutionary and mass segregation effects (see Meyer et al. [17] for a recent review).

The current situation is summarized in Figure 2 in the same way and with the same units used in Figure 1. Represented here are the two best studied very young clusters (ages < 3 Myr) with the most complete MFs obtained so far, namely the Orion Nebula Cluster (ONC) [11], and IC348 [14], [10] together with the most recent determinations of the MFs for the three classical nearby open clusters: the Pleiades (reported by Bouvier et al. [1], the Hyades [20] and M67 [5]. The latter are $\sim$ 125, 625 and 10^4 Myr old respectively. The latest determination of the MF of the nearest globular cluster NGC 6397 [4], [18] is also plotted in Figure 2. As in Figure 1, I have distilled the reported MFs in Figure 2 in power law line segments in the appropriate mass ranges. I have also deliberately resisted the temptation to extrapolate the stellar MF to the substellar mass regime to avoid injecting preconceived notions in the discussion of the stellar MF.

As in the case of the field stars shown in Figure 1, it is difficult to escape the conclusion that, at least to the level of accuracy of the present data, all the MFs schematically represented in Figure 2 come from basically the same underlying distribution function that increases with mass from the substellar limit to a peak somewhere between 0.2 and 0.5 solar masses and then drops steeply beyond $\sim$ 1 solar mass. It is somewhat tempting to see the peak shifting to higher masses with increasing age possibly due to dynamical evolution but small number statistics and large measurement errors contribute to considerable uncertainty but it is indisputable that some such effect must be at work at least on some clusters. Only more and better data will allow this fascinating possibility to become reality and a vital tool in our understanding of these important effects. In the meantime, we are content to point out that there seems to be no reason to think that the IMF is much different from a log-normal with m_c somewhere in the range ~ 0.2–0.4 and $\sigma \sim 1$ since there is *no gross age dependence* in the results shown in Figure 2.

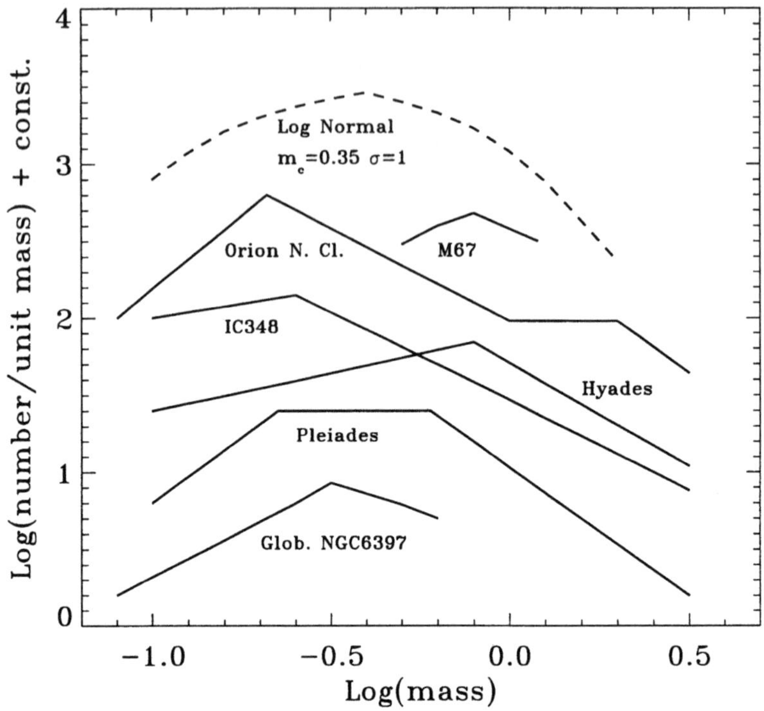

Fig. 2. The mass function of selected clusters in logarithmic units. The various curves represent from top to bottom: a log-normal distribution function with $m_c = 0.35\ m_\odot$ and $\sigma = 0.8$, the Orion Nebula cluster [11], M67 (from 0.5 to 1.2 $m_\odot$ [5]), IC348 [14], the Hyades [20], the Pleiades [1] and NGC 6397 [4].

4 Discussion and Conclusions

There are several points that can be stressed in the context of our currently incomplete but quickly improving view of the stellar IMF. First is the, by now, robust result that the steep MF of high mass stars ($> 1\ m_\odot$) is not maintained in the low mass star regime below 1 $m_\odot$. Thus the overly simplified recipe used in the vast majority of galaxy evolution calculations of a $\alpha = -2.35$ power law extending from 100 $m_\odot$ all the way down to the H burning limit is simply untenable and any result that depends on this assumption has to be viewed with suspicion especially those concerning expected dark masses of low mass stars. Second, it is highly unlikely that the IMF has a simple power law structure even in the decade between 1 and 0.1 $m_\odot$. Certainly the present day MFs do not and the only way to reconcile the two is to suppose that the former has been modified by dynamical evolution to resemble the latter. From the present results on clusters, however, this modification needs to have been impressed very early in the life of the cluster, certainly in the first Myr or so. A more likely possibility is that the clusters actually started with a log-normal type of IMF already peaking around 0.2 solar masses with erosion of the lowest mass stars possibly moving the mean mass upwards in time. The erosion rate would be expected to depend sensitively on the characteristics of their orbit in the galaxy. Theoretical calculations of this effect using a log-normal IMF have yet to be carried out but the tools to do so are rapidly improving [23].

Third, the physical reality of an actual drop in the MF below $\sim$ 0.2 solar masses rather than a flattening has always been in dispute up to now. There are quite a few observations that do not show this feature but they are all suspect since they do not have anywhere near the sensitivity and resolution required to detect it. Those that do, do so systematically as shown dramatically in Figure 2. This brings up the correlated issue of the connection between the stellar MF and its substellar counterpart. Must we insist on a smooth transition as one might expect from stellar formation theories that ignore the existence of the boundary or is some other mechanism at work to decouple the two populations? If it indeed turns out that the substellar IMF rises as m^{-1} with decreasing mass [15] and the two populations are drawn from the same sample, the question might arise as to what happened to the very low mass stars. Have they all been stripped from the clusters? If so, they should appear in abundance in the field MFs shown in Figure 1 but they do not seem to.

Finally, there are the fundamental questions of the existence and universality of the IMF alluded to earlier. If the picture I have just described is correct, the underlying IMF, at least in our galaxy and in the mass range 0.08–3 $m_\odot$ exists but whether it is universal or not simply cannot be decided unambiguously with the present data mainly because of the uncertain effects of dynamical evolution. But there seems to be no reason to believe that the effect cannot be measured and understood both with wider and more precise observations and with improving theoretical calculations. Having then removed or at least understood this effect, we could move on to probing differences in MFs due to environment that are now quite likely completely masked by sample pollution in the field and dynamics in clusters. All of this will definitely require deep unbiased surveys of the type that the VLT is designed to carry out owing to its very high sensitivity coupled to very wide field capabilities.

To conclude, it seems that from a purely observational standpoint there is no basis for pessimism regarding this fundamental parameter. It will not be easy for sure but unveiling the stellar IMF and exploiting the resulting cornucopia of information may be a feasible objective in the VLT and NGST era and no effort should be spared to meet it either because we think the *IMF does not exist* or because we think all is already clear.

References

1. Bouvier, J. et al., 1998, A&A, 336, 490
2. Chabrier, G. and D. Mera, 1997, A&A, 328, 83
3. Dahn, C. et al., 1995, Proceedings of ESO Workshop "The Bottom of the Main Sequence and Beyond", C.G. Tinney (ed.), Springer, Berlin, p. 239
4. De Marchi, G., F. Paresce and L. Pulone, 1999, ApJ, in press
5. Fan, X., et al., 1996, AJ, 112, 628
6. Fuchs, B. and H. Jahreiss, 1998, A&A, 329, 81
7. Gilmore, G., 1998, in "The Stellar Mass Function" Proc. 38th Herstmonceux Conference, eds. G. Gilmore, I. Parry, and S. Ryan, editors preface
8. Gizis, J.E. and I.N. Reid, 1999, AJ, 117, 508
9. Graff, D.S. and K. Freese, 1996, ApJ, 467, L65
10. Herbig, G., 1998, ApJ, 497, 736
11. Hillenbrand, L.A., 1997, AJ, 113, 1733
12. Holtzman, J.A. et al., 1998, AJ, 115, 1946
13. Kroupa, P., 1998, ASP Conf., 134, 483
14. Luhman, K., G.H. Rieke, C.J. Lada and E.A. Lada, 1998, ApJ, 507, 347
15. Martin, E.L., M.R. Zapatero-Osorio and R. Rebolo, 1998, in "Brown Dwarfs and Extrasolar Planets", eds. Rebolo, Martin and Zapatero-Osorio, ASP Conf. Ser., 134, p. 507
16. Martini, P. and P. Osmer, 1998, AJ, 116, 2513
17. Meyer, M.R., F.C. Adams, L.A. Hillenbrand, J.C. Carpenter and R.B. Larson, 1999, in Protostars & Planet IV, eds. V. Mannings, A. Boss, S. Russell (Tucson: The University of Arizona Press), in press
18. Paresce, F. and G. De Marchi, 1999, ApJ, in press
19. Reid, I.N. and J.E. Gizis, 1997, AJ, 113, 2246
20. Reid, I.N. and Hawley, 1999, AJ, 117, 313
21. Scalo, J., 1998, in "The Stellar Mass Function" Proc. 38th Herstmonceux Conference, eds. G. Gilmore, I. Parry, and S. Ryan, p. 201
22. Scalo, J., 1999, astro-ph/9811341, proceedings for The Birth of Galaxies, Blois
23. Takahashi, K. and S.F. PortegiesZwart, 1999, astro-ph/9903366

Aspects of the Substellar Mass Function

Gibor Basri

Univ. of California, Berkeley CA 94720

Abstract. The best method we have at the moment for finding the substellar mass function is the study of young clusters. These have the advantages that 1) the stellar mass function for the same environment is known, 2) the brown dwarfs are all at the same (known) age as are the stars, 3) effects of dynamical evolution are minimized the younger the cluster (and the brightness of brown dwarfs is maximized). Rather than concentrate on this method, I consider issues relevant to two other methods that are also being employed to study the substellar mass function: brown dwarfs as companions to stars, and brown dwarfs in the field. I argue that the impression of a brown dwarf "desert" that arose from radial velocity studies of solar-type stars is misleading (although there is a "desert island" associated with this class of object). I also discuss the difficulties in finding a mass function in the field, and the use of lithium and luminosity distributions to help constrain it.

1 Introduction

Until 1995, there was no empirical information at all on the substellar mass function. I have summarized the development of the field since then in a recent review ([3]; cf. [12]). The paper here by Paresce discusses the stellar MF, and the fact that it seems to flatten out at low stellar masses. The stellar data should not be construed as much of a constraint on the substellar MF; it is poorly enough known at low masses that one is better off trying to understanding the substellar MF and joining it to the stellar MF rather than trying to extrapolate the stellar MF to the substellar domain. The best data on the substellar MF to date comes from work on the Pleiades (see the paper here by Zapatero-Osorio). It has the advantage that the stellar MF for the same environment is known, and the brown dwarfs are all at the same age. Recent work by Tamura [14] on Taurus, and work presented here by Zapatero-Osorio on the σ Ori cluster, suggests that the substellar MF can reach down all the way to the deuterium burning limit.

Besides the work on clusters, there are two other sources of data which in principle give us a handle on the substellar mass function. One of these is the search for BDs as binary companions to stars. This method had great favor in the time before BDs were found, because it was felt that at least one knows where to look for them in this case. The other is the search for BDs in the solar neighborhood - freely floating in the nearby field. Both of these methods have been successful in finding certified BDs. But neither of them is as useful as clusters for actually studying the substellar mass function. I discuss each of them in turn below.

2 Brown Dwarfs in Binary Systems

The hunt for BDs as companions can proceed on two parallel tracks. One can find them by indirect means, just as with extrasolar planets. In particular, precision radial velocity (PRV) searches will automatically turn them up, even more easily than planets. It now seems fairly likely that the first BD was in fact found this way: HD114762 was found as a PRV companion to a solar type star by [6]. The difficulty with these detections is that they are only lower mass limits. For a particular case, one is never quite sure that the correction may not push it into the stellar mass range. On statistical grounds one can argue that all the inclination corrections cannot be large. The extent to which this argument can be made, however, depends on the intrinsic MF of binary companions.

The main source of BD candidates from PRV studies has been the work of Mayor and collaborators [10]. They, and [1] before them, concluded that the number of BD candidates was consistent with a flat or slowly rising mass function into the substellar domain. It was surprising when Mayor et al. and collaborators from the Hipparcos project found that the companions are actually primarily stellar. The question can now be recast: what distribution of stellar companions can produce the large number of false BD candidates? Of course, the stellar companion distribution must also be compatible with the known results in the stellar regime. I have redone the analysis in [1] using lognormal distributions. I was able to find a solution which produces the right number of false positives in the BD domain, and matches reasonably with the stellar results (Fig. 1). The peak occurs at a mass of 0.175 $M_\odot$, and the number of companions drops to half the peak value at about the substellar limit.

The cluster work implies there could easily be as many BDs in the Universe as there are stars. The PRV searches, furthermore, are not subject to the worst selection effect - that BDs fade out with time. Why then, the rapidly falling substellar MF in binaries contrasted with the rising substellar MF in clusters? The likely cause is that when one looks at binary companions, the mass-ratio of the system is a very important parameter. The binary formation mechanism may care more about the mass-ratio than the absolute mass of the companion. Thus, searching for BD companions to solar-type stars could yield a rather different result than a search for BD companions to low mass stars (or even other BDs). The evidence to date is certainly in favor of this. We cannot be sure of *any* of the PRV BD candidates at the present time; the remaining candidates must have their orbital inclinations determined.

On the other hand, there are a number of certified BD candidates to low mass stars already known. Gl229B is certainly the most well-known of these; it is a very cool companion to an early M star [11]. G196-3 is a lithium-certified L-type BD companion to a young early M star [13]. The companion to GG Tau B [15] is an even younger example of such a system (here we rely on models as well as lithium, but the overall system gives confidence in

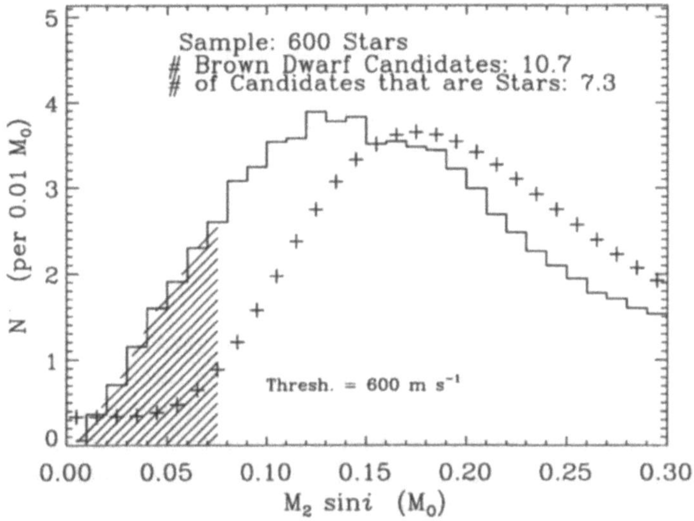

Fig. 1. A revised estimation of the companion mass function, based on the latest data from the Mayor et al. PRV surveys, and the analysis method of [1]. A log normal companion distribution peaking at 0.175 $M_\odot$ is assumed (plusses), which generates an "observed" distribution of candidate mass limits (histogram). The number of candidates, and fraction of them which are stars with orbital inclinations that make them look substellar, is approximately what is observed

the result). LHS 102B [5] is an L-type companion to an early M star too, although the lithium test does not give certification in this case. That does not exclude the companion from being a BD, but it must have a mass greater than 60 M_J . There are further unconfirmed candidates in Taurus [14] and the TW Hya association [7]. Finally, there are at least two double BD systems known: PPl15 [4] which is a close spectroscopic binary, and DENIS 1228-15 [9] which is an imaged system with a few AU separation. What is striking is that the surveys which found these systems are all substantially smaller than the surveys for BD companions to solar-type stars (or around white dwarfs, which would have been higher mass primaries). In particular, the double BD systems were found right at the beginning of very small surveys.

Thus, it is clearly easier to find systems containing BDs if the mass-ratio is not too large. The distribution of companions in Fig. 1 can be reinterpreted as a distribution with a peak at a mass-ratio of about 1:6 and a cutoff for mass-ratios greater than 1:10 or so. If the same distribution applied for M stars, one might expect many companions in the BD range (though it is not given that the same mass-ratio distribution holds for all primary masses). At any rate, this means that binary companions are probably not a good means of addressing the general substellar MF. They probably tell us more about the binary formation mechanism (itself a very interesting topic), than the general likelihood of forming substellar objects. And the metaphoric brown dwarf "desert", which expresses the result that BD companions to solar-type stars are substantially rarer than either stellar or planetary companions [8], should now be seen as merely a "desert island" which occurs in high mass-ratio systems. There are "verdant fields" of BDs in clusters and in the actual field, and BDs are more common as companions if one sticks with low mass primaries.

3 Brown Dwarfs in the Field

Since 1997, the new IR all-sky surveys (DENIS and 2MASS) have been uncovering nearby young BDs in the field at an increasing rate. The interpretation of field survey data requires two separate and difficult steps. The first (which I do not really discuss), is the correction of the survey for observational biases and effects. A survey with given sensitivity will sample smaller and smaller volumes for objects of cooler temperatures. There is also a correction for completeness effects as a function of observed brightness in the survey colors, and the conversion of observed intensities to luminosity or effective temperature. Binaries would also have to be accounted for.

The second, overriding problem lies in the nature of the BDs themselves. By definition, they never come onto the main sequence and so are continually fading with time. There is a complete degeneracy in the relations between luminosity/temperature, mass, and age. Both VLM stars and BDs will start at mid-M, then cool off. A star settles at its main sequence temperature – BDs do not. Different mass objects will pass through the same temperature at different ages. Observations, unfortunately, generally don't yield mass and age separately.

To deal with the mass/age degeneracy for a MF, one must resort to a model for star formation history; there is no way to disentangle the mass function at a given time from the history of the production rate of objects at various masses. I have made a simulation purely for illustrative purposes, using models kindly provided me (for other purposes) by G. Chabrier and I. Baraffe, in the mass range 45-100 M_J . I assume a constant star formation rate from 10 Gyr ago to the present, and a flat MF (equal numbers of objects in 5 M_J bins). The models provide the luminosity history in each mass bin. The results are shown in Fig. 2 (*without including any observational biases*). The histogram of current luminosities shows a pronounced dip at the bottom of the main sequence. Brighter than that, the number of objects is influenced primarily by the pileup of VLM stars at their main sequence luminosities.

The sudden lack of objects which stabilize in luminosity accounts for the sharp dip at the substellar boundary (which these models place at $\log(L_\odot)$=-3.9). The BD numbers increase again as we see the effects of larger age bins contributing to them at very low luminosities. Of course, the observational biases will convolve this intrinsic luminosity distribution with a function which drops rapidly towards lower luminosities as the effective survey volume de-

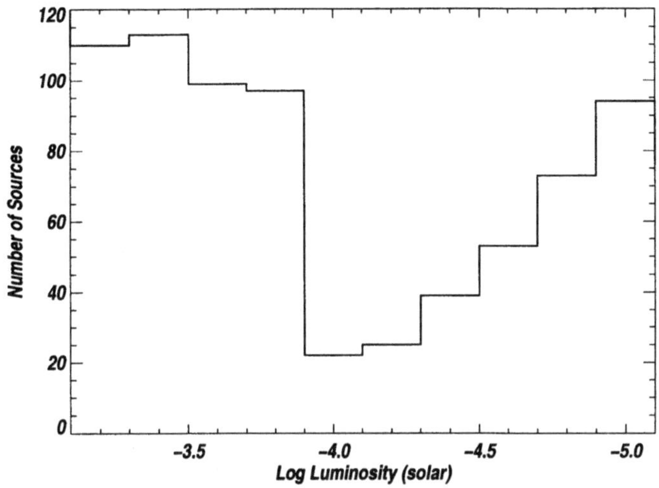

Fig. 2. A histogram of current observed luminosities generated by assuming a flat MF and constant star formation rate, for objects between 45–100 M_J and ages between 0–10 Gyr. The luminosity histories of the objects are from models by Baraffe and Chabrier

creases for such objects. Nonetheless, we might hope to see the sudden drop in the final survey data and so empirically locate the bottom of the main sequence. Changing the assumed star formation rate, or the MF, alters the shape of the predicted histogram.

It is easier to find the relative numbers of VLM stars and BDs. If one finds an object which cooler than M6 with lithium, that object must necessarily [2] be substellar (regardless of its age). The ambiguity between stars and BDs is also gone for objects cooler than the minimum main sequence temperature. Figure 3 shows the variation of the fraction of lithium-bearing objects with luminosity. The combined results from 2MASS and DENIS seem to indicate that the fraction of lithium-bearing L stars is around 0.2 in the early L subclasses, and increases to perhaps 0.4 in the later L subclasses.

Above the substellar boundary, the fraction is basically that between objects below and above 60 M_J (which is dominated by the number of stars which reach each subclass on the main sequence). It will suddenly rise below the substellar boundary to the ratio of BDs above and below 60 M_J . It will increase further below that luminosity, as one begins to run out of BDs above 60 M_J which have had time to fade enough, until it becomes 100% lower than the luminosity which the oldest 60 M_J BD will have reached. Of course, we have not sampled such low luminosities at all, yet. The point is, the behavior of the lithium fraction as a function of luminosity contains information on the star formation rate and mass function together, and can be used to help

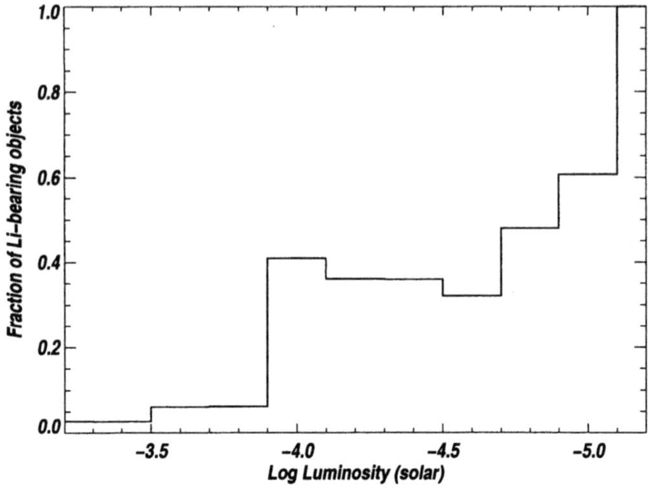

Fig. 3. A histogram of the current fraction of lithium-bearing objects at current luminosities for the same set of assumptions as in Fig. 2

constrain them (even better if the star formation rate can be taken as given from other lines of evidence).

Finally, though, one must conclude that both of the methods I have discussed here in detail are difficult to use for getting the substellar MF. They rely on models for the MF and star formation rate together. There are very substantial observational selection effects to be accounted for (which are harder to account for in advance of the answer). My real point is that the study of cluster MFs, where there is no ambiguity arising from different ages, and the effects of binaries are muted and can be studied, will serve as our best diagnostic of the substellar MF for the next stage of study. It is also likely that our understanding of the bottom end of the substellar mass function will come from studying very young clusters, which have the additional advantage that very low mass objects are still relatively bright. This does not mean that we should drop the study of BDs in binary systems or in the field; each has significant other advantages and provides complementary information in the overall effort to understand brown dwarfs and substellar objects. In particular, the study of cooler brown dwarfs will come primarily from these two areas, since suitable targets can be found much closer to us than in clusters.

References

1. Basri, G.; Marcy, G.W. (1997) Early hints on the substellar mass function. Star Formation Near and Far. Seventh Astrophysics Conference. A.I.P. Conf. Proc. **393**, 228–240
2. Basri, G. (1998) The lithium test for young brown dwarfs. Brown Dwarfs and Extrasolar Planets Workshop. A.S.P. Conf. Ser. **134**, 394–404
3. Basri, G. (1999) Brown Dwarfs: The First Three Year. Reviews in Modern Astronomy **12**, 101–113
4. Basri, G., Martín, E. (1999) PPl15: The first brown dwarf spectroscopic binary. Astr. J. (submitted)
5. Goldman, B., Delfosse, X., Forveille, T., Afonso, C., et al. (1999) EROS 2 proper motions: a brown dwarf, and an L dwarf companion to LHS 102. Astr. Ap. (submitted)
6. Latham, D.W., Mazeh, T., Stefanik, R.P., Mayor, M. et al. (1989) The unseen companion of HD 114762: a probable brown dwarf. Nature **339**, 38–40
7. Lowrance, P.J.; McCarthy, C.; Becklin, E.E.; Zuckerman, B. et al. (1999) A candidate substellar companion to CD -33 degrees 7795 (TWA 5). Ap. J. **512**, L69–L72
8. Marcy, G.W., Butler, R.P. (1998) Detection of extrasolar giant planets. Ann. Rev. Astr. Ap. **36**, 57–97
9. Martín, E., Brandner, W., Basri, G. (1999) A search for companions to nearby brown dwarfs: The binary DENIS-PJ1228.2-1547. Science **283**, 1718–1720
10. Mayor, M.; Udry, S.; Queloz, D. (1998) The mass function below the substellar limit. Cool Stars, Stellar System and the Sun. Tenth Cambridge Workshop. A.S.P. Conf. Ser. **154**, 77–87
11. Nakajima, T.; Oppenheimer, B.R.; Kulkarni, S.R.; Golimowski, D.A. et al. (1995) Discovery of a cool brown dwarf: Gliese 229B. Nature **378**, 463–465
12. Oppenheimer, B.R., Kulkarni, S.R., Stauffer, J.R. (1999) Brown Dwarfs. Protostars and Protoplanets IV. preprint (astro-ph/9812091)
13. Rebolo, R.; Osorio, M.R.Z.; Madruga, S.; Bejar, V.J.S. et al. (1998) Discovery of a low-mass brown dwarf companion of the young nearby star G196-3. Science **282**, 1309–1312
14. Tamura, M., Yoichi, I., Oasa, Y., Nakajima, T. (1998) Isolated and companion young brown dwarfs in the Taurus and Chamaeleon molecular clouds. Science **282**, 1095–1097
15. White, R., Ghez, A., Schulz, G., Reid, I.N. (1999) The quadruple PMS system GG Tau: evidence for a substellar companion. Ap. J. (accepted)

Planetary Microlensing: Present Status and Long-Term Goals

Penny D. Sackett

Kapteyn Astronomical Institute, 9700 AV Groningen, The Netherlands
psackett@astro.rug.nl

1 Introduction

Massive gravitational microlensing programs were begun about a decade ago as a means to search for compact baryonic dark matter in the Galaxy [1], but before the first events were detected [2–4] the technique was also proposed as a means of detecting extra-solar planets in our Galaxy [5]. Current microlensing

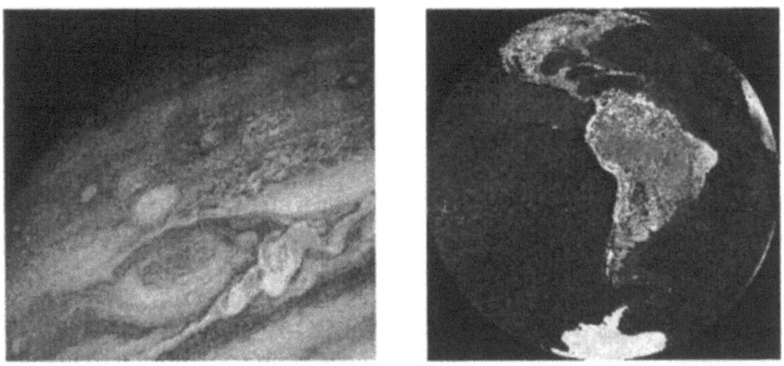

Fig. 1. Present microlensing planet detection programs are sensitive to planets similar to the one on the left; theoretical and observational capabilities must be increased by an order-of-magnitude before the long-term goal of reliably detecting planets similar to the one the right can be achieved.

planet searches, which have been underway for four years, are sensitive to jovian-mass planets orbiting a few to several AU from their parent Galactic stars. Within two years, sufficient data should be in hand to characterize or meaningfully constrain the frequency of massive planets in this range of parameter space, nicely complementing information about planets at smaller orbital radii now being provided by radial velocity searches. In principle, the technique could be pushed to smaller planetary masses, but only if a larger number of faint microlensed sources can be monitored with higher precision and temporal sampling. The VST on Paranal, with spectroscopic follow-up with the VLT, may be the ideal instrument for such an ambitious program.

2 Point Lenses

As light rays from a distant background source S pass a distance ξ from a gravitational point lens L of mass M, they are bent by an angle $\alpha = 4\,G\,M/c^2\,\xi$. Simple geometric arguments reveal that the two resulting images have an angular separation on the sky $\Delta\theta_I = \sqrt{\theta_S{}^2 + 4\,\theta_E{}^2}$, where θ_S is the angular distance between the lens and the observer-source sight line and θ_E is the angular Einstein ring radius, a characteristic size for the lensing geometry defined by

$$\theta_E \equiv \sqrt{\frac{4GM\,D_{LS}}{c^2\,D_{OL}\,D_{OS}}} \quad . \tag{1}$$

Here D_{OL}, D_{OS}, D_{LS} are the observer-lens, observer-source, and lens-source distances, respectively (Fig. 2).

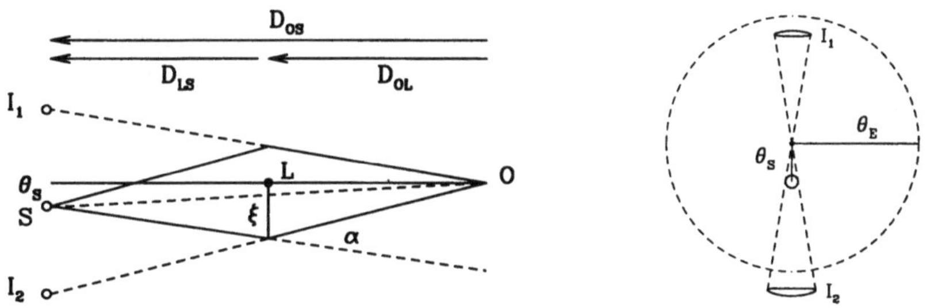

Fig. 2. Point-lens microlensing geometry as seen from the side (left) and projected onto the sky (right). The images I_1 and I_2 straddle the source position on the sky.

Since the specific intensity of each ray remains unchanged, the magnification of the images is just the ratio of the image area to the source area, which can be found by evaluating at the each image position the determinant of the Jacobian mapping J describing the lensing coordinate transformation:

$$\text{Magnification Image}_{\,i} = \left.\frac{1}{|\,det\,J\,|}\right|_{\theta_I=\theta_i} = \left|\frac{\partial\,\theta_S}{\partial\,\theta_I}\right|^{-1}_{\theta_I=\theta_i} \quad . \tag{2}$$

The largest magnifications occur when the determinant is near zero. For a point lens this occurs when $\theta_S \approx 0$ and the source lies almost directly behind the lens; the images then lie close to the Einstein ring θ_E. For lenses with stellar masses and distances typical for stars within the Milky Way, $\theta_E \sim 1$ mas, so that whenever the images are significantly magnified, they are too close together to be resolved by traditional imaging. Only the combined magnification A of both images can be measured. For a point-lens, Eq. 2 can be used to show that $A = (u^2+2)/(u\sqrt{u^2+4})$, where $u \equiv \theta_S/\theta_E$ is the instantaneous source-lens separation in units of the angular Einstein radius.

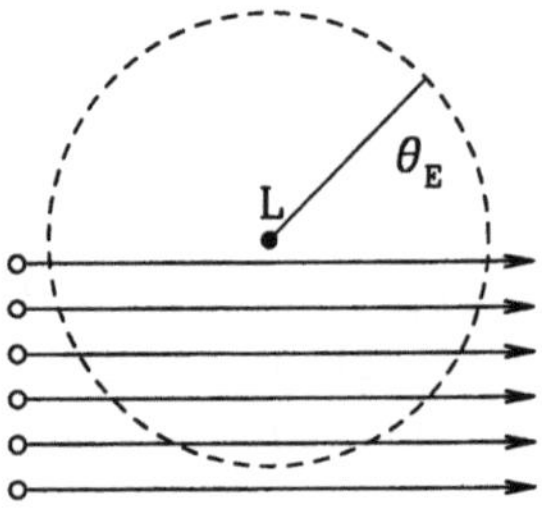

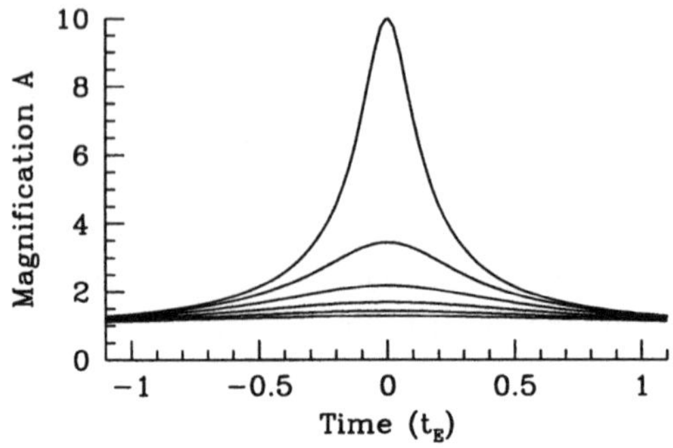

Fig. 3. Sample source trajectories (left) and the resulting light curves (right).

Because the source, lens, and observer are in relative motion, u is a function of time. When u is at its minimum, the observed light curve of the microlensing event undergoes its maximum magnification. Microlensing events for which the source has a small minimum impact parameter $u_{\rm MIN}$ will have the largest peak amplifications (Fig. 3). The characteristic time $t_{\rm E} \equiv \theta_{\rm E} D_{\rm OL}/v_\perp$ is the time taken by the lens, moving at speed $v_\perp$ across the sight line to the source, to travel one Einstein radius, and is a several days to months for most Galactic microlensing events.

3 Binary Lenses

Point lens light curves are symmetric because they are one-dimensional cuts through two-dimensional circularly-symmetric magnification patterns on the sky. Double lenses can generate more complicated light curves because this symmetry is broken. In particular, the loci of points in the source plane for which $\left| det\, J \right| = 0$ have quite complicated structure (Fig. 4).

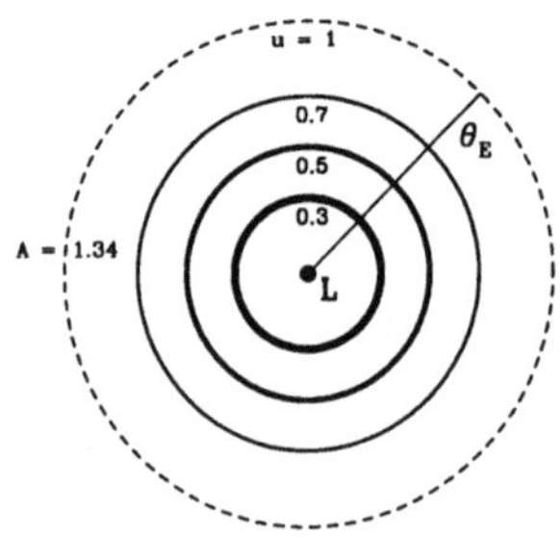

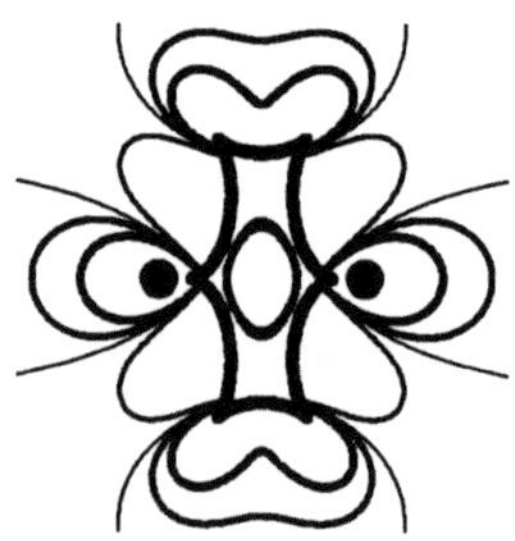

Fig. 4. Single lens magnification map (left) and a binary lens fractional deviation map (right). The binary components (heavy dots) are separated by 1 $\theta_{\rm E}$ and have a total mass equal to the single lens. Regions where the binary magnification is depressed or increased by 1% and 5% relative to that of the single lens are marked by light and bolder contours, respectively. The very bold contour is the caustic.

These loci — called *caustics* — mark positions at which point sources would experience infinite image magnification. The observed flux remains finite as a caustic crosses a real source; integration over the source size θ_* is required to compute the observed total magnification. Sources passing near binary-lens caustics will exhibit light curves that deviate strongly from those of single lenses. The smaller the source and the closer the caustic approach, the larger the deviation will be. Unlike single lens curves, the shape of a binary light curve depends on the angle of the source trajectory through the magnification pattern.

Any static, unblended binary light curve is described by eight parameters: four quantities relevant to single lenses ($u_{\rm MIN}$, $t_{\rm E}$, time at peak t_0, and baseline flux F_0) and four additional parameters, namely, the binary mass ratio $q \equiv m_1/m_2$, the instantaneous angular separation b of the components in units of $\theta_{\rm E}$, the ratio $\rho_* \equiv \theta_*/\theta_{\rm E}$ of source radius to Einstein radius, and the angle ϕ between the source trajectory and the binary axis.

4 Planetary Microlensing

A lens orbited by one or more planets is a multiple lens and thus may have a magnification pattern that differs measurably from that of a single lens. Measuring and characterizing light curve deviations induced by planetary lensing companions is the goal of microlensing planet detection programs. The number, size and relative positions of the caustics depend on the mass ratio q and separation b of the double lens (Fig. 5).

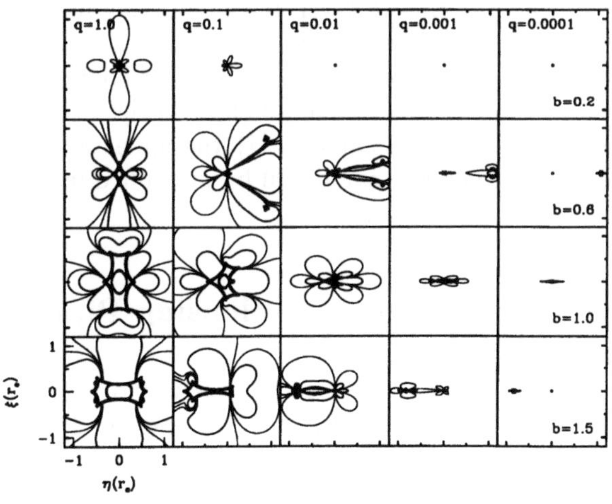

Fig. 5. Fractional deviation patterns for binary lenses with different mass ratios q and separations b. Contours are the same as in Fig. 4; heavy lines are caustics. A "super-jupiter" with mass equal to 10 $M_{\rm J}$ orbiting a late M dwarf would have $q \approx 0.1$; $q \lesssim 0.01$ would almost certainly correspond to a planetary rather than stellar binary system. Adapted from Gaudi & Sackett 1999.

The duration, amplitude and placement of a planetary anomaly atop an otherwise normal microlensing light curve depends on the mass ratio $q = m_p/M$, instantaneous separation b, and the source trajectory. Well-sampled light curves thus allow the determination of q and b if an anomaly is detected, but not all source trajectories will generate a detectable anomaly even if the lens has a massive planet in its lensing zone (Fig. 6).

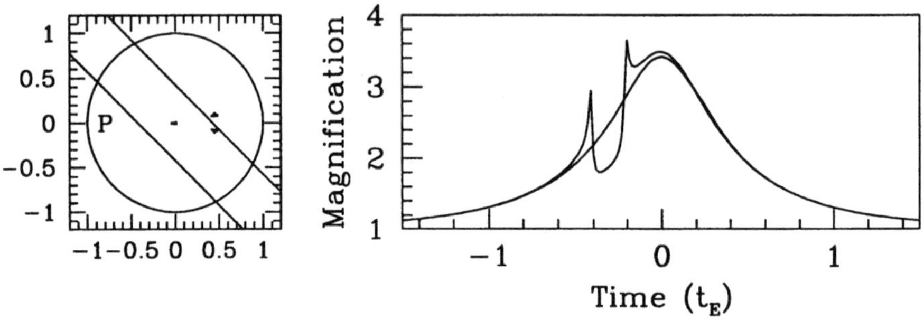

Fig. 6. Light curves (right) generated by two source trajectories (left) through the same magnification pattern due to a stellar lens (center) orbited by a planet ("P") with $q = 0.003$ at a separation $b = 0.8\theta_E$. One light curve reveals the planet; the other does not. Note the three tiny caustics. The central caustic is always located close to the primary. The planetary caustics do not coincide with the planet, but their number and location do depend on the planet's position and mass ratio.

Each planet in a lensing system will generate isolated planetary caustics that will influence the overall magnification pattern in a nearly independent manner (Fig. 6); for most events, a given planet will be detected only if the source passes near one of these caustics. The central caustic, on the other hand, is affected by any planet in the system [6], so that high magnification events — in which the source always passes close to the primary lens — are sensitive to the presence of *multiple planets* [7], though attaching a unique set of multiple planets to a given deviation will be difficult due to the increased complexity of the caustic structure.

5 Capabilities of Current Microlensing Planet Searches

Light curve morphologies are quite varied [8], but broadly speaking both the caustic cross section presented to a (small) source and the duration of a planetary anomaly are proportional to $\sqrt{q}$ [9]. Larger planets are easier to detect both because it is more likely that the source will pass near a caustic and because the perturbation lasts longer. For Galactic stellar lenses, jovian-mass planets are expected to have durations of 1-3 days; terrestrial-mass anomalies will last only a few to several hours. As Fig. 5 illustrates, stellar binaries ($1 \lesssim q \lesssim 0.1$) in the Lensing Zone nearly always create perturbations larger than 1%, regardless of source trajectory. Planetary companions ($q \lesssim 0.01$)

can escape detection much more easily since it likely that the source will not cross any of the deviation contours that are above the photometric noise. The efficiency with which planets can be detected in a given microlensing data set will depend therefore not only on the separation and mass ratio of the planet, but also on the temporal sampling and photometric precision. Studies of simulated data sets using different detection criteria and assuming light curves monitored continuously with the best precision currently possible in crowded fields have estimated that planets like our own Jupiter ($q = 10^{-3}$) may be detectable with 15–40% efficiency [10,11]. Efficiencies for observed data sets with erratic photometric quality and sampling must be computed separately for each light curve [12]. Detection efficiencies of most light curves observed today are effectively zero for earth-mass ($q = 3 \times 10^{-6}$) planets.

Planets too close to their parent lenses on the sky will closely resemble a single combined lens; planets too widely separated will behave like isolated single lenses. Planet-star systems with separations $0.6 \lesssim b \lesssim 1.6$ — that is, planets inside the so-called *Lensing Zone* — generate the most prominent binary caustic structure inside the Einstein ring $\theta_{\rm E}$ of the primary (Fig. 5). Since microlensing events are seldom alerted and monitored unless the source is inside $\theta_{\rm E}$ (i.e, $u < 1$), current surveys are most sensitive to planets in the Lensing Zone. Planets outside this zone could be detected by microlensing if the light curve is monitored for source positions outside $\theta_{\rm E}$ in order to have sensitivity to distant, outer planetary caustics [13] or, a in very high amplification events which bring the source close to the central caustic generated by all planets [6,7]. A 1 $M_\odot$ lens positioned halfway to source stars in the Galactic Bulge (the location of the overwhelming majority of events) has a physical Einstein ring radius of 4 AU. Most lenses will be somewhat less massive and closer to the Bulge, yielding Lensing Zones between 1 and 6 AU. Depending on their orbital inclination, planets with larger orbital separations may be brought into this zone for a certain fraction of their orbit. In sum, current microlensing searches are sensitive to massive ($q > 10^{-4}$) planets orbiting 1 – 10 AU from their parent stars, a region rich with planets in our own Solar System.

Three teams, PLANET [14], MPS, and MOA, now routinely use longitudinally distributed networks of southern 1m telescopes to monitor events discovered by microlensing surveys [2–4], with the detection of planetary anomalies as one of their primary goals. Of the three, PLANET currently enjoys the most extensive network of semi-dedicated telescopes, and performs intensive, nearly continuous (every 2 hours, weather permitting) photometric monitoring of several events per night with $\sim 1 - 5\%$ precision [14].

No convincing planetary signal has yet been detected, though not all data have been analyzed thoroughly. However, by comparing models with and without planets for all possible source trajectories [12], the presence of massive planets within certain zones of angular separation from their parent lenses can be *ruled out* in very well sampled, high magnification events [15,16]

(Fig. 7). The accumulation of many such events will lead in the next few years to the detection of planets like our own Jupiter — or to meaningful constraints on the abundance of such planets in the Galaxy (Fig. 8).

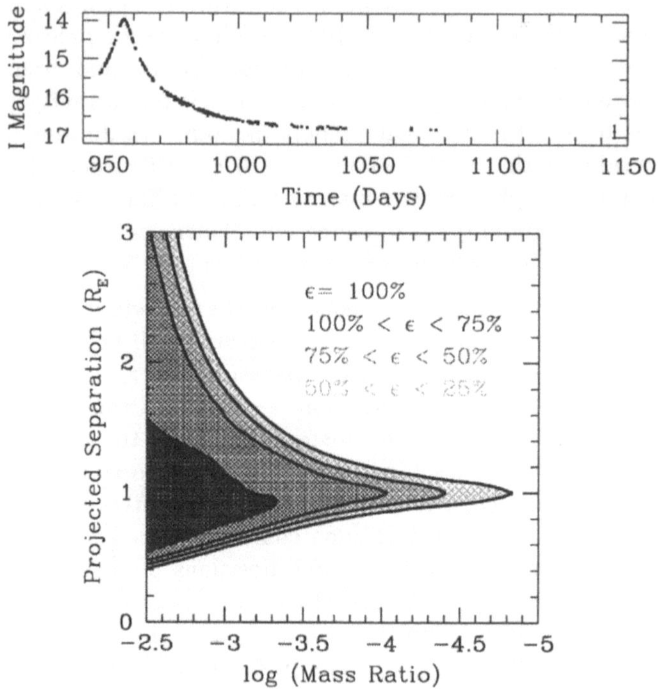

Fig. 7. *Top:* The PLANET collaboration I band light curve containing ~400 points for event OGLE 98-BLG-14 does not differ obviously from that of a single lens. *Bottom:* Exclusion probability contours for planetary companions of given mass ratio and instantaneous separation (in Einstein radii) orbiting this microlens [15,16]. Within the solid black region, the best model with a planet differs from that without a planet by $\Delta\chi^2 > 100$ for all source trajectories.

6 Planetary Microlensing in the VLT Era

Small planets are difficult to detect with any method; microlensing is no exception. Earth-mass planets are especially elusive because the angular size of the planetary caustics are smaller than a giant star in the Bulge. Thus, even if an earth-mass caustic directly transits a background giant, only a piece of source is significantly magnified; source resolution greatly dilutes the signal [17]. In order to detect such small planets, several obstacles associated with earth-mass microlensing anomalies must be overcome simultaneously, namely their: (1) rarity, (2) short duration, (3) typically small amplitude, and (4) near invisibility against giant sources. A successful program would thus need to monitor hundreds of dwarf or turn-off stars undergo microlensing in the Bulge, and to do so frequently and with high precision.

The 2.5m VST equipped with OmegaCAM, a 16K × 16K CCD detector spanning a 1 degree field of view, is expected to see first light on Paranal in 2001 as a survey telescope for the VLT. The excellent weather and median seeing ($0''.65$), large field, and possibility of immediate VLT follow-up could make the VST the most formidable microlensing machine of its era; *simultaneous detection and monitoring of 10 – 20 on-going microlensing events in every bulge field* would be possible. Detailed observing strategies and simulations for the VST remain to be worked out, but first estimates are encouraging. Assuming that 1% photometry on V=20 (I=19) turn-off stars can be achieved with 4-5 minute integrations in these very dense fields, and allowing for 30-50% lost time due to weather or poor seeing, continuous observations with the VST during "bulge season" could yield as many as ~20 jovian and ~2 terrestrial-mass planets per year — if every lens has one of each sort of planet in its lensing zone [18,19]. Since this is unlikely to be the case, detected numbers will be smaller, but it is only by measuring how much smaller that we can determine the abundance of such planets in our Galaxy.

Since microlensing "selects" lensing stars by mass, not luminosity, very distant and dim stars can be probed for planetary systems. Furthermore, the selection is a weak function of mass, so all types of Galactic stars can be studied. Unfortunately, because the parent star is unseen, its mass and distance are unknown: generally stellar lenses are too close (mas) to the source to be resolved with normal imaging and too faint to be detected in a combined spectrum. Lens mass M and distance $D_{\rm OL}$ are the scaling parameters that are required to translate the planetary mass ratio q into a planetary mass m_p and the normalized instantaneous separation b into physical units such as AU (see Eq. 1). Without knowledge of the lens mass and distance, ensembles of events must be fit with reasonable Galactic models to derive statistical estimates for these quantities.

On the other hand, 8-10m telescopes offer the first hope to spectrally type the otherwise unseen microlenses so that their distance and mass can be determined directly. Because the source and lens star are likely to differ in both spectral type and radial velocity, high resolution spectra ($\lambda/\Delta\lambda$ = 40000) with large apertures are expected to detect the lens signal in composite spectra, even for lens-source contrasts of 4 magnitudes — allowing dwarf lenses to be discerned [20]. In order to make best use of the potential for scientific gain, the microlensing events should be identified in real time at the VST so that VLT spectra can be taken both at peak, when the magnified source spectral energy distribution will dominate, and nearer baseline, when the composite spectrum will reflect the unlensed fraction of light from source and lens. This ambitious goal of direct lens detection would open a new era for microlensing, and is a task that the VLT is especially well suited to tackle due to its flexible instrumentation and scheduling, proximity to the VST, and the excellent seeing conditions on Paranal. The speed with which VLT instrumentation can be made ready and the availability of service observing will make it ideal for other target-of-opportunity science as well, such as the monitoring of (1) caustic crossings for proper motion and limb-darkening measurements, (2) supernovae, and (3) gamma ray burst optical counterparts.

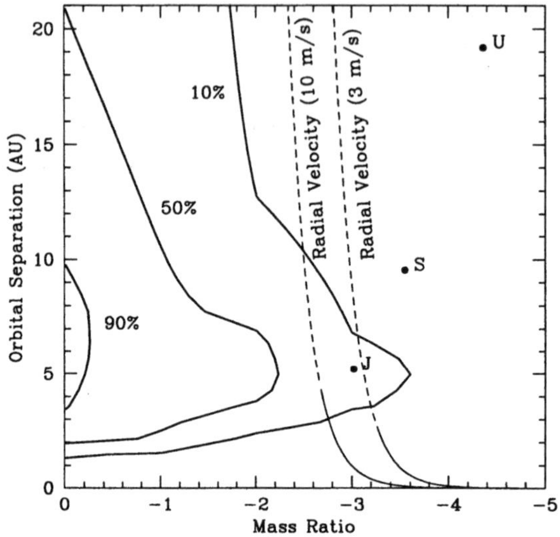

Fig. 8. Estimated detection efficiency contours of 10, 50, and 90% for current microlensing searches for planets of given mass ratio and true orbital separation a in units of AU. (The Einstein ring radius is taken to be 3.5 AU, appropriate to a solar type Bulge lens.) Also shown are expectations for a radial velocity planet search running for 5 and 10 years (solid lines) and requiring 3-σ detections at limiting sensitivities of 3 and 10 m s^{-1}, respectively. Positions of Jupiter, Saturn and Uranus are shown for reference. Adapted from Sackett (1999).

As Fig. 8 makes evident, if a sufficient number of events can be monitored sufficiently well, current microlensing searches will contribute to our knowledge of jovian mass planets orbiting with true orbital separations comparable to and somewhat larger than that of our own Jupiter, complementing current radial velocity searches which are sensitive to (and finding!) jovian planets at smaller orbital radii ($\lesssim$3 AU). The VLT (and VST) era may see a widening of the parameter space to which microlensing can contribute valuable information, especially by pushing the technique forward to lower masses while allowing an improved characterization of detected planetary systems through better photometry and — possibly — direct spectroscopic microlens detection.

Acknowledgements

It is a pleasure to thank my colleagues in the PLANET collaboration for permission to display our results for OGLE 98-BLG-14 prior to publication, and Scott Gaudi for assistance in preparation of Fig. 8.

References

1. Paczyński, B. (1986) Gravitational microlensing by the galactic halo. Ap.J. **304**, 1-5.
2. Alcock, C., et al.: The MACHO Collaboration (1993) Possible gravitational microlensing of a star in the Large Magellanic Cloud. Nature **365**, 621-22.
3. Aubourg, E., et al.: The EROS Collaboration (1993) Evidence for gravitational microlensing by dark objects in the galactic halo. Nature **365**, 623-24.
4. Udalski, A., et al.: The OGLE Collaboration (1993) The optical gravitational lensing experiment. Discovery of the first candidate microlensing event in the direction of the Galactic Bulge. Acta Astron. **43**, 289-294.
5. Mao, S. & Paczyński, B. (1991) Gravitational microlensing by double stars and planetary systems. Ap.J. **374**, L37-40.
6. Griest, K. & Safizadeh, N. (1998) The Use of High-Magnification Microlensing Events in Discovering Extrasolar Planets. Ap.J. **500**, 37-50.
7. Gaudi, B.S., Naber, R.M. & Sackett, P.D. (1998) Microlensing by Multiple Planets in High Magnification Events. Ap.J. **502**, L33-37.
8. Wambsganss, J. (1997) Discovering Galactic planets by gravitational microlensing: magnification patterns and light curves. MNRAS **284**, 172-188.
9. Dominik, M. (1999) The binary gravitational lens and its extreme cases. Submitted to Astron. Astrophys. (astro-ph/9903014)
10. Gould, A. & Loeb, A. (1992) Discovering planetary systems through gravitational microlenses. Ap.J. **396**, 104-114.
11. Bolatto, A. & Falco, E. (1994) The detectability of planetary companions of compact Galactic objects from their effects on microlensed light curves of distant stars. Ap.J. **436**, 112-116.
12. Gaudi, B.S. & Sackett, P.D. (1999) Detection Efficiencies of Microlensing Data Sets to Stellar and Planetary Companions. Submitted to Ap. J. (astro-ph/9904339)
13. Di Stefano, R. & Scalzo, R.A. (1999) A New Channel for the Detection of Planetary Systems through Microlensing. II. Repeating Events. Ap.J. **512**, 579-600.
14. Albrow et al.: The PLANET Collaboration (1998). The 1995 Pilot Microlensing Campaign of PLANET: Searching for Anomalies through Precise, Rapid, Round-the-Clock Monitoring. Ap.J. **509**, 687-702.
15. Gaudi et al.: The PLANET Collaboration (1998). Limits on Planetary Companions in Microlensing Event OGLE-BUL-98-14. AAS, **193**, 108.07
16. Albrow et al.: The PLANET Collaboration (1999). In preparation.
17. Bennett, D. & Rhie, S. H. (1996) Detecting Earth-Mass Planets with Gravitational Microlensing. Ap.J. **472**, 660-664.
18. Sackett, P.D. (1997) Planet Detection via Microlensing. Final Report of the ESO Working Group on the Detection of Extrasolar Planets. ESO SPG-VLTI-97/002, Appendix C (astro-ph/9709269)
19. Peale, S. J. (1997) Expectations from a Microlensing Search for Planets. Icarus **127**, 269-289.
20. Mao, S., Reetz, J. & Lennon, D.J. (1998) Detecting luminous gravitational microlenses using spectroscopy. Astron. Astrophys. **338**, 56-61.
21. Sackett, P.D. (1999) Searching for Unseen Planets via Occultation and Microlensing. Planets outside the Solar System: theory and observations, NATO-ASI Series, Kluwer, 189-227 (astro-ph/9811269)

Approaching the Deuterium Burning Limit in the σ Orionis Young Cluster

M.R. Zapatero Osorio[1], V. Béjar[1], E.L. Martín[2], R. Rebolo[1], and G. Basri[2]

[1] Instituto de Astrofísica de Canarias, E-38200 La Laguna, Tenerife, Spain
[2] Astronomy Department, University of California, Berkeley, CA 94720, USA

Abstract. We have obtained J photometry for a sample of 12 objects in the interval $16.0 < I \leq 20.5$ which are member candidates in the young σ Orionis cluster. Our data confirms that 11 of them do fit the photometric sequence expected for the cluster implying their very likely membership in σ Orionis. Their masses span from $0.09\,M_{\odot}$ down to the very low-mass brown dwarf regime at the very young age (1–5 Myr) of the cluster. We have also obtained low-resolution spectroscopy for the faintest object in our sample, SOri 47 ($I = 20.5$). We classify it as an early L-type young brown dwarf with a very likely mass around the deuterium burning limit. SOri 47 provides evidence that the initial mass function extends down to masses of the order of $0.015\,M_{\odot}$ and serves as a guide for the study of the *new* L-class discoveries in the field.

1 Introduction

Given its very young age ($\leq$ 5 Myr) the σ Orionis cluster is ideally suited for the detection of very low-mass brown dwarfs (BDs) and subsequently for investigating the initial mass function in the substellar regime. In this paper we present follow-up IR photometry for a sample of 12 member candidates in the σ Orionis cluster taken from the RIZ survey of Béjar et al. (1999) [1]. These candidates have I magnitudes in the range 16.0–20.5 corresponding to a mass interval of $0.09\,M_{\odot}$ down to the deuterium burning mass limit at $\sim 0.01\,M_{\odot}$. The membership confirmation of the faintest candidates in the sample could unambiguously contribute to the comprehension of stellar formation processes. IR data have been collected in 1998 August and December at the 1.5 m Carlos Sánchez Telescope (Observatorio del Teide at Tenerife). We also present the low-resolution (20 Å) spectrum for the faintest candidate, SOri 47, and the intermediate-resolution (3 Å) spectrum for SOri 27 taken with the Keck II telescope (Mauna Kea Observatory at Hawaii) in 1998 December. The final J magnitudes are given in Table 1. The rms scatter in the calibration is $\sim$0.1 mag.

2 Analysis

The I vs $(I - J)$ color-magnitude diagram for our sample is shown in Fig. 1. For reference we also plot previously known Pleiades cool members shifted to

Table 1. Optical and IR photometry for σ Orionis cluster member candidates.

Object	J_{UKIRT}	$I - J$	Sp. Type
SOri 12	14.28 ± 0.05	2.00 ± 0.09	M6
SOri 15	14.53 ± 0.05	2.02 ± 0.07	
SOri 16	14.51 ± 0.06	2.09 ± 0.08	
SOri 17	14.72 ± 0.03	2.09 ± 0.06	M6
SOri 21	14.82 ± 0.10	2.08 ± 0.11	
SOri 25	14.55 ± 0.06	2.49 ± 0.08	M6
SOri 27	14.65 ± 0.06	2.42 ± 0.09	M7
SOri 31	15.14 ± 0.10	2.17 ± 0.11	
SOri 40	15.43 ± 0.06	2.50 ± 0.08	M7
SOri 42	16.68 ± 0.20	2.33 ± 0.22	
SOri 45	16.79 ± 0.09	2.80 ± 0.11	M8.5
SOri 47	17.13 ± 0.20	3.40 ± 0.21	early L-type

the distance of the σ Orionis cluster. We have adopted a distance of 352 pc for the multiple star σ Orionis as measured by Hipparcos. This star is associated to the cluster of the same name and its age is estimated in the interval 1–7 Myr [2]. All of our σ Orionis candidates, with the only exception of SOri 42, appear to be more luminous than their Pleiades color counterparts, laying above the upper envelope of this older cluster. This location in the diagram is expected for true cluster members. Our candidates clearly define a photometric sequence parallel to that of the Pleiades but about 2 mag brighter in the I-band. The IR photometry presented in this paper confirms their membership in σ Orionis (except SOri 42), being 6 young BDs and 5 very low-mass stars and objects in the stellar/substellar transition region of the cluster.

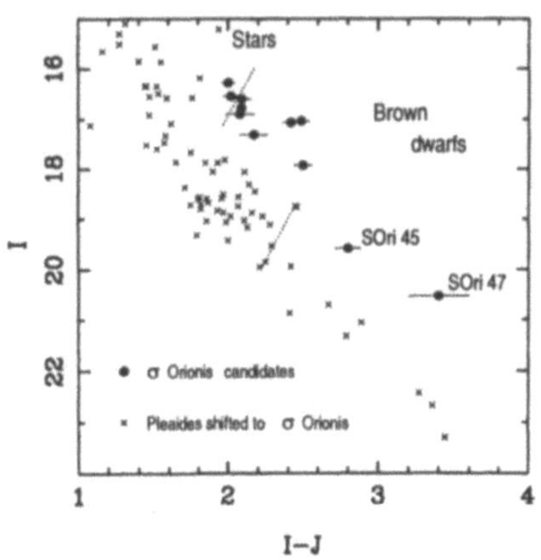

Fig. 1. Color-magnitude diagram for σ Orionis cluster member candidates compared to known Pleiads. The location of the stellar/substellar boundary is indicated for each cluster *with dotted* lines.

Among the objects in our sample, SOri 47 deserves special attention given its very likely young age ($\leq$ 5 Myr), low luminosity and red $(I - J)$ color. The optical spectrum of SOri 47 is shown in Fig. 2 in comparison to the early L-type BDs Roque 25 in the Pleiades and Kelu 1 in the field, which were observed with the same instrumental configuration. The $(I - J)$ color of the three objects is very similar and so, their spectral types. Hα is not seen in strong emission in any of them. We can set an upper limit of 7Å to the equivalent width for the cases of SOri 47 and Roque 25, and 2Å for Kelu 1. The NaI lines at 818.3, 819.5 nm appear weaker in the cluster objects than in the field one. The *new* "L" classification [5] major spectroscopic signature is the progressive replacement of the metallic oxides (which characterize the M-class) by metallic hydrides. This is very likely due to the effects of dust condensation in very cool atmospheres [7]. A detailed inspection of the relative strength of some molecular bands reveals some differences among the three objects. For example, the average strength of TiO and VO at 710–730 nm, 735–752 nm, 795–805 nm, 820–860 nm is larger in SOri 47 and Roque 25. This could be due to a light difference in $T_{\rm eff}$. It has been [6] argued that the more intense absorption due to these molecules seen in the Pleiades late-M BDs compared to similar spectral type stars in the field as well as the weakness of the NaI lines is a consequence of the smaller gravity of the young objects. This works in favor of the youth both for SOri 47 and Roque 25.

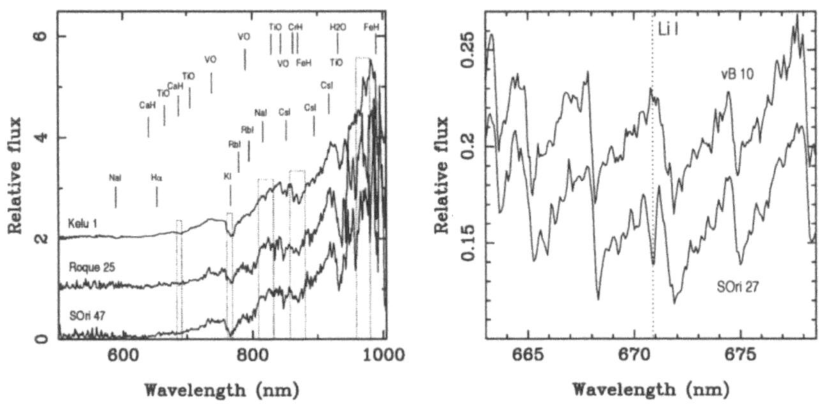

Fig. 2. *(Left panel)* Keck II spectra for early L-type BDs: SOri 47 (~0.015 $M_\odot$) in σ Orionis, Roque 25 (~0.035 $M_\odot$) in the Pleiades, and Kelu 1 (0.035–0.065 $M_\odot$) in the field. Some spectral features are identified. The telluric absorption bands are marked with dotted lines. *(Right panel)* Lithium detection in SOri 27 (M7). By comparison with the most recent theoretical models the mass of this young BD is in the interval 0.03–0.04 $M_\odot$ for the adopted age range 1–5 Myr of the cluster.

We show in Fig. 3 our sample of objects in σ Orionis and the comparison of their location in the HR diagram to various sets of isochrones from [4]

Chabrier & Baraffe (1997) (Next Gen models) and [3] Burrows et al. (1997). These models match the observations nicely and provide an upper limit to the age of the cluster which is in accord with previous measurements. On the other hand, mass estimates depend on the choice of evolutionary tracks and the conversion between observables and theoretical parameters. Adopting an age interval of 1–5 Myr for the σ Orionis cluster we derive a mass for SOri 47 of $0.015 \pm 0.005\, M_{\odot}$. The deuterium burning mass limit is around 0.013–$0.015 M_{\odot}$. Because of its low mass and young age, SOri 47 has still preserved deuterium, and together with SOri 45, these two substellar members of σ Orionis are the most promising candidates for detecting deuterated hydrogen in their spectra. If the age of SOri 47 were as young as 1 Myr, its mass would be below the deuterium limit. In that case and following the proposed rule based on the deuterium depletion for separating planets (which never burn deuterium) from BDs (which efficiently deplete it), SOri 47 might be then an example of a *free-floating super-giant planet*. Nevertheless, the finding of SOri 47 provides evidence that the initial mass function extends down to masses of the order of $0.015\, M_{\odot}$ and serves as a guide for the study of the *new* L-class discoveries in the field.

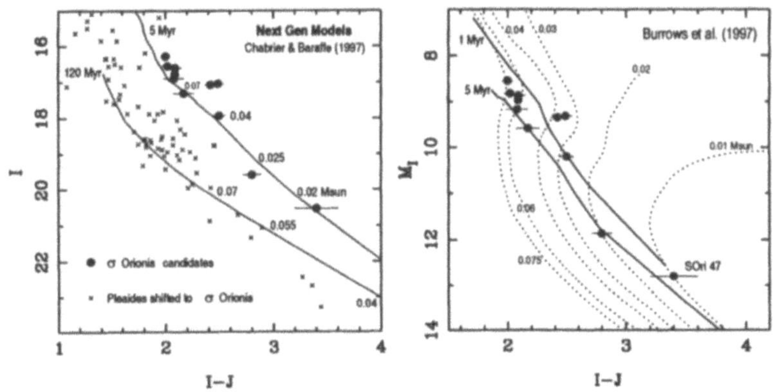

Fig. 3. Color-magnitude diagram for our sample in σ Orionis (filled circles) compared to the theoretical evolutionary models of *(left panel)* Chabrier & Baraffe (1997, Next Gen), and *(right panel)* Burrows et al. (1997). Masses are labelled in solar units. For symbols see Fig. 1

References

1. Béjar, V.J.S., Zapatero Osorio, M. R., Rebolo, R., 1999, ApJ, in press
2. Brown, A. G. A., de Geus, E. J., de Zeeuw, P. T., 1994, A&A, 289, 101
3. Burrows, A., et al., 1997, ApJ, 491, 856
4. Chabrier, G., Baraffe, I., 1997, A&A, 327, 1039
5. Martín, E. L., et al., 1997, A&A, 327, L29
6. Martín, E. L., Rebolo, R., Zapatero Osorio, M. R., 1996, 469, 706
7. Tsuji, T., Ohnaka, K., Aoki, W., 1996, ApJ, 305, L1

From PTI to the Keck Interferometer

Gerard T. van Belle

Jet Propulsion Laboratory, California Institute of Technology,
Pasadena, CA 91109, USA

Abstract. The Keck Interferometer project will link together the Keck 1 and Keck 2 telescopes as a single instrument, building the technical and scientific achievements of the Palomar Testbed Interferometer (PTI). Specifics of PTI research and the Keck design are presented herein.

1 Introduction

The Keck Interferometer project will link together the Keck 1 and Keck 2 telescopes as a single instrument. The addition of four 1.8 m 'outrigger' telescopes will further enhance the capabilities of this array. Specific science capabilities of this interferometer will include direct detection of zodiacal shells about nearby stars, direct detection of warm giant planets, indirect astrometric detection of cool giant planets, and interferometric imaging of general science targets, such as protoplanetary disks. The California Association for Research in Astronomy (CARA) and The Jet Propulsion Laboratory (JPL) are partners in this NASA-funded venture, which expects to have first light between the two Kecks by the end of 2000, and the fully operational 6 element array by 2002.

Much of the technology that will go into the Keck Interferometer is already operation in the form of the Palomar Testbed Interferometer (PTI). PTI is a technology pathfinder for Keck, and a productive scientific tool in its own right.

2 Interferometer Basics

An astronomical instrument does three basic functions: 1) Collect light, 2) Equalize the path differences, and 3) Combine and detect the light. The basic result from any such instrument is resolution on the scale of λ/B, where λ is the wavelength of light being observed, and B is the basic scale of the instrument. If step 2 has been correctly done, then step 3 represents the special case where the collected light is combined in a coherent fashion. Implicit in each of these steps are a great many more tasks: for example, wavefront correction in the form of adaptive optics is assumed to fall under step 1, where necessary. Conventional telescopes achieve these tasks through static optics and structures, and B is replaced by D, the telescope primary diameter. For a "standard" Michelson interferometer, these three tasks are accomplished

Fig. 1. The Palomar Testbed Interferometer, located atop Palomar Mountain, as seen from the catwalk of the Hale 5 m telescope.

by the basic pieces of an interferometer: conventional telescopes collect light, delay lines equalize path differences, and back end optics combine and detect the light. Examples of such interferometers are PTI, COAST, NPOI, IOTA, and GI2T. Development of nanometer precision delay line technology over the past 20 years has enabled a key tool for the construction of these instruments.

3 The Palomar Testbed Interferometer

3.1 Instrument

Current status. PTI is a two-element interferometer located atop Palomar Mountain in California. PTI uses the feed from two of three 40 centimeter apertures, located on a N-S baseline of 110 m, or a NE-SW baseline of 85 m. Current limiting magnitude for the instrument is $m_K < 5$, and sources must fall within a declination range of $0^o < \delta < 50^o$ on the sky. One of the basic purposes of PTI is to demonstrate dual-star astrometry, a narrow ($\Delta\theta < 30''$) field approach to determining stellar positions to tens of microarcsecond (μas) precision as first discussed in Shao & Colavita (1992). At present, dual-astrometry has been demonstrated down to the 300 μas level on a relatively bright pair of objects (both $m_K < 4.0$). A detailed description of the instrument has been published in Colavita et al. (1999), and a specific discussion of fringe visibility estimators may be found in Colavita (1999).

Near-term upgrades. A number of upgrades will be incorporated over this observing season. An expanded fringe tracking integration window will push the limiting magnitude down to $m_K < 6.5$. An H band filter has been added to the dewar, and it is expected to have a limiting magnitude between $m_H < 3.5$

and 5.0. Additional delay line range is being considered which would expand the sky coverage to $-10^o < \delta < 65^o$. The first two of these improvements have already been demonstrated on engineering nights and be implemented as standard facility tools shortly.

Two significant improvements are slated for the dual-star mode. First, identification and control of systematics in the optical beam train will enable 100 μas astrometry. Second, a "bright-dim" mode will enable the secondary star to be much dimmer than before; $m_K < 10$ is expected, in contrast to the "bright-bright" mode which required $m_K < 4$. Demonstration of both of these advances will ultimately enable 30 $\mu as/\sqrt{\mathrm{hr}}$ astrometry with the Keck Interferometer, with secondary stars that are brighter than $m_K < 17.3$.

3.2 Science

Even with the operational limitations listed above, a considerable science program has been undertaken with PTI and has begun to produce a number of results, including:

- A vigorous program of observing spectroscopic binaries has already resulted in a 10 milliarcsecond (mas) orbital solution for the double-lined spectroscopic binary ι Pegasi (Boden et al. 1999a). Possible eclipses in the ι Peg system suggested by Fekel & Tomkin (1983) were ruled out by interferometric orbit. A second binary system, 64 Piscum, will be addressed in a forthcoming publication (Boden et al. 1999b).
- Basic stellar parameters (effective temperature, linear radius) as derived from angular sizes measured for over 100 giant and supergiant stars are beginning to be published as well (van Belle et al. 1999). These observations build on the previous work of Ridgway et al. (1980), DiBenedetto & Rabbia (1987), and Dyck et al. (1996, 1998).
- A protoplanetary disk about FU Ori has been resolved at a scale of 2 AU (Malbet et al. 1998); the data are consistent with an accretion disk inferred from earlier spectral and spectrophotometric data.
- An orbit for the RS CVn binary star TZ Tri has produced independent confirmation of spectroscopic results (Koresko et al. 1998).
- An upper limit to 51 Peg B mass has been established by interferometric observations of 51 Peg A (Boden et al. 1998). Measured from the visibility data is a magnitude difference no less than $\Delta K > 4.27$, implying a maximum mass of $M_B < 0.22 M_\odot$.
- Dual star astrometry of 16 Cyg A & B to be discussed in a forthcoming publication (Boden et al. 1999c). 16 Cyg is a pair of stars separated by $\Delta\theta = 40''$; these two stars are nearby ($d \approx 3$ pc) and in a 600 year orbit about each other. For the 1998 observing season, narrow angle astrometry at a precision of 300 μas exhibited proper motion easily discernable on a day-to-day basis and has yielded significant improvement to the original orbital solution by De Caro-Veca (1948).

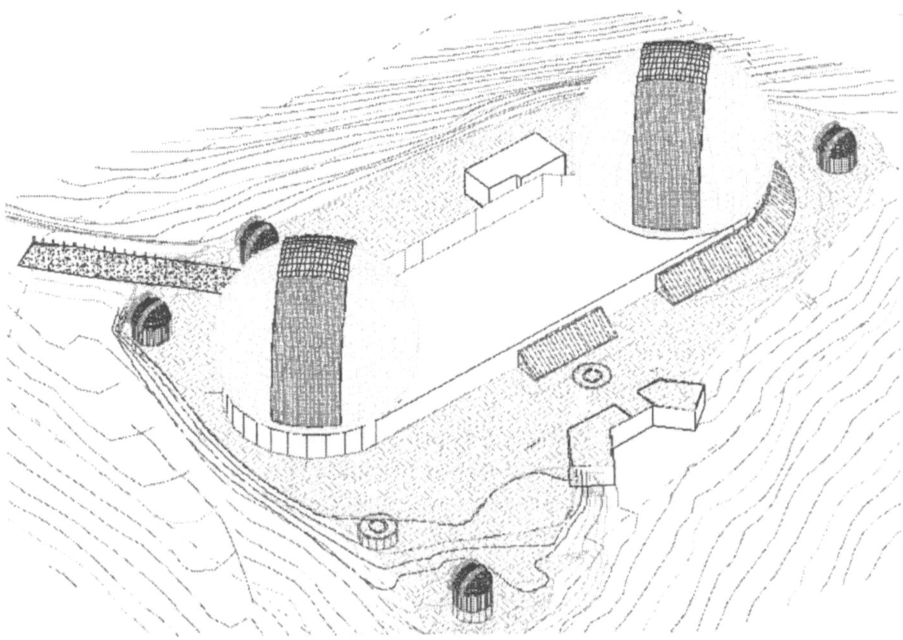

Fig. 2. Bird's eye view of the Keck Observatory as it will appear in the fall of 2002, from the north looking south. Four outrigger telescopes are currently under construction and will be added to the site, with pads for two additional outriggers for future expansion as well.

4 The Keck Interferometer

4.1 Telescopes

The apertures for the Keck Interferometer include the two 10 m Keck telescopes, as well as four proposed 1.8 m "outrigger" telescopes, which are scheduled to begin installation in late 2000. The observatory site is located at the summit of Mauna Kea, at an altitude of 4,200 m. The two Kecks are on a 85 m baseline which runs NNE-SSW. The outriggers are used for the astrometric program, and with the Kecks for synthesis imaging. The various outrigger telescope baselines measure from 25 to 140 m in length, in a variety of orientations; as a result of the requirements of the astrometric program, there are two orthogonal baselines that use separate pairs of outriggers, each in excess of 100 m in length. Another requirement of that program is that the outriggers have tight specifications on their pivots in order to allow high accuracy measurements. For testing, two 40 cm siderostats with fixed compressor telescopes will be installed to debug the interferometer systems prior to integration with the Kecks or the outriggers. These siderostats will be similar to those used by the Palomar Testbed Interferometer (PTI), which serves as a *testbed for a number* of aspects of the Keck Interferometer.

4.2 Wavefront Correction

The Keck Observatory has developed an adaptive optics system for the Keck 2 telescope. As part of the interferometer project, a second adaptive optics system will be developed for Keck 1, providing phased pupils at near- and mid-IR wavelengths. For the smaller outrigger telescopes, fast tip/tilt compensation will be implemented using an active secondary and a 1.2 μm wavefront sensor at the interferometer back end. For 1.8 m telescopes at Mauna Kea, tip/tilt compensation is adequate for observations at 1.6 μm and longer wavelengths.

4.3 Dual-Star Module

To enable cophasing using an isoplanatic reference, a dual-star module is installed at each telescope. The dual-star module selects two objects in the telescope field of view, and collimates and directs the light into the interferometer beam train. One of these objects will be bright (primary star) and serve as the phase reference for the other object (secondary star). On the Kecks, the dual-star module will be located at an output port of the adaptive optics bench. The dual-star module will support a field of view of up to 1 arcmin radius.

4.4 Coude Train and Beam Transport

To propagate the light from the dual-star module on the Nasmyth platform of each Keck telescope, the Keck coude train will be completed with the addition of coude mirrors M4 – M7 (M4 is the output mirror of the dual-star module, while M7 is the fixed mirror at the base of the telescope). Since two beams are propagated from the dual-star feed, each coude mirror is actually a mirror pair. For the primary star, these mirrors are located along the coude centerline; for the secondary star, these mirrors are offset from the centerline and are actuated to compensate for azimuth rotation. A similar (but simpler) coude system is needed for the outrigger telescopes. After mirror M7, the light is directed into the beam-combining lab located in the basement of the Keck facility. For the outriggers, the light is directed using buried pipes; for the Kecks, the light from the M7 mirrors is directed into the coude tubes which pass through the telescope footings into the coude tunnel which connects the two telescopes.

4.5 Delay Lines

The delay lines on the interferometer will be implemented in two stages. The long delay lines will be located in the coude tunnel between the two Kecks. They use adjustable range mirrors that move along tracks in the tunnel. The range mirrors are flat, and delay both the primary and secondary beam identically. The are repositioned to select an area of the sky, and remain fixed during an observation. The fast delay lines are located in the beam-combining lab. There are separate fast delay lines for the primary and secondary beams from each telescope. These delay lines move continually during the observation to track sidereal motion and atmospheric turbulence; separate delay lines for the two stars enable phase referencing with feedforward, and different pathlength modulations for fringe demodulation. The fast delay lines will be similar to those used at PTI; a 20 m delay range per delay line will allow significant sky coverage without repositioning the long delay lines.

4.6 Pathlength Sensing

Pathlength sensing on the interferometer includes:

- Local metrology of the fast delay lines. This is implemented with a conventional heterodyne metrology system, and will be used primarily for delay-line servo control.
- End-to-end, or constant-term, metrology of the complete optical path from beam combiner to dual-star module. This metrology references the primary and secondary beam paths to identical fiducials in order to implement cophasing and narrow-angle astrometry. The metrology will also monitor vibrations in the beam-transport optics.
- Accelerometer sensing of coude optics. Accelerometers on key optics not monitored by laser metrology will sense vibrations which could affect fringe visibility. Vibrations measured with the accelerometers will be fed forward to the delay lines for high-bandwidth compensation.

4.7 Angle Tracker

Angle tracking will be implemented at 1.2 μm using an infrared array in a fast-readout mode. For the primary beams from the Kecks, the angle tracker will provide low-bandwidth track offsets to the adaptive-optics systems. For the primary beams from the outriggers, centroids determined from the angle sensor will control active secondary mirrors on the telescopes. For the secondary beams from both telescopes, the angle tracker will provide low-bandwidth track offsets to the dual-star modules.

4.8 Fringe Tracker

The fringe tracker will provide five identical two-way beam combiners; each of these will be similar to those in use at PTI. The detectors will be HgCdTe infrared arrays used in fast-readout mode, sensing at 1.52.4 μm. The detectors for all of the combiners will be remotely located in two fiber-fed dewars, fed via post-combination single-mode IR fibers. The fringe trackers support all of the observing modes of the interferometer. For cophasing, fringes are tracked on up to 5 baselines to cophase the 6 telescopes. For astrometry, two of the beam combiners will track the bright primary star on orthogonal baselines, while two others will track the faint secondary star.

4.9 Multi-Way Combiner

The multi-way combiner will provide pair-wise fringe measurements on up to 15 baselines simultaneously for imaging. The proposed configuration is a non-redundant cross-dispersed design, which provides unique spatial frequencies along one dimension of the infrared array to code baselines, with frequency information provided along the other dimension. The primary detector would be a 5 μm cutoff HgCdTe array.

4.10 Nulling Combiner

The nulling combiner will implement two stages of achromatic nulling interferometers feeding a Si:As BIB detector. A laboratory nulling testbed has been assembled at JPL to test various aspects of nulling; recent results include visible light nulls of 1 part in 1,000.

4.11 Control System

Integrating the various subsystems of the interferometer will be a real-time control system. Based on the PTI control system, the system will use synchronized VME single-board computers running VxWorks to control time-critical systems such as servos and trackers. User interface, display, and overall coordination of subsystem functions will be hosted on workstations.

4.12 Other Subsystems

An optical switchyard will provide automated reconfiguration among the various observing modes. A stimulus will provide white-light and laser sources for boresighting, calibration, and testing. An automated alignment system will provide low-bandwidth sensing of alignment beacons at the periphery of the beam-transport optics and long delay lines to control the mirror tilts.

5 Keck Interferometer Key Science Projects

5.1 Characterization of Stellar Zodiacal Dust

For a typical G star at a distance of 10 pc, its apparent diameter would be about 1 mas. A 1 AU radius zodiacal dust cloud around the star subtends a full angle at 10 pc of 200 mas. The detection problem has two parts. For an exozodiacal dust cloud with a density equal to 10 times our solar system, the intensity ratio between the star and the integrated exozodiacal signal is 1000:1. Furthermore, at 10 μm, which is the preferred wavelength for this observation, both of these signals are deeply embedded in the 10 μm thermal background. Nevertheless, with the high sensitivity provided by the 10 m apertures of the Kecks, the photometric signal-to-noise ratio is high, allowing in principle 1σ detections of exozodiacal dust clouds as faint as one solar-system equivalent.

The key to the measurement technique proposed for the Keck Interferometer is to use interferometric nulling to distinguish the exozodiacal signal from the star, and to provide an AC signal for synchronous detection. An achromatic rotational shearing nulling interferometer is implemented for each aperture separately, followed by an achromatic nulling interferometer that combines the outputs of the two aperture nullers. The achromatic nullers use polarization rotation, rather than a phase shift, in order to provide a null with matched internal optical paths. The two stages provide nulling on two different spatial scales: the aperture nullers provides nulls at a characteristic scale of 200 mas at 10 μm, corresponding to the aperture diffraction limit, while the interferometer nuller provides a null at a characteristic scale of 25 mas, corresponding to the 85 m separation between the two Kecks.

While the interferometer null primarily attenuates the star, the aperture null attenuates both the star and the exozodiacal dust cloud. These nulls can be modulated on and off by changing internal optical paths in order to implement source chopping while remaining pointed at the source. By modulating both the aperture and interferometer nullers, AC signals are provided to allow separate detection of the exozodiacal dust and the star in the presence of the thermal background.

5.2 Planet Detection

Direct: Phase Difference Interferometry The field of exoplanet research has exploded in the last several years with the discovery of massive planets by radial velocity techniques. Planets with a orbital radii less than 0.3 AU (Mercury is at 0.4 AU) are often referred to as "hot Jupiters". 51 Peg B, the first of these discovered, is the prototypical hot Jupiter. These planets are so close to the parent star that their surface temperature is quite high, 900K or hotter, and they can be detected from their direct emission. A Jupiter-sized planet at a temperature of 900K is about 10,000 times dimmer than a solar type star at 2.2 μm. Jupiter itself, on the other hand, is about a million times fainter than the Sun in the thermal infrared and a billion times fainter in the visible and near infrared. Direct detection of a cold Jupiter can only be done from space, but direct detection of a hot Jupiter can be done with the Keck Interferometer.

The principle of the observation exploits the temperature difference between the planet and the parent star. At short wavelengths, the intensity ratio between the star and the planet is quite large. Thus the fringe position of the system essentially the center of light will lie very close to the star. However, at longer wavelengths the intensity ratio between the star and the planet, while still large, is more favorable to the planet, and the center of light will be displaced toward the planet. Thus, simultaneous measurements of a planetary system at two different wavelengths can be used to detect the planet; rotational synthesis can be used to determine a planetary orbit.

As the center of light shift is quite small, and must be detected in the presence of atmospheric turbulence, the multi-wavelength measurements must be made simultaneously with a single beam combiner. For some objects, especially the brown dwarfs, which may have significant structure in their near-infrared spectra, detection could use the $1.5 - 2.4$ μm coverage of the fringe-tracking detectors. For other objects, the detector of choice would be the $1.5 - 5.0$ μm multi-way combiner. These observations will use the large apertures of the Kecks to provide high sensitivity.

Indirect: Dual-Star Astrometry Astrometry looks for the transverse wobble of a star's position caused by a planet orbiting the star. For a solar system like our own, viewed from a distance of 10 pc, the motion of the Sun would appear as shown below. The scale of the motion is easily calculated: a Jupiter around a Sun at 10 pc would produce a wobble with an amplitude of 0.5 mas.

Astrometry with a ground-based telescope with a diameter of 1 m, measuring the relative position of stars over a 0.5 degree field of view, is limited to an accuracy of a few mas in a 1 hour measurement. The limitation is attributable to atmospheric turbulence, which causes the positions of stars to fluctuate with time. However if two stars are sufficiently close together in the sky, the motion of the two stars will be almost identical because the light from the two stars will traverse almost identical paths through the atmosphere.

For small telescopes or short-baseline interferometers, the amount of atmospherically induced motion is nearly independent of the size of the instrument. But when the size of the telescope or interferometer baseline grows to be larger than the separation of the two stellar beams at the top of the troposphere (roughly a 10 km altitude), the behavior changes, and the accuracy rapidly increases with the size of the instrument. In this regime, the accuracy also increases as the size of the field decreases. Interferometers allow the very long baselines needed to exploit this behavior. Differential astrometric measurements with tens of mas accuracy are possible over a 20 arcsecond field of view with an interferometer with a baseline of 100 m at an excellent site like Mauna Kea.

If one wishes to make measurements with significantly higher accuracy than 10 mas, the only solution is to go to space. The need to do this comes from the very small astrometric signatures of Earth-like planets. While the effect of Jupiter on a Sun at 10 pc produces a wobble with an amplitude of 0.5 mas, the signature of an Earth is only 0.3 mas. In its narrow-angle mode, the Space Interferometry Mission (SIM) will be able to look for planets with a mass equal to several Earth masses.

Astrometry with the Keck Interferometer measures the position of the target star with respect to one or two reference stars. While the target star will generally be nearby and bright, the reference stars will generally be distant and relatively faint. The Keck Interferometer will use phase referencing, which tracks the bright target star to stabilize the interferometer, to obtain high sensitivity in order to detect the fainter reference stars. Two reference stars are needed in most cases to resolve ambiguities in the astrometric signature: with a single reference star, it is not possible to distinguish the signature of a planet around the nearby star from that of a more massive object around the more distant reference star. With two reference stars, two measurements are available, and signatures in the astrometric signature are attributed to the target star if it is seen in both astrometric measurements. To allow detection of systems with any orientation on the sky, the Keck Interferometer will make simultaneous measurements with orthogonal baselines.

A number of planet search programs can be considered. The number of target stars depends on the desired accuracy. For Jupiter-mass planets, shorter integration times are needed because of the large signature; for Uranus-mass planets, longer integrations are needed. One proposed program would search 80 stars for planets down to Uranus mass (25 mas signature) and 250 stars for planets down to Jupiter mass (125 mas). This program would use approximately 50% of the outrigger time, assuming 4 observations per star per year.

5.3 Protoplanetary Disk Imaging

The Keck Interferometer with its outrigger telescopes will be capable of imaging in the infrared with unmatched sensitivity and angular resolution. The imaging beam combiner for the Keck Interferometer will be designed to combine light from the two Kecks and the four outriggers and measure the complex fringe visibility for all 15 baselines simultaneously. As with radio interferometry, Earth rotation will be used to provide a more uniform coverage of the (u, v) plane. One advantage of including the Keck Telescopes in the array is, or course, their large collecting area. This not only improves sensitivity on the target star, but also helps with sky coverage in order to image very faint sources. When the target is very faint, a nearby reference star is used in order to stabilize the interferometer to provide long integration times on the target. Inclusion of the Keck Telescopes improves the ability to find reference stars around faint targets, improving the sky coverage of the interferometer. The interferometer can be configured for imaging using four outriggers alone, or for five or six telescopes with either or both of the Kecks.

One specific target of interest to the NASA Origins program is interferometric mapping of protoplanetary disks. Recent HST images and Keck images of objects such as *HR* 4796A and HD 141569 (Weinberger et al. 1999) have

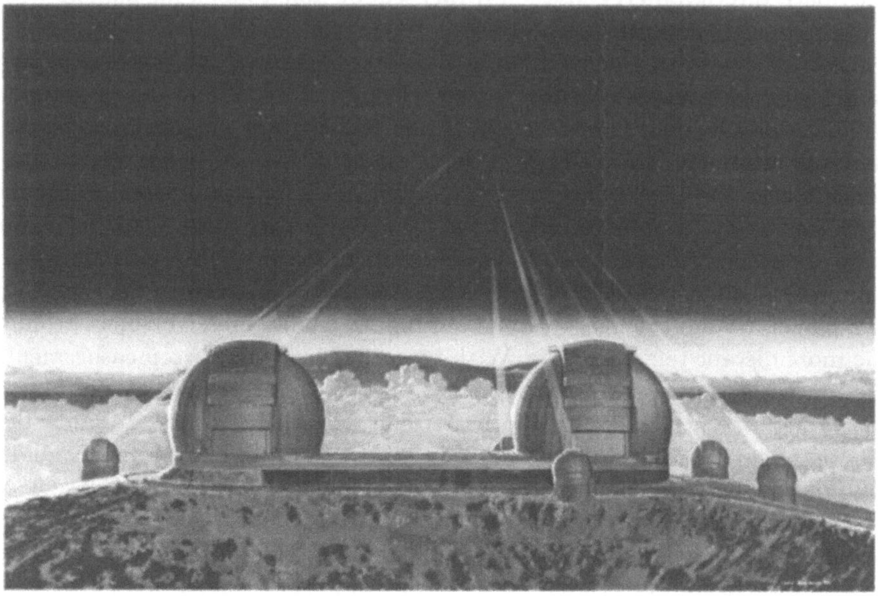

Fig. 3. Artist's concept of the fully operational Keck Interferometer.

provided evidence of disks surrounding the young stellar objects. The HD 141569 disk shows evidence of a gap in the disk on a scale of roughly 100 AU; one of the goals of the Keck Interferometer is to map protoplanetary disks and search for gaps at much smaller spatial scales, on the order of 0.5 AU. It should be noted that in addition to protoplanetary disks, the interferometric mapping capability of the instrument will be designed to be applicable to a general range of astronomical targets.

6 Conclusion

The Keck Interferometer will build upon the current success of the Palomar Testbed Interferometer.

References

1. Boden, A.F., et al., 1999a, in press.
2. Boden, A.F., et al., 1999b, in preparation.
3. Boden, A.F., et al., 1999c, in preparation.
4. Boden, A.F., et al., 1998, ApJL, 504, L39.
5. Colavita, M.M., et al., 1999, ApJ, 510, 505.
6. Colavita, M.M., 1999, PASP, 111, 111.
7. DiBenedetto, G.P., Rabbia, Y., 1987, A & A, 188, 114.

8. Dyck, H.M., Benson, J.A., van Belle, G.T., Ridgway, S.T., 1996, AJ, 111, 1705.
9. Dyck, H.M., van Belle, G.T., Thompson, R.R., 1998, AJ, 116, 981.
10. Fekel, F. and Tomkin, J, 1983, PASP, 95, 100.
11. Koresko, C.D., 1998, et al. ApJL, 509, 45.
12. Malbet, F., et al., 1998, ApJL, 507, L149.
13. Ridgway, S. T., Jacoby, G. H., Joyce, R. R., Wells, D. C., 1980, AJ, 85, 1496.
14. Shao, M., & Colavita, M.M., 1992, A&A, 262, 353.
15. van Belle, G.T., et al., 1999, AJ, 117, 521.

Astrometry with the VLT Interferometer

Andreas Quirrenbach

University of California, San Diego; Physics Department; Center for Astrophysics and Space Sciences; Mail Code 0424; La Jolla, CA 92093-0424; USA

1 Introduction

The VLTI was originally conceived as an imaging instrument, providing a resolution of a few milliarcseconds at near-infrared wavelengths for studies of stars, circumstellar matter, and extragalactic objects. However, following the pioneering work on interferometric narrow-angle astrometry by Shao and Colavita (1992) and Colavita (1994), it was proposed that the VLTI could also be used for astrometric planet detection (Quirrenbach 1995). It was envisaged that the astrometric mode of the VLTI could be implemented by taking advantage of the large unvignetted field-of-view foreseen at that time (von der Lühe, Quirrenbach, & Koehler 1995). The idea of using the VLTI for narrow-angle astrometry was embraced by ESO's Interferometry Science Advisory Committee (Paresce et al. 1996), but the technical concept for the delay lines has changed. The current plan for narrow-angle astrometry is based on dual star feeds at the telescopes and comprehensive internal laser metrology, as described in the PRIMA (Phase-Referenced Imaging and Microarcsecond Astrometry) study (Quirrenbach et al. 1998).

2 Astrometric Precision of the VLTI

The precision of the VLTI for narrow-angle astrometry will ultimately be limited by atmospheric turbulence. For sufficiently small angles ϑ on the sky, the atmospheric error for a baseline b scales with $b^{-2/3} \cdot \vartheta$ (Shao & Colavita 1992). For typical turbulence and wind velocity profiles at Cerro Paranal, the atmospheric error on the longest baselines of the VLTI approaches $10\,\mu$as over a $10''$ arc for a half-hour integration (von der Lühe, Quirrenbach & Koehler 1995). However, several challenges have to be met to realize this stunning potential for precise measurements. First, the differential instrumental errors have to be controlled to a precision of about 5 nm. Second, the baseline vector has to be known to $\sim 50\,\mu$m, for all orientations of the telescope. Third, differential atmospheric effects have to be taken into account and modeled precisely (Quirrenbach et al. 1994, Gubler & Tytler 1998). In addition to these technical requirements, suitable reference stars — preferably stable at the $10\,\mu$as level themselves — have to be found in a very small patch of the sky. It is therefore likely that astrometric observations with the VLTI will start at a somewhat more modest level, perhaps with a $50\,\mu$as precision goal, and progress towards the ultimate atmospheric limit over time. Finally, a long-term effort is needed to detect the planets at large orbital separations to which the astrometric method is most sensitive.

3 Observing Extrasolar Planets with the VLTI

The planet observing programs that can be carried out with the astrometric mode of the VLTI fall into two broad categories. (1) Follow-up observations of planets known from radial velocity searches will provide the inclination of the systems and thus the planet masses without $\sin i$ ambiguities. (2) Astrometric surveys can target stars that are not accessible to the radial velocity technique (e.g., early-type stars), and thus establish a comprehensive census of planetary systems. Table 1 lists the amplitude of the astrometric signature for a number of representative cases. It is immediately obvious that there is an enormous range of amplitudes even for giant planets; it is therefore instructive to evaluate the potential of astrometry with the VLTI more closely.

Table 1. Astrometric signature from different planet / parent star combinations. Note that the numbers given in the last column have to be divided by the distance to give the astrometric signature. 15 $M_{\rm Jup}$ was used for the mass of the brown dwarf.

Planet	Orbit [AU]	Star	Amplitude $\cdot$ d [μas $\cdot$ pc]
Earth	1	G2	3
Jupiter	5	G2	4 800
Uranus	20	G2	880
Jupiter	0.1	G2	96
Brown Dwarf	0.1	M5	7 500
Jupiter	5	A5	2 200
Jupiter	5	M5	24 000

In Table 2, the numbers from Table 1 have been converted to the maximum distance at which the planet is detectable, for different assumptions about the astrometric precision. This table can be used to outline some of the observing programs that will be feasible with the VLTI, both initially and with its ultimate precision. Astrometry on the 50 μas level will allow the following projects:

- Detection of all brown dwarfs in orbits of at least 0.1 AU around all late M dwarfs that are sufficiently bright to serve as fringe-tracking reference for the interferometer ($m_K \approx 12$).

Table 2. Maximum distance to which different planets can be observed with the VLTI in astrometric mode, with either 50 μas or 10 μas precision. 1 μas can probably be attained only from space. It is assumed that a 4 σ peak-to-peak variation is required for secure detection of a planet.

Precision	Planet	Orbit [AU]	Star	max. distance [pc]
50 μas	Jupiter	5	G2	48
	Uranus	20	G2	9
	Brown Dwarf	0.1	M5	75
	Jupiter	5	A5	22
	Jupiter	5	M5	240
10 μas	Jupiter	5	1 $M_{\odot}$	240
	Uranus	20	G2	44
	Jupiter	5	A5	110
	10 Earths	1	G2	1.5
1 μas	Jupiter	0.1	G2	48
	Earth	1	G2	1.5

- Census of Jupiter-like planets (in 5 AU orbits) along the main sequence. This may be our first view of planet formation as a function of stellar mass.
- Detection of systems with multiple planets down to Uranus masses around nearby G dwarfs.

At the 10 μas level, additional observing programs will become possible. These include:

- Searches for Jupiter-like planets around pre-main-sequence stars in nearby star-forming regions. This will help to establish the epoch of planetary system formation, and the relation between pre-main-sequence disks and planets.
- Surveys for multiple systems and long-period planets with better signal-to-noise.
- Detection of massive terrestrial planets ($M \gtrsim 10\,M_{\oplus}$) around a few selected nearby stars.

Finally, Table 2 shows that some interesting projects such as astrometric detection of giant planets in close orbits, or searches for true Earth twins around nearby stars, require a precision of 1 μas or better, which probably requires an interferometer in space.

4 Reference Stars

The discussion in the previous section has tacitly assumed that there are reference stars against which the motion of the target star can be measured. However, as mentioned earlier, identifying suitable references is an important and non-trivial task by itself. It is possible to get a feeling for the difficulty of this task by cross-correlating a catalog of potential targets (e.g. the HIPPARCOS catalog, ESA 1997) with a sufficiently deep catalog of possible reference stars (USNO-A1.0, Monet et al. 1996). This exercise gives the following results:

- The HIPPARCOS catalog contains 4760 stars with $\delta \leq +20^\circ$ and parallax $\pi \geq 20\,\text{mas}$.
- For 1762 of these stars, 3 reference stars within 50″ are found.
- For 734 of these, 3 reference stars within 30″ are found.
- The HIPPARCOS catalog contains 1018 stars that have $\delta \leq +20^\circ$ and $\pi \geq 40\,\text{mas}$, and 130 stars with $\delta \leq +20^\circ$ and $\pi \geq 100\,\text{mas}$.
- The Gliese catalog contains 2381 stars with $\delta \leq +20^\circ$.
- The proportion of these samples for which potential references within 30″ or 50″ are found is about the same as for the full HIPPARCOS sample.

There are a few caveats about the results from these catalog cross-correlations, which may lead to an overestimate of the number of potential reference stars. For example, some of the "reference stars" may actually be galaxies that were not recognized as such during the compilation of the USNO catalog. In some cases, a target star with high proper motion is found as a potential "reference" for itself because the USNO catalog lists the stellar positions at the epoch of the underlying Schmidt plates. (It would be necessary to look up the plate epoch for each target star, compute its position at that epoch, and compare it to the positions of the "references" found by the cross-correlation. This is even more important when reference stars within 10″ from the target are sought.) These cautionary remarks notwithstanding, it is apparent from the above numbers that the requirement of finding nearby reference stars reduces the number of potential targets considerably, especially for the highest precision when the search radius is only 10″. The problem is exacerbated by the high proper motion of many of the nearby target stars: a star that has three references today may move far enough away from them during the course of a long-term program that the astrometric accuracy gets severely degraded. It is obviously possible to increase the number of potential targets by requiring only one or two references, but then the risk gets high that the references turn out to be members of multiple systems, which can lead to unusable data, or worse, to false "planet" detections if the motion of the reference is ascribed to the target. It is possible, however, to take this strategy to the extreme and search for planets in double stars. The Washington Double Star Catalog (Worley & Douglass 1997) contains 745 F, G, and K main sequence stars with $\delta \leq 20^\circ$ and $V \leq 10$ in pairs with separation d between

5″ and 20″; 23 of these are G main sequence stars with $V \leq 7.5$. Most of these are members of wide physical binaries, and searching for planets in these systems may be scientifically interesting and technically somewhat less challenging than searches around single stars. The downside of this approach is the difficulty of determining to which of the two system components any detected planet belongs.

5 Spectroscopy of Extrasolar Planets

The near-infrared spectra of extrasolar planets should be characterized by extremely deep absorption bands of water and methane (Burrows et al. 1997), somewhat similar to the spectrum of the cool brown dwarf Gl 229 B (Oppenheimer et al. 1998). This opens the possibility of using wavelength-dependent astrometric information for the detection of extrasolar planets, and perhaps even to obtain their spectra.

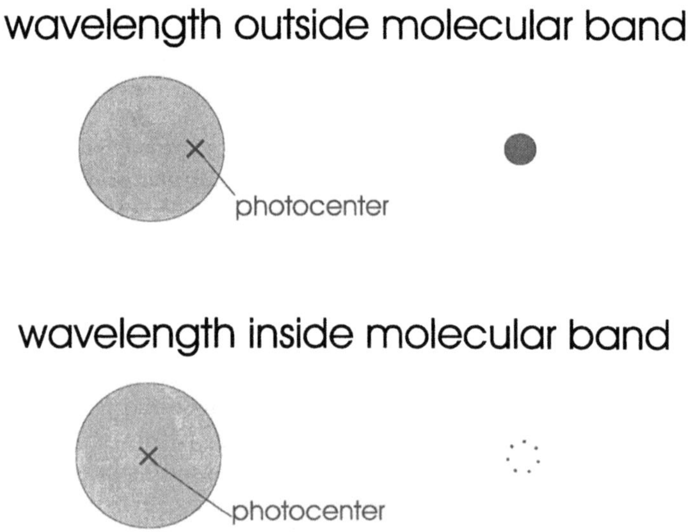

Fig. 1. The shift of the star – planet photocenter with wavelength gives rise to an interferometric phase shift that can be exploited to obtain a spectrum of the planet.

Figure 1 illustrates that the photocenter of a star – planet system is slightly different between molecular bands, where the planet is relatively bright, and within bands, where it is much dimmer. Actually, the shift of the photocenter is proportional to the planet / star brightness ratio and can thus be used as a proxy for the planet spectrum. The shift of the photocenter gives rise to a corresponding wavelength dependence of the interferometer phase, which can be measured if the signal-to-noise is sufficient and systematic effects are kept small. In the case of the planet around 51 Peg, which is quite favorable because the planet is close to the star and therefore hot and bright, the expected effect on the interferometer phase is $\sim$ 0.5 mrad on the longest baselines of the VLTI. This could be measured with a signal-to-noise ratio of $\sim$ 3000, but it remains to be seen whether the systematic instrumental and atmospheric effects can be overcome.

6 Conclusions

The VLTI provides exciting new capabilities for the astrometric detection of extrasolar planets. At the 50 μas level, which may be a reasonable goal for the first observations with PRIMA, a number of very interesting programs aimed at gas giants will be possible. Improvements in the astrometric precision will substantially increase the scientific capabilities by extending the reach to more distant objects and less massive planets. However, it must not be overlooked that these programs require careful preparation and selection of target and reference stars. It is not too early to begin with comprehensive preparatory programs.

References

1. Burrows, A., Marley, M., Hubbard, W.B., Lunine, J.I., Guillot, T. Saumon, D., Freedman, R., Sudarsky, D., & Sharp, C. (1997). ApJ 491, 856-875
2. Colavita, M.M. (1994). A&A 283, 1027-1036
3. ESA (1997). *The Hipparcos and Tycho Catalogues*, ESA SP-1200
4. Gubler, J., & Tytler, D. (1998). PASP 110, 738-746
5. Monet, D., Bird, A., Canzian, B., Harris, H., Reid, N., Rhodes, A., Sell, S., Ables, H., Dahn, C., Guetter, H., Henden, A., Leggett, S., Levison, H., Luginbuhl, C., Martini, J., Monet, A., Pier, J., Riepe, B., Stone, R., Vrba, F., Walker, R. (1996). *USNO-A1.0*, (U.S. Naval Observatory, Washington DC)
6. Oppenheimer, B.R., Kulkarni, S.R., Matthews, K., & van Kerkwijk, M.H. (1998). ApJ 502, 932-943
7. Paresce, F., Mourard, D., Bedding, T., Beletic, J., Haniff, C., Leinert, C., Malbet, F., Mariotti, J.-M., Mozurkewich, D., Mundt, R., Petitjean, P., Quirrenbach, A., Reinheimer, T., Richichi, A., Röttgering, H., von der Lühe, O., Waters, R. (1996). ESO Messenger 83, 14-21
8. Quirrenbach, A. (1995). In *Science with the VLT*. Eds. Walsh, J.R., & Danziger, I.J., Springer-Verlag, p. 33-37
9. Quirrenbach, A., Coudé du Foresto, V., Daigne, G., Hofmann, K.-H., Hofmann, R., Lattanzi, M., Osterbart, R., le Poole, R., Queloz, D., & Vakili, F. (1998). In *Astronomical Interferometry*. SPIE Vol. 3350, p. 807-817
10. Quirrenbach, A., Mozurkewich, D., Buscher, D.F., Hummel, C.A., & Armstrong, J.T. (1994). A&A 286, 1019-1027
11. Shao, M., & Colavita, M.M. (1992). A&A 262, 353-358
12. von der Lühe, O., Quirrenbach, A., & Koehler, B. (1995). In *Science with the VLT*. Eds. Walsh, J.R., & Danziger, I.J., Springer-Verlag, p. 445-450
13. Worley, C.E., & Douglass, G.G. (1997). A&AS 125, 523

PRIMA – The VLTI Dual Feed System

Andreas Glindemann and Samuel Lévêque

ESO, Karl-Schwarzschild-Str. 2, 85748 Garching, Germany

Abstract. The PRIMA (**P**hase **R**eferenced **I**maging and **M**icro-arcsec **A**strometry) facility is the VLTI dual feed system adding a faint object imaging and an astrometry mode to the VLTI. The science drivers and the concept have been discussed previously [6]. In this paper, we describe the performance requirements for the subsystems of PRIMA, *i.e.* the star separator, the differential delay line, the laser metrology system and the fringe sensor unit. In addition, the principle of operation will be explained.

1 Introduction

The Very Large Telescope Interferometer (VLTI) is located on top of Cerro Paranal in the Atacama Desert, Chile, at an altitude of 2640 m. The interferometer coherently combines the light from up to four 8-m telescopes ('Unit Telescopes', UT's) or of the three 1.8-m telescopes ('Auxiliary Telescopes', AT's) of the VLT observatory.

The field of view of the VLTI is 2 arcsec. The objective of the Phase Referenced Imaging and Micro-arcsecond Astrometry facility (PRIMA) is to enable simultaneous interferometric observations of two objects – each with a maximum size of 2 arcsec – that are separated by up to 1 arcmin, without requiring a large continuous field of view[3]. One object will then be used as a reference star for fringe tracking whilst the other object will be the science target. PRIMA is the key to access:

1. Higher sensitivity, the limiting magnitude will be about K = 20,
2. imaging of faint objects with high angular resolution (<10 milli arcsec),
3. high precision astrometry (≈10 μarcsec over a 10 arcsec field).

The principle of operation relies on finding within the isoplanatic angle (≈1 arcmin) of the science target a sufficiently bright star that can be used as a reference star for the stabilisation of the fringe motion induced by atmospheric turbulence. Controlling additionally all optical path lengths of the reference star and of the science star inside the interferometer with a laser metrology system introduces the capability of imaging faint objects and of determining the precise angular separation between the two stars.

The scientific potential of PRIMA is huge and covers stellar as well as extra-galactic topics (see [4]): **the Galactic center** – providing evidence for the presence of a massive black hole in the center of our Galaxy, and **substellar companions and planet detection** – the anticipated performance allows us to detect Jupiter and Saturn-like planets up to 500pc, **Star formation** – determining the duplicity of massive stars in young open clusters, **stellar dynamics** – measurements of differential proper motions between stars in globular clusters, **quasars and active galactic nuclei** – investigating the nature of the relative contributions of a circumnuclear starburst and a compact nuclear source, **extra galactic astrometry** – determining the so-called AGN Dance-centroid due to SN explosion.

2 Principle of Operation

The need for a reference star in interferometry is similar to the need of a reference star in adaptive optics. Due to atmospheric turbulence the fringe pattern in a telescope interferometer is smeared out after a few 100 msec depending on the observing wavelength. In order to freeze the fringe pattern the maximum exposure time is limited to a few 10 msec, corresponding to limiting magnitudes of K = 10 in the near infrared and N = 5 in the mid infrared. If a bright reference star within the isoplanatic angle of the science object is used to measure the fringe position and then stabilise the fringes in a closed loop system, the scientific camera can integrate for several minutes on the science object, extending the sensitivity by many magnitudes.

The first reason to implement PRIMA is thus to increase the sensitivity. The second reason is to produce images using a phase reference technique. The observable is the so called complex visibility function whose modulus and phase are respectively given by the contrast and position of the recorded fringe pattern. For a point like object the visibility is one and the phase equals zero whereas for an extended and non-symmetric object, the complex visibility varies in contrast and phase depending on the choice of interferometer baseline. By measuring the complex visibility for many different baselines the image can be reconstructed by Fourier transforming the complex visibility (as a function of the baseline vector). Determining the phase of the fringe pattern in addition to their contrast is the key to interferometric imaging.

The random motion of the fringe pattern due to atmospheric turbulence can be frozen by using a reference star for fringe tracking. The problem remains to determine the zero OPD position, i.e. to define the default position of the white light fringe for zero phase. This position cannot be determined in absolute terms when using two telescopes but only with respect to the reference star, hence the term phase referenced imaging (Fig. 1 illustrates the situation). The reference star is assumed to be symmetric, i.e. the phase of the visibility function is zero.

Four terms determine the difference of the white light fringe positions of the object and the reference star:

- The differential external OPD, defined as the product $\boldsymbol{\Delta S B}$, with $\boldsymbol{\Delta S}$ the angular distance vector between object and reference star, and $\boldsymbol{B}$ the baseline vector,

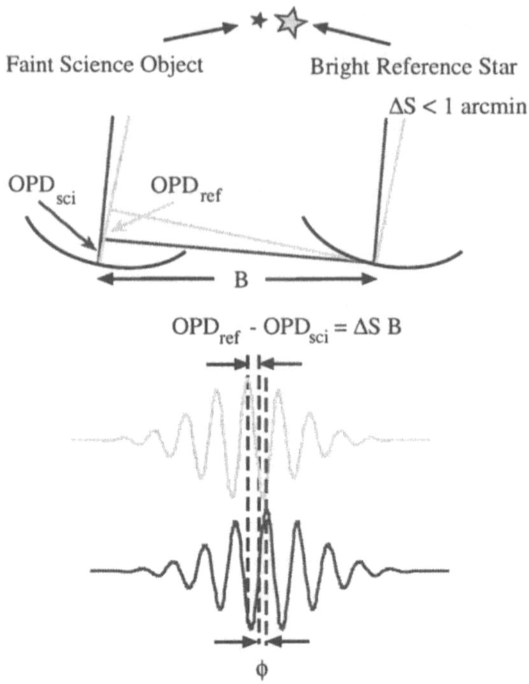

Fig. 1. Principle of phase referenced imaging and astrometry with an interferometer. $\Delta \boldsymbol{S}$ is the angular separation vector of the stars, $\boldsymbol{B}$ is the baseline vector, and ϕ is the phase of the visibility function of the science object.

- The phase ϕ of the complex visibility of the object,
- The internal optical path difference (ΔIOPD) between object and reference stars (not shown in Fig. 1), and
- Random differential fringe motion due to atmospheric turbulence (anisoplanatism).

The latter introduces an rms error $\sigma_{\mathrm{OPD}} \propto B^{-2/3}\Theta/\sqrt{t}$ (see Fig. 2) with B the baseline, Θ the distance of the two stars and t the observing time[7]. Thus, increasing the observing time decreases the astrometric error.

The Internal Optical Path Difference (ΔIOPD) between object and reference star can be measured inside the interferometer with a laser metrology system. The first two terms cannot be distinguished with a single measurement. Measuring the phase ϕ for many different baselines the (constant) angular separation $\Delta \boldsymbol{S}$ between the object and reference star can be determined and subtracted. In an astrometric measurement it is exactly the angular distance that one is looking for. Assuming that the object star and the reference star are symmetric with zero phase of the visibility, the difference of the white light fringe positions together with the measured ΔIOPD give the exact angular separation of the two objects.

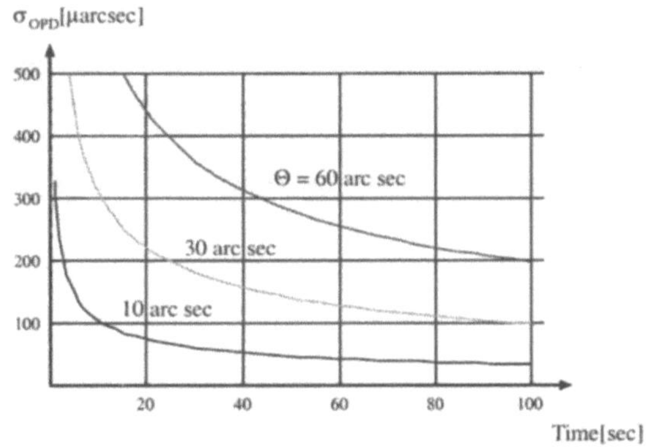

Fig. 2. Astrometric error due to atmospheric turbulence as a function of observing time and distance of the two stars for a baseline of 200m. To achieve 10μarcsec accuracy the exposure time has to be increased to 30min for a separation of 10arcsec.

3 The Subsystems of PRIMA

PRIMA can be subdivided into four sub-systems (see Fig. 3):

– **Star Separator**

An optical system for the UT's and AT's located in the Coudé foci feeding two arbitrary objects from the Coudé field of view into the delay lines of the VLTI. The star separator will allow the transmission of two fields up to 1 arcmin in separation and 2 arcsec in diameter towards the delay lines and the interferometric laboratory. The optical design of the delay lines allows guiding two beams in parallel from the star separator to the beam combination lab. This means that the optical paths of the two stars are separated over about 200m. If a multiplexing scheme is chosen for the star separator only a single beam containing both fields is transferred through the delay line tunnel. However, problems with cross-talk between the two objects, the possible increase of the background radiation in the mid infrared due to transmission through beam splitters and the lack of a convincing multiplexing technique make this solution rather unattractive.

– **Laser Metrology System**

The internal OPD between object and reference star has to be monitored with high accuracy (5 nm rms over 30 minutes) using a laser metrology system. The laser beam is fed into the VLTI optics in the beam combination laboratory and travels along the same path as the star light up to a retro reflector at the secondary mirror sending the light back to the beam combination laboratory.

– **Differential Delay Lines**

The difference of the white light fringe position of the object and the reference star can be expressed as an OPD (maximum 65mm) that has to be adjusted with the differential delay line with an accuracy of 5nm. The differential delay lines will be installed in the beam combination laboratory.

– **Fringe Sensor Unit**
An infrared camera measuring the position of the white light fringe of the reference star providing the error signal for the fringe tracking system. The fringe sensor unit has to be able to track on a H = 13 star (on the UT's) with an accuracy of 30nm rms. The fringe sensor unit is located in the beam combination laboratory.

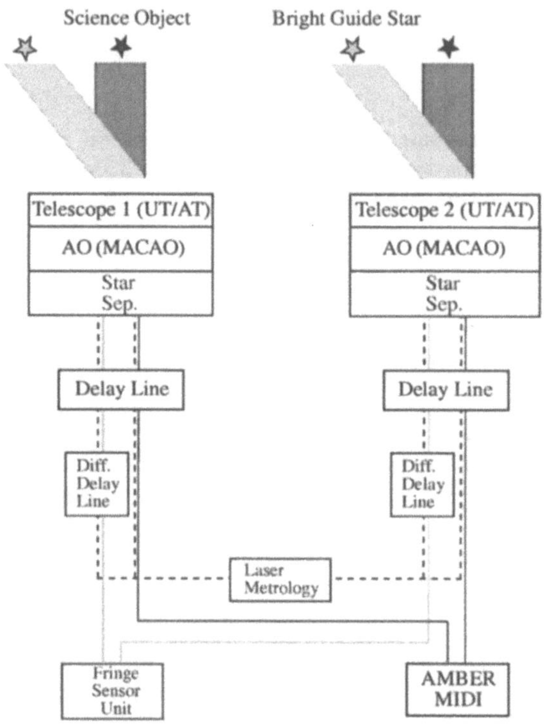

Fig. 3. Functional block diagram of the VLTI and the PRIMA components. The VLTI has the following subsystems: Tel. 1 and 2 – two UT's or two AT's, AO – Adaptive optics, Delay Lines, AMBER/MIDI – Science Instruments. The PRIMA subsystems are: Star Separator, Differential Delay Line, Laser Metrology, Fringe Sensor Unit.

In the beam combination laboratory (see Fig. 3), the reference beam will be fed into a Fringe Sensor Unit and the science beam into one of the VLTI instruments. For observations with the UT's the fringe sensor unit and all near infrared instruments will take advantage of the MACAO adaptive optics system[1]. MACAO has a 60-actuator bimorph mirror and a curvature wavefront sensor. The anticipated performance is a Strehl ratio of 30% in the K-band with a V = 16 magnitude star.

Two scientific instruments are currently being designed to make use of the stabilised and calibrated fringe measurements for phase reference imaging: MIDI in the mid infrared (10 μm)[2] and AMBER in the near infrared (1-2.2 μm)[5]. MIDI and AMBER will see first light around 2001/2002. It is also planned to build a dedicated PRIMA camera for micro-arcsecond astrometry.

The PRIMA facility contains a number of technological challenges, in particular in the area of laser metrology, that require careful analysis of the possible technical solutions. Therefore, feasibility studies for the four subsystems are being performed by Dornier Satellite Systems, Friedrichshafen, and by Onéra, Paris, to find the best technical solutions and to obtain a financial estimate for the manufacturing of the subsystems.

4 Conclusions

Dual feed observations with PRIMA can start as soon as the star separator, the fringe sensor unit and the differential delay lines are ready. Then, a reference star can be used for fringe tracking whilst integrating on the fainter science object as described above. If the fringe pattern can be stabilised over 10-100 sec the expected limiting magnitudes are about K $\approx$ 16 and N $\approx$ 8.

Phase information required for imaging and astrometry becomes available if the laser metrology system is installed. metrology provides An OPD measurement accuracy of 500nm rms over 10 min sets the limiting magnitudes to about K = 19 and N = 11. The Strehl ratio in the reconstructed image can be as good as 30% in the K-band and 80% in the N-band depending on the uv coverage. Reaching the final goal of 5nm rms over 30min allows 10μarcsec astrometry. PRIMA should be operational by 2002.

References

1. Bonaccini D., Rigaut F., Glindemann A., Dudziak G., Mariotti J.-M. and Paresce F. (1998) Adaptive Optics for ESO VLT-Interferometer. Proc. SPIE 3353, 224–232
2. Leinert Ch. and Graser U. (1998) MIDI – the Mid-infrared interferometric instrument for the VLTI. Proc. SPIE 3350, 389–393
3. von der Lühe O., Quirrenbach A., Koehler B. (1995) Narrow-angle astrometry with the VLT Interferometer. In *Science with the VLT*, Eds. Walsh, J.R. and Danziger, I.J., Springer-Verlag, 445–450
4. Paresce F., Ed. (1997) *Science with the VLT Interferometer*. Springer-Verlag
5. Petrov R., Malbet F., Richichi A. and Hofmann K.-H. (1998) AMBER, the Near-Infrared/Red VLTI Focal Instrument. The ESO Messenger, **92**, 11–14
6. Quirrenbach A., Coudé du Foresto V., Daigne G., Hofmann K.-H., Hofmann R., Lattanzi M., Osterbart R., Le Poole R., Queloz D., Vakili F. (1998) PRIMA - Study for a Dual-Beam Instrument for the VLT Interferometer. Proc. SPIE 3350, 807–817
7. Shao M. and Colavita M.M. (1992) Potential of long-baseline infrared interferometry for narrow-angle astrometry. Astron. Astrophys, **262**, 353–358

High Precision Interferometry with the NPOI

David Mozurkewich

Naval Research Laboratory, Remote Sensing Division,
4555 Overlook Avenue SW, Washington, DC 20375 USA

1 Introduction

Now that we have seen successful operation of the first 8 meter VLT unit telescopes and the 10 meter Kecks, we are all looking forward to what will be accomplished with interferometry using these large telescopes. For this, it is reasonable to use the existing ground based interferometers as a guide. Although these instruments have produced an extensive literature of interesting results, their promised potential has not yet been fully realized. The problem is with our ability to calibrate the data. We need to increase the aperture and integration time to increase sensitivity. As a result, we end up working with apertures and integration times comparable to or somewhat larger than the coherence length and time imposed on the system by the Earth's turbulent atmosphere. The resulting loss of coherence means the fringes we observe fluctuate wildly. Some efforts at improving calibration have been successful. For example, the SUSI team [1] has shown that by averaging the data with several integration times during post processing they can successfully determine and correct for the finite coherence time. A similar technique using several aperture sizes is not currently feasible, but may be a useful technique for interferometers with larger apertures such as the CHARA array at visible wavelengths, or the VLTI. The use of single mode fiber optics, notably the work by de Foresto with the FLUOR experiment at IOTA, has show impressive repeatability of the fringe amplitude calibration. Another potentially very powerful approach is to identify data products that are either immune to calibration errors or are self-calibrating. For example, the use of closure phases allows radio interferometry to produce very high dynamic range images.

The Navy Prototype Optical Interferometer (NPOI) is a long baseline optical interferometer located at the Lowell Observatory dark sky site on Anderson Mesa, Flagstaff AZ USA. The NPOI was built for use both as an imaging and as a wide-angle astrometric interferometer. Currently, it has baselines with lengths up to 38 meters which are combined three at a time. It operates at visible wavelengths making simultaneous observations at 20 wavelengths between 550 and 850 nm.

We have started a program of looking at self-calibrating data products which explicitly exploit the NPOI's wide bandpass. Both atmospheric and time varying instrumental effects that limit our calibration accuracy have well known wavelength dependencies. We can use this knowledge to provide very accurate relative calibration of data taken at different wavelengths. In this talk, I explore two of these possibilities and discuss their applicability to the study of exo-solar planets.

2 High Precision Phase Measurements

For three baseline observations, the NPOI uses pair-wise beam combination. Consequently, a change of some path lengths within the beam combiner produce errors in the closure phase. Typically, we monitor the drift in closure phase on an unresolved star and transfer that calibration to the program stars assuming the phase varies smoothly with time. This is done separately for each wavelength and produces closure phases that are good at the 3 to 5 degree level.

However, we can do better. We know the calibration of the closure phases varies because of changes in the optical path through the beam combiner. Since this path is partly in air and partly in glass (fused silica), we can model the phase variation with three parameters:

$$\phi_c(\lambda, t) = \frac{2\pi}{\lambda}(n_{Air} L_{Air}(t) + n_G L_G(t) + \phi_0(t)). \tag{1}$$

I solve for the path lengths in air, L_{Air} and glass L_G as well as a wavelength independent phase offset, ϕ_0. The phase offset is required in order to obtain a good fit, but I have not yet investigated the origin of this term. I suspect it arises from inaccuracies in the adopted values for the index of refraction of fused silica. Instead of fitting the time variation of each wavelength separately, we can fit all the wavelengths with one model. This reduces the number of parameters required to obtain a good fit and can potentially decrease the required amount of time observing calibration stars.

An example of this type of fit is shown in Figure 1. The observation consisted of 90 seconds of data on an unresolved star so that the source closure phase should be identically zero. The upper panel shows the raw data (points) and the model fit (line). The lower panel shows the residuals. The error bars are the deviation of the mean of 90, one second averages. L_G, which is due to a mismatch in the thickness of the beam splitters, should vary slowly, if at all. I assumed a constant value of $32\mu m$. Fitted values for L_{Air} in nm are shown in the second column of Table 1 for eight observations during a night. These data are very well behaved. The RMS error bars are on the order of one degree and the median $\chi^2_n = 1.5$. Due to our alignment procedures, the beam combiner is thermally unstable at the beginning of the night. But for the last 6 observations spanning more than 3 hours, L_{Air} grows smoothly, increasing my confidence in the appropriateness of the model.

Having demonstrated the accuracy of the model, I can now exploit the real advantage of this technique: A model involving both the calibration parameters and the source structure can be fit directly to observations of the source without the use of a calibration star. Figure 2 shows similar data

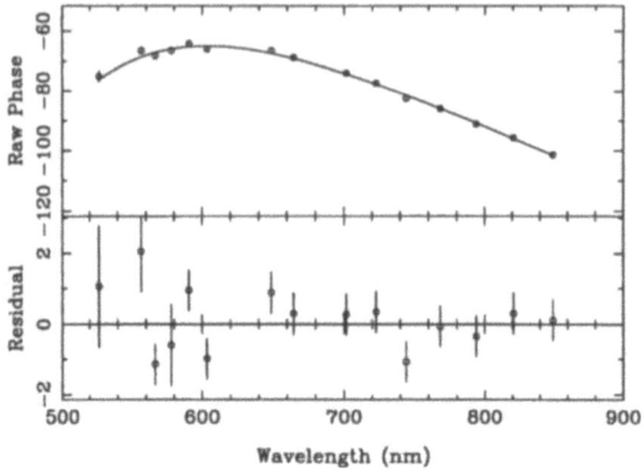

Fig. 1. Closure phase in degrees versus wavelength on a calibration star. The solid line is the model fit. The difference between the data and the model is shown in the lower panel.

Table 1. Model Parameters

UT(seconds)	$L_{Air}nm$	χ^2_n	Flux Ratio
13670	15718.	1.1	
14963	15735.	1.5	0.997 ± 0.002
16199	15713.	2.3	
17301	15719.	2.9	0.998 ± 0.002
18266	15722.	0.7	
19260	15719.	1.6	0.998 ± 0.003
25017	15734.	1.4	
27871	15740.	3.1	

and a model fit for the binary star system Mizar A taken on the same night. I modified the model to include the separation and position angle of the two stellar components and the brightness ratio as well as the calibration parameters. The separation and position angle of the two components are not very well constrained by the phases. This is typical of systems with two almost equally bright components. The flux ratio, however, is very well constrained. The Mizar data is also shown in Table 1 where it can be separated from the calibration star data by the presence of a flux ratio in the fourth column. These values are consistent with each other and have formal errors nearly an order of magnitude smaller than anything in the literature.

There is no evidence in this data for systematic effects in that the residuals to the model fits do not correlate between observations. If this continues to be true for longer observations, 3 hours of data will produce a flux ratio with

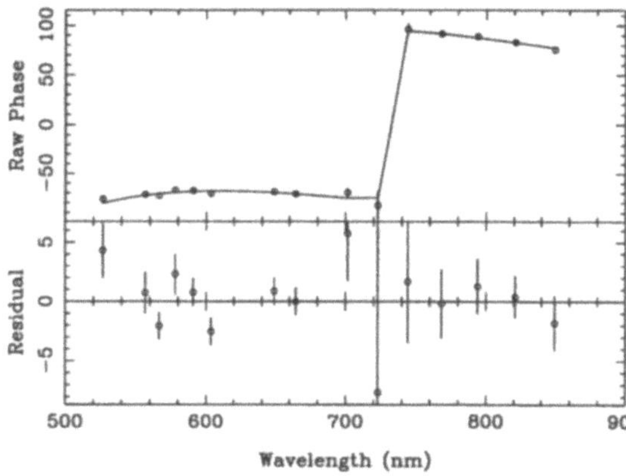

Fig. 2. Closure phase in degrees versus wavelength on a double star. The solid line is the model fit. The residuals are shown in the lower panel.

signal to noise of 3000, or a 3σ detection of a system with $\Delta m = 7.5$, large enough to detect large, gaseous planets in the thermal infrared.

In binary star systems with non unity flux ratios, the high precision phases also provide a good constraint on the separation and position angle. Separations measured with an accuracy of better than 1000 to 1 provide a sensitive astrometric search for planets in binary star systems.

3 Equivalent Diameters, Spots and Transits

If we measure the visibility amplitude of a star at a number of wavelengths, we can determine the wavelength where the amplitude drops to zero. Dividing this wavelength by the baseline gives a measure of an equivalent stellar diameter that can be determined with extremely high precision. Hajian et al, [2] presented data where this diameter can be determined with a precision of better than a part in 1000 from 90 seconds of data. This measurement is not affected by the multiplicative calibration errors which limit conventional interferometric measurements which leads me to believe that the ultimate accuracy of this type of measurement will be significantly better than what has been demonstrated. To date, measured stellar diameters have been used for determining effective temperatures and surface brightnesses where accuracies better than about 0.5% are of no value because at that level the accuracy of the results are limited by photometry. However very small changes in a star's diameter are interesting; this extreme precision provides an exciting new tool for studying stellar variability.

Another application of high precision stellar diameters is for the detection of stellar spots and planetary transits. These change the observed flux distribution from a star resulting in a change in the equivalent diameter that is large compared to the precision of the measurement. Figure 3 shows the results of a simulation of a stellar transit. The gaseous giant was centered on the stellar disk obscuring 1% of the flux. The squared visibility amplitude, V^2, is plotted versus wavelength. With a change in the equivalent diameter of the star by almost 2%, this event could be easily detected.

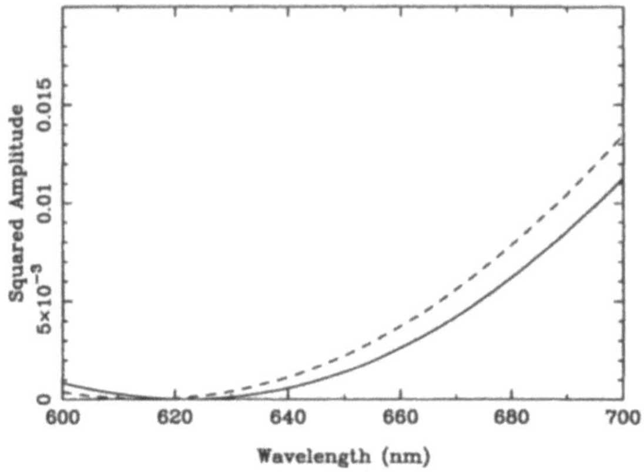

Fig. 3. Squared visibility amplitude as a function of wavelength for a naked star (dashed line) and for the same star with a transit by a Jupiter sized planet (solid line).

4 Conclusion

Interferometric observations provide very high angular resolution. To date, however, most of the published measurements do not have very high precision. Angular diameters typically have 1% to 3% errors. Binary star separations are better, with errors in the 0.5% range. However, with the success of FLUOR, and with techniques similar to those presented here, we should expect to be seeing one to two orders of magnitude improvement in precision for properly designed observations. These improvements provide exciting new tools for stellar astronomy and for the detection of planets.

Interferometry research at NRL is funded through the Office of Naval Research.

References

1. Davis, John and Tango, William J. (1996) Publ. Astr. Soc. of Pacific, **108**, 456
2. Hajian et al (1998) Astrophysical Journal, **496**, 484

Extra-Solar Planets with GAIA

Mario G. Lattanzi[1], Alessandro Sozzetti[2,1], and Alessandro Spagna[1]

[1] Osservatorio Astronomico di Torino, 10025 Pino Torinese, Italy
[2] Space Telescope Science Institute, Baltimore, MD, 21218, USA

Abstract. GAIA, like Hipparcos, is a global astrometric mission which will survey the sky several times during the 5 years of its nominal lifetime. GAIA is being actively studied by ESA as a possible candidate for the next Cornerstone-class mission to be launched around 2009 and possibly earlier.

Results from detailed simulations show that GAIA would detect giant planets ($M_p \sim M_J$) around normal stars up to a distance of $\sim$ 200 pc. This horizon includes over 500,000 dwarf stars earlier than K5 to be screened for planets. The same simulations predict that the main orbital parameters (semi-major axis, eccentricity, period, and inclination) of a significant fraction of the detected planets should be measurable to better than 30% up to 200 pc.

Our results indicate that GAIA's contribution would uniquely complement the expectations from other spectroscopic and astrometric surveys for ground breaking science in the fields of formation and evolution of planetary systems.

1 The Present Scenario

In less than four years since the discovery [12] of the first Jupiter-mass planet orbiting a normal star other than the Sun, some twenty candidate extra-solar planets have been discovered, thanks to spectroscopic techniques (http://www.obspm.fr/encycl/encycl.html). However, these discoveries have raised new questions in our understanding of the properties of planetary systems. Several of these candidate planets present characteristics that are hard to explain within the context of current theoretical models describing the formation and evolution of planetary systems. The fundamental tenets upon which present theories are based include nearly circular orbits and giant planets formed several AU from the central star, in contrast with the very short orbital periods [12] and high eccentricities [9,5,13] found for several candidate planets. Their interpretation as *bona-fide* planets largely rests on our understanding of correlations shown by their orbital and physical parameters, as recently discussed by Black [2] and earlier by Duquennoy & Mayor [7] in their work on solar-type binary stars.

The inadequacy of the current models, when confronted with the first actual observations of extra-solar planets, clearly indicates that our knowledge of the complicated physical phenomena governing the formation and evolution of planetary systems is still partial. The interplay between additional theoretical work and more observational data is instrumental for continued improvement in our theoretical understanding of how planets form and evolve, and where Earth-like planets could eventually be found.

However, simply adding a few tens of additional detections of giant extra-solar planets is not enough. A better understanding of the conditions under which planetary systems form and of their general properties requires sensitivity to less massive planets (down to ~10 Earth-mass), better characterization of known systems (mass, and orbital elements), and complete samples of planets, with useful upper limits on Jupiter-mass planets up to several AU from the central star. Astrometric measurements good to 2–10 micro-arcsec (μas), and in particular those made from space by the NASA mission SIM and by GAIA, will contribute substantially to these tasks. SIM will be able to study in detail targets detected by other methods, including microlensing, and to search for low mass planets [16]. GAIA's strength will be its discovery potential, following from the astrometric monitoring of all of the several hundred thousands bright stars out to distances of ~ 200 pc.

2 The Role of GAIA

The first firm discoveries of substellar companions around normal stars in the neighborhood of our Sun have shifted the perspectives in the science of extra-solar planets: the most basic issue to be addressed is no longer their *Existence*, but rather their *Classification*, which will help clarify the nature and diversity of substellar companions. Also, decisively addressing the other two key questions of *Formation* and *Evolution* of planetary systems will ultimately require statistically firm observational evidence.

Substellar companions can be divided in two classes, namely planets and brown dwarfs. There exist three major genesis indicators that can help classify substellar objects as either brown dwarfs or planets: *1)* mass, *2)* shape and alignment of the orbit, and *3)* composition and thermal structure of the atmosphere. Recently, Oppenheimer & Kulkarni [14] have tried to establish from theory both the lower limit to the mass of a brown dwarf and the upper bound to the mass of a planet. However, giant planets and brown dwarfs may still populate a common mass range. Therefore, mass alone cannot be decisive for the classification of a low mass companion to a star as a planet, unless this mass is close to that of the Earth [2].

The formation mechanisms of low mass stars and giant planets are thought to be quite different. Stars form from large-scale dynamical instabilities, while planets form by core accretion or dynamical instability of protoplanetary disks. Consequently, it is reasonable to expect that if correlations among stellar orbital parameters exist, such correlations might be significantly different for planetary companions. It has recently been shown [2] that correlations between eccentricity and the logarithm of orbital period for PMS (and main-sequence) binaries and for objects thought to be the result of accretion in a disk (like the giant planets in the solar system) are significantly different.

The majority of the candidate planets discovered so far by the radial velocity programs appear to follow the (e, logP) relation that can be established for PMS binaries. Low eccentricity alone is not sufficient to classify as planet a newly discovered low mass companion, as low eccentricity is expected for stellar companions orbiting close to the other star.

Different correlations among orbital parameters (eccentricity, period or semi-major axis) and measurable differences in planetary frequency are likely to be generated by diverse planetary formation scenarios (core accretion and disk instability are the two known to date) and evolution mechanisms as well as different formation and evolution processes of the parent star (binarity, spectral type, metallicity, age).

Orbital evolution mechanisms through the stellar disk (like gravitational migration) could generate differences in the distribution of orbital parameters (e.g., Jupiter-size planets on large orbits with very high eccentricity) with age. If disk instability is the preferred mode for the formation of giant planets, the gravitational pull induced by them on the parent star should already be measurable in very young (age < 1 Myr) PMS stars (see Table 1 for a list of the closest star forming regions), and planetary frequency is probably enhanced in binary stars [3,4].

Table 1. The closest star forming regions

Name	Distance (pc)
Chamaeleon complex	100-180 pc
Corona Austrina	140 pc
Lupus	140 pc
ϱ Ophiuchi	125-150 pc
Scorpio AB	180 pc
Serpens	150-200 pc
Taurus-Auriga	135 pc

Indications on the history of planetary frequency will come from the observations of the local old populations. Galaxy models [1,15] predict $\sim$ 4000 F–G–K dwarfs and sub-dwarfs to 200 pc, brighter than $V = 13$ mag, and with metallicity [Fe/H] < -1.0 (cfr. Fig. 1 right). By looking at stars in the age range $\sim$1–8 Gyr (cfr. Fig. 1 left), i.e., up to values comparable with the age of the Galaxy, it will be possible to evaluate if metal enrichment has made planetary formation more likely.

The ongoing spectroscopic work in conjunction with ground based astrometric searches will probably have more than 10 years of continuous observations on a number of substellar companions, so that astronomers are likely to begin basic *classification* before the sky is surveyed by GAIA.

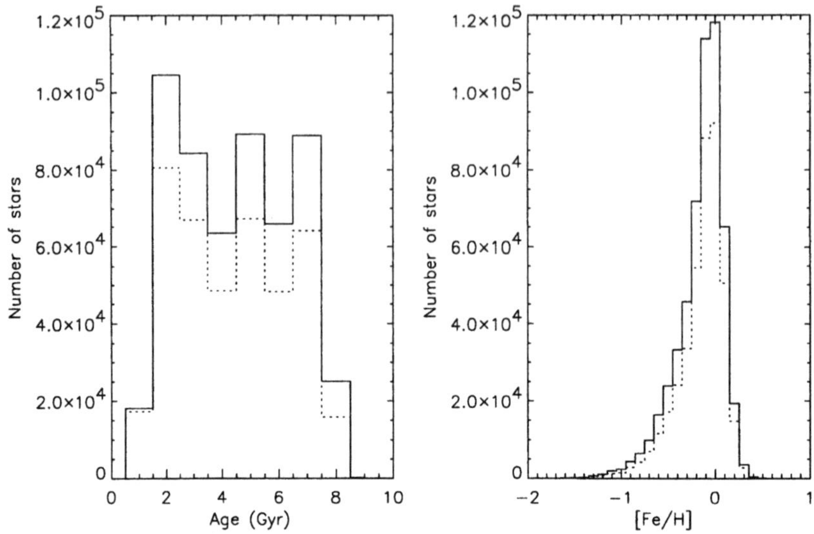

Fig. 1. Stellar distribution in the solar neighborhood ($d < 200$ pc), as function of age and metallicity, for $V < 13$ (*solid line*) and $V < 12$ (*dotted line*)

As for the range of masses between 1 Earth-mass and a few Earth-masses (Neptune-class planets), which are probably marginally accessible to GAIA anyhow, existence might be established by the SIM mission [6].
On the other hand, the major ground based programs and space missions employ pointed instruments and are therefore in need of input lists of candidate stars. Although these lists can be carefully optimized to yield much more than just the existence of extra-solar planets of different masses, the target selection criteria will depend on current understanding of formation and evolution mechanisms. Also, the "effectiveness" of these pre-selected lists will depend on actual planetary frequencies (to date largely unknown) which, in turn, are likely to be functions of several other parameters, as discussed above.

The ability of simultaneously and systematically determining planetary frequency and distribution of orbital parameters for the stellar mix in the solar neighborhood, stems out as a fundamental contribution that GAIA will uniquely provide, the only limitations being those intrinsic to the mission, i.e., to the actual sensitivity of the GAIA measurements to planetary perturbations. In the next section we discuss those limits and provide quantitative estimates of sample sizes accessible to GAIA.

Finally, two more issues are worth mentioning here. First, under the sole assumption that the portion of the orbital parameter space probed by GAIA is adequate, a magnitude limited survey enhances the probability of uncovering new facts or more details in the way planets form and/or evolve. Second, GAIA could play an important, although indirect, role in the quest for habitable Earth-size planets. These should be found around solar-type stars with a Jupiter-mass planet orbiting at a distance ≥ 3 AU, effectively protecting terrestrial planets from cometary impacts. GAIA measurements are particularly sensitive to such systems thanks to the mission duration and the favorable magnitude of normal dwarf stars (earlier than K5) up to $\sim$ 200 pc from the Sun ($V < 13$ mag). Therefore, GAIA could provide us with a quantitative estimate of the likelihood that inhabitable Earths exist in the solar neighborhood.

3 How GAIA Does It

The magnitude of the gravitational perturbations induced by a single planet on the parent star, as seen by an astrometric mission, can be quantified through its ***astrometric signature*** α defined as:

$$\alpha = \frac{M_{\rm p}}{M_{\rm s}} \frac{a_{\rm p}}{d} \qquad (1)$$

where $M_{\rm p}$, $M_{\rm s}$ are the masses of the planet and star respectively, $a_{\rm p}$ the semi-major axis of the planetary orbit, d the distance of the system from us. If $a_{\rm p}$ is in AU and d in parsec, then α is expressed in arcsec.

GAIA's scanning law is such that any given object on the sky is regularly re-observed (approximately every month, on average) throughout the 5-year mission. Thanks to the partial overlap of consecutive scans, these periodic observations are each comprised of several FOV crossings spread over a time interval of approximately half a day (i.e., the duration of 4-5 consecutive scans), for a total observing time of $\sim$160 sec. The elementary one-dimensional measurements made during these FOV crossings can be combined together to estimate a *normal* one-dimensional position. This is the longitude ψ relative to a local great circle (GC) located in the region of the sky surveyed by the satellite during the few consecutive scans of an observing period.

Then, sensitivity to the astrometric signature α depends on the longitude error σ_ψ, which was fixed to 10 μas in our simulations (see Lattanzi et al. 1999 [10], and references therein), a value $\sim$ 3 times larger than the expected final astrometric error. This value applies to stellar magnitudes $V < 13$ mag and it is consistent with the adopted mission profile and the level of residual system errors of both the baseline and interferometric backup designs [8]. Also, it was verified through Monte Carlo tests with different values of σ_ψ that, as could be expected, the planetary detection probability depends on σ_ψ and α only through their ratio, the "astrometric" ***signal-to-noise ratio***

$$S/N = \alpha/\sigma_\psi, \qquad (2)$$

so that the results described below can be straightforwardly rescaled to different measurement errors.

At magnitudes $V \lesssim 12.5$ mag, the astrometric error reaches its saturation limit, i.e., the photon error becomes negligible compared to the magnitude-independent residual system errors intrinsic to the adopted design. Therefore, that is the magnitude range where GAIA will have maximum sensitivity to planetary perturbations acting on normal stars. However, many other planetary systems, as those formed around late type stars, with larger astrometric signatures can also be detected at fainter magnitudes.

GAIA's potential is effectively assessed by gauging, on a homogeneous set of extra-solar systems and as function of the most relevant orbital parameters (semi-major axis, period, and eccentricity) and of distance from the Sun, both its sensitivity to the presence of planets and its limitations in accurate orbital parameters estimation. A natural choice is to experiment with planet-star combinations grossly resembling our own solar system. Consequently, all of the sources considered in the simulation were made of single Jupiter-mass planets orbiting 1-$M_\odot$ stars. In practice, this choice for the mass of the parent star encompasses the spectral class range from $\sim$ F0 to early K type dwarf stars, whose masses are within a factor of ~ 1.5 that of the Sun. This, in combination with the $V < 13$ magnitude limit, translates into a distance cutoff of ~ 200 pc.
To this distance, F-G-K type dwarfs dominate the star counts at bright magnitudes.

3.1 Detection of Giant Planets

The detection method [10] applies a standard χ^2 test (with the confidence level set to 95%) to the residuals $\psi - \psi_\mathrm{r}$, where ψ represents the actual measurements and ψ_r the GC abscissae recomputed after fitting the observations of each star with a single-star model, i.e., solving only for the five astrometric parameters appropriate for a single star (position, parallax, and proper motion). Being χ^2_o the value provided by the null model (no planet), the test fails when $\Pr(\chi^2 \leq \chi^2_\mathrm{o}) \geq 0.95$. In this case the planet is considered *detected.*

The χ^2 test was utilized on a set of 160 000 planetary systems, uniformly distributed on the sky, as function of period P and astrometric signature (or apparent semi-major axis) α, which we expect to be among the major contributors to detection efficiency. The simulated values ranged from 10 to 100 µas for α, and between 0.5 and 20 years for P. The remaining orbital elements were distributed randomly in the ranges: $0° \leq i \leq 90°$, $0 \leq e < 1$, $0 \leq \Omega$ (position angle of the ascending node) $\leq 2\pi$, $0 \leq \omega$ (argument of periastron) $\leq 2\pi$, $0 \leq \tau$ (periastron epoch) $\leq P$. With the observation error set to $\sigma_\psi = 10$ µas, the astrometric signal-to-noise ratio varied in the range $1 \leq S/N \leq 10$.

The results are summarized in Fig. 2. The solid lines were derived from the simulated dataset, and express the empirical relationship between the

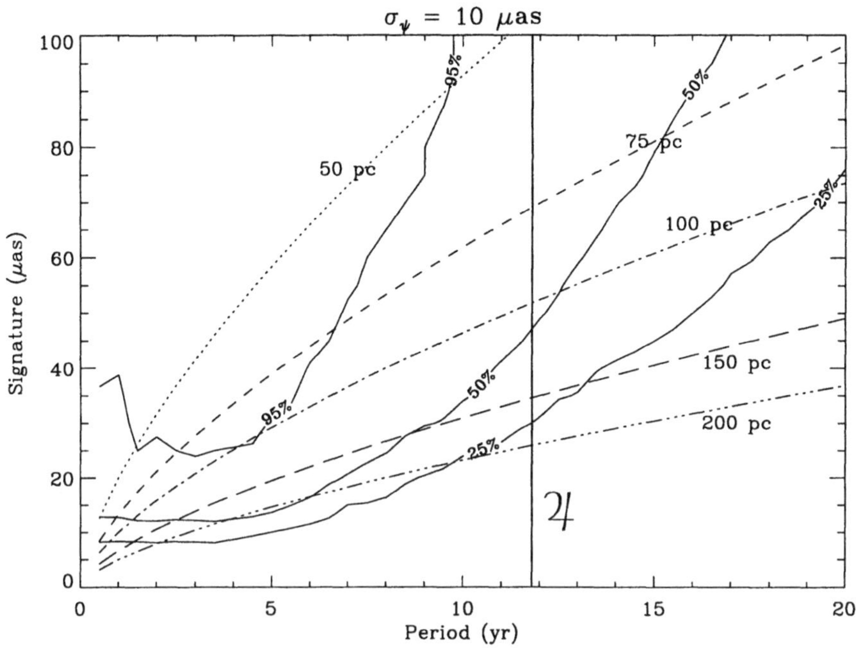

Fig. 2. Iso-probability contours (*solid lines*) for 25 per cent, 50 per cent and 95 per cent of detection probability, compared with Kepler's third laws (*dotted/dashed lines*) for systems with Jupiter-Sun masses at D=50, 75, 100, 150, and 200 pc

amplitude of the astrometric perturbation and the orbital period for detection probabilities of 25%, 50%, and 95%.

The over-plotted curves show the behaviour of α as function of the orbital period for a Jupiter-mass planet around a solar-mass star at different distances from the observer, obtained by substituting Kepler's third law in the expression defining α:

$$\alpha \simeq \frac{M_{\rm p}}{M_{\rm s}^{2/3}} \frac{P^{2/3}}{d} \tag{3}$$

Practically all of the existing Jupiter-mass planets (95% detection probability) within 50 pc and with periods between 1.5 years and 9 years would be revealed by GAIA. The range of periods narrows with increasing distance. At 100 pc statistical certainty is possible only for those Jupiters with orbital periods clumped around the mission duration.

For the remainder of this section and the following, we will consider the value of 50% as the threshold for significant probability of planetary detection. Then, Jupiter-like planets (same orbital period of Jupiter) appear detectable up to a distance of 100 pc. Jupiter-mass planets with shorter periods (roughly between 2.5 and 8 years) can be detected up to 150 pc, and detectability is still effective for a short range of periods (4-6 years) around

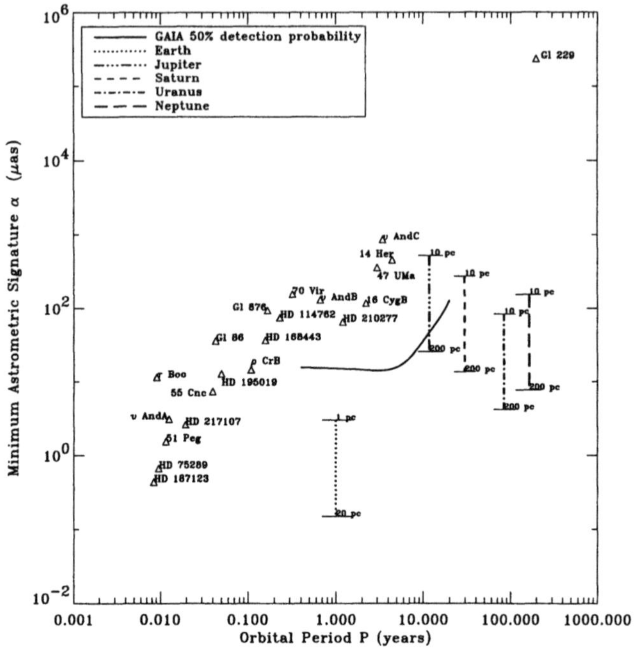

Fig. 3. Minimum astrometric signature, α, induced on the parent star by presently known planetary mass objects ($0.5 M_J < M < 11 M_J$), as a function of orbital period [10]

the mission lifetime at 200 pc. These results confirm the existence of a good overlap with the range of periods probed by spectroscopy. This is illustrated in Fig. 3, which compares the GAIA 50% detection probability curve (solid line) with the minimum astrometric signatures of the known planetary systems.

To evaluate the number of potential planetary systems within GAIA's detection horizon one needs to know the stellar content in the solar neighborhood and the planetary frequency distribution.

Star counts to 200 pc were obtained from simulated catalogs produced with the Besançon model of stellar population synthesis [1,15] constrained to $V < 13$ mag (Fig. 4). According to this Galaxy model, and for spectral types earlier than K5, we should expect $\sim 61\,000$ stars within 100 pc from the Sun; the number increases to $\sim 175\,000$ and $\sim 470\,000$ stars, for distances of 150 pc and 200 pc, respectively. Note that the contribution from early type stars and giants is negligible.

To date, only coarse attempts at establishing planetary frequency have been made, which are based on the radial velocity programs concluded so far. A recent estimate [11] yields an integral planetary frequency $F_p \sim 4\%$ (~ 300 stars surveyed) for giant planets (defined as having masses in the range 0.5-5 M_J) orbiting within 3 AU from the parent star. Clearly, revising this estimate

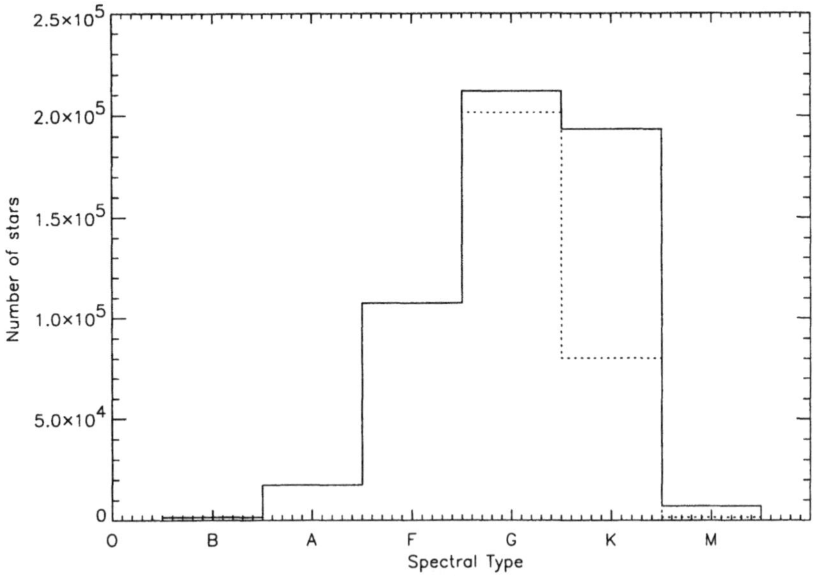

Fig. 4. Stellar content to $d < 200$ pc, as function of the spectral type, for $V < 13$ (*solid line*) and $V < 12$ (*dotted line*)

and extending it to wider ranges of orbital radii is one of the objectives of the current and future surveys. Nonetheless, we can utilize this estimate of $F_{\rm p}$ to predict the number of giant planets GAIA could reveal. Under the assumption that the actual planetary frequency is uniform with semi-major axis, $F_{\rm p}$ can be turned into an estimate of the planetary frequency ($f_{\rm p}$) by simply dividing it by the limiting orbital distance of 3 AU; this yields the value $f_{\rm p} \sim 1.3\%$ per 1-AU bin.

The *lower limit* to the number of detected giant planets $N_{\rm d}$ at a given distance d (in pc) predicted for GAIA is then obtained from the relation:

$$N_{\rm d} > 0.5 \cdot f_{\rm p} \cdot \Delta a \cdot N_{\star}, \tag{4}$$

where $N_{\star}$ is the total number of stars within a sphere of radius d centered on the Sun (for given limiting magnitude and spectral type), while the ranges (Δa's) of semi-major axis were computed from converting the orbital period intervals generated by the intersections of the 50% iso-probability curve with the Kepler's third law plots for different distances (Fig. 2). The 0.5 factor explicitly accounts for the adopted 50% detection probability threshold. The results are summarized in Table 2.

The numbers for N_{150} and N_{200} correspond to actual *new detections*, once the fractions of detected giant planets in common with the overlapping semi-major axis regions at lower distances have been properly subtracted.

Table 2. Number of giant planets that could be revealed by GAIA, as function of increasing distance. A uniform frequency distribution of 1.3% planets per 1-AU bin is assumed

Δd (pc)	$N_\star$	Δa (AU)	$N_{\rm d}$
0-100	~61 000	1.3 - 5.3	$\gtrsim$ 1600
100-150	~114 000	1.8 - 3.9	$\gtrsim$ 1600
150-200	~295 000	2.5 - 3.3	$\gtrsim$ 1500

Therefore, the total number of giant planets GAIA could discover orbiting around normal stars within the distance limit of 200 pc from the Sun is then *greater* than 4700.

3.2 Measuring New Giant Planets

Once a planet is "detected", the goal is to reliably determine its orbital characteristics and mass.

For a complete reconstruction of the orbital geometry we implemented an analytic model in which the observation residuals $\psi - \psi_{\rm r}$ are evaluated with the recomputed abscissa $\psi_{\rm r}$ function of both the standard astrometric parameters (λ, β, μ_λ, μ_β, π) and the parameters describing the stellar relative orbit with respect to the barycenter of the planetary system: period P, periastron epoch τ, eccentricity e, semi-major axis a, inclination i, argument of periastron ω, position angle of the ascending node Ω. Both orbital and astrometric parameters are solved for with a Least Squares method applied to the linearized version of the observation equations.

The planet mass can be derived from the familiar expression defining the mass function of binary systems as applied to the case of a planet mass secondary, i.e.:

$$M_{\rm p} \simeq \left(\frac{a_{\rm s}^3}{\pi^3} \frac{M_{\rm s}^2}{P^2} \right)^{1/3} \tag{5}$$

where the planetary ($M_{\rm p}$) and stellar ($M_{\rm s}$) masses are given in units of $M_\odot$, π the parallax in arcsec, P the orbital period in years, and $a_{\rm s}$ the semi-major axis of the primary, expressed in arcsec. Equation 5 shows that the mass of a planet cannot be determined directly from the orbital parameters alone, but reliable independent estimates of the mass of the parent star must be provided by some other means (e.g., spectroscopy).

Since our purpose is to estimate GAIA's ultimate potential in measuring giant planets, it is not crucial to have at this stage a suitable strategy for choosing appropriate initial guesses for the unknown parameters of the

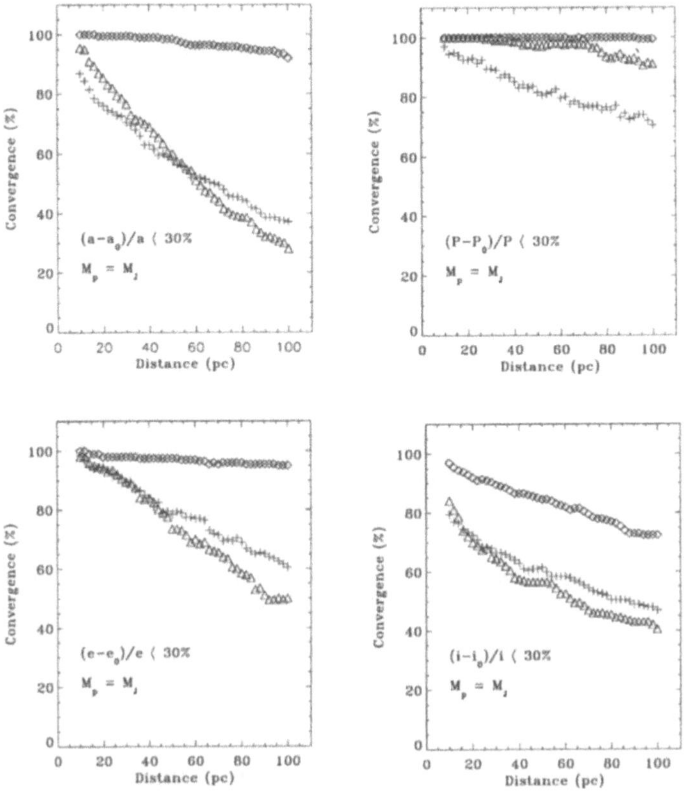

Fig. 5. Convergence probability to a 30% accuracy for a, P, e, i, as a function of the distance from the Sun

fit. Instead, the orbital elements were initialized to their "true" values [10]. We then evaluated the convergence probability for each parameter to a well defined accuracy level, i.e., we computed the percentage of stars for which the solution of each parameter converged within a given fraction of its true value.

Figure 5 shows convergence probability for systems with Sun-Jupiter masses and given orbital period (triangles for 0.5 years, diamonds for 5 years, and crosses for 11.8 years), expressed as a function of distance from the observer, within the 30% accuracy level for the parameters a, P, i, and e.

As expected, the 5 year-period case is particularly favorable, due to the optimal sampling of the orbit. The 11.8 year period, although it corresponds to a larger signal (α, as we have seen, increases with period), suffers from poorer sampling of the orbit during the mission duration, while in the short-period case (0.5 years) both the smaller signal amplitude and the generally non-optimal timing of the observations contribute to worsen the accuracy of the solution.

Figure 5 also highlights how we may expect different accuracy levels for different orbital parameters. Orbital period and eccentricity come out to be easier to measure, compared to semi-major axis and inclination.

Finally, the results obtained allow us to provide a lower bound, consistent with the assumptions above, of the fraction of giant planets detected by GAIA for which it would be possible to derive orbital elements within well defined accuracy levels.

As shown in Fig. 5, the distance horizon was fixed at 100 pc, as the 30% convergence fraction for the orbital parameters a and i would quickly drop to values exceedingly low for long periods (P = 11.8 years). At 100 pc, the fraction of acceptable solutions reaches its lowest value of ~40% for the parameter a. If we retain this as the probability value appropriate for the entire set of orbital parameters, a lower limit to the number of detected planets for which there would be reliable orbits follows from the data in Table 2. More than 640 of the 1600 detected planets would have accurate estimates of the orbital parameters. The plot in Fig. 5 for P = 5 years is still useful for estimating what the prediction would be at a distance of 200 pc. A rough extrapolation suggests that at least 50% of the planets would have orbital parameters good to 30% or better, adding more than 750 planets to the previous count, and a similar number of accurately measured planets should come from the 150-pc distance bin, bringing the total to over 2000 planets. The statistical value of such a sample would be instrumental for critical testing of theories on planet formation and evolution. Finally, one immediately realizes the uniqueness of the GAIA sample of *measured* planets, as its size is *at least* comparable to that of the *observing lists* of the largest ground based surveys, and to the size of the planet-finding program which is reasonable to expect for SIM.

Acknowledgements

We wish to thank M.A.C. Perryman for his comments and continuing support to this investigation. Over the course of this work, we have also benefited from discussions with Stefano Casertano, Gerry Gilmore and Lennart Lindegren. The authors gratefully acknowledge partial financial support from the Italian Space Agency.

References

1. Bienaymé, O., Robin, A. C., Crézé, M. (1987), A&A **180**, 94.
2. Black, D. C. (1997), ApJL **490**, L171
3. Boss, A. (1998), Nature **393**, 141-143
4. Boss, A. (1998), Nature **395**, 320-321
5. Cochran, W. D., Hatzes, A. P., Marcy, G. W., Butler, R. P. (1996), ApJ **483**, 457

6. Danner, R., Unwin, S. (1999), SIM: Taking the Measure of the Universe, NASA/JPL
7. Duquennoy, A., Mayor, M. (1991), A&A **248**, 485
8. Gilmore, G., et al. (1998), in Proc. SPIE 3350, Astronomical Interferometry, Ed. R. D. Reasenberg & M. Shao
9. Latham, D.W., et al. (1989), Nature **339**, 38
10. Lattanzi, M. G., Spagna, A., Sozzetti, A., Casertano, S. (1999), MNRAS, in press
11. Marcy, G., Cochran, W. D., Mayor, M. (1999) in Protostars and Planets IV, Ed. V. Mannings, A. P. Boss & S. S. Russell (Tucson: University of Arizona Press), in press
12. Mayor, M., Queloz, D. (1995), Nature **378**, 355
13. Mazeh, T., Latham, D. W., Stefanik, R. P. (1996), ApJ **466**, 415
14. Oppenheimer, B. R., Kulkarni, S. R. (1999), in Protostars and Planets IV, Ed. V. Mannings, A. P. Boss & S. S. Russell (Tucson: University of Arizona Press), in press
15. Robin, A. C., Crézé, M. (1986), A&A **157**, 71
16. Unwin, S. (1999), this volume

Searching for Planets with the Space Interferometry Mission

Stephen Unwin

Jet Propulsion Laboratory, California Institute of Technology,
4800 Oak Grove Drive, Pasadena, CA 91109, USA

Abstract. The Space Interferometry Mission (SIM) will be the first space-based long baseline Michelson interferometer designed for precision astrometry. It will address a wide range of problems in stellar astrophysics and Galactic structure, delivering precision astrometry of stars down to 20 magnitude throughout the entire Galaxy. SIM uses a 10-m Michelson interferometer in Earth-trailing solar orbit to provide 4 microarcsecond (μas) precision astrometry. With a 5-year mission lifetime, SIM will be a powerful tool for discovering planets around nearby stars, through detection of the stellar reflex motion. The astrometric method complements the radial velocity technique which as already yielded many new planets, with an important benefit of directly measuring planetary masses. SIM will have a single-measurement precision of 1 μas in a frame defined by nearby reference stars, enabling searches for planets with masses as small as a few earth masses around the nearest stars. More massive planets will be detectable to much larger distances. In addition to precision astrometry SIM will also serve an important role as a technology precursor for future astrophysics missions using interferometers. Two technologies demonstrated will be high dynamic-range aperture synthesis imaging at 10-milliarcsec resolution in the optical, and fringe nulling to 10^{-4}.

1 Introduction

The Space Interferometry Mission (SIM) is designed as a space-based optical interferometer operating in the visible waveband. This mission will open up many areas of astrophysics, via astrometry with unprecedented accuracy during its 5-year mission. In its wide-angle mode, SIM will be capable to provide a 4 μas precision absolute position measurements of stars, with parallaxes to comparable accuracy [1]. The expected proper motion accuracy is around 2 μas/yr, corresponding to a transverse velocity of 10 m/s at a distance of 1 kpc. It is expected to achieve an accuracy of 1 μas in one hour, over a narrow field of view. In this mode, SIM will search for planetary companions to nearby stars, by detecting the astrometric 'wobble' relative to nearby reference stars.

The SIM architecture comprises three 10-m baseline Michelson interferometers with parallel baselines. Collectors at each end of the structure contain afocal telescopes which compress the input beam from a diameter of 0.3 m. Two telescope pairs observe bright guide stars, to stabilize the optical system, and a third pair observes the science target. A fourth set of optics are carried for redundancy. Light from each pair is directed to a beam combiner through as series of fold mirrors. One light path passes through a movable delay line which tracks the white-light fringe. The position of this delay line is the fundamental astrometric quantity measured by SIM. An internal metrology beam is used to measure the path difference between the two arms of the interferometer. The SIM Project is currently evaluating two designs for the layout of the three interferometers. In one of the designs, a deployable metrology boom measures the positions of the individual collectors, and the baseline lengths are triangulated. In the alternate design, the three baselines are coincident, and a single metrology beam measures the baseline length; this design has a more complicated optical layout, but a simpler metrology system.

The telescope field of regard is 15°; this is the sky area accessible while the instrument is inertially pointed. Astrometry is performed by measuring the change in white-light fringe delay between successive targets. Global astrometry is accomplished by 'tiling' the sky with grid stars which serve as positional references for the astrometry. A summary of the key instrument and mission parameters is given in Table 1.

While SIM is primarily an astrometric instrument, it will serve as a technology pathfinder for future astrophysics missions, such as the Terrestrial Planet Finder (TPF). Key technologies needed by future missions will be demonstrated by SIM, then carried over directly or readily adapted. For TPF the nulling beam combiner, which operates by applying a polarization flip to one arm of the interferometer will demonstrate a null depth (suppression of on-axis light) to 10^{-4}. A scientific goal of this observing mode for SIM is imaging of reflected light from dust disks around main-sequence stars. Nulling mode greatly increases the effective sensitivity to a disk, by canceling most of the light from the bright central star.

SIM will also operate as an imaging instrument, using aperture synthesis to provide a fully diffraction-limited aperture of 10 m in the optical band. The image resolution will be an unprecedented 10 milliarcsec (mas) in the optical. High dynamic range images will be possible because of the instrument stability and accurate fringe calibration. An example of an imaging target for SIM is a nearby active galactic nucleus with strong Hα line emission near its center. By imaging the line-emitting gas, SIM will measure the mass distribution in the very dense central region.

SIM is a key element of the NASA Origins Program. The long-range goal of this program is to seek answers to questions of the origins of galaxies, stars, and planets, and ultimately life itself. The program has a phased approach to the technical challenges of the proposed space missions. More information on SIM and Origins is available on the Web at: `http://sim.jpl.nasa.gov`.

Table 1. SIM Instrument and Mission Parameters

Baseline	10 m
Wavelength range	0.4 - 1.0 μm
Telescope aperture	0.3 m diameter
Astrometric Field of Regard	15°
Imaging field of view	0.3 arcsec
Detector	Si CCD
Orbit	Earth-Trailing solar orbit
Mission Duration	5 years (launch June, 2005)
Astrometry (wide-angle)	4 μas mission accuracy
Astrometric Sensitivity	20 mag in 4 hours
Astrometry (narrow-angle)	1 μas in 1 hour
Imaging Resolution	10 milliarcsec
Imaging Sensitivity (point source)	25 mag in 1 hour
Interferometric Nulling	Null depth 10^{-4}

2 Nearby Planetary Systems

The discovery of planets around other stars [2], [3] has brought into focus a long standing, but poorly understood problem in astronomy, namely the formation of planetary systems. These new data are the first solid pieces in a puzzle which will take some time to unravel. SIM will contribute strongly to the observational database available for testing of solar-system formation theories, by searching for planets. It will be capable of detecting planets down to Earth mass around the nearest stars, and Jupiter-mass planets out to a kiloparsec.

An important goal for the mission is to detect the presence of earth-sized planets orbiting nearby solar-type stars and study their characteristics in order to learn about the formation and evolution of planetary systems that might include habitable planets. SIM will search for earth-sized planets orbiting in or near the habitable zones of 50 of the most suitable solar-type stars.

Because planets are extremely faint compared to the star around which they orbit – typically 10^9 times fainter in the visible waveband, indirect detection techniques must be used. In a two-body system, the components orbit around the center of mass of the system, so that the presence of a comparatively small body, such as a planet, around a star induces a 'wobble' on its motion which can be detected by tracing the motion of the center of light of the system. The well-known radial velocity method utilizes the Doppler shift resulting from the radial motion of the star. With a resolution of better than 10 m/s, many giant planets around several solar-type stars have been found [3]. With this technique, the mass of the planet is unknown to a factor ($\sin i$) where i is the inclination of orbit plane to the line of sight.

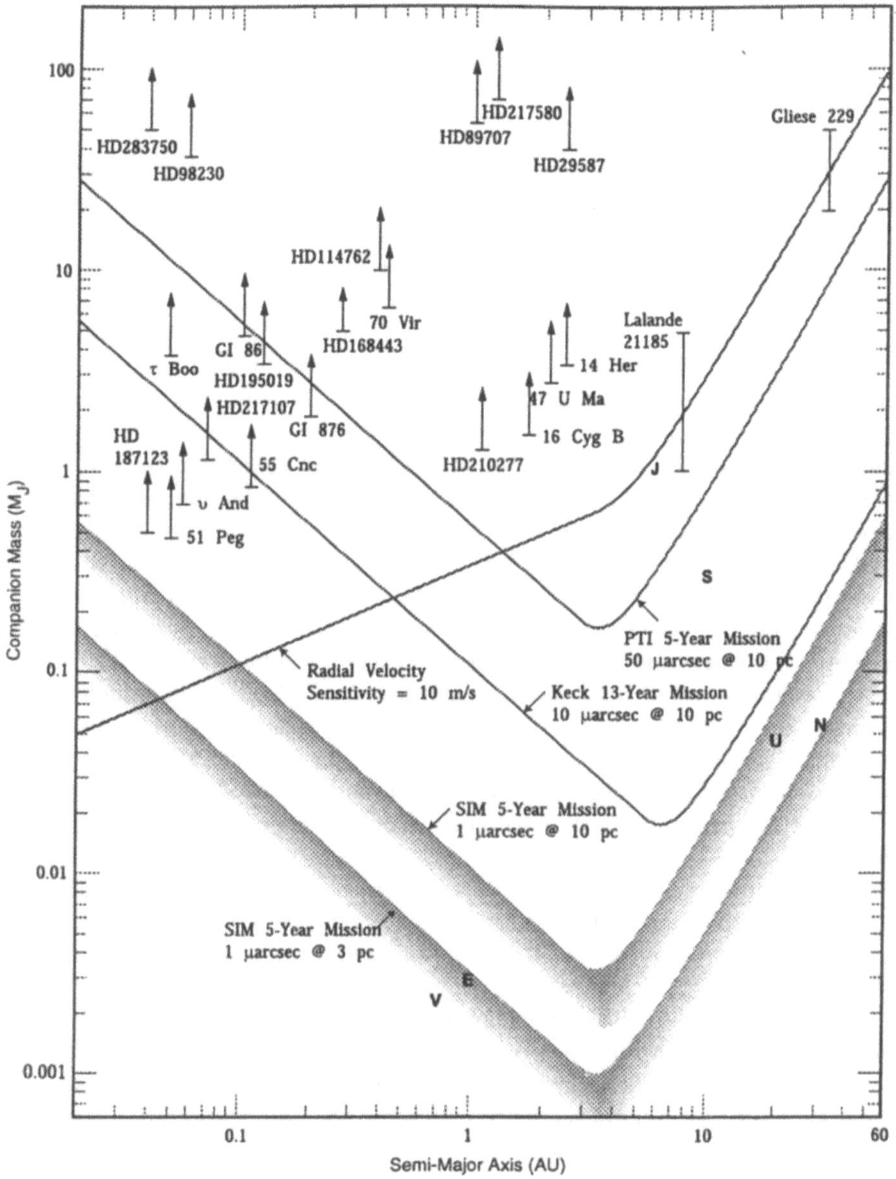

Fig. 1. Masses and orbital semi-major axes for known substellar objects. Planets in our own solar system are also shown. Diagonal lines represent the detection sensitivities of the several instruments: radial velocity with a 10 m/s sensitivity; Palomar Testbed Interferometer and Keck Interferometer for stars at 10 pc; and SIM. For SIM, the curves are for stars at 3 pc and 10 pc distance. Note the reduced sensitivity of all instruments to very long-period orbits

Figure 1 shows where the known planets (including our own solar system) lie in the mass - orbit-radius plane, together with the sensitivity limits of the two main detection methods.

SIM detects the the *transverse* reflex motion of the target star with respect to a reference frame of nearby ($\leq 1^\circ$) stars. Our own Sun would reveal the presence of its planetary system to an observer at 10 parsec away with a signal of ± 500 μas. The dominant sources of the wobble are Jupiter and Saturn, and these would be easy SIM targets. In its narrow-angle mode, SIM will reach a precision of about 1 μas in a 1-hour measurement. Repeated visits during the mission will allow the characterization of orbits with signatures smaller than this value. This is of the same order as the reflex motion induced by an Earth-mass planet around one of the nearest few G-type stars.

The recent announcement[4] of a triple planetary system around Upsilon Andromedae provides additional impetus for planet searching for SIM. Now that a multiple system has been found, one can begin asking more detailed questions of planetary searches. What range of orbit periods and masses are typical? Are the orbits circular? Are they coplanar? Continued radial velocity searches will provide many answers, and probably several new multi-planet systems. SIM has at least two unique contributions to make. It can measure all the astrometric parameters, and test the coplanarity of the orbits, which cannot be done with radial velocities. And it will reach much lower in mass than the radial velocity method, exploring the Earth–Uranus mass range for a large number of systems. Unlike Hipparcos, and all the proposed future astrometry missions, SIM is a *pointed* instrument. This capability will allow SIM to perform very detailed studies of the most interesting planetary systems.

SIM will detect large (5 to 10 earth masses) terrestrial planets orbiting nearby solar-type stars and study their characteristics in order to learn about the role of rocky cores in the formation and evolution of planetary systems. It will search a large nearby sample for planets which formed as rocky cores in the region closer to their parent stars than the inner boundary of the ice condensation zones in their proto-planetary disks.

SIM will explore the prevalence of planetary systems as a function of stellar type along the main sequence, since the spectral characteristics of the parent star do not affect the sensitivity of the astrometric detection method.

3 SIM Astrometric Science

The discovery and characterization of planetary systems around nearby stars is a prime objective for SIM, as indicated above. But the science program for SIM is very broad, and will impact many areas of astrophysics including fundamental astrophysics, the cosmic distance scale, and Galactic dynamics and structure [5]. SIM offers the ability to make astrometric measurements with an accuracy far in excess of ground-based observations, or any other space mission in the near term. As well as contributing strongly to known astrophysical problems, it has the potential for completely unexpected results. With an absolute positional accuracy of 4 μas, SIM will improve on the best currently available measures by 2 to 3 orders of magnitude, providing parallaxes accurate to 10% and transverse velocities accurate to 0.2 km/s anywhere in the Galaxy, to stars as faint as 20 magnitude.

The following is a brief list of some of the important areas of astrophysics in which SIM will play an important role:

- *Stellar Dynamics of the Galaxy.* Using precision astrometry, SIM will address many fundamental questions concerning the mass distribution in our Galaxy, the dynamics of its stars, and the evolution of its stellar populations.
- *Masses and Evolution of Stars in Close Binary Systems.* By detecting the astrometric signature of the binary star orbit, SIM can determine the masses and orbits of a large number of ordinary binary star systems, as well as more exotic systems: white dwarf/CVs, neutron stars and black holes.
- *Rotational Parallaxes.* Distances to nearby spiral galaxies can be measured directly, without use of luminosity-based indicators, using proper motions of their brightest stars due to rotation of the disk about the center of the galaxy.
- *Stellar Luminosities and Calibration of Distance Indicators.* SIM will calibrate the luminosities of Cepheid and RR Lyrae 'standard candle' variable stars, and planetary nebulae. These sources are used as standard candles in distance determinations for calculating the Hubble constant, and in distance measurements to globular clusters.
- *MACHO Gravitational Microlensing Events.* SIM will be able to detect the astrometric signature of microlensing events caused by massive compact halo objects (MACHOs) along the line of sight [6]. Masses for these objects can be inferred, providing important clues to the nature of dark matter in the Galaxy.
- *Ages of Globular Clusters.* Stellar evolution models for clusters apparently conflict with the age of the universe inferred from the Hubble expansion. SIM will measure the distances to globular clusters directly using trigonometric parallax, and hence the luminosities of the oldest main-sequence stars can be determined. Subtle effects due to reddening and metallicity effects can be studied.
- *Dynamics of Small Stellar Systems.* By measuring the proper motions of globular clusters, the mass distribution of the Galaxy can be determined. This will help our understanding of the formation of the Galactic halo and the globular cluster system. Tidal tails from disrupted dwarf spheroidal galaxies [7] also provide a powerful means of tracing the Galactic potential.

The work described in this paper was performed at the Jet Propulsion Laboratory, California Institute of Technology, under a contract with the National Aeronautics and Space Administration.

References

1. Unwin S.C., Shao M., Peterson D.M., Jones C.P. (1999) Space Interferometry Mission Science Requirements Document. JPL internal document
2. Mayor M., Queloz, D. (1995) A Jupiter-mass companion to a solar-type star. Nature 378, 355
3. Marcy G.W., Butler R.P. (1996) A planetary companion to 70-Virginis. ApJ 464, L147
4. Butler R.P., Marcy G.W., Fischer D.A., Brown T.W., Contos A.R., Korzennik S.G., Nisenson P., Noyes R.W. (1999) Evidence for Multiple Companions to Upsilon Andromedae. ApJ (submitted)
5. Peterson D. M. (1995) Scientific Support for Space Interferometry, Spaceborne Interferometry II. Proc. SPIE 2477, 2
6. Boden A., Shao M., Van Buren D. (1998) Astrometric Observation of MACHO Gravitational Microlensing. ApJ 502, 538
7. Ibata R.A., Wyse R.F.G., Gilmore G., Irwin M.J., Suntzeff N.B. (1997) The kinematics, orbit, and survival of the Sagittarius dwarf spheroidal galaxy. AJ 113, 634.

Extrasolar Planets Transits: Detection and Follow-Up

Jean Schneider

CNRS - Observatoire de Paris, 92195 Meudon, France
schneider@obspm.fr

Abstract. The detection of transits has a rich potential of information about extrasolar planets. This contribution gives an overview and new perspectives on the subject:
- constraints on the internal structure from radius measurement
- detection of rings and satellites of planets
- determination of the albedo of planets in close orbit ('Pegasids')
- spectroscopic investigation of transits

In addition, the high precision photometry of transits has several interesting by-products such as stellar surface investigation or microlensing in binary systems. Ground-based and space-based projects dedicated to these searches are reviewed and some potential artefacts are discussed.

1 The Planet Transits Method

The aim of the transit method, first investigated by Rosenblatt [1] and Borucki and Summers [2], is to search for extrasolar planets by the luminosity drop of a star produced by a planet transit. The relative drop $\Delta F/F$, given by the surface ratio R_P^2/R_*^2, runs from 10^{-4} for an Earth-like planet to 10^{-2} for a Jupiter. The low probability $p = R_*/a$ of a transit ($p = 0.5\%$ at $a = 1AU$) governs the search strategies discussed in section 2.

1.1 Expected Scientific Results

Assuming that a transit has been detected, it can be exploited for several science purposes.

1/ If the stellar spectrum allows for high precision radial velocity (RV) measurements, giving the product $M_P \times \sin i$, there is no ambiguity in the planet mass determination since the inclination i of the orbit is exactly $\pi/2$.

2/ The most obvious, and unique, output is the radius of the planet. Indeed, the imaging gives only the product $A \times R_P^2$ where A is the planet albedo . The value of the radius is inferred from the value of $\Delta F/F$, assuming a value for R_*; the latter can be estimated from the spectrum, the brightness and the distance of the star. If the planet mass can be known from RV or astrometric measurements, it gives constraints on its internal structure and in particular, for massive planets, on its solid or gaseous nature.

3/ From the shape of the transit one can infer if the planet is surrounded by rings [3,4]. The investigation of the precise shape gives clues for the inclination of the ring on the sky and on its orientation relative to the planet orbit [5]. It must be noted that the presence of a ring degrades the precision on the planet radius measurement, and thus on the conclusions one can draw on its internal structure.

4/ The study of the transit can also reveal the presence of giant satellites of the planet, either by the shape of the transit or by the timing of repeated transits [6]. From the shape of the light curve $L(t) = L_P(t) + L_S(t + \Delta T)$, where $L_{P,S}(t)$ are the light curves of the planet and the satellite and ΔT the shift of $L_S(t)$ relative to $L_P(t)$, one can infer the satellite radius and its distance projected onto the planet orbit. If the transit is seen several times, one can infer from the time variation of ΔT from one transit to another the orbital period of the satellite around the planet. Even if the photometric precision is not sufficient to see $L_S(t)$, one can infer the presence of the satellite from the timing of $L_P(t)$ if there are at least three transits: in that case there is a periodic modulation in the time of occurrence of transits with an amplitude $\Delta T = 16(M_S/M_\oplus)(M_P/M_{Saturn})^{-1}(a_S/0.001AU)$ min. for a planet at 1 AU from the star.

5/ When the planet is nearly behind the star, it has a relative illumination, relative to the stellar flux, $I/F = (A/4) \times (R_P/a)^2$. When the planet passes behind the star, the observer sees a dip $\Delta F/F = I/F$ in the planet + star light curve with the same duration than the planet transit. For a Jupiter-size planet at 0.05 AU from the star and $A = 0.7$, $\Delta F/F = 2N_T\ 10^{-5}$ when cumulated over N_T transits, giving thus an access to A.

6/ Spectroscopy of transits.

For planets in close orbits, the stellar wind can produce a cometary-like tail [7] which, when there is a transit, must produce absorption lines in the stellar spectrum.

For spectral lines having a width W due to stellar rotation, the transit makes in the line centered on λ_o a bump with a width $d\lambda = W(R_P/R_*)$ and a depth $dF/F = R_P/R_*$ at a position $\lambda = \lambda_o + Wx/R_*$, x being the planet distance to the star's axis ('spectro-transits'). This effect can be used for the confirmation of the transit, the determination of the transit latitude and puts constraints on the orientation of the star axis with respect to the planet orbital plane [9].

7/ During the transit, the star photocenter experiences a uniform displacement [8] with an amplitude $\Delta PC = R_P^3/R_*^2$ corresponding to 20μas for a Jupiter and a star with $R_* = 0.3R_\odot$ at 5 pc. The combination of $\Delta F/F$, of the duration and period of transits and of ΔPC gives an unambiguous determination of both R_P and R_*.

8/ Further in the future, it will be possible to have an image of a transit: if the stellar disc is resolved by interferometric imaging into $N \times N$ elements of size r, the transit will make in such an element a luminosity drop $(R_P/r)^2$

= 100% for a Jupiter and for $N = 10$. It will be possible to sound spectroscopically the planet atmosphere by absorption of the stellar light and to have a direct determination of the transit latitude and of the relative orientation star axis/planet orbit.

9/ Additional by-products:

If the planet passes in front of a bright (resp. dark) spot there is an extra dip (resp. bump) in the light curve whose detection reveals the corresponding surface feature of the star. In addition, for a typical stellar rotation period of 30 days and a planet orbital period of 4 days, the planet can cross the same spot 3 to 4 times (if the the planet orbit is perpendicular to star's rotation axis). The detailed investigation of this effect can 1) give informations on the stellar surface brightness irregularities 2) help to disentangle the planet transit from the stellar activity light curve and 3) impose constraints on the relative orientation star axis/planet orbit [9].

If the star is eclipsed by a white dwarf (WD) companion at a distance a, the Einstein radius $R_E \approx \sqrt{4GM_{WD}a/c^2}$ of the WD can be larger than its physical radius (but $< R_*$); there is then an amplification by microlensing with an amplitude $\approx (R_E/R_*)^2 \approx 10^{-4}(M_{WD}/M_\odot)(a/0.1AU)$ for $R_* = 1R_\odot$ [10].

In addition to these scientific aspects, the transit method has the observational advantage to allow the detection of planets around those binary stars (in general eclipsing binaries [11]) for which the radial velocity, astrometric and imaging methods are inappropriate. For eclipsing binaries, the transit is double with a very characteristic shape [11].

1.2 Background Noise, Ambiguities and Artefacts

1/ The main source of background noise is stellar activity. It can in principle be disentangled from a planet transit since, contrary to stellar activity (as confirmed by Virgo results) is achromatic. A multicolor photometric monitoring thus reduces this background noise (but without reducing it to zero) [1].

2/ One of the main ambiguities comes from binary stars (which represent 70% of the stellar population). From the lightcurve alone it is impossible to decide around which of the components in the binary the planet is circling. Here again, a multicolor photometry can help to remove the ambiguity when the two stars are different in color. Of course, a subsequent follow up by 'spectro-transits' makes the removal even easier.

3/ Several artefacts can mimic planet transits:

A faint background eclipsing binary with a magnitude difference Δm with the target star and superposed to it would mimic a transit by a planet having a radius $R_P \approx R_* 10^{-0.4\Delta m}$. If the background star has a spectral type different from the target, the artefact can be eliminated with the help of multicolor photometry.

Eclipses by brown dwarfs and white dwarfs mimic respectively transits by giant and Earth-sized planets. In both cases the ambiguity is removed by radial velocity measurements.

2 Detection Strategies: Ongoing Programmes and Future Projects

2.1 Systematic Search for Transits

The systematic search of transits is governed by the low probability of transits which must be compensated by large samples (several thousands) of stars. The choice of such samples is very different for ground-based and for space-based surveys: from the ground, one can explore the full sky while from space the high energy consumption of pointing forbids to explore more than a very few sky regions; the small aperture of telescopes ($< 10^o$) then obliges to select far away ($>$ 100 pc) and therefore faint ($m_V > 12$) stars in order to grasp thousands of them. From the ground on the contrary, the samples can contain several thousands of bright stars (e.g. brighter than $m_V = 12$).

From the ground, low photometric precision ($>$ 0.2%) surveys of samples of several thousand stars with small telescopes (diam. = 0.3 to 0.8 m) have already started [12,13] or are in preparation [14–16]. They can essentially detect Jupiters and large circumplanetary rings.

In 1999, a planet detected by radial velocity measurement [17,18] near the star HD 209458 turned out to transit across the stellar disc [19,17]. The orbital period is 3.52 days (corresponding to a distance to the star of 0.045 AU), allowing for frequent observations of the transit which has since been observed many times by different observers. The planet radius inferred from the magnitude of the brightness drop is, taking the weighted mean of the values given by Charbonneau *et al.* and by Henry *et al.*, $1.30 \pm 0.03\ R_J$, compatible with the models of hot Jupiters [20]. This discovery, allowing many of the follow-up studies described above, opens a new field in exoplanetology.

From space, high photometric precision ($\approx 10^{-4}$) surveys are in preparation: 1) the HST had a 9×24 h run in July 1999 for the monitoring of 20,000. [23]; this duration allows the detection of planets in orbits closer than 0.07 AU. 2) the COROT mission (to be launched in 2002) will monitor 12,000 stars brighter than m = 15 - 16 in five regions of the sky for 5 months each [21], allowing thus the detection of Earth-like planets in the habitable zone of M-type and K-type star and of planets around eclipsing binaries. Both programmes should detect several tens of planets. The Kepler project, submitted to NASA, aims to monitor 160,000 stars and could detect several hundreds of Earth-like planets [22].

2.2 Subsequent Follow-Up of Transits

1/ From the ground, low photometric precision can be used to perform the timing of transits and their spectroscopic follow-up.

2/ From space, high photometric precision is suited for the investigating of the shape of transits to search for satellites and rings, to measure the albedo of the planet when it is eclipsed by the star and for the detailed investigation of the stellar surface. The high astrometric precision of space missions is well adapted to measurement of the photometric center displacement.

The perspectives and programmes listed above make this field of research evolving rapidly; updates of ideas, projects and results are found in www.obspm.fr/planets.

References

1. Rosenblatt F. (1971) A two-color photometric method for detection of extrasolar planetary systems Icarus **14**, 71
2. Borucki W. and Summers A. L. (1984) The photometric method of detecting other planetary systems Icarus **58**, 121
3. Schneider J., Auvergne M. et al. (1998) THE COROT Mission: From Structure of Stars to Origin of Planetary Systems. in *Proc. NASA Origins Conference*, Colorado 19 - 23 May 1997. Woodward, Schull and Thronson Eds. ASP Conf. Series no. 148.
4. Schneider J. (1999) C.R. Acad. Sci. Paris The study of extrasolar planets: methods of detection, first discoveries and future perspectives. **327** Série II b,
5. Schneider J. (1999) On the detection and investigation of rings of extrasolar planets (submitted)
6. Sartoretti P., Schneider J. (1999) On the detection of satellites of extrasolar planets with the method of transits. Astron. and Astrophys. Suppl., **134**, 553
7. Schneider J., Rauer H. et al. (1998) The cometary-like tail of giant extrasolar planets in close orbits. in *Proceedings of "Brown Dwarfs and Extrasolar Planets"* (Rebolo et al. Eds). ASP Conference Series 134, p. 241
8. Mozurkewich D. (1999) (these proceedings)
9. Laurent E. and Schneider J. (1999) Spectrophotometric detection and study of exoplanets. Poster at the NATO ASI Summer School *Planets Outside the Solar System : Theory and Observations* (Cargèse, 1998) and paper in preparation.
10. Schneider J. (1990), Possible gravitational amplification in the binary pulsar 1957+20. Astron. and Astrophys. **214**, 1
11. Schneider J. and Chevreton M. (1990), The photometric search for earth-sized extrasolar planets by occultation in binary systems. Astron. and Astrophys., **232**, 251
12. Caldwell L., Borucki W., et al. (1999) Photometric Search for Exoplanets with the NASA Ames Vulcan Camera. Bull. Amer. Astron. Soc. **30** n0. 4, 97.04
13. Esquerdo G. et al. (1999) The ASP Project. See http://www.psi.edu/ esquerdo/asp/asp.html
14. Brown T. and Charbonneau D., (2000). The STARE Project; A Transit Search for Hot Jupiters. in 195th AAS Meeting

15. Dunham E. private communication
16. Hatzes A. private communication
17. Henry G., Marcy G., Butler P. and Vogt S., (1999). A Transiting 51 Peg-like Planet. ApJ. Letters, submitted
18. Mazeh T. *et al.*, (2000). ApJ. in preparation.
19. Charbonneau D., Brown T., Latham D. and Mayor M., (1999) Detection of Planetary Transits Across a Sun-like Star ApJ. Letters, accepted
20. Guillot T., Burrows C., Hubbard W., Lunine J. and Saumon D., (1996) Giant Planets at Small Orbital Distances. ApJ. Letters, 459, L35
21. Rouan D. et al. (1999) The detection of extrasolar planets by transits with the Corot mission. Physics and Chemistry of the Earth. (In press)
22. Borucki W., Koch D. et al. (1996) The Kepler Mission: A Photometric Mission to Determine the Frequency of Inner Planets Near the Habitability Zone of Solar-like Stars.
    in "Planets Beyond the Solar System and the Next Generation of Space Mission," STScI, Baltimore 16 - 18 October 1996
23. Gilliland (private communication)

2MASS Brown Dwarfs and a First Estimate of the Substellar Mass Function

James Liebert[1], I. Neill Reid[2], J. Davy Kirkpatrick[3], A. Burrows[1], J. E. Gizis[4], A. Burgasser[2], C.C. Dahn[5], D.G. Monet[5], R. Cutri[3], C. Beichman[3], and M. Skrutskie[4]

[1] Steward Observatory, Univ. of Arizona, Tucson AZ 85721
[2] Palomar Observatory, Caltech, Pasadena CA 91125
[3] IPAC, M/S 100-22, Caltech, Pasadena CA 91125
[4] Dept of Physics and Astronomy, Univ. of Massachusetts, Amherst MA 01003
[5] U. S. Naval Observatory. P.O. Box 1149, Flagstaff AZ 86002

Abstract. More than 70 L dwarfs have now been found in followup work on a small portion of the sky surveyed by the Two Micron All Sky Survey. It is estimated that the L dwarfs span a temperature range of 1400 to 2000K. Several have trigonometric parallax determinations, and luminosities between $10^{-3.5}$ and $10^{-4.3}$ $L_{\odot}$. These properties, together with the detection of lithium in one third of the objects, are consistent with the majority having substellar masses. Evolutionary models are utilized to derive estimates of the space density and mass function of substellar objects, predicting a majority that are not detectable with 2MASS. In a given volume the brown dwarfs may outnumber the stars, though contributing only modestly to the local disk mass.

1 Brown Dwarfs of the Local Galactic Disk

After searching in vain for nearly three decades after Kumar (1963) proposed their existence, literally dozens of likely substellar objects have now been discovered in several ways: (1) as resolved companions to nearby stars (eg. Gliese 229B); (2) as inferred radial velocity companions to nearby stars; (3) as sequences in young stellar clusters and star forming regions; and (4) as isolated field objects found in the infrared sky surveys, DENIS and 2MASS. With this last method, one lacks initially any kind of direct knowledge of the distance, luminosity and age. Hence, it is difficult to establish more than a crude mass for a field object, although the presence or absence of lithium in the spectrum is a valuable clue.

In comparison with surveys of young clusters, searching for field brown dwarfs must confront the difficulty of finding cooler and fainter objects, because they are generally older. However, surveys of large areas of sky should enable us to find the nearest such neighbors to the Sun, assuming they form as single entities in the field. At a given temperature or luminosity, these will be the brightest such objects, and the most amenable to detailed followup studies. The *older* brown dwarfs are indeed cooler, yet it turns out that the

1-1.5μ wavelength region – roughly the J and H bands – are relatively transparent intervals of the spectrum. Models of Burrows et al. (1997) show that brown dwarfs down to below 1,000 K emit a relatively large amount of flux there.

Fortunately, two groups have recently initiated the first substantial sky surveys at near-infrared wavelengths. The DEep Near Infrared Sky survey (DENIS) was started in January 1996 by a consortium of European investigators for the southern hemisphere, and their recent results are reported in the previous presentation. The Two Micron All Sky Survey (2MASS, Skrutskie et al. 1997) began data acquisition in May 1997 at Mt. Hopkins AZ for the northern sky, and at Cerro Tololo Interamerican Observatory in April 1998 for the southern sky. The nominal 10 sigma detection limiting magnitudes for 2MASS are $J<15.8$, $H<15.1$ and $K_s < 14.3$. Rapid progress has been made in both hemispheres, with more than 60% of the sky scanned (though not all of these will be accepted) at the time of this writing.

2 L Dwarfs and the Lithium Test

In the paper by Delfosse et al. (1997), DENIS announced the discovery of the first brown dwarf candidates from an infrared sky survey. Those objects show spectra indicating they are cooler than stars assigned the M9-M9.5 V subtype. The 2MASS project has analyzed a sample of 25 of these objects (Kirkpatrick et al. 1999, K99). As of this writing, more than 70 have been found in this followup work. Both groups have proposed that the objects later than M9.5 V should be called L dwarfs. The spectroscopic properties of the L dwarfs are discussed in the papers by Tinney et al. (1998) and K99. A complete classification system has been developed in the latter reference. We give here only a brief summary of the principal spectroscopic features.

In the latest M dwarfs, the dominant TiO absorption features of the red spectrum have begun to weaken. The slightly cooler T_{eff} where VO also begins to weaken is defined as the beginning of the L sequence (L0 V). By the latest L subtype, they disappear entirely. It is believed that the molecular constituents begin to precipitate out as dust at temperatures above 2000 K, as first proposed by Tsuji et al. (1996ab) – see also Burrows and Sharp (1998), Allard (1999).

With the removal of these principal sources of opacity, dwarfs a few hundred degrees cooler than late M stars have much more transparent atmospheres at optical wavelengths. Among the remaining absorbers are hydride bands with limited wavelength coverage. Otherwise, there is little continuum opacity, due to the extremely low electron supply (for H^- and H_2^-). Thus, the atomic resonance lines of the neutral alkalis become quite strong for an abundant species like K I, and easily detected for rare species like Rb I, Cs I, and if undepleted, Li I.

With these red spectra featuring pressure-broadened K I, the late L dwarfs may resemble cool white dwarfs more than they do M dwarfs. On the other hand, the L dwarfs show strong CO and H_2O like M dwarfs in the near infrared. The effective temperatures of the coolest L dwarfs are not known, but in any case they are not cool enough for the CO molecule to give way to CH_4, as it is predicted to occur near 1500 K (Burrows and Sharp 1998). Indeed, our infrared spectra of the latest objects (K99) show no evidence of CH_4 bands. The only known "methane dwarf" remains GL 229B. A series of arguments are given in K99 suggesting that the L dwarfs inhabit a temperature range like 1400-2000 K.

Martin et al. (1997) reported the detection of lithium in a DENIS L dwarf. Passing the "lithium test" (Rebolo et al. 1992; Magazzu, Martin & Rebolo 1993) proves unambiguously that DENIS-P J1228.2-1547 and many 2MASS objects like it with T_{eff} below about 2,000 K are substellar. Not all the L dwarfs observed to high signal-to-noise ratio at 6700Å show a lithium feature, but the detected fraction is very sensitive to the slope of the mass function, and the extent to which it penetrates the substellar range. Note that, since Li undergoes nuclear reactions at a lower central temperature than hydrogen, the highest mass brown dwarfs ($\sim$0.065-0.075$M_{\odot}$) are expected to deplete lithium. The establishment of a lithium luminosity boundary in the Pleiades (Stauffer, Schultz and Kirkpatrick 1998; Martin et al. 1999) – where ages and luminosities are known – lends credence to the model-dependent interpretations of the field objects that will be discussed here.

3 Towards a Substellar Mass Function

Reid et al. (1999, R99) use a sample of L dwarfs drawn from about 1% of sky to make first estimates of the luminosity and mass functions, and an inference as to the space density of substellar objects. For this purpose simulations of the observed 2MASS sample were produced, assuming a constant "star" formation rate over a 10 Gyr Galactic disk age, with a fixed mass function characterized by the Salpeter exponent α. Evolutionary models (Burrows et al. 1997) were employed to estimate luminosity and age as functions of mass, for solar composition. In addition to a T_{eff} vs. spectral type scale, we have to estimate the absolute magnitude (M_K) as a function of spectral type. Thanks to the efficiency of the U.S. Naval Observatory team, we have available preliminary but good trigonometric parallaxes for several L dwarfs. An HR Diagram of these L dwarfs, overlapping in luminosity with a few benchmark latest M dwarfs, is shown as Fig. 1. To derive total luminosities for comparison with the models, bolometric corrections for unprecedentedly cool objects had to be estimated (and are a principal uncertainty). Finally, a mass-luminosity relation had to be employed, though the ages of the objects are not known. For lack of space, the reader is referred to R99 for the important

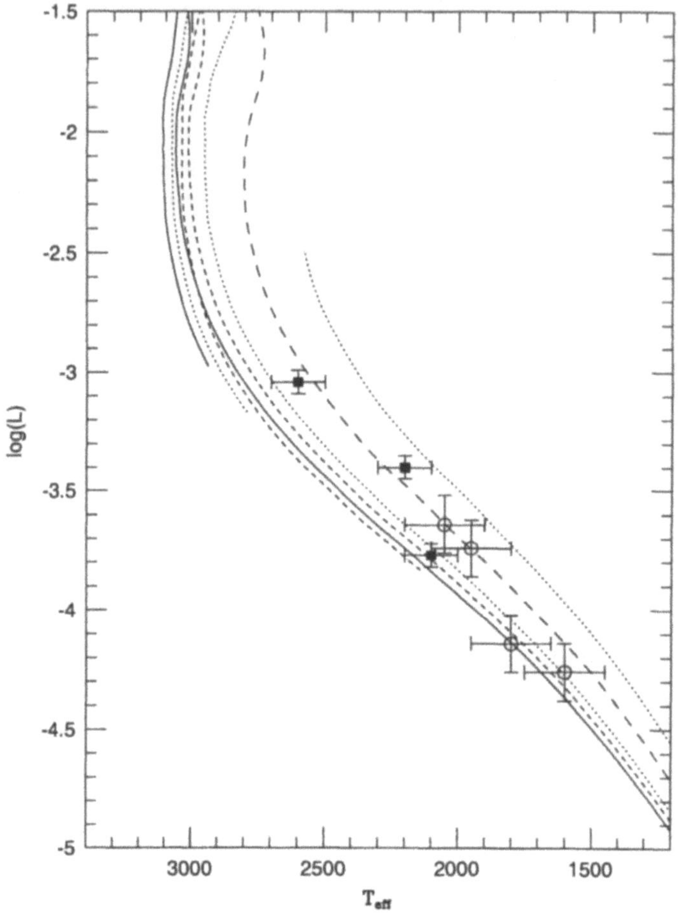

Fig. 1. The HR diagram from VLM/brown dwarfs: we plot Burrows (1993, 1997) models for masses 0.015, 0.03, 0.05, 0.06, 0.07, 0.08, 0.09 and $0.10 M_\odot$, in order of decreasing luminosities. The solid points indicate the positions of VLM dwarfs GL406, vB10 and LHS 2924. The open circles mark the four L dwarfs.

details of these assumptions. The reader will appreciate that this exercise is merely a first attempt!

The 1/3 or so of the L dwarfs with lithium detections must be interpreted as brown dwarfs below about $0.065 M_\odot$. The Burrows models also indicate that they are quite young – of the order of 10^8 to possibly 10^9 years. If one can assume a uniform "star" formation rate, the inference is that, for every lithium dwarf, there are ten times as many much cooler objects of similarly low mass.

This implies that we are detecting only the luminous, young, high mass tip of a brown dwarf "iceberg." The best fitting simulations suggest a mass function steeper than M^{-1} and predict a space density of substellar objects

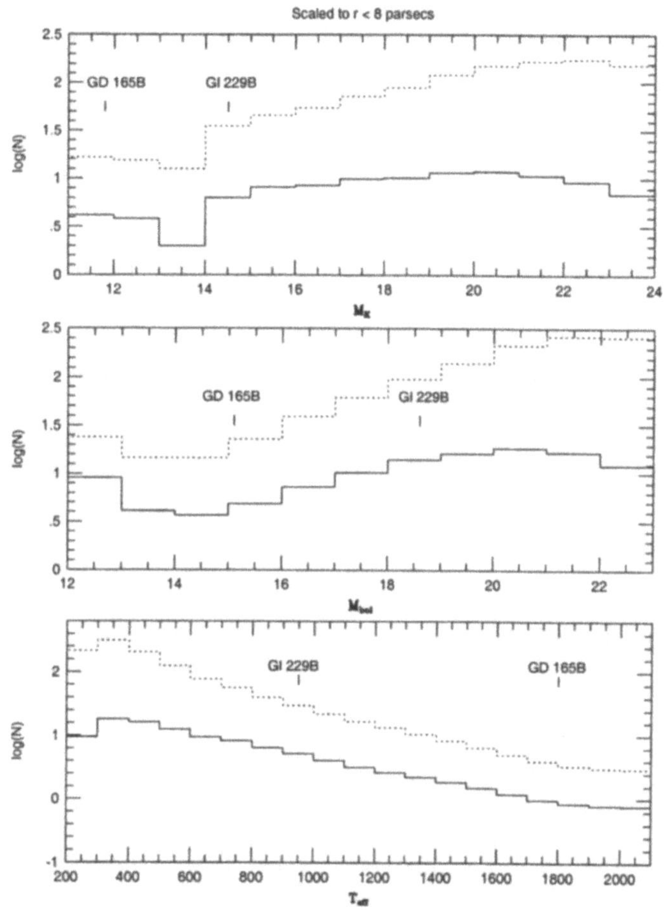

Fig. 2. The 2.2μm and bolometric luminosity functions and the effective temperature distribution predicted by our models with constant formation rate and power laws $M^{-\alpha}$, $\alpha = 1$ (solid line) and 2 (dotted histogram). All results are scaled to a volume with radius 8-parsec around the Sun.

of nearly 0.1 per cubic parsec. That is, 50 brown dwarfs could reside within 5 pc of the Sun, more than the number of stars! Such a population would add only about 15% to the mass of the local disk, but indeed this matter would be pretty dark. The predicted luminosity function is shown in Fig. 2. The bottom panel depicts a majority of objects much cooler than the 950 K temperature of the only known methane dwarf, GL 229B. Moreover, the fact that the absolute M_K peak is near 20th means that the 2MASS survey, complete only to an average $K_s = 14.5$ or so, could detect such objects only out to about one parsec!

At the time of writing, the 2MASS search has uncovered no methane dwarfs with *blue* *J-K* colors like GL 229B, nor have we found any between the temperature of an L8 V dwarf and GL 229B which might just be forming methane, and have neutral J-K colors overlapping the vast majority of 2MASS point sources. However, this null result is not inconsistent with the predictions of the simulations, and underscore the lack of sensitivity to objects cooler than the L dwarfs. *If* these predictions are correct, a much deeper JHK survey than 2MASS or DENIS may be necessary to find them in significant numbers. Another interesting possibility would be a thermal IR sky survey probably conducted above the atmosphere, especially covering 4-6μ where the energy distributions of objects with $T_{eff} < 1{,}000$ K are expected to peak.

We conclude by reemphasizing all the assumptions of the simulations listed earlier, and discussed at length in R99, on which these precarious predictions depend. Perhaps the number of brown dwarfs envisioned here will not turn out too large. Still, the search for them should be an exciting, challenging task, since we know that entities in the mass range between Jupiter and a low mass star are an important constituent of the Galactic disk.

References

1. Allard, F. 1999, in *Very Low Mass Stars and Brown Dwarfs in Clusters and Associations*, La Palma, Spain, 11-15 May 1998, eds. R. Rebolo and M.R. Zapatero-Osorio, Cambridge Univ. Press (U.K.), in press.
2. Burrows, A. & Sharp, C.M. 1998, *Ap.J.*, in press.
3. Burrows, A., Marley, M., Hubbard, W.B., Lunine, J.I., Guillot, T., Saumon, D., Freedman, R., Sudarsky, D., & Sharp, C. 1997, ApJ, 491, 856.
4. Delfosse, X., Tinney, C.G., Forveille, T., Epchtein, N., Bertin, E., Borsenberger, J., Copet, E., de Batz, R., Fouque, P., Kimeswenger, S., Le Bertre, T., Lacombe, F., Rouan, D., & Tiphene, D. 1997, Astr. Ap., 327, L25.
5. Kirkpatrick, J.D., Reid, I.N., Liebert, J., Cutri, R.M., Nelson, B., Beichman, C.A., Dahn, C.C., Monet, D.G., Skrutskie, M.F., & Gizis, J. 1999, ApJ July 10 (K99)
6. Kumar, S. 1963, Ap.J., 137, 1121.
7. Magazzu, A., Martin, E.L., & Rebolo, R. 1993, Ap.J.Let., 404, L17.
8. Martin, E.L., Basri, G., Delfosse, X., & Forveille, T. 1997, Astr. Ap., 327, L29
9. Martin, E.L., Basri, G., Zapatero Osorio, M.R., Rebolo, R., & Garcia Lopez, R.J. 1999, in *Very Low Mass Stars and Brown Dwarfs in Clusters and Associations*, La Palma, Spain, 11-15 May 1998 eds. R. Rebolo and M.R. Zapatero-Osorio, Cambridge Univ. Press (U.K.), in press.
10. Rebolo, R., Martin, E.L., & Magazzu, A. 1992, Ap.J.Let., 389, L83.
11. Reid, I.N., Kirkpatrick, J.D., Liebert, J., Burrows, A., Gizis, J., Burgasser, A., Dahn, C.C., Monet, D.G., Cutri, R.M., Beichman, C.A., & Skrutskie, M.F., & 1999, Ap.J., 521, in press (R99)
12. Skrutskie, M.F., et al. 1997, in *The Impact of Large Scale Near-IR Sky Surveys*, eds. F. Garzon et al., Kluwer (Netherlands), pp 25-32.
13. Stauffer, J.R., Schultz, G.000, & Kirkpatrick, J.D. 1998, Ap.J.Let., 499, L199.
14. Tinney, C.G., Delfosse, X., Forveille, T. & Allard, F. 1998, A&A, 338, 1066
15. Tsuji, T., Ohnaka, K., & Aoki, W. 1996, Astr. Ap., 305, L1.
16. Tsuji, T., Ohnaka, K., Aoki, W., & Nakajima, T. 1996, Astr. Ap., 308, L29.

Brown Dwarfs in Open Clusters

María Rosa Zapatero Osorio

Instituto de Astrofísica de Canarias, E-38200 La Laguna, Tenerife, Spain

Abstract. Over the last two decades, many efforts have been devoted to the search for brown dwarfs, and various ways of finding them have been developed. Deep photometric surveys in nearby open clusters is one approach that has provided a large amount of candidates. These clusters offer a number of advantages (a single known distance, metallicity and age) that make feasible the identification of such once elusive objects. Moreover, deep searches constitute one of the most direct means of measuring the mass function through the whole stellar and substellar mass range. In this paper the progress is presented of recent work on several young open clusters which leads to the finding of unambiguous brown dwarfs beyond the substellar mass limit. These discoveries, particularly in the Pleiades, imply a mass function in the substellar domain down to 0.05 $M_\odot$ which can be fitted by a power law with $\alpha = -0.8 \pm 0.2$ ($\mathrm{d}N/\mathrm{d}M \sim M^{\alpha}$). The detection of reliable candidate members with estimated masses of only 0.02–0.015 $M_\odot$ in younger clusters, such as σ Orionis, provides substantial evidence for the initial mass function extending down to the deuterium-burning mass limit.

1 Introduction

Since the theoretical prediction of the existence of brown dwarfs in the early '60s [10], many efforts have been devoted to their search, both in the field and in open clusters. Although various candidates could be found in the literature, none was convincingly assigned a substellar status. It was in the mid-nineties when the first bona fide brown dwarfs were unambiguously identified orbiting around a low-mass field star [18], and free-floating in the Pleiades [23], [22], [2]. In only four years a flurry of brown dwarf discoveries (see [24] for a summary) has proven that these once elusive objects do exist and that their population in the Galaxy is numerous. These findings have led to an increasing scientific interest in the comprehension of the atmospheres of brown dwarfs, their activity, internal structure and evolution. However, one of the most attractive topics is indeed the substellar mass function (MF). Nearby, young open clusters offer a unique scenario for its study. Clusters present a single population of known distance, age and metallicity. Their youth and proximity ensure that brown dwarfs are relatively luminous and thus, easy to detect, even those with low masses. Additionally, the MF in young clusters must resemble the *initial* MF, as no strong dynamical evolution is expected to have taken place within the cluster except perhaps for the least massive members. Another significant advantage of this kind of analysis being performed in the field is that there is no Malmquist bias in a cluster sample.

This paper provides a summary of current knowledge of the Pleiades brown dwarf regime. The efforts to characterize the substellar limit and the brown dwarf population of this cluster will be described briefly.

2 The Pleiades Cluster

The Pleiades, and occasionally a few more young clusters like α Persei, Praesepe, and Hyades, have so far been the most preferred targets (see [7] for an overview) for the hunting of brown dwarfs because of their age, proximity, richness in numbers, and the fact that their internal extinction is almost negligible. This, in addition to the availability of extensive combined proper motion, photometric and spectroscopic studies of the low-mass stellar population of these clusters, makes them ideal grounds for investigating the MF through the whole stellar and substellar mass range. The largest, deepest surveys aimed at finding brown dwarfs have been carried out in the Pleiades, and subsequently this is the only cluster for which a relatively numerous brown dwarf population has been identified. Photometric candidates are selected by fitting the sequence defined by other cluster members of higher mass, and the comparison with theoretical evolutionary tracks can provide a mass estimate. However, given the uncertainties in the derivation of temperatures and luminosities from observables and the differences among various sets of models, this is not enough to asses the substellarity of candidates whose location in the HR diagram is close to the faint end of the stellar cluster sequence.

2.1 Lithium in Pleiades Brown Dwarfs: the Substellar Limit

The detection of lithium in the photospheres of cool dwarfs ($T_{\rm eff} \leq 2800$ K) is indicative of a brown dwarf nature [21], [1]. The lithium test was applied unsuccessfully to a number of substellar candidates in the field and belonging to young clusters (α Persei and the Pleiades) [16], [13], with the result that these candidates are stars rather than brown dwarfs. The discovery of this element in the Pleiads PPl 15, Teide 1 and Calar 3 [2], [22] showed that the frontier between stars and brown dwarfs in this cluster had been finally crossed (at the age of the Pleiades the substellar mass limit, $0.075\,M_{\odot}$, and the mass at which lithium is preserved do coincide; see [1] for further details). More recent work [20], [14], [25] has proven that there is a drastic transition from the complete destruction of lithium in low-mass and very low mass stars to its preservation in brown dwarfs (Fig. 1, left panel), and that this transition takes place in a narrow range of luminosity and effective temperature. These observations allow the location of the substellar limit of the Pleiades to be determined with high precision [14], [25] at $I = 17.80$, $R - I = 2.20$, $I - J = 2.25$, $I - K = 3.35$, and M6.5–M7 spectral type. Figure 1 (right panel) displays the colour-magnitude diagram showing those Pleiads (with K photometry available) for which a good assessment of their lithium abundance can be made. In the future, objects with fainter photometric magnitudes and with radial velocity and proper motion that guarantee membership in the Pleiades should be considered as bona fide brown dwarfs.

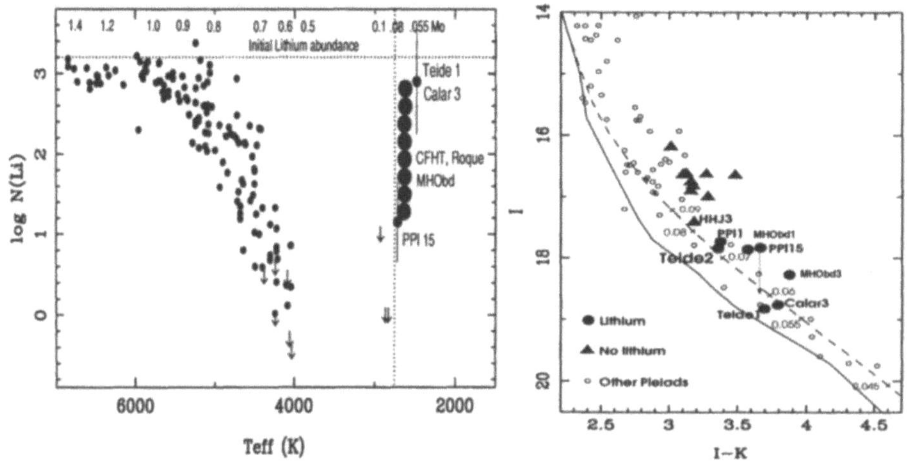

Fig. 1. *(Left panel)* Lithium abundances versus effective temperatures for Pleiades members. Abundances are given in the usual scale $\log N(H) = 12$ for stars and the brown dwarfs PPl 15, Teide 1 and Calar 3 as taken from the literature (see [22] and references therein). With large-sized filled circles in the substellar region we aim only to indicate the number of lithium detections so far available in Pleiades brown dwarfs [14], [25]. Given the luminosity, temperature and lithium equivalent width of the Roque, CFHT and MHObd objects we expect them to show lithium abundances between those of PPl 15 and Teide 1. *(Right panel)* Color-magnitude diagram showing the location of the substellar limit in the Pleiades cluster. Superimposed are the Next Gen 120 Myr-isochrone [5] with masses labelled in solar units (dashed line) and the field main sequence (full line) shifted to the cluster distance and reddening.

Adopting an evolutionary model, it is possible to evolve the location of the substellar mass limit from the age of the Pleiades (120 Myr) to other ages of interest, for example, younger and older clusters and the field, and infer expected magnitudes and spectral types at which the stellar/brown dwarf borderline is expected. This exercise provides $I \sim 17.7$, 18.2, ~M6.5, M9 types for the α Persei and Hyades, respectively, and $M_I \sim 15.8$ and types well within the new L-class for ages older than 900 Myr. This can be useful for studying the nature of the recent findings of L-type dwarfs in the field [6], [9].

2.2 Photometric and Spectroscopic Sequences

Several photometric searches [27], [4], [26], [8] have gone far beyond the substellar limit in the Pleiades and have covered significant areas within the cluster. As a result of these surveys, and with the membership assessment given by follow-up observations of the candidates, the optical and IR photometric Pleiades sequence is defined down to masses of $0.035\,M_{\odot}$. Figure 2 (left panel) summarizes our current knowledge of the I vs. $I-K$ colour-magnitude diagram for the cluster. The Next Gen isochrone of 120 Myr [5] (which does not incorporate the treatment of dust) is overplotted on the data. This theoretical model fits the observations nicely until $\sim 0.045\,M_{\odot}$ ($I-K=4.4$, $\sim$2200 K), where dust condensation starts to play an important role in the atmospheres of these cool objects. It is worth noting that less massive Pleiads appear red in optical/IR colours (e.g. $I-K, I-J$), but blue in optical colours (e.g. $R-I$), very probably due to an effect of dust. This implies that future searches for low-mass cluster brown dwarfs should not be based on optical colours.

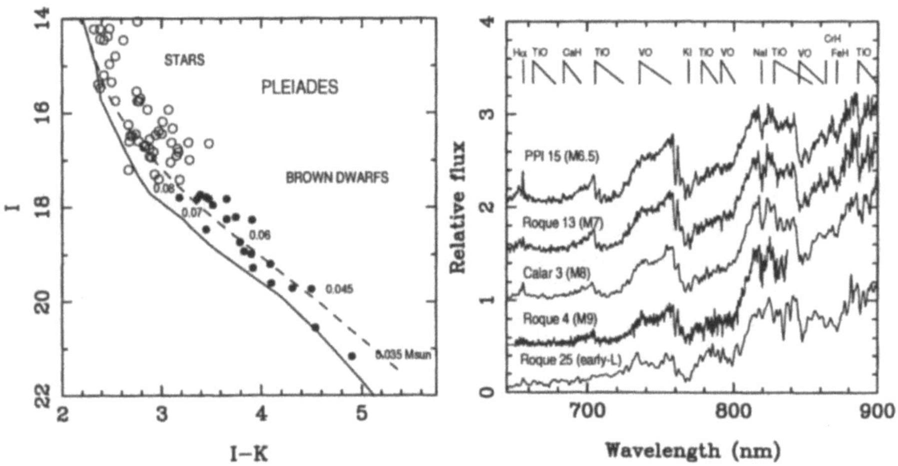

Fig. 2. *(Left panel)* Color-magnitude diagram showing the photometric sequence of the very low mass stars and brown dwarfs in the Pleiades. Superimposed are the Next Gen 120 Myr-isochrone [5] with masses labelled in solar units (dashed line) and the field main sequence (full line) shifted to the cluster distance and reddening. *(Right panel)* Some Pleiades brown dwarf low-resolution spectra ordered by decreasing temperature. Roque 25 ($\sim 0.035\,M_{\odot}$) is so far the least massive Pleiad found. The most relevant molecular and atomic features are indicated.

Low-resolution spectroscopy is available for all the brown dwarfs displayed in the previous figure, and a sample of these spectra is shown in Fig. 2 (right panel) [17], [15]. Pleiades brown dwarfs in the mass range 0.075–$0.035\,M_{\odot}$ cover spectral types from M6.5 to early L-class. So far Roque 25 is the only example of an L-type brown dwarf in the cluster. It is observed that VO bands strengthen from mid-M to M9, while the TiO bands become weaker. For cooler temperatures, VO molecular absorption also weakens. Remnants of TiO in Roque 25 are barely visible, but VO is still detectable. Compared to the field, it seems that precipitation of TiO and VO (first TiO and then VO) into grains in the atmospheres of Pleiads takes place at later spectral types, which could be a hint of lower gravities. The Pleiades spectroscopic sequence will allow an accurate temperature classification to be derived, as it is made up of objects of the same age and metallicity.

2.3 The Mass Function

The two most relevant questions regarding the MF are *(i)* its form across the stellar/substellar boundary and through the brown dwarf domain, and *(ii)* whether there is a minimum mass for cloud fragmentation which should lead to a turnover of the function. So far the best constraints on our knowledge of the shape of the MF are provided by the Pleiades. Figure 3 presents the MF for the central $1\,\mathrm{deg}^2$ of this cluster as derived from only confirmed members of the survey of Zapatero Osorio et al. [28]. Both the linear and log-normal representations are shown. The theoretical 120 Myr Next Gen mass-luminosity relationship has been used. The low-mass stellar domain, the stellar/substellar boundary, and the brown dwarf regime down to $0.05\,M_\odot$ can be approximated by a power law MF with $\alpha=-0.8\pm0.2$ ($\mathrm{d}N/\mathrm{d}M \sim M^\alpha$). This is in fairly good agreement with what is found in other outer regions of the cluster [4], [8], [19]. There are some hints that the MF decreases drastically at lower masses, but much deeper surveys are needed in order to address this point. It is interesting to note that the number of brown dwarfs in the Pleiades is comparable to that of M-type stars; however, the decreased power-law index implies that the substellar members would not make a significant contribution (a few per cent) to the mass of the cluster.

Recent discoveries of less massive brown dwarfs in younger clusters (σ Orionis, ρ Oph, IC 348, [3], [11], [12]) shed new light on the substellar MF and thus, on the theories of substellar formation. Objects like SOri 47 in σ Orionis (see Zapatero Osorio et al., these proceedings) with a mass estimated at around $0.015\,M_\odot$, show evidence for the initial MF extending down to the deuterium-burning mass limit. Brown dwarfs of these masses in the Pleiades have low luminosities and are therefore very faint. Their detection will be a challenge to astronomers. In order to carry out an exhaustive study of the initial MF down to very low mass brown dwarfs and planets, nearby compact clusters rich in members and with ages $\leq$20 Myr are ideal targets.

3 Conclusions

Solar-metallicity, free-floating brown dwarfs have been found as members of young open clusters (Pleiades, σ Orionis, IC 348, ρ Oph), and with masses ranging from the substellar boundary ($0.075\,M_\odot$) down to the deuterium burning limit ($\sim0.015\,M_\odot$). The Pleiades substellar population in the interval *0.075–0.035*$\,M_\odot$ *is so far formed by* ~30 confirmed members (about the same

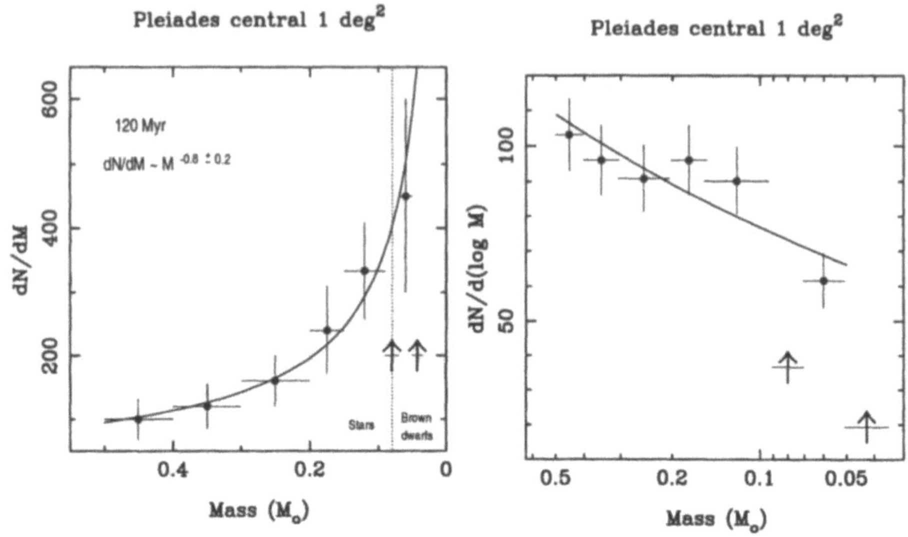

Fig. 3. The mass function (linear representation – *left panel*; log-normal representation – *right panel*) for the central area of the Pleiades. The power law fit (full line) is overplotted on the data. Error bars in the number of objects are poissonian count errors, and the horizontal bars only indicate the mass bins for which each circle symbol is representative.

number still remains as candidates), and only a few per cent of the total area of the cluster has been investigated down to such small masses. This sample allows the Pleiades substellar limit to be determined with high accuracy and also makes possible the characterization of the transition between the late M-types and the *new* L-classes.

Young open clusters provide the most direct means of measuring the mass function. The Pleiades substellar mass function down to $0.05\,M_\odot$ can be fitted by a power law with $\alpha = -0.8 \pm 0.2$ ($\mathrm{d}N/\mathrm{d}M \sim M^\alpha$). There are some hints for the mass function decreasing drastically at lower masses. The detection of reliable candidate members with estimated masses of only 0.02–$0.015\,M_\odot$ in younger clusters like σ Orionis, does provide substantial evidence for the initial mass function extending down to the deuterium burning mass limit.

References

1. Basri, G., 1998, ASP Vol. 134, eds. Rebolo, Martín, Zapatero Osorio, p. 394
2. Basri, G., Marcy, G.W., Graham, J.R., 1996, ApJ, 458, 600
3. Béjar, V.J.S., Zapatero Osorio, M.R., Rebolo, R., 1999, ApJ, in press
4. Bouvier, J., et al., 1998, A&A, 336, 490
5. Chabrier, G., Baraffe, I., Plez, B., 1996, ApJ, 459, L91
6. Delfosse et al., 1997, A&A, 327, L25
7. Hambly, N.C., 1998, ASP Vol. 134, eds. Rebolo, Martín, Zapatero Osorio, p. 11

8. Hambly, N.C., et al., 1999, MNRAS, 303, 835
9. Kirkpatrick, J.D., et al., 1999, ApJ, submitted
10. Kumar, S., 1963, ApJ, 137, 1121
11. Luhman, K.L., Liebert, J., Rieke, G.H., 1997, ApJ, 489, L165
12. Luhman, K.L., et al., 1998, ApJ, 508, 347
13. Marcy, G.W., Basri, G., Graham, J.R., 1994, ApJ, 428, L57
14. Martín, E.L., et al., 1998, ApJ, 499, L61
15. Martín, E.L., et al., 1998, ApJ, 507, L41
16. Martín, E.L., Rebolo, R., Magazzù, A., 1994, ApJ, 436, 262
17. Martín, E.L., Rebolo, R., Zapatero Osorio, M.R., 1996, ApJ, 469, 706
18. Nakajima, T., et al., 1995, Nature, 378, 463
19. Pinfield, D.J., Jameson, R.F., Hodgkin, S.T., 1998, MNRAS, 299, 955
20. Oppenheimer, B., Basri, G., Nakajima, T., Kulkarni, S., 1997, AJ, 113, 296
21. Rebolo, R., Martín, E.L., Magazzù, A., 1992, ApJ, 389, L83
22. Rebolo, R., et al., 1996, ApJ, 469, L53
23. Rebolo, R., Zapatero Osorio, M.R., & Martín, E.L. 1995, Nature, 377, 129
24. Rebolo, R., Zapatero Osorio, M.R., 1999, in "Very Low Mass Stars and Brown Dwarfs in Stellar Clusters and Associations", Cambridge University Press, in press
25. Stauffer, J.R., Schultz, G., and Kirkpatrick, J. D., 1998, ApJ, 499, L199
26. Stauffer, J.R., et al., 1998, ApJ, 504, 805
27. Zapatero Osorio, M.R., et al., 1997, ApJ, 491, L81
28. Zapatero Osorio, M.R., et al., 1999, A&AS, 134, 537

Adaptive Optics Imaging of Faint Companions: Test Case of MWC480

Laird Close[1], Dan Potter[2], and Alan Tokunaga[2]

[1] ESO, Garching, Germany *lclose@eso.org*
[2] University of Hawaii, Honolulu, USA

Abstract. We briefly describe how diffraction-limited imaging with adaptive optics (AO) can detect young massive brown dwarfs by direct imaging. We will review how current curvature AO systems can already detect point sources 1 million times fainter at separations of 3 arcsec in median seeing (0.65 arcsec). We will show real examples of such detections made with the University of Hawaii AO system located at CFHT on Mauna Kea around the young (2 Myr) nearby (132 pc) Herbig Ae/Be star MWC480. Moreover, we show that the 4 faint (H=18-19) companions within 6 arcsec of MWC480 are unlikely to be physical since they are non-common proper motion objects, and so are likely just background stars.

1 AO Sensitivity to Massive Planets

Current curvature AO systems (like the University of Hawaii AO system Hokupa'a: Graves et al. 1998) can already detect point sources 1 million times fainter at separations of 3″ in median seeing (0.65″). Real examples of such detections made with the University of Hawaii AO system located at CFHT on Mauna Kea around the young (~2 Myr) nearby (132 pc) Herbig Ae/Be star MWC480 can be found in Close et al. (1998).

Based on these sensitivities and the Brown Dwarf atmospheric models of Burrows et al. (1997) detection limits with such an AO system can be predicted. For example, current AO systems could detect an older 5 Gyr object in the mass range of 5-80 Jupiter masses with separations of 8-60 AU (assuming a M4 primary at a system distance of 3 pc). Another example would be that of younger objects (~1 Gyr) of 10-80 Jupiter masses could be detected around a M1 star at 10 pc from 24-180 AU (in a 6 min. exposure at a 4m telescope with AO at J band).

The potential for AO systems to detect these wide systems (with periods > 50 years) is the perfect compliment to the radial velocity or astrometric searches that require shorter periods for detection.

2 Procedure to Detect Faint Companions with AO

1) Take both short unsaturated, and then deep saturated AO images of the target primary star. Make sure to dither (or jitter) the star around the science camera FOV.

2) Replace the saturated core of the image with the scaled unsaturated data. Median filter all the data to remove scattered light ghosts and flat field errors, as well as removing bad pixels and poorly corrected frames.

3) Use Fourier filtering to remove the large 1–2″ uncompensated AO halo.

4) search for diffraction-limited point sources in the final image.

5) One should then be able to detect objects of several (>4 M_J) Jupiter masses at 1″ - 10″ from the central star in the near IR. Or in other words objects that are ~1 million times fainter at ~ 3″ from the central star. A detailed account of this performance (AO sensitivity VS separation) is given in Close et al. (1998).

6) Follow-up observations should be taken one year later to determine if these faint objects are both real and physically associated with the target star. *Only through confirming that the faint objects are common proper motion objects that one can have some confidence that these are true physical companions.*

3 Case Study MWC480: Planetary System or Just Background Stars?

In 1997 we discovered MWC480 to have 4 faint "companions" located at distances of 4–6″ with magnitudes of 17.67, 18.80, 19.24, & 19.37 at H (since MWC480 has H~7.0 these are all extremely faint companions). Since a Gl229B has H=19.98 (at MWC480s distance of 132 pc) these would likely be true planetary companions -*if they are physically associated with MWC480.* If they are physical then these faint objects should share MWC480's proper motion (of 6.25 mas/yr East and 28.3 mas/yr South) on the sky.

Follow up AO observations in 1998 and 1999 with the CFHT PUEO curvature AO system suggested that in fact that these stars appear to have systematically less proper motion based on 1.5 years of observation. Indeed the brightest of these 4 stars (located 6″ East of MWC480) has only a 10±6 mas/yr motion East and 5±7 mas/yr North, which is inconsistent with that of MWC480's motion. A fainter closer star (located 4″ to the south-east) has a motion of ~15±10mas/yr East and 5±10 mas/yr North which is also inconsistent with MWC480.

The last 2 stars (of the 4 companions) had large errors in their centroids and so their proper motions were difficult to measure, but the knowledge that 2 out of 4 of these objects were background in nature strongly suggests that all 4 of these objects are background stars (which is possible since MWC480 is near the galactic plane). Hence, we can likely reject the hypothesis that these are physically associated with MWC480 and are in fact most likely background M type stars directly behind MWC480. However, future observations will be required to unambiguously eliminate these other 2 stars as physically related.

4 Conclusions

The case of MWC480 shows how important it is to use proper motion to confirm physical association with faint companions. However it is clear that AO is a powerful direct imaging technique to detect massive planets and Brown Dwarfs. In particular, the potential for AO systems to detect wide systems (with periods > 50 years) is the perfect complement to the radial velocity or astrometric searches that require shorter periods for detection.

References

1. Burrows, A., et al. 1997, ApJ, 491, 856
2. "First light for Hokupa'a: the 36 element Curvature AO system at UH", Graves, J.E., Northcott, M., Roddier, F., Roddier, C., & Close, L.M 1998, Proc SPIE 3353 *download from -http://www.ifa.hawaii.edu/ao/publi.html*
3. "Adaptive Optics Imaging of Circumstellar Disks and low Mass Companions around Young Stars", Close, L.M., Roddier, F., Potter, D., Roddier C., Graves, J.E., & Northcott, M. 1998 ESO/OSA Astronomical Adaptive Optics Sept. 6-11th Sonthofen, Germany, Ed. D. Bonaccini page 109-120 *download from-http://www.ifa.hawaii.edu/ao/publi.html*

Observations of Disks and Faint Companions with NAOS

Anne M. Lagrange[1] and David Mouillet[1]

Laboratoire d'Astrophysique de Grenoble
Observatoire de Grenoble, BP 53,
F-38041, Grenoble Cedex 9

Abstract. NAOS on the VLT will provide diffraction limited images and high Strehl ratio over a wide wavelength domain. Together with CONICA, a 1-5 micron spectro-imager equipped with coronographic masks and also, possibly, other instruments, they are expected to allow the detection and detailed studies of circumstellar disks and faint companions around stars.

In this paper, we present the instruments, their expected performances, and some associated simulations of imaging of disks and faint companions.

1 Introduction

One of the today instrumental challenges is to image planetary systems, disks or planets around neighbour stars. A few disks of planetary origins have been found and studied in the recent years (see [1] this volume), but there is still a strong need to observe many other systems, of various ages, to allow an analysis in terms of planetary system evolution. Concerning extrasolar planets, there are only indirect detections so far (see [5])

In both cases, the challenge is to detect very faint structures very close to bright objects. This translates into two instrumental constrains: a high dynamical range(typ. 10^9 for "old" giant planets at optical wavelengths, much smaller in the case of still forming planets), and very high angular resolution ($\leq$ 0.1"). High angular resolution is achievable with large telescopes either equipped with Adaptive Optics (hereafter AO) systems or in space (0.05" for a 10m telescope at K band) or with interferometers (a few mas). High dynamical range requires the use of anti-blooming detectors or coronographic technics; the latter technic being so far much more developed. An additional improvement in getting high dynamics in coronographic observations is provided by dark speckle technics (see [7]).

The Nasmyth Adaptive Optics System (NAOS) on the VLT will provide diffraction limited images (FWHM $\simeq$ 60mas) and high Strehl ratio (hereafter Sr) over a wide wavelength domain.

We present here briefly NAOS and the associated focal instrumentation, their expected performances, and give simulations of observations of disks and faint companions around stars.

2 NAOS and Focal Instruments

NAOS basic design will feed the spectro-imager CONICA and/or a parallel instrument, and provide high Sr ($\geq 70\%$) on bright ($V \leq 13$) reference sources as well as diffraction limited images with Sr $\geq 10\%$ on faint sources up to V=16. It will also offer the possibility to implement later-on a Laser Guide Star. Detailed technical informations may be found in [8], [2], [3].

CONICA is a spectro-imager working between 1 and 5 μm ([4]). The detector will be an SBRC, with a well capacity of 200000 e- and a read-out-noise of about 25 e-. Different cameras will be available to ensure proper sampling of the images in this wavelength range. CONICA will provide several observing modes: broad and narrow band imaging, spectroscopy at various resolutions: 50–500 plus 1800 (Fabry-Pérot mode at K band), polarimetry and coronography with various sets of masks adapted to the J, H, and K wavelengths. The latter mode is obviously very important for the present topic.

The overall NAOS+CONICA sensitivity is given in Fig. 1. As concerns the planet detection, we stress on the fact that these detection limits cannot be directly used to determine whether a faint object could be detected or not, as the detection limit in such a case is limited by the dynamical range and not by the sensitivity.

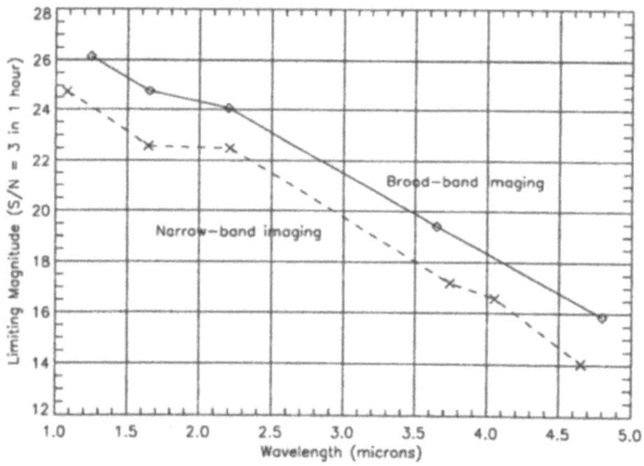

Fig. 1. NAOS+CONICA limiting magnitudes for a diffraction-limited source

NAOS and CONICA are currently being built respectively in France and in Germany. Both will be available to the community on UT1 in early 2001.

3 Expected Performances and Consequences

The high performance expected in terms of Sr (Fig. 2) achievable on bright objects directly translates into high performance in terms of detection limit.

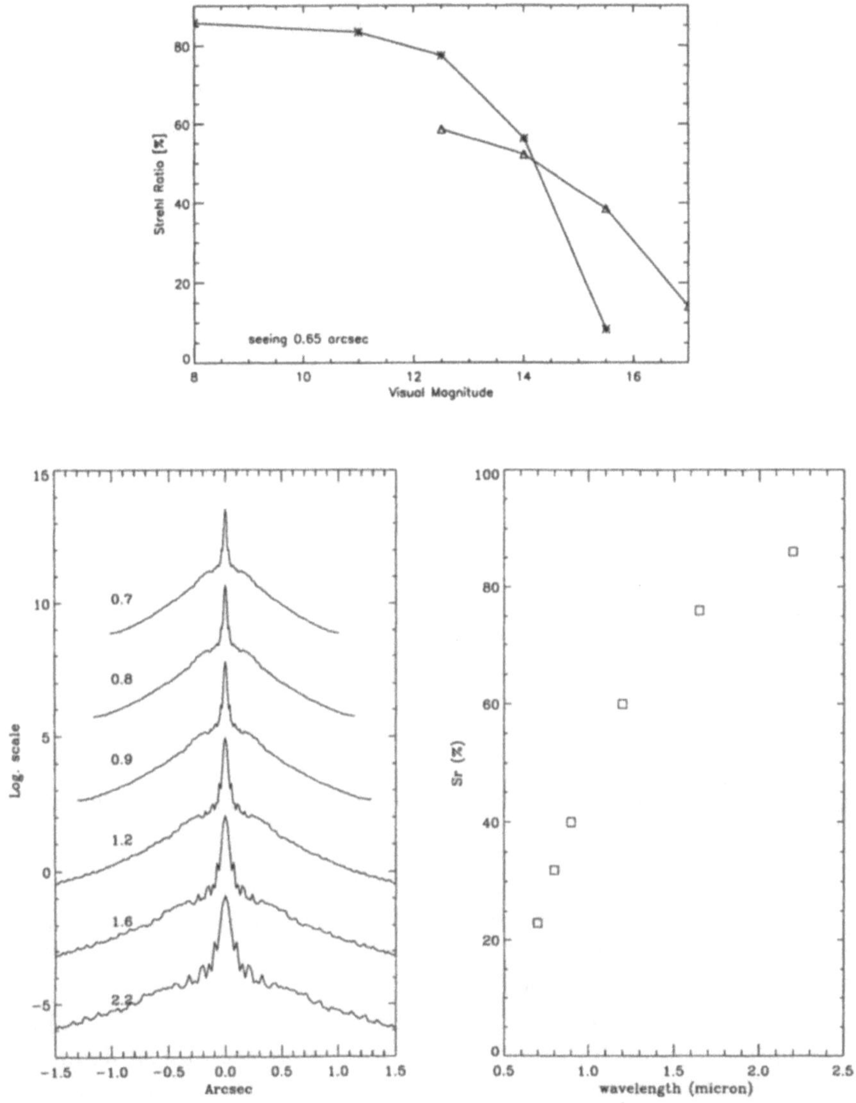

Fig. 2. Top: Expected Sr vs V magnitude at K band, with the NAOS Visible Wavefront Sensor, and with a 0.65" seeing; stars and triangles represent values obtained with different subapertures. Bottom: simulated PSFs for different wavelengths (left) and Sr as a function of wavelength (right) for a 11th magnitude reference object and a 0.85" seeing.

As an illustration, Fig. 3 gives examples of two simulated Point Spread Functions (herefater PSF) with Sr of 70% and 56%. In the region 0.25–0.5" from the center of the PSF, the halo is a factor of 2 higher with the 56 % Sr PSF; this translates into a loss of 0.5–1 magnitude in terms of detection limit.

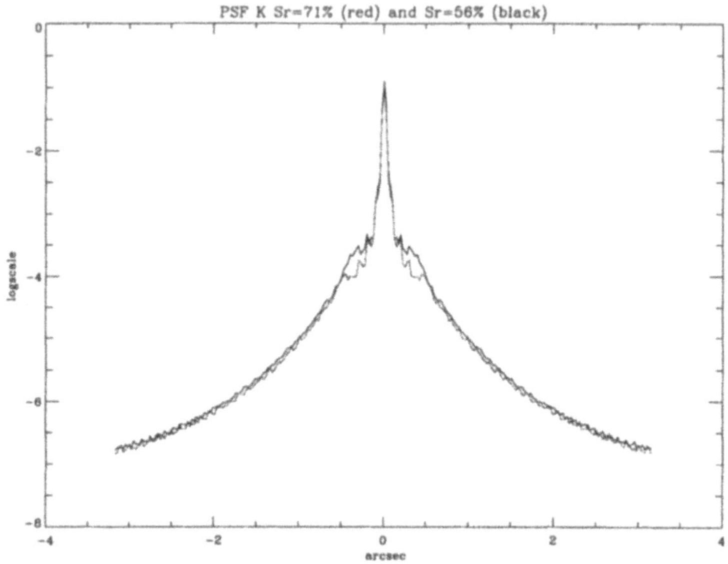

Fig. 3. Superimposition of two PSFs in K band, one with Sr=0.71 (solid) and one with Sr=0.56 (dashed).

An example of the impact of Sr improvement is given in Fig. 4 in the case of a binary system with a 0.3" separation and a ΔK of 6.0, observed with NAOS and CONICA with both Sr.

Also importantly, the good performances in terms of Sr (see Fig. 2) *down to the optical domain* for bright objects show that the optical domain will also benefit from the use of AO. This justifies the implementation in NAOS of an extra output channel to feed other instruments, working in the visible domain.

4 Examples of Disk Observations with NAOS

It is expected that NAOS will allow the detection of disks 1) fainter than those detected so far, and 2) associated to stars further away. Fig. 5 and Fig. 6 give examples of such detections, using the β Pictoris disk (see [1]) as a comparison. Noticeably, gaps due to the presence of giant planets in the disks should be detectable.

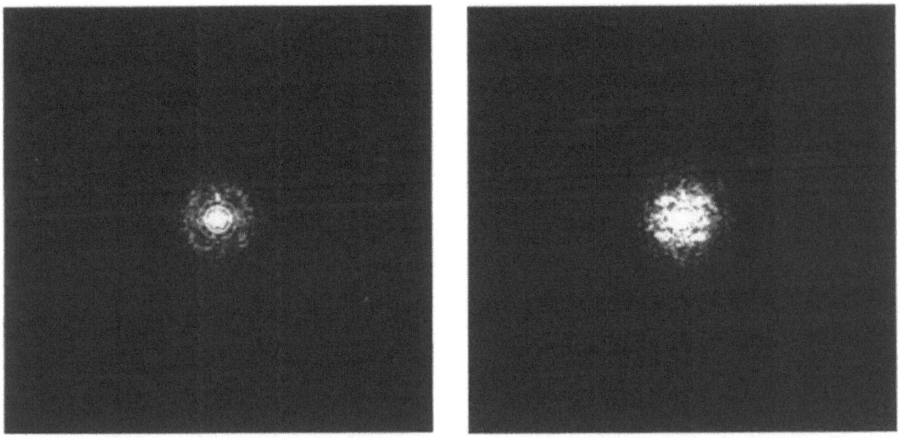

Fig. 4. Simulation of K images of a binary system with a separation of 0.35" and a ΔK of 6.0 with NAOS+CONICA and Sr=0.71 (left) and Sr=0.56 (right). Same cuts are used in both figures.

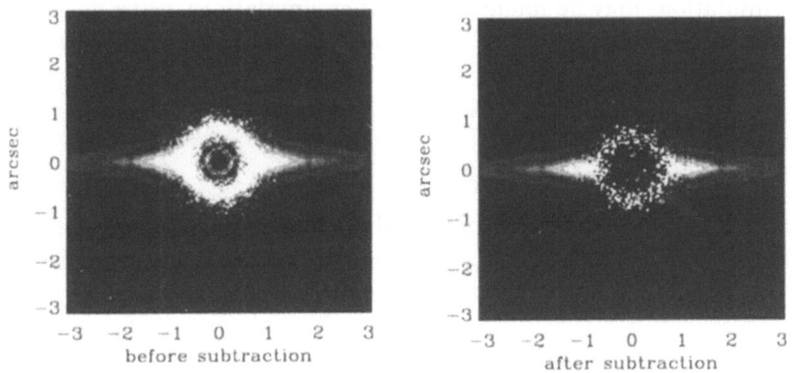

Fig. 5. Simulation of a disk similar to that of β Pic, with a 1 AU gap at $\simeq$ 30 AU. K band. Left image shows the disk before removal of the scattered light with a comparison star; right image shows the disk after removal. The gap is clearly detected.

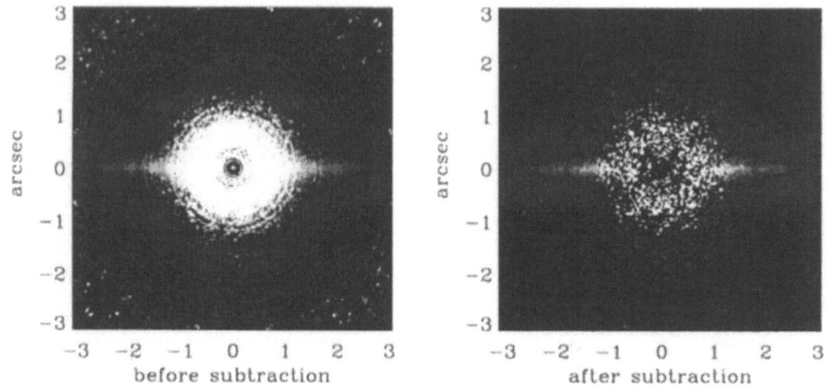

Fig. 6. Simulation of a disk $\simeq$ 100 times fainter than β Pic with a 1 AU gap at $\simeq$ 30 AU. Same lay-out as in former figure.

5 Examples of Observations of Faint Companions

NAOS will obviously allow to increase the detection limit of faint companions over a wide spatial domain, compared to present AO systems working on smaller telescopes. Noticeably, the very high level of correction achievable will offer *unique possibilities* for the detection faint objects in the *immediate vicinity of the central stars* ($\leq$ 0.5"). Fig. 7 illustrates such possibilities when observing in K band or at shorter wavelengths. The example could represent a still hot giant planet surrounding a young star in the solar neighbourhood. Other simulations may be found in [6]. Other simulations, using dark speckle technics can be found in [7]. A still higher dynamics may be reached with this method.

6 Conclusion

Major progress in the detection of faint disks and companions close to bright objects are expected with NAOS. In particular, giant planets could be found, either directly for very young planets or indirectly through gaps in disks.

References

1. Kallas, P. (1999) this volume
2. Lagrange, A.M. et al (1998) Astronomical constrains for the design of the VLT-NAOS adaptive optics system. in *Infrared Astronomical Instrumentation*, A. M. Fowler, ed., *Proc. SPIE* **3353**, 591–599
3. Lagrange, A.M. et al (1999) Astrophysical drivers for the VLT/NAOS in the proceedings of the 1998 ESO/OSA conference

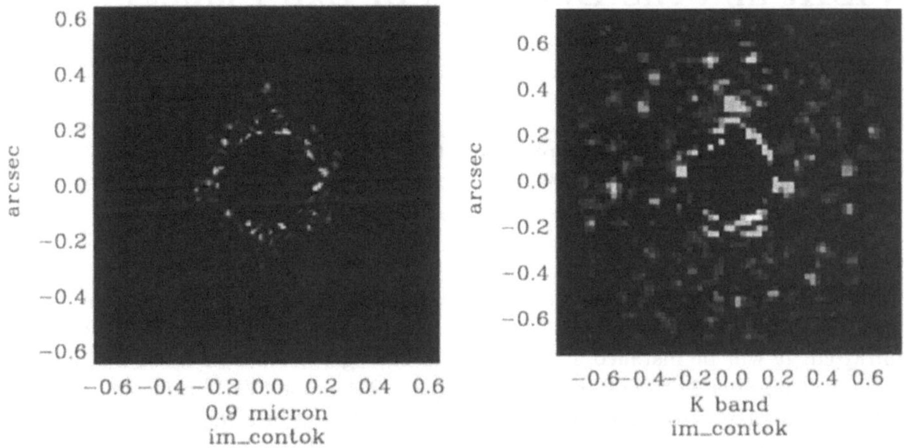

Fig. 7. Simulation of a binary system with Sep = 0.35" and Δ K = 10, at 0.9 micron (left image) and at K band (right image).

4. Lenzen, R., Hofmann, R., and Bizenberger, P. (1998) in *Infrared Astronomical Instrumentation*, A. M. Fowler, ed., *Proc. SPIE* **3354**
5. Mayor, M. (1999) this volume
6. Mouillet, D. and Lagrange, A.M. (1999) Observations of disks and faint companions with AO systems on large telescopes. in preparation
7. Moutou, C. et al (1999) this volume
8. Rousset, G., et al (1998) Design of the Nasmyth adaptive optics system (NAOS) of the VLT," in *Adaptive Optical System Technologies*, D. Bonaccini and R. K. Tyson, eds., *Proc. SPIE* **3353**, 508–516

VISIR and the Detection of Exo-Planets

Pierre-Olivier Lagage

DSM/DAPNIA/Service d'Astrophysique, CEA/Saclay,
F-91191 Gif-sur-Yvette Cedex, France
E-mail: lagage@cea.fr

Abstract. In this paper, we discuss the detection of exo-planets with an instrument like VISIR. VISIR stands for VLT Imager and Spectrometer for the mid-InfraRed. It will be installed in 2001 at the Cassegrain focus of Melipal, the telescope unit number 3 of the European Very Large Telescope (VLT) program. Indirect detection of planets by looking for footprints of planets in a dust disk will be one of the key programs conducted with the instrument. The direct detection of giant (10 Jupiter mass) exo-planets as cool as 350 K appears feasible for a limited, but plausible, range of parameters: planet orbiting at more than 5 AU from a M dwarf located at less than 5 pc from us.

1 Introduction

The study of disk and planets around stars is one of the fields in Astrophysics which has undergone huge progresses the last years. Protoplanetary disks, a key ingredient of star-formation theories, have been imaged, (see the review by M. McCaughrean, these proceedings). More than a dozen of exo-planets have been discovered (see the reviews by M. Mayor, D. Queloz, these proceedings). Debris disks have been imaged, (see the review by P. Kalas, these proceedings).

In this paper, I shortly describe VISIR, the VLT instrument for the mid-infrared, (Sect. 2). Then, I discuss the prospects in using the instrument to detect exo-planets, either indirectly by imaging debris disks (Sect. 3) or directly (Sect. 4).

2 VISIR, the VLT Mid-InfraRed Instrument

Since the beginning of the Very Large Telescope (VLT) project, the European Southern Observatory (ESO) has set a comprehensive instrumentation plan which provides a wide range of imaging and spectroscopic facilities from near-ultraviolet to mid-InfraRed (IR). The two mid-IR atmospheric windows, the N-band (7.7-13.5 μm) and the Q-band (16.5-25 μm), will be covered by VISIR, the VLT Imager and Spectrometer for the mid-InfraRed. VISIR is being built by a consortium of two institutes: CEA/Saclay (France) and NRFA/Dwingeloo (The Netherlands). The instrument combines:

1) imaging capabilities over a field up to about 1 arcmin with a choice among three scales (0.075″, 0.127″, 0.200″ per pixel) and with a choice among 27 narrow- and broad-band filters;
2) long-slit (>30″) grating spectroscopy capabilities with various spectral resolutions (350, 3200, 25000 at 10μm; 1600, 12500 at 20μm).

The design of the instrument is completed and the manufacturing phase has started (see Fig. 1). For details about the instrument, the reader is referred to the article by Lagage et al. (1998). VISIR will be mounted at the Cassegrain focus of Melipal, the VLT telescope unit number 3, at the beginning on 2001.

Fig. 1. Picture of the VISIR test cryostat and cable wrap mounted on the integration support.

In terms of performances, the two key issues related to the observing programs discussed hereafter are: the angular resolution and the sensitivity.
In the mid-infrared, the image quality is limited by telescope diffraction effects rather than seeing effects. The diffraction effect on the angular resolution, defined here as the wavelength (λ) over the telescope diameter (D), is 0.25″ at 10 μm. The smallest pixel field of view of VISIR (0.075″) will allow to oversample the diffraction pattern. Thanks to image deconvolution techniques, we can expect to reach an angular resolution in the 0.1″ range.
A typical sensitivity of 1 mJy (10 sigma 1 hour) is expected in the N-band and of 10 mJy (10 sigma 1 hour) in the Q-band. The sensitivity is limited by the noise associated with the huge photon background generated by the at-

mosphere and the telescope. Note that a large gain in sensitivity is obtained when going from a 3-m class telescope to a 8-m class telescope. Indeed, the observing time needed to reach a given signal over noise ratio when observing a point source, decreases as D^4. One hour of point source observations with VISIR will be equivalent to 2 nights of observations with a mid-IR instrument on a 3.6-m telescope.

3 Perspectives in Debris Disk Observations

Disks and planets are intimately connected (see D. Lin, P. Artymowicz, these proceedings). Planets are made from protoplanetary disks. Interactions between newly formed planets and "protoplanetary" disks can modify the orbit of the planet, which may "migrate" towards the star. Protoplanetary disks disappear with time, so that main-sequence stars are not expected to be surrounded by a disk. However, collisions between planetesimals or comet sublimation can generate a "second generation" dust disk (for example Weissman 1984); these disks are usually quoted as debris disks. The presence of a planet in such a disk can deeply modify the morphology of the disk through gravitational perturbations of the dust orbit (Paresce 1992, Roques 1994). Such effects have been observed in the solar system (Dermott et al. 1994; Reach et al. 1995). Thus, studying the morphology of debris disks is a way to search for "footprints" of planets.

3.1 Debris Disk Imaging

The study of debris disks has started with the IRAS discovery of an IR excess around many main-sequence stars (Aumann 1985). The best-studied dust disk is the disk around the star β-Pictoris (see reviews by Artymowicz 1997, by Vidal Madjar et al. 1998 and references there-in). The outer parts of the dust disk were imaged in the visible soon after the IRAS discovery (Smith & Terrile 1984). The inner parts of the disk were imaged 10 years later, thanks to mid-IR observations (Lagage & Pantin 1994, see also Fig. 2). Both the mid-IR images and recent observations in the visible or near-IR (Mouillet et al. 1997) show features (asymmetry, deficit of matter near the star, warp), which indicates the possible presence of planets in the system.

The origin of the disk emission observed in the mid-IR is dust thermal emission; at shorter wavelengths, the origin of the emission is star-light dust scattering; (see the contribution by A.M. Lagrange, these proceedings, for a discussion of the prospects in detecting dust disk in the near-IR). Observations in the mid-IR and in the near-IR (or visible) are complementary. On one hand, a better angular resolution can in principle be obtained at shorter

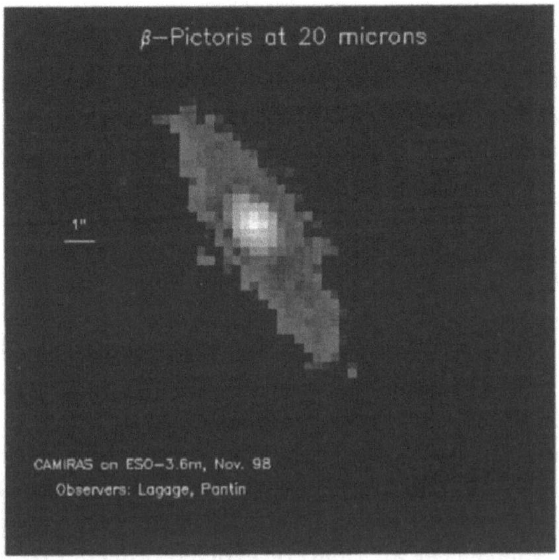

Fig. 2. β-Pictoris dust disk as observed at 20 μm (raw image). North is up; east is left. The pixel field of view is 0.3 arcsec, which means 5.7 AU at the distance of β-Pictoris (19.3 pc). At this wavelength, the disk contribution dominates over the star contribution (not removed here).

wavelengths, at least when using adaptive optics. On the other hand, only mid-IR observations can reveal the innermost parts of a disk (closest to the star). Indeed, given that the disk emission often dominates over the star emission in the mid-IR (especially in the Q-band), the resolution is not limited by the need of a coronograph, as required at shorter wavelengths. Second, dust thermal emission is isotropic, when dust scattering is anisotropic (forward scattering). Then, different parts of a disk are seen according to the origin of the observed emission and, in the case of an edge-on disk similar to the β-Pictoris disk, the scattered emission is dominated by the outer part of the disk (let's say greater than 30 AU from the star), so that the inner parts are inaccessible, whatever the angular resolution is (Pantin 1996).

Up to now, only two disks have been resolved in the mid-IR, the disk around the β-Pictoris star and recently the disk around HR4796 (Koerner et al. 1998, Jayawardhana et al. 1998). The list should extend rapidly thanks to the improved sensitivity obtained with mid-IR imagers mounted on large telescopes. For example, VISIR will be about 100 times more sensitive than TIMMI, the instrument which was used to image for the first time the β-Pictoris dust disk at 10 μm.

3.2 Dust Composition

Spectroscopic studies of debris disks in the mid-IR are also crucial. Indeed, mid-IR dust features, such as silicate features, can help in identifying the nature of the dust orbiting the star and hopefully in understanding the origin of the dust.
Comparison of mid-IR spectra of the dust around β-Pictoris (Knacke et al. 1993, Aitken et al. 1993) with spectra of comets, have led to the claim that the β-Pictoris dust is generated from the sublimation of comets. Of special interest was the presence of crystalline olivine in both comets and β-Pictoris. The presence of crystalline olivine was first identified in comet Halley (Campins & Ryan 1989); crystalline olivine was also present in large quantities in comet Hale-Bopp and was identified as forsterite (Crovisier et al. 1997). However, it should be pointed out that not all comets feature silicate dust (Hanner et al. 1994). Furthermore, ISO/SWS observations have shown that crystalline olivine dust was present in various environments, such as young stellar objects (Waelkens et al. 1996) or evolved objects (Waters et al. 1998); the origin of such a dust is not yet well understood. Thus, the link between the dust around β-Pictoris and comets is not as straightforward as usually thought.

Additional studies of dust both in comets and in debris disks is definitively needed to progress. The spectroscopic mode of VISIR will be well adapted to such studies.

4 Giant Exo-Planet Direct Detection

We have to face two main problems when trying to detect directly an exo-planet.
The first issue is a sensitivity issue. The flux expected from a giant exo-planet is now well calculated, (F. Allard these proceedings). In the following, we only consider the detection of an exo-planet at a temperature in the range 450-100 K, typical of the mid-IR radiation. Given the sensitivity discussed in Sect. 2, only giant planets (7-12 M_{jup}) at a distance of less than 5-10 pc and at a relatively high temperature in the range 450-350 K can be detected by VISIR. For example, the flux expected at 10 μm for a 7 M_{jup} exo-planet at a temperature of 450 K and located at 10 pc from the Earth is of 1/3 of mJy (Burrows et al. 1997, Marley private communication); the age of such a planet is 200 Millions years. A 12.6 M_{jup} exo-planet will be at 300 K at an age of 4.2 Billions years and will emit a flux of 1/30 mJy, if located at 10 pc.
The second issue is the contrast in brightness between the hosting star and the planet. As already mentioned in Sect. 2, the main limitation to the angular resolution is the diffraction by the telescope. The signal at respectively 1″ and 2″ from a point star is respectively 1/300 and 1/10000 of the central signal. These numbers have to be compared with the intensity contrast between

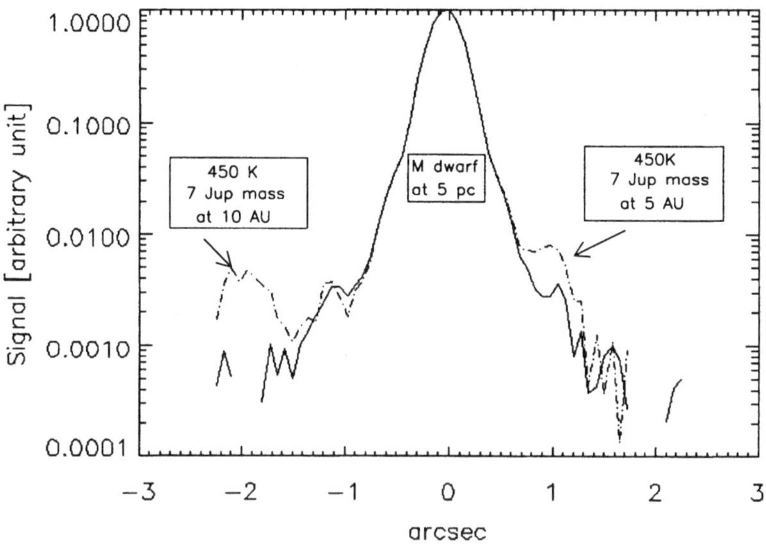

Fig. 3. Cut along an axis of an image simulating the 11 μm signal expected in VISIR from a system made of two 7 M_{jup} exo-planets at 450 K, orbiting respectively at 5 AU and 10 AU from a M-dwarf located at 5 pc from us. The diffraction by the telescope and a seeing of 0.4″ in the visible have been taken into account, as well as the background noise generated by the atmosphere and the telescope. Solid line: star alone; dashed-dotted line: star + companions.

a star and its planet. If we consider a solar type star and a 7 M_{jup} planet at 450 K, the contrast is 1/4500. A more favourable case is obtained when the hosting star is a M-dwarf; in this case, the contrast is only 1/200 and the exo-planet can be easily detected (see Figure 3). Note that a giant planet (2/sin(i) M_{jup}) orbiting close by (semi-major axis of 0.2/sin(i) AU) Gl876, a M dwarf located at 4.7 pc from us, has already been detected indirectly, by the radial-velocity technique (Delfosse et al. 1998).

Thus, the direct detection of giant exo-planets with an instrument like VISIR appears possible for a limited, but plausible, range of parameters.

References

1. Aitken D.K. et al. (1993) MNRAS **265**, L41
2. Artymowicz P. (1997) Ann. Rev. Earth & Planets Sci. **25**, 175
3. Aumann H.H. (1985) PASP **97**, 885
4. Burrows A. et al. (1997) ApJ **491**, 856
5. Campins H. & Ryan E. (1989) ApJ **341**, 1059
6. Crovisier J. et al. (1997) Science **275**, 1904

7. Delfosse X. et al. (1998) A&A **338**, L67
8. Dermott S.F. et al. (1994) Nature **369**, 719
9. Hanner M.S. et al. (1994) ApJ **425**, 274
10. Jayawardhana R. et al. ApJ ApJ **503**, L83
11. Koerner D.W. et al. ApJ **503**, L79
12. Knacke R.F. et al. (1993) ApJ **418**, 440
13. Lagage P.O. & Pantin E. (1994) Nature **369**, 628
14. Lagage P.O. (1998) Messenger **91**, 17
15. Mouillet D. et al. (1997) MNRAS, **292**, 896
16. Pantin E. (1996) PhD, Orsay University, France
17. Paresce F. (1992) Adv Space Res **12**, 208
18. Reach B. et al. (1995) Nature **374**, 521
19. Roques F. (1994) ICARUS **108**, 37
20. Smith B.A. & Terrile R.J. (1984) Science **226**, 1421
21. Vidal Madjar A., Des Etangs A.L. & Ferlet R. (1998) Planetary and Space Science, **46**, 629
22. Waelkens C. et al. (1996) A&A **315**, L245
23. Waters R. et al. (1998) Nature **391**, 868
24. Weissman P.R. (1984) Science **224**, 987

The Darwin Space Interferometer

Alan Penny[1], Alain Leger[2], Robin Laurance[3], and Malcolm Fridlund[3]

[1] Rutherford Appleton Laboratory, Didcot, U.K.
[2] Institute d'Astrophysique Spatiale, Paris, France
[3] ESTEC, Nordwijk, The Netherlands

Abstract. Darwin is a mid-IR (5-30 micron) interferometry observatory with baselines greater than 50 metres. As a long-duration general purpose observatory, ESA is studying it as part of its Horizons 2000 programme for a possible launch in the 2009 – 2016 timeframe. This paper describes some of the technical, scientific and programmatic aspects of the Darwin mission as presently conceived by the ESTEC study, the Alcatel industrial study and by the members of the Darwin Team.

1 Introduction

Darwin will have two science aims, the study of nearby planetary systems and general astrophysics. The search for planets orbiting nearby stars will be done in an imaging and spectroscopic mode. Its ultimate goal will be to search for the signature of presently active life on any terrestrial planets found. In this study, the interferometer will operate in its nulling mode. The general astrophysics will be done using high angular resolution imaging together with spectroscopy, with the interferometer operating in its imaging mode.

The development of Darwin will also drive technological capability for interferometry in space. This will form part of a long-term trend to have high-resolution telescopes in space.

The detection of active life on a planet orbiting another star would have implications far beyond the planetary science and exobiology involved.

2 The Mission

Darwin will operate at the L2 Sun-Earth Lagrangian point, some 1.5 million km from the Earth in the anti-Sun direction, and consist of five or more 1.5-3.5 m-class IR telescopes with separations of up to some hundreds of metres. It will have a beam-combining hub and a separate command and control module. Each of these components will be mounted on a separate spacecraft. Thrusters and laser metrology will maintain their relative positions to within accuracies of micrometres, and perhaps even nanometres. The beam-combining hub will have separate optical benches for the imaging and nulling modes. Passive cooling will maintain the temperatures of the optical components below 40K, so that they will not contribute significant noise to the input signal.

Fig. 1. Artist's impression of Darwin in action. (Courtesy Alcatel)

The ideas for this kind of mission have origins in a number of papers [[1], [2], [3], [4], [5]] 2,3,4,5], starting with Bracewell's original paper in 1979 [6].

3 Observing Modes

3.1 Nulling Mode

The nulling mode is used to remove a bright on-axis source. By joining together all the beams from the different interferometer elements into one output element, and with suitable phase differences imposed on the beams, an on-axis source can be placed at the null of the response. Further, the width of this null can be made broader then the 'natural' width due to the spacing of the interferometer elements, and the depth of the null can be made very deep. These advantages are very important when the on-axis source is a nearby star, which is both extremely bright and has significant angular extent.

The gain in the central null is achieved at a price. One loses the separate angular information on the source which would normally come from having a number of elements, providing many baselines. Further information may be had by rotating and expanding the array, which keeps the deep null. Further information may be had by changing the phase shifts in the different beams ([7], (Mariotti and Mennesson, in preparation). A number of arrangements of the interferometer elements are possible, each with advantages and disadvantages. For a simple target, such as a planetary system, which has a small

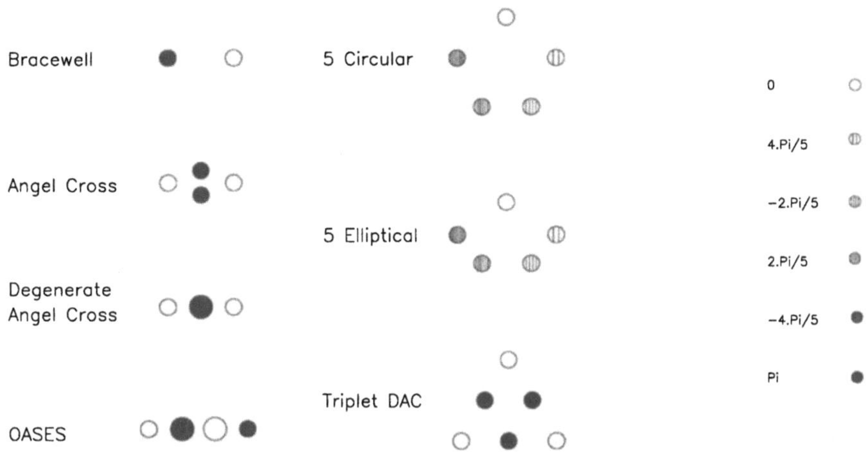

Fig. 2. Various interferometer element arrangements for nulling, with the phase shifts applied to each element.

number of point sources, together with an axisymmetric elliptical zodiacal dust light, the target can be extracted with reasonable confidence.

The performance depends on the optical throughput, the background noise and the noise introduced by the system, and on the complexity of the image. The dominant source of noise will be the target zodiacal dust light. Léger et al. [8], together with Mennesson and Mariotti simulated the performance of an array of five 1.5 m telescopes looking at a solar system twin at 10 pc. A 30 hour integration would lead to the detection of all the three major telluric planets. Giant planets within 10 AU from the central star could also be detected, with a spectral range of 7 to 18μm, with R=20.

3.2 Imaging Mode

For imaging, visibility measurements are made for all the individual baselines of the interferometer, and the image of the target reconstructed as is done for present-day radio interferometers. However, as the interferometer elements are at controlled positions, no S/N performance need be lost in performing phase closure. Calculations show that significant performance is possible for many astrophysically interesting objects. With baselines of up to hundreds of metres, Darwin will complement the NIR capability of the Next Generation Space Telescope.

4 Science Objectives

For planetary systems, some 100-300 nearby stars may be surveyed looking for planetary systems and in particular those with terrestrial planets. Darwin will tell us how frequent such objects are, and it will perform spectroscopy on any found, searching for the CO_2 band and for the H_2O and O_3 bands in the 7-20 μm range. It will simultaneously obtain images and spectra for all planets present. The study of the orbits of the planets orbiting stars of different masses and ages will give information on the formation and evolution of planetary systems, and the spectra will give information on the physical properties of the planets.

For general astrophysics, an angular resolution ranging up to 2 milli-arcseconds at 10μm, will have a sensitivity much higher than that possible for ground-based interferometers observing through the warm Earth atmosphere, and will be clear of atmospheric lines, Darwin will be a very powerful tool for general astrophysics. Observations of protostars, protoplanetary disks, and the disks around local stars will trace the formation of stars and planetary systems. Observations of red giants, AGB stars, novae and supernovae trace the process of mass-loss and grain nucleation. Studies of the cores of galaxies, will separate the non-thermal and thermal radiation. For AGNs, the 10-pc obscuring torus may be resolvable. The structures of the earliest galaxies in their rest-wavelength infrared emission will be observable.

Table 1. A 5-year observing programme: Fractions of time allocated to different activities

Year	1	2	3	4	5
In-orbit checkout	0.2	0.0	0.0	0.0	0.0
Planet					
Survey	0.4	0.4	0.2	0.1	0.1
Medium spectroscopy	0.1	0.2	0.4	0.3	0.3
Deep spectroscopy	0.1	0.1	0.1	0.2	0.2
General astrophysics	0.2	0.3	0.3	0.4	0.4

References

1. Leger, A. et al. (1993) "The Darwin Mission Concept", proposal to the ESA "Horizon 2000" planning process
2. Angel, Cheng and Woolf. (1986) Nature **322**, 341
3. Shao, M. & Colavita, M. (1992) Ann. Rev. Ast. Ast. 31
4. Owen, T. (1980) in "Strategies for the search for life in the universe", ed Papagiannis, Reidel, p177
5. Hitchcock and Lovelock (1967) Icarus, **7**, 149
6. Bracewell and MacPhie, (1979) Icarus **38**, 136
7. Mennesson, B., Mariotti, J.-M. (1997) Icarus, **128**, 202-212
8. Leger A. et al. (1996) Icarus **123**, 249

Radial Velocity Searches for Extra Solar Planets from Keck and McDonald Observatories

Artie P. Hatzes and William D. Cochran

McDonald Observatory, The University of Texas, Austin, TX, 78712, USA

Abstract. We review several programs using precise stellar radial velocity measurements to search for extra solar planets. Two programs that are conducted from the Keck Observatory using the HIRES spectrograph are discussed. The Cochran & Hatzes Hyades program is searching for extra-solar planets among members of the Hyades cluster. These stars represent a homogeneous population that allows us to probe the characteristics of planetary systems primarily as a function of stellar mass. The Marcy et al. Keck program is searching for planets among 400 of the brightest G-M dwarfs. The McDonald Observatory Planet Search Program is a multi-faceted program that not only searches for planets around a sample of bright F–M dwarfs, but also uses high resolution spectral data to place constraints on companion masses as well as confirm extra-solar planet discoveries.

1 Introduction

In this contribution we discuss several programs searching for extra-solar planets using precise stellar radial velocity (RV) measurements. Two of these programs are conducted at the Keck Observatory: the Cochran & Hatzes survey of members of the Hyades cluster and the Marcy et al. program to search for planets around nearby F–K stars. The McDonald Observatory Planet Search Program is a multi-faceted extra-solar planet search program using the 2.7-m telescope at McDonald Observatory.

2 The Keck Hyades Program

One of the main goals of extra solar planet searches is to understand how the characteristics of planetary systems correlate with the properties of the central star. A fundamental understanding of the process of planet formation requires a knowledge of how the stellar mass, metallicity, magnetic activity, and other stellar properties influence the process of planet formation. Field stars represent a random mix of stars with a wide range of properties (age, metallicity, activity). Stars in clusters represent a much better controlled homogeneous sample of stars that are coeval and with the same metallicities. They thus present an ideal environment for probing the characteristics of planetary systems primarily as a function of the stellar mass.

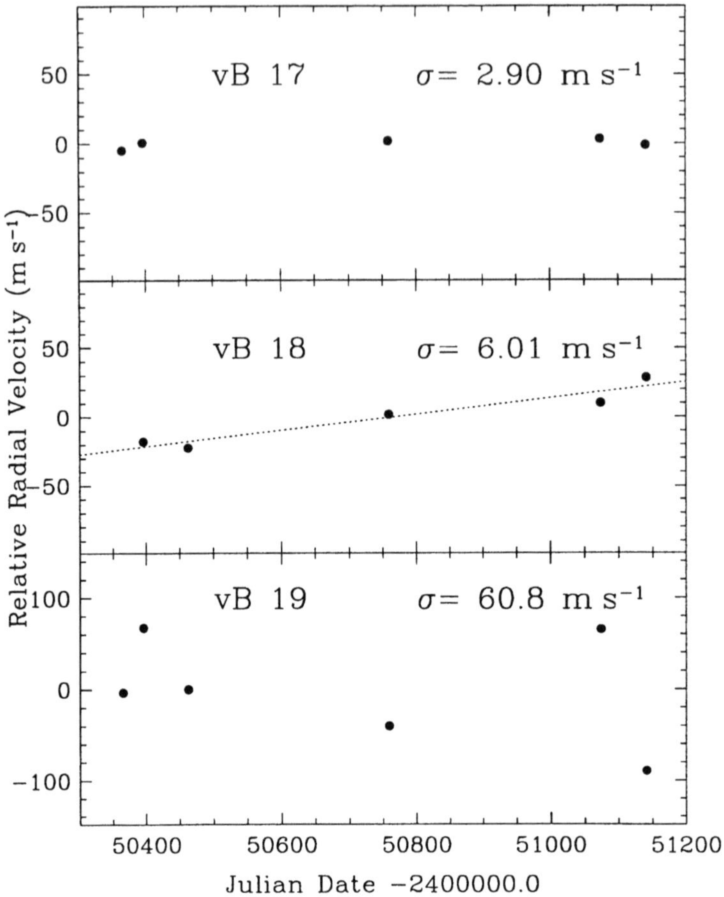

Fig. 1. Radial velocity measurements taken with Keck HIRES for 3 members of the Hyades: vB 17, vB 18, and vB 19.

To investigate the properties of planetary systems in clusters we began a search in late 1996 for extra-solar planets around members of the Hyades. This program uses the HIRES spectrograph of the Keck Observatory and an iodine absorption cell for the velocity metric. The Hyades cluster is the nearest and best studied cluster and has an age of $\sim$ 800 Myrs and a metallicity of [Fe/H]=+0.18. The brightest isolated F-M dwarfs known to be members of the cluster from proper motion and radial velocity studies were chosen for the target list and this includes 14 F dwarfs, 25 G dwarfs, 42 K dwarfs, and 19 M dwarfs. Each year we typically have 3 runs separated by a month and strive to observe each program star at least twice a year. To date we have, on average, about 3 total observations per program star.

There was some concern that because of the relatively young age of the Hyades, activity related RV variations would hinder our ability to make these measurements with sufficient precision to detect planetary companions. Saar & Donahue (1997) have demonstrated that high levels of stellar activity can induce RV variations with amplitudes as high as 30 $m\,s^{-1}$. Our preliminary analysis of the RV variations of three Hyades members indicate that stellar activity is not compromising the RV precision for our sample of Hyades dwarfs.

Figure 1 shows the RV variations for 3 members of the Hyades: vB 17 (G5 V), vB 18 (G1 V), and vB 19 (F8 V). The RV variations of vB 17 are constant to within about 3 $m\,s^{-1}$ for measurements spanning 3 years. vB 18 shows a linear trend of increasing RV over this same period with an rms of 6 $m\,s^{-1}$ about the trend. On the other hand, vB 19 shows considerable more RV scatter of about 60 $m\,s^{-1}$. The variations for vB 19 may be due to a planetary companion and, in fact, the data can be fit by a sine curve (rms scatter of 9 $m\,s^{-1}$ about the curve) with a period of 100 days. This conclusion, however, is premature due to the small number of measurements. With sparse data sampling one can find sinusoidal variations in pure noise and there are a variety of reasons why the RV scatter for vB 19 should be higher than for the other two stars. First, vB 19 shows appreciable rotation and for a given signal-to-noise ratio the RV error scales linearly with $V \sin i$. The rotational velocity of vB 19 is about a factor of two greater than for the other two stars, which is still insufficient to account for the observed RV scatter. Second, the higher rotation rate means that this star may exhibit enhanced magnetic activity and thus more intrinsic activity-related RV variability For very active stars this can be as high as several tens of $m\,s^{-1}$. Finally, vB 19 is a late F star and RV surveys have demonstrated that these stars can show significant RV scatter due to intrinsic stellar variability of an unknown nature, possibly activity or pulsations (Cochran & Hatzes 1994).

3 The Marcy et al. Keck Program

The Marcy et al. RV program is the most extensive one conducted from Keck Observatory and it is searching for planets around a large sample of the brightest ($V < 8.5$ mag) G and K dwarfs as well as a modest sample of bright ($V < 11$ mag) M dwarfs. The target list contains 220 G dwarfs, 150 K dwarfs, and 50 M dwarfs. The key selection criterion for this program was Ca II H and K emission. The Mt. Wilson S value was measured for about 550 original sample stars and those objects having an age less than 3 Gyr were excluded from the sample. By selecting only the older stars the Marcy group hoped to bias the sample towards those stars with the smallest amount of *expected* intrinsic RV variations due to stellar activity. Since the RV scatter for these stars is less likely to be set by the intrinsic RV variations, they hope to maximize the likelihood of detecting planets. This program was initiated in mid-1996 and currently there are 2.5 yrs of data on all stars with an average of 5 observations per star.

4 The McDonald Program Planetary Search Program

The McDonald Observatory Planet Search Program (MOPS), which began in September 1987, uses precise stellar RV measurements to search for Jovian planets in a modest sample of solar-type stars. Observations were made using the coudé spectrograph of the McDonald Observatory's 2.7-m Harlan J. Smith Telescope. The initial survey had only limited wavelength range as it isolated a single order of the echelle grating. The survey began using the telluric O_2 absorption lines at 6300 Å as the velocity metric, but switched to a stabilized molecular iodine (I_2) gas cell in October 1990. This technique achieved a velocity precision of 10-15 $\mathrm{m\,s^{-1}}$ using a 9 Å bandpass centered on 5190 Å.

The initial MOPS program comprised of 36 bright dwarfs with spectral type ranging from late F to early K (8 F-, 12 G-, and 13 K-dwarfs). The small target list was driven by the efficiency of the coudé spectrograph and our desire to concentrate intensively on a small sample of stars with high quality, well-sampled observations. Observations were made with monthly sampling and typically 6 observations were made of each star per year. The observing strategy was primarily geared to finding Jovian planets in long-period orbits as this reflected our understanding of planet formation at the time. The MOPS program successfully co-discovered the planetary companion in the highly eccentric orbit around 16 Cyg B (Cochran et al. 1997).

To date, an analog to the Sun-Jupiter system (i.e. a Jovian mass at 5 AU) has yet to be found. Undoubtedly this is a selection effect since the shortest period planets will be the easiest and thus first planets to be discovered. As RV planet searches extend their time base the Jovian planets will certainly be discovered. One advantage of the MOPS program is that is has a long time base which may eventually reveal long-period Jovian planets.

Several stars in the MOPS show interesting trends which may be indicative of such a long period planet. Figure 2 shows the seasonal averages of our RV measurements for ϵ Eri. The solid line represents the best fit sine wave with a period of 11.5 years and a K-amplitude (semi-amplitude) of 15 $\mathrm{m\,s^{-1}}$. The rms scatter about the sine curve is 5.8 $\mathrm{m\,s^{-1}}$. These variations are consistent with a 1.2 Jupiter mass planet in orbit 5.1 AU from the star. Intriguingly, Walker et al. (1995), in their 12 year-long RV survey of solar type stars found evidence for a 9.9 year period with comparable amplitude in ϵ Eri. This period is consistent with the one we find given the rather short time span of each data sets with respect to the period.

It may be premature to attribute these RV variations to a planetary companion due to the relatively high level of activity in this star. Gray & Baliunas (1995) found a period of 5 yr, or about 1/2 the RV period, in the Ca

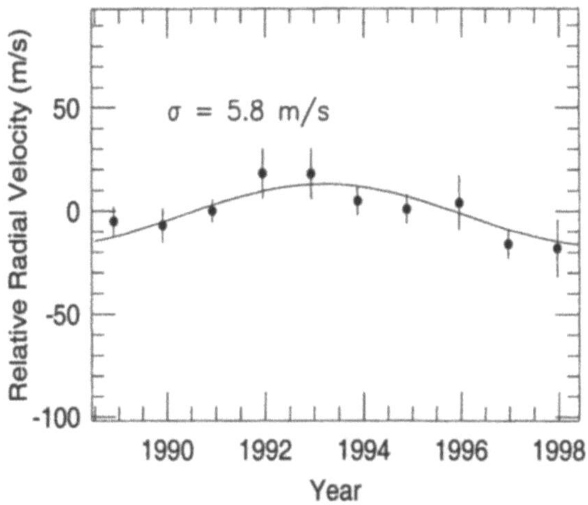

Fig. 2. Seasonal averages of the RV measurements for ϵ Eri taken at McDonald Observatory's 2.7-m telescope. The curve represents a sine fit with a period of 11.5 years and an amplitude of 15 $\mathrm{m\,s^{-1}}$.

II H&K emission and in the velocity span for spectral line bisectors of this star. McMillan et al. (1996) also found RV variations in ϵ Eri and suggested that they may instead be due to changes in the convection pattern on the star. Clearly, ancillary measurements, particularly line shape measurements, are crucial for establishing whether ϵ Eri indeed has a long-period Jovian companion.

In 1998 the MOPS program was expanded to a survey of 121 stars (15 F-, 79 G-, 21 K-, and 6 M-dwarfs). This expanded program uses the new 2dcoude cross-dispersed spectrograph of the 2.7-m telescope (Tull et al. 1995) which provides a wavelength coverage of 3500-10000 Å at a resolving power of 60,000 (an R=240,000 mode is also available, but with more limited spectral coverage). The greatly increased efficiency of this instrument allows us to survey more stars and the longer wavelength coverage enables us to achieve an RV precision of about 5 $\mathrm{m\,s^{-1}}$.

4.1 High Resolution Line Profile Measurements

An important part of the MOPS program is the use of high resolution line profile profiles to measure the projected rotational velocities (V sin i) of stars as well as the bisectors of photospheric lines. These can be used to place constraints on the inclination of the star and to confirm planet discoveries.

Because of the sin i ambiguity in the orbital determination, RV measurements cannot yield accurate companion masses so there is a possibility that we have actually detected a relatively high mass companion in a highly inclined orbit. Assuming that orbital and stellar rotational axes are approximately aligned (the case in the solar system), then measurements of the

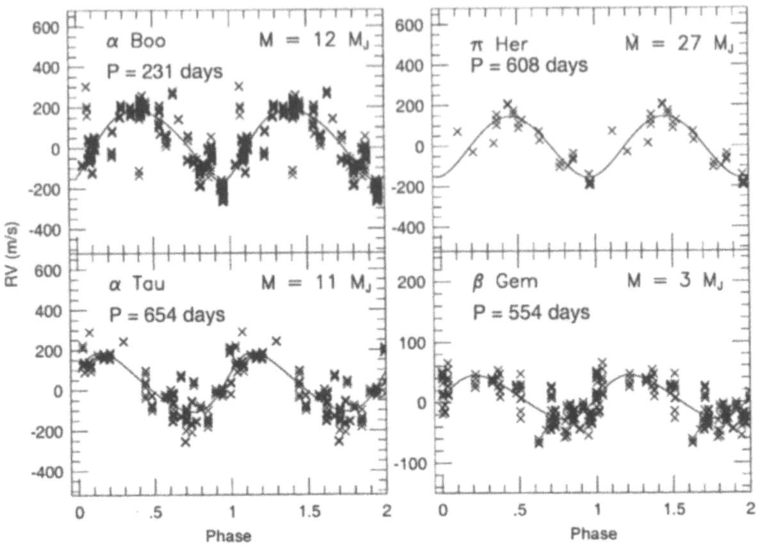

Fig. 3. Long period RV variations for four K giant stars. The curves represent orbital solutions with the companion mass (× sin *i*) given in each panel.

V sin *i* may help place constraints on the companion mass. If accurate radii determinations (via interferometry) and rotational periods are known (via Ca II H&K monitoring), then reasonably accurate stellar inclinations can be derived.

HD 114762 (F9V) is one example of using *V* sin *i* to constrain a companion mass. This star has an 11 $M_{Jupiter}$/sin *i* companion in a 83.9 day orbit (Latham et al 1987). Cochran et al. (1991) measured the *V* sin *i* for this star using high spectral resolution (R = 220,000) and found that the best fit to the data was provided by V sin $i = 0$. The mean rotational rate for stars of this spectral type is about 5 $\mathrm{km\,s^{-1}}$, so there is a high probability that we are viewing HD 114762 nearly pole-on which implies an orbital inclination of ≈ 0. The true mass for the companion can easily be a factor of 10 higher which would put it in the brown dwarf to stellar mass range.

On the other hand, *V* sin *i* measurements for 51 Peg (G2.5V) yield a value of 2.35 ± 0.1 $\mathrm{km\,s^{-1}}$ (Hatzes et al. 1997), which is greater than the solar rotational velocity. This solar-type star is expected to have a radius and rotational period comparable to those of the Sun, yet the projected rotational velocity is slightly higher. The true mass of the planetary companion is probably not much different than the 0.5 $M_{Jupiter}$ inferred from the RV amplitude.

High resolution measurements of the spectral line shapes can also provide important clues as the the nature of the RV variations. Both nonradial pulsations (Hatzes 1996; Brown et al. 1988) and surface structure (Toner & Gray 1988) should produce changes in the spectral line bisectors along with the

RV variations. To first order low-mass companions should not affect the line shapes, although the reflected light from a nearby companion may alter the line shapes at selected orbital phases (Charbonneau et al. 1998).

These types of measurements were the source of the short-lived controversy concerning the planetary companion to 51 Peg. Gray (1997) reported spectral line bisector variations in the photospheric spectral lines of the star with the same period as the planet's orbit which implied that the RV variations were due to intrinsic variability of the star and not due to a planet. The semi-amplitude of changes in the velocity span of the bisector was about 40 $m\,s^{-1}$, comparable to the RV amplitude. We used the 2dcoude spectrograph of the 2.7-m telescope at McDonald Observatory in its high resolution mode (R = 240,000) to obtain over 120 spectra of 51 Peg spanning 18 nights. These data had twice the resolution and more wavelength coverage than Gray's data. By averaging the bisector variations of several spectral lines we were able to show that the bisector span variations were no more than 1.3 $m\,s^{-1}$, or a factor of 40 smaller than the variations claimed by Gray (Hatzes et al. 1998). These data firmly established that 51 Peg had no spectral variations with the 4.23-d orbital period of the planet.

5 Selection Effects

Precise stellar RV measurements have had spectacular success at finding the first extra-solar planets. However, many of these surveys are introducing selection effects which may be giving us a biased view of planet formation. The planet around 51 Peg, a discovery which may have altered the paradigm for planet formation, should warn us of the dangers of selection effects. These systems eluded discovery for years in spite of the fact that some RV surveys, with measurement precision comparable to current detection methods, started more than a decade ago. Furthermore, due to the relatively high RV amplitude and short period of 51 Peg systems these would have been an easy detection. One of the reasons for the long delay in their discovery was that observing strategies were not geared to searching for Jovian planets in short-period orbits. Our own solar system biased the initial search strategies.

Although we are now more aware that planets can come with a diverse range of orbital elements, current surveys still introduce biases. Most RV searches concentrate on G-type stars. The Marcy et al. program, in order to achieve the highest RV precision, excludes F-type stars, which may be intrinsically variable, as well as stars younger than 3 Gyrs. In order to obtain a more fundamental understanding of the planet formation process we must become more "equal opportunity" planet hunters and search for planets throughout the H-R diagram in spite of the reduction in RV precision.

There is evidence that dusty disks, possibly the remnants of protoplanetary disks surround many A-type stars (see Kalas, these proceedings). The disk around β *Pic* is warped, possibly due to a planetary companion (Mouillet et al. 1997). Unfortunately, RV searches have excluded A-type stars, and for good reason; the high rotation rates and paucity of stellar lines result in an RV measurement precision of several $\mathrm{km\,s^{-1}}$ for these stars, woefully inadequate for detecting planetary companions. The next generation of interferometers as well as photometric transits may offer the best chance for detecting planetary companions around A-type stars.

We also should not shy away from stars because they may be intrinsically variable. F-type stars have more intrinsic RV scatter, yet one of the planet discoveries was around the F8 star τ Boo. It is important to understand the dynamical evolution of planetary orbits, but this can only come by searching for planets around young stars. This would require removing the possible RV signature due to the high activity in these stars.

Giant stars may also provide important clues to planet formation. K giant stars were included as part of the MOPS program and these were shown to have short-term variations ($\sim$ days) which are due to p-mode oscillations, and long-term variations (200-700 days) the nature of which is unknown. One possible explanation for the long period (LP) variations are low-mass companions with masses in the range of 3–27 $M_{Jupiter}$ (Fig. 3). Although these variations are often dismissed as being due to rotational modulation or pulsations, there is tantalizing evidence that these may actually be due to low-mass companions. The LP variations in three stars (α Boo, α Tau, and β Gem) are long-lived and coherent for more than a decade of observations (Hatzes & Cochran 1993). In α Tau, spectral line bisector measurements reveal no variations with the period of the LP variations (Hatzes & Cochran 1998). Possibly these stars may tell us something about the formation of planets around the progenitors of these stars (some of which may be A-type stars). We should point out that the long period signal was extracted from these stars in spite of the increased "noise level" due to the presence of short period variations with amplitudes comparable to the LP variations.

We thank G. Marcy for "briefing" us on the status of his Keck Program. This work was supported by NASA grant NAGW-4384.

References

1. Brown, T. et al. (6 other authors) 1998, ApJS, 117, 563.
2. Charbonneau, D. Jha, S., & Noyes, R.W. 1998, ApJ, 507, L153.
3. Cochran, W.D., Hatzes, A., Hancock, T. 1991, ApJ, 380, 35.
4. Cochran, W.D. & Hatzes, A. 1994, BAAS, 184, 608
5. Cochran, W.D., Hatzes, A., Butler, R.P., & Marcy, G.W. 1997, ApJ, 483, 457.
6. Gray, D.F. & Baliunas, S.L. 1995, ApJ, 441, 426.
7. Gray, D.F. 1997, Nature, 385, 795.
8. Hatzes, A.P. & Cochran, W.D. 1993, ApJ, 413, 339.
9. Hatzes, A.P. 1996, PASP, 468, 391.
10. Hatzes, A.P., Cochran, W.D., & Johns-Krull, C.M. 1997, ApJ, 478, 374.
11. Hatzes, A.P. & Cochran, W.D. 1998, MNRAS, 293, 469.

12. Hatzes, A.P., Cochran, W.D., & Bakker, E.J. 1998, 508, 380.
13. Latham, D.W., Stefanik, R.P., Mazeh, T., Mayor, M., & Burki, G. 1989, Nature, 339, 38.
14. McMillan, R.S., Moore, T.L., Perry, M.L., & Smith, P.H. 1996, DPS, 28, 1205.
15. Mouillet, D., Larwood, J.D., Papaloizou, J.C.B., & Lagrange, A.M. 1997, MNRAS, 292, 896.
16. Saar, S.H & Donahue, R.A. 1997, ApJ, 485, 319.
17. Toner, C.G. & Gray D.F. 1988, ApJ, 334, 1008.
18. Tull, R.G., MacQueen, P.J., Sneden, C., & Lambert, D.L. 1995, PASP, 107, 251.
19. Walker, G.A.H., Walker, A.R., Irwin, A.W., Larson, A.M., Yang, S.L.S., & Richardson, D.C. 1995, Icarus, 109, 2698.

Extrasolar Planets in the Southern Hemisphere: The CORALIE Survey

D. Queloz[1,2], M. Mayor[2], D. Naef[2], N. Santos[2], S. Udry[2], M. Burnet[2], and B. Confino[3]

[1] Jet Propulsion Laboratory, Mail-Stop: 306-388, 4800 Oak Grove Drive, Pasadena, CA 91109, USA
[2] Observatoire de Genève, 51 ch. des Maillettes, CH–1290 Sauverny, Switzerland
[3] Observatoire de Saint-Luc, Saint-Luc, Switzerland

Abstract. We are conducting a large survey to detect extrasolar planets by high precision Doppler spectroscopy using the new swiss Euler telescope and the spectrograph CORALIE. The sample includes 1600 nearby G and K stars from the southern hemisphere. The simultaneous thorium technique is used to reach very high precision radial velocity measurements. Less than 6 months after the setup of the instrument two planetary candidates have been detected. Almost 20 planetary candidates have been discovered by various high precision Doppler spectroscopy surveys. The semi-major axis of the planetary candidate orbits ranges from very small (0.05 AU) to 3 AU. Some have eccentric orbits, others have a more massive secondary companion, some have both. This large spread in orbital characteristics is difficult to fit into a single formation scenario. Current results also suggest a planet maximum mass around 7-9 M_J.

1 Introduction to CORALIE

CORALIE is the fiber fed echelle spectrograph installed in April 1998 at La Silla on the new 1.2m alt-azimuth Euler Swiss telescope. It is an improved version of the ELODIE spectrograph [3] of the Observatoire de Haute Provence. The CORALIE front-end adaptor is located at the Nasmyth focus of the Euler telescope. Two sets of two fibers can alternatively feed the spectrograph which is located in an isolated temperature controlled room. The set of fibers used for high precision radial velocity measurements includes a double scrambler device [15] to improve the stability of the input illumination of the spectrograph. Thanks to a slightly different optical combination at the entrance of the spectrograph and the use of a 2k by 2k CCD camera with smaller pixels (15μm), CORALIE has a larger resolution than ELODIE. A resolution power of 50'000 ($\lambda/\Delta\lambda$) is observed with a 3 pixel sampling. Similarly to the ELODIE spectrograph, CORALIE is provided with on-line reduction software that computes the radial velocity of stars several minutes after their observation.

The main scientific goal of CORALIE is to search for planets through very precise radial velocity measurements by looking for stellar reflex motions (changes in the radial velocity of the star). In June 1998 we began a systematic survey on a sample of 1600 G and K dwarfs in the southern hemisphere. For details about the sample see Udry et al. in this volume.

2 CORALIE Achievements

2.1 Precision and Limitations

The long term precision in radial velocity measurements is ensured by the thorium lamp reference, the "scrambling effect" of fibers and of the double-scrambler, as well as the high thermal and mechanical stability of the instrumental environment. The simultaneous thorium technique provides very precise monitoring of zero-point drifts during measurements (see [15] for details about the technique). On Fig. 1, the result of the Allan analysis[1] [1], made on a 6 hour measurement sequence taken with CORALIE on the star HD160691, indicates that we reach almost $1\,\mathrm{m\,s^{-1}}$precision level. We believe that the deviation from the random statistic, visible after two hours time scale on Fig. 1, is the remnant of the slow thermal relaxation stemming from the filling up of the cryostat 12 hours before. On the long term, using the results from the planetary search survey stars, we measure, with the $P(\chi^2)$ statistic, $7\,\mathrm{m\,s^{-1}}$instrumental (or systematic) error. For young and active stars, spot effects is known to produce radial velocity "jitter" of the order of few $10\,\mathrm{m\,s^{-1}}$[19]. For older stars one expects some instabilities with an amplitude of few $\mathrm{m\,s^{-1}}$. Such effect could lead us to overestimate the systematic error of CORALIE.

2.2 Two Planets Already Detected

So far, two planetary candidates have been detected with the CORALIE planet search survey: a $4\,M_J$ minimum mass planet with a 15.8 day period for the early K star Gliese 86 [15] and a 0.42 minimum mass planet with a 3.5 days period for the G0 star HD75289 [13] . On Fig. 2 are displayed the phased orbits. In Tables 1 the orbital parameters of both planets are indicated. The residual to the fit of the orbital solutions is respectively $7\,\mathrm{m\,s^{-1}}$and $8\,\mathrm{m\,s^{-1}}$suggesting very low CORALIE systematic errors during the period of time of these measurements. For Gliese 86 a drift of the γ point is observed. This drift, observed as well by 20 year CORAVEL radial velocity measurements, suggests the presence of a more distant low mass stellar companion.

The cross-correlation technique allows us to study the mean bisector of (non-saturated) spectral lines "selected" by the template. No bisector variation has been observed with an upper limit of a few $\mathrm{m\,s^{-1}}$on the amplitude

[1] Measurement of the best precision that one can reach by averaging radial velocity measurements over a period of time Δt

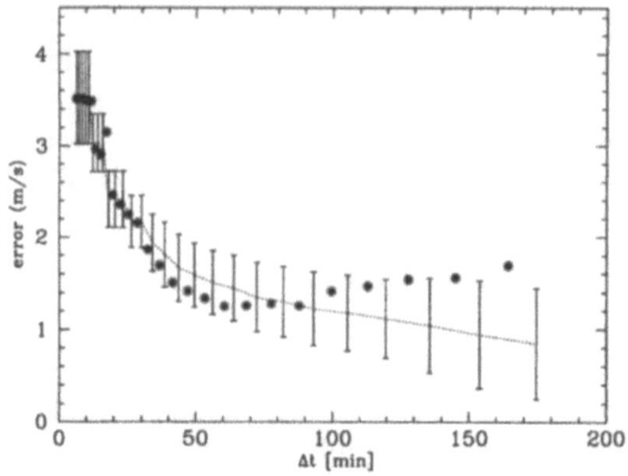

Fig. 1. Allan analysis on a sequence of 5 minute exposures on the star HD160691 (stars) with CORALIE. The dotted line with error bars indicates a random set of observations with the same accuracy and same sampling but no systematic errors from the instrument. The deviation to the random statistic after 2 hours is the result of systematic errors probably arising from long term thermal relaxation initiated at the filling up of the cryostat.

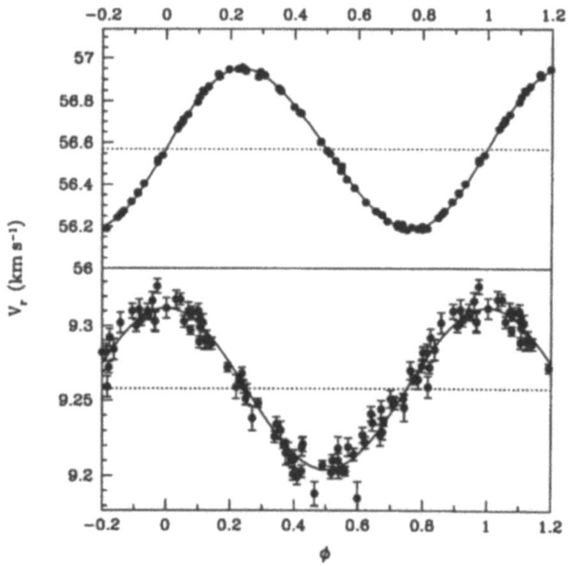

Fig. 2. Phased orbital motion for the two stars: Gliese 86 (top) and HD75389 (bottom). See orbital elements in Table. 1

Table 1. Orbital elements of Gliese 86 and HD75289

	Gliese 86*	HD75289	
P	15.78 ± 0.04	3.508 ± 0.001	d
T	2451146.7 ± 0.2	2451196.6 ± 0.5	d
e	0.046 ± 0.004	0.0	
V_r	$56.57^{\dagger} \pm 0.01$	9.26 ± 0.01	$\mathrm{km\,s^{-1}}$
ω	270 ± 4	0	^{0}
K_1	380 ± 1	54 ± 1	$\mathrm{m\,s^{-1}}$
$f_1(m)$	$8.9 \cdot 10^{-8}$	$5.7 \cdot 10^{-11}$	$M_\odot$
N	61	82	
$(O-C)$	7	8	$\mathrm{m\,s^{-1}}$

(†) at $T_0 = 2451150$ d
(*) after correction of the $0.36\,\mathrm{m\,s^{-1}\,d^{-1}}$ linear drift of the γ-point.

of the bisector span of HD75289 phased with the 3.5 d period. We can then rule out any surface feature phenomena significantly affecting the observed radial velocity curve of the star (activity, spots). Moreover HD75289 is not detected as a photometric variable star. The short period radial velocity variation is best explained by the presence of a planet orbiting the star.

3 The Search for Planets: Overview of the Results

The number of detected planetary candidates orbiting nearby stars has never stopped growing since the detection of 51 Peg, the first of them (see [11] for a review and [4] and [16] for recent results). In the last four years about 20 stars with radial velocity orbits suggesting the presence of planetary companions have been detected. From these results we get a first estimate (lower limit) of the amount of stars with giant planets: about 3%. But more important, the observed distribution of orbital parameters and planetary masses ($m \sin i$) may be use to get some insight into the planetary formation mechanism.

3.1 Orbital Parameters

The main orbital characteristics of planetary companion detected from radial velocity measurements are displayed on Fig. 3: the distance to the star or semi-major axis (a), the eccentricity of the planetary orbit (e) and the minimum mass of the planet ($m \sin i$). The large shaded region in the mass versus distance diagram represents the zone of non-detectability in the current radial velocity surveys. We then clearly see that caution must be used when interpreting any parameters related to the semi-major axis of the orbit or the mass of the planet, since these two parameters are tightly connected. For example, the fact that all detected low mass planets ($m \sin i$ about

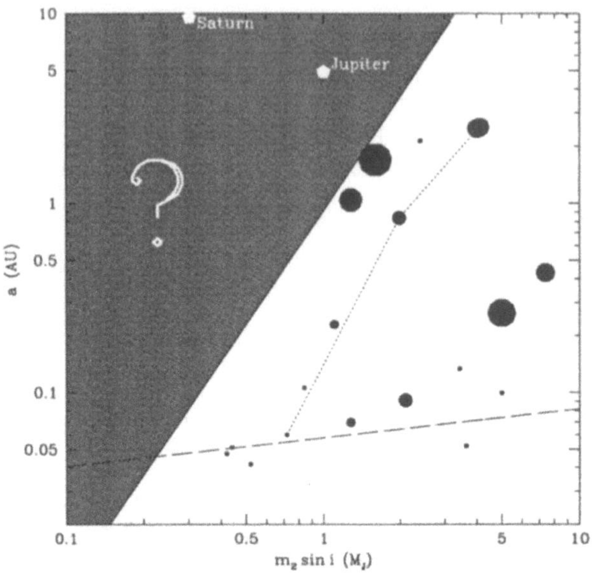

Fig. 3. Orbital parameters of planetary companions detected so far. The size of the dots is scaled to the eccentricity of the orbit (the bigger the dot, the larger the eccentricity). The shaded area indicates the non-detectable mass and distance domain considering typical precisions and observing strategies of surveys from which planets have been found ($30\,\mathrm{m\,s^{-1}}$ detection sensitivity). Planets below the dashed line are likely to be circularized after a few billion years by the tidal effect between the star and the planet [17]. Dotted lines connect the 3 planets orbiting the star Ups And.

$0.5 M_\mathrm{J}$) have circular orbits, is a direct consequence of the semi-major axis detection limit ($a < 0.1$ AU) since all these short period systems are likely to be tidally circularized. The same planets ($m \sin i \sim 0.5 M_\mathrm{J}$) but with a much larger semi-major axis would be undetectable.

The standard paradigm of planet formation model (accretion of planetesimal) tells us that we expect to find most of the planets on circular orbits. The current data don't support this prediction. If one excludes planets that may have been circularized by tidal interactions with the star, the planetary orbits show a large range of eccentricities. Various late processes in planet formation can divert the planet from its initial eccentricity: secular effects by a remote companion [8][14], disk-planet interactions [2], orbit scattering by gravity interactions with other planets [22][18][9] or planetesimals [10]. However, given the current data, most of the planets far enough from their stars to be free of any tidal interaction with them have eccentric orbits. If one believes that most of the detected planets migrated somehow during their formation, we don't expect an eccentricity growth during that process. Other processes may possibly need to be accounted for to explain such high number of eccentric orbits (see [12]).

The planet mass distribution can be studied only on a restricted range of masses and semi-major axes to avoid the mass-distance sensitivity threshold. We see on Fig. 3 that planets with semi-major axes smaller than 1 AU and more massive than $1 M_J$ are free of such observation constraint. On Figure 4. a mass distribution histogram with logarithmic bin is displayed for this selection. For this semi-major axis range a drop in the rate of planet detection is observed around 7-9M_J, suggesting a kind of planet mass upper limit.

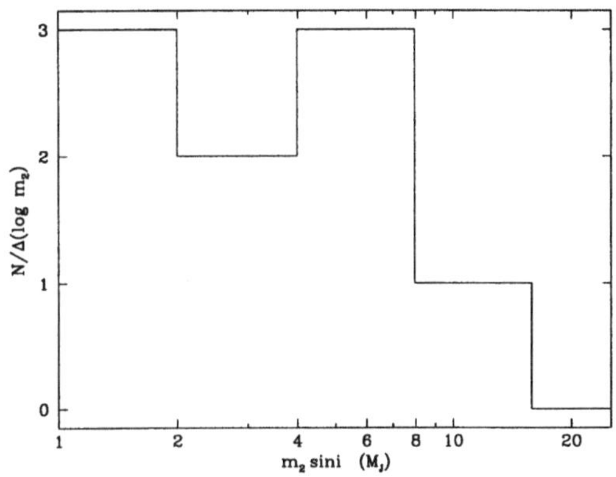

Fig. 4. Planet mass function per logarithm bin of the mass for planets with semi-major axis smaller than 1 AU

3.2 The Star Metallicity

The recent spectroscopic studies of stars around which planets have been detected have shown that the host stars may also bear marks from some processes occurring during the planetary formation [6][7][20]. More specifically, a large number of planets with short period orbits have surprisingly metal rich hosting stars. If we take a closer look at the relationship between the semi-major axis of planets and the metallicity content of host stars, we observe two remarkable behaviors (see on Fig. 5.) First, all planets with semi-major axes smaller than 0.08 AU have stars with an unusual high metal content compared to the other stars with planets. Second, the stars having planets on more remote orbits spread out on a broader range of metallicity, but they are in average more metal rich than field G dwarfs. The unusual metal rich content of stars with planets had already been pointed out earlier [6]. Figure 5 suggests in addition that there may be two kinds of metallicity behavior amongst the stars with planets. If confirmed, such a systematic high stellar metallicity content of very close orbit planetary systems compared to the larger orbit ones ought to somehow be related to specific processes governing the formation and evolution of these systems. This could be related to the way the migration process is halted, or the way the disk material feeds the star. The intrinsic metallicity content of the proto-planetary nebula may also have some effects on the efficiency and the likelihood of any migration and possible mass transfer between the planet and the star [21].

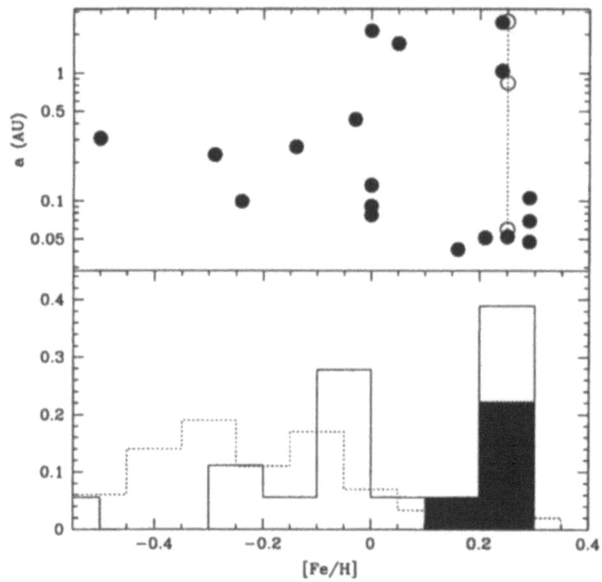

Fig. 5. Top [Fe/H] content of stars versus semi-major axis of the planetary orbit for all known planetary candidates with a $m_2 \sin i < 12\,M_J$. [Fe/H] measurements are from [6] and [7]. The dotted line connects the 3 planets orbiting the star υ Andromedae.
Bottom Distribution of the metallicity for field G dwarfs (hatched histogram)[5] and stars with a planet (solid line). The shaded area indicates stars with a planet closer than 0.08 AU. The υ And system is not included in the histogram.

4 Conclusions and Perspectives

A large number of planet detections is to be expected from the CORALIE survey. The simultaneous thorium technique has demonstrated that on the short term a $2\,\mathrm{m\,s^{-1}}$intrinsic precision could be achieved. Significant efforts are under way to improve the long term precision of the instrument. If successful, this may move the detection threshold closer to planets like our own Jupiter or hypothetic Saturn mass objects on earth like orbits. We may hope that, at some point, we shall get a better knowledge of the formation processes to understand the diversity of planetary systems. Foreseen ultra-high precision radial velocity instruments like HARPS (see Pepe et al. this issue) is the very next step to reach that goal.

Acknowledgments

We are deeply indebted to all the people who participated in the building and the rapid commissioning of the EULER telescope and CORALIE spectrograph, as well as to the observers who carry on the many observations dedicated to the Geneva planet search program. The two first planets are a direct outcome of their efforts. D.Q. thanks S. Loiseau for his careful reading of the manuscript.

References

1. Allan, D. W., 1966, Proc IEEE, 54, 221.
2. Artymowicz P., 1993, ApJ, 419, 166.
3. Baranne A., Queloz D., Mayor M., Adriansyk G., et al. 1996, A&A Suppl. Ser 119, 1.
4. Butler R.P., Marcy G.W., Fischer D.A., et al. 1999, ApJ submitted.
5. Favata F., Micela G., Sciortino S., 1997, A&A 323, 809.
6. Gonzalez G., 1998, A&A 334, 221.
7. Gonzalez G., Wallerstein G., Saar S.H., 1999, ApJ 511, L111.
8. Holman M., Touma J., Tremaine S., 1997, Nature, 386, 254.
9. Lin D.N.C., Ida S., 1997, ApJ, 477, 781.
10. Levison H. F., Lissauer J. J., Duncan M. J., 1998, AJ, 116, 1998.
11. Marcy G.W., Cochran W.D., Mayor M., 1999, Protostars and Planets IV, eds. V. Mannings, A. Boss, S. Russell, in press.
12. Marcy G., Butler R.P., Vogt S.S., Fischer D., Liu M.C., 1999, ApJ, in press
13. Mayor M., et al., 1999, in prep.
14. Mazeh T., Krymolowski Y., Rosenfeld G., 1997, ApJ, 447, L103.
15. Queloz D., Casse M., Mayor M., 1999, IAU Colloquium 170, "Precise stellar radial velocities", Victoria BC Canada, eds J.B. Hearnshaw and C.D. Scarfe, ASP Conference Series (1999), in press.
16. Queloz D., Mayor M., Weber L., et al. 1999, A&A submitted
17. Rasio F.A., Tout C.A., Lubow S.H., Livio M., 1996, ApJ, 470, 1187.
18. Rasio F.A., Ford E.B., 1996, Science, 274, 954.
19. Sarr, S. H., Butler, R. P., Marcy, G. W., 1998, ApJ, 498, L153.
20. Sandquist E., Taam R.E., Lin D.N.C., Burkert A., 1998, ApJ 506, L65.
21. Trilling D.E., Benz W., Guillot T., et al., 1998, ApJ, 500, 428.
22. Weidenschilling S.J., Marzari F., 1996, Nature, 384, 619.

Use of the 3.6m Telescope at La Silla for Planetary Detection

Martin Kürster

European Southern Observatory, Casilla 19001, Santiago 19, Chile

Abstract. An overview of the ongoing planet search programmes via precise RVs is given for comparison with ESO's efforts at the 3.6m. ESO's announcement of opportunity for a spectrograph for the 3.6m capable of a measurement precision of 1 m/s (HARPS) is calling for the best achievable precision worldwide. I touch on the instrumental requirements, reasons for setting the goal to 1 m/s, and the complementary role of ESO's highest resolution instrument, the refurbished CES at the 3.6m, to obtain important ancillary data.

1 Introduction

To date precise differential radial velocity (RV) measurements have been the most successful technique to discover extra-solar planets. Not only have they produced 18 discoveries during the past 3.5 years, they have also revealed the difference of the majority of these objects from solar system planets. The best known example is the short period (4.2 d) orbit of the first extrasolar planet discovered, the companion to 51 Peg [11], an object of roughly Jupiter mass that has become the prototype of a whole class of similar planets. Also noteworthy are the high mass of the superplanets around HD 114762 [9] ($m \sin i = 11\ \mathrm{M_{Jup}}$) and 70 Vir [10] ($m \sin i = 7.4\ \mathrm{M_{Jup}}$) and the high eccentricities found for many of these objects with the companion to 16 Cyg B [2] being the extreme ($e = 0.67$).

In order to understand planet formation scenarios and observational selection effects a larger statistical sample is now required. To this end large surveys are currently underway that monitor a few thousand late-type stars for RV variations (see Table 1). For the discovery of 'true Jupiters' within a few years' time also an increased measurement precision is indicated. Owing to the intrinsic variability of late-type stars due to activity (star spots, convection patterns) a natural limit of the RV method appears around 1 $\mathrm{ms^{-1}}$; at this level virtually every late-type star seems to be variable [13] with the implication that any further improvement in RV measurement precision would not lead to a similar improvement in detection probabilities.

ESO has issued an announcement of opportunity for the construction of a new spectrograph for the 3.6m telescope in La Silla capable of achieving a long-term RV precision of 1 $\mathrm{ms^{-1}}$. Project HARPS (High Accuracy Radial Velocity Planet Search) was initiated following the recommendations of the ESO Working Group On the Detection of Extrasolar Planets [12] and the

Table 1. Ongoing planet search programs via precise RVs

Observatory/instrument (Investigators)	# of stars	Types (dwarfs)	Start	Accuracy
Keck HIRES		late F to	July	
(Marcy, Butler et al.)	400	late K	1996	≈ 6 ms^{-1}
Lick 3m Shane/CAT				
+ Hamilton Ech.		late F		
(Marcy, Butler et al.)	120	to M	1987	≥ 3 ms^{-1}
AAO 3.9m		late F	Sept.	
(Butler, Marcy et al.)	150	to K	1997	≥ 3 ms^{-1}
Keck HIRES+AFOE/CORALIE		late F	mid-	
follow-up (Latham et al.)	1000	to G	1997	≈ 10 ms^{-1}
Haute Pr. 1.93m + ELODIE				upgraded to
(Mayor, Queloz, Sivan et al.)	350	G and K	1994	7 ms^{-1}
Swiss 1.2m+CORALIE (La Silla)		late F to	June	
(Mayor, Queloz, Udry et al.)	800	late K	1998	6 ms^{-1}
Whipple 1.5m + AFOE		solar-	mid-	
(Noyes et al.)	100	type	1996	10 ms^{-1}
Hobby Ebberly Tel. + HRS		mid-F to		
(Cochran, Hatzes)	300	late G	2000	$3-10$ ms^{-1}
Keck HIRES		late-F to M	fall	
(Cochran, Hatzes)	100	Hyades	1996	$3-6$ ms^{-1}
McDonald 2.7m + Coudé		late F to		
(Cochran, Hatzes)	30	late K	1987	20 ms^{-1}
ESO CAT/3.6m + CES				
(Kürster, Hatzes, Cochran,		late F	Nov.	
Dennerl, Döbereiner, Endl, Els)	35	to M	1992	$10-20$ ms^{-1}

ESO OPC. As a large stellar sample must be monitored over a long time baseline the amount of 3.6m time to be allocated to the project will be on the order 100 nights per year over five years.

2 Main Issues on the Road to 1 m/s

Complex instrumental requirements will have to be met when a long-term precision of 1 ms^{-1} shall be reached. Opening the possibilities for innovative approaches ESO has explicitly not made any design preselection. Therefore, only two major points are shortly addressed here, the issues of stabilization of the beam that feeds the spectrograph and of the spectrograph itself.

Possibilities to reach a stable input beam include fibre feeding the spectrograph, using image scramblers, and a tip-tilt and/or adaptive optics system. To stabilize the spectral measurement the spectrograph itself can be stabilized (temperature and pressure control) and parallel wavelength calibration used. An alternative is presented by the self-calibration approach where the star light is sent through a (temperature and pressure controlled) gas absorption cell that superimposes stable reference absorption lines onto the stellar spectrum. When combined with the detailed modelling of the instrumental profile this latter approach has produced the highest RV precision to date of 3 ms^{-1}[1]. The most frequently used device for self-calibrating, and iodine gas absorption cell, is technically easy to produce and install.

3 ESO's Facilities for Measuring Precise RVs

Several instruments will be available at ESO for measuring precise RVs:

- UVES at UT2 on Paranal, self-calibrated with an iodine cell, will be particularly useful for faint objects (e.g. M dwarfs).
- CRIRES at UT4 on Paranal, possibly with an CH_3I cell for self-calibration (G. Wiedemann, priv. comm.), holds the prospect for the direct discovery of planetary spectral signals [14].
- FEROS at the 1.52m on La Silla has been demonstrated to achieve an RV precision of at least 21 ms^{-1} using simultaneous calibration [6].
- CES at the 3.6m on La Silla, self-calibrated with an iodine cell, has been shown to achieve an RV precision in the $10-25$ ms^{-1} range [4] [5][7][3].
- HARPS at the 3.6m with a goal to achieve a precision of 1 ms^{-1}.

4 The Role of the CES

ESO's highest resolution spectrograph at the 3.6m ($R = 230,000$), the fibre-fed CES will be in a position to provide important ancillary data to control planetary candidates for intrinsic stellar variability via the method of line bisector analysis. With an iodine cell already in place it can also be used for an independent confirmation of planetary candidates.

To illustrate the potential of this instrument with a recent result (from [8]) Fig. 1 shows the outcome of four years of precise RV monitoring of Prox Cen with the CES fed previously with the CAT telescope and with the previous Long Camera ($R = 100,000$). Despite of the lack of a clear detection of a substellar companion to Prox Cen the data could be used to determine (via Monte Carlo simulation) stringent upper limits to the mass of any possible objects around the star as a function of orbital period and/or separation.

Similar analyses on non-detection data for the ongoing surveys will provide a wealth of information for constraining planet formation models.

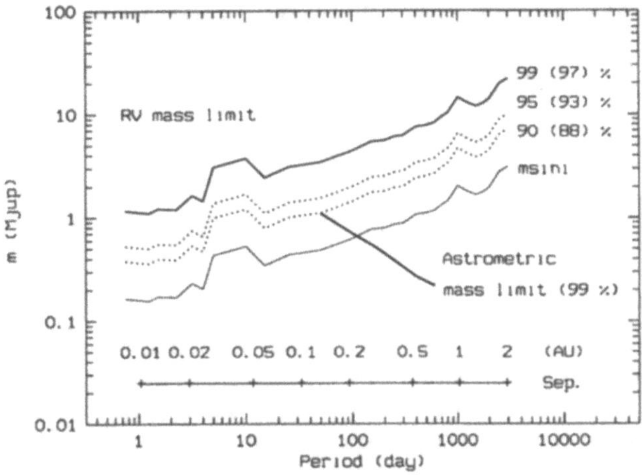

Fig. 1. From precise RV measurements with the ESO 1.4m CAT+CES: Upper limits to the mass of companions to Prox Cen as a function of orbital period and separation. The four parallel curves represent the $m \sin i$ and several confidence intervals for the mass limit. The single bold line is the mass upper limit from HST FGS astrometry (Benedict, priv. comm.)

References

1. Butler R.P., Marcy G.W., Williams E., McCarthy Ch., Dosanjh P., Vogt S.S.: 1996, PASP 108, 500
2. Cochran W.D., Hatzes A.P., Butler R.P., Marcy G.W.: 1997, ApJ 483, 457
3. Hatzes A.P., Kürster M., Cochran W.D., Dennerl K., Döbereiner S.: 1996, J. Geophys. Rev. (Planets), 101, 9285
4. Kürster M., Hatzes A.P., Cochran W.D., Pulliam C.E., Dennerl K., Döbereiner S.: 1994, The Messenger, No. 76, 51
5. Kürster M., Hatzes A.P., Cochran W.D., Dennerl K., Döbereiner S., Endl M., Vannier M.: 1998, Workshop Science with Gemini, B. Barbuy, E. Lapasset, R. Baptista, R. Cid Fernandes (eds.), IAG-USP & UFSC 1998
6. Kaufer A., Stahl O., Tubbesing S., Nørregaard P., Avila G., Francois P., Pasquini L., Pizzella A.: 1999, The Messenger 95, 8
7. Kürster M., Hatzes A.P., Cochran W.D., Dennerl K., Döbereiner S., Endl M.: 1999, IAU Colloq. 170, J.B. Hearnshaw, C.D. Scarfe (eds.), PASP Conf. Ser., in press
8. Kürster M., Hatzes A.P., Cochran W.D., Döbereiner S., Dennerl K., Endl M.: 1999, A&A 344 L5
9. Latham D.W., Mazeh T., Stefanik R.P., Mayor M., Burki G: 1989, Nature 339, 38-40
10. Marcy G.W., Butler R.P.:1996, ApJ 464, L147
11. Mayor M., Queloz D.: 1995, Nature 378, 355
12. Paresce F., Renzini A.: 1997, The Messenger 90, 15
13. Saar S.H., Butler R.P., Marcy G.W.: 1998, ApJ 498, L153
14. Wiedemann G.: 1996, The Messenger 86, 24

Direct Observation of Hot Jupiters and Other Faint Companions with the VLTI

Vincent Coudé du Foresto

Observatoire de Paris (DESPA), 92195 Meudon Cedex

Abstract. The existence of Jupiter-size planets very close to their parent stars (two examples of which are 51 Peg and τ Boo) creates a challenge for theories of planetary formation and structure. Only a direct measurement of their mass and luminosity (not accessible via the current radial velocity techniques) can help discriminate between the different models that have been proposed for this class of objects.

Direct observation of such extrasolar planets with an infrared interferometer is a difficult task because of the high dynamic range required (around 10^4), but not impossible if the seeing noise is taken care of by spatial filtering and fringe tracking.

A method is proposed for the VLTI, analogous to a synchronous detection using radial velocity orbital phase information. Simulations show that its sensitivity would be sufficient to detect hot Jupiters with a small spectral resolution.

1 Introduction

Giant planets are known to orbit in the near proximity of a small percentage of nearby solar-type stars (such as 51 Peg or τ Boo). Their existence is inferred from indirect observations with radial velocity techniques, which provide firm information on orbital parameters only (maximum elongation, period, eccentricity and ephemeris). They revolve around a quasi-circular orbit within a few days. Their mass can be hinted at through the $M \sin i$ product, where i is the unknown orbital inclination with respect to the plane of the sky. Beyond this, however, the physical nature of these objects remains elusive. They are expected to be hot because they are close to their parent star (0.05 AU or even less). A rudimentary model, assuming the planet to be a grey body, gives their equilibrium temperature T_{eq}:

$$T_{eq} = \frac{T_\star}{\sqrt{2}}(1-A)^{\frac{1}{4}}(\frac{R_\star}{a})^{\frac{1}{2}} \tag{1}$$

where $T_\star$ is the temperature of the parent star, $R_\star$ its radius, a the orbital radius and A the planet albedo. Assuming $A = 0.1$, the equilibrium temperature would be about 1260 K and 1620 K for the companions of 51 Peg and τ Boo, respectively.

A spectrophotometric analysis of these exotic "hot Jupiters" requires a direct observation. At 15 parsecs, the parent star and its planetary companion form a binary pair separated by 2 to 10 milliarcseconds, and interferometric aperture synthesis is required to achieve that angular resolution. The resolving power of the VLTI is adequate, since with a 200 m maximum baseline in its sub-array (with the 1.8 m auxiliary telescopes) it ranges from 1 mas at $\lambda = 1\,\mu$m to 10 mas at $\lambda = 10\,\mu$m. The challenge, of course, comes from the very high contrast between the star and its faint companion. From this point of view, however, hot Jupiters are –by far– the most favorable class of exoplanets, because of their thermal emission and also (depending on albedo) a non negligible stellar flux reflection at shorter wavelengths. Of all hot Jupiters discovered so far, 51 Peg B and τ Boo B feature the smallest expected contrast, with a brightness ratio α of the order of 10^4 for 51 Peg B and 3×10^3 for τ Boo B at $\lambda = 2.2\,\mu$m. Further in the infrared around 10 μm, the brightness ratio is 5 times lower, but the maximum angular separation of the systems (3.6 mas for 51 Peg and 2.4 mas for τ Boo) make them marginally resolved at those wavelengths.

The issue, then, is to achieve interferometric observations of relatively bright stars with a dynamic range of at least 10^4. A solution is proposed here, which involves accurate determinations of fringe visibility moduli and a synchronous detection scheme. Another method exists which uses differential phase measurements; it is detailed elsewhere in these Proceedings by Lopez and Petrov.

2 Proposed Method

Although a challenging problem, the observation of a planet orbiting its parent star is ideally suited for interferometry: the system can be well described by a simple model (a small central disk with an orbiting point source), and useful astrophysical information can be retrieved from a reduced set of parameters. In this case there are only three unknown parameters: the orbit inclination i (which removes the ambiguity on the mass of the companion in the $M \sin i$ product), its orientation Ω (which can be compared to the orientation of a circumstellar disk, if any), and the brightness ratio α which, if obtained at several wavelengths, provides the spectroscopy. The other parameters in the model are known from radial velocity observations: they are the maximum elongation, the orbital period P, the eccentricity, and the ephemeris.

The model can be used to predict fringe visibilities at any time, with any pair of telescopes. The visibility modulus function of the binary system shows, on top of the standard stellar Airy function, a modulation pattern whose peak to peak amplitude is $2/\alpha$, and whose periodicity and orientation in the (u, v) plane depend on the instantaneous position of the planet, i.e. the known phase ϕ of the radial velocity oscillations (Figure 1). Thus any visibility point can be located into a (u, v, ϕ) data cube (or, rather, a data tore, since ϕ is periodic), which can be filled in progressively as the source is observed during several (not necessarily consecutive) nights (using diurnal rotation to explore the (u, v) plane and with different baselines. Once enough data have been accumulated, they can be solved for a, i and Ω.

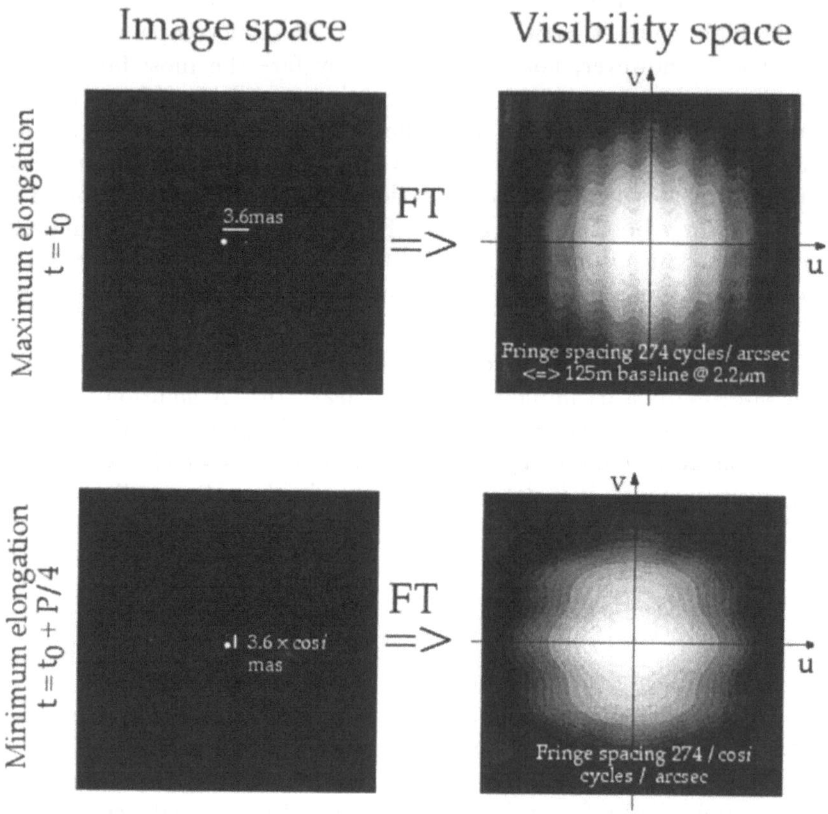

Fig. 1. Modelization of the 51 Peg binary system in the image and visibility spaces, at two different epochs. For clarity the brightness ratio in this example has been lowered to the unrealistic value of 20.

In practice the target source is observed in alternance with a reference star of a similar spectral type (and known to have no planet) to calibrate the interferometric point source response. In the data cube, the planet signature builds up as a coherent signal above the noise constituted by instrumental and external errors.

Instrumental errors include any drift in the point source response, and statistical errors on fringe visibility measurements. Statistical errors, for bright sources, are dominated by seeing noise, and their amplitude depends on how atmospheric turbulence is treated:

- In a multimode instrument the point source response has a random atmospheric term and the precision reaches a few percent;

- If a spatial filter (and/or good adaptive optics) is used to remove the spatial modes of the atmospheric turbulence, the precision (standard error) has been demonstrated to be 0,3 % [1] or lower in 5 mn of observing time. It is dominated by the differential piston between the two interferometric sub-pupils (the only non-spatial mode of the turbulence);
- If, in addition, a fringe tracker is used to remove the differential piston, the standard error should be further reduced drastically since all atmospheric effects have in principle been eliminated. However, the amount of residual statistical noise is yet to be determined since the combination of perfect spatial filtering and fringe tracking (that the VLTI will offer) does not exist in current instruments.

External errors include wrong estimates of the source and reference diameters (constant bias in the fringe visibilities, to which the data fitting procedure is insensitive). Starspots may generate 0.1 % visibility signatures for solar-type stars [2] but much less in the upper part of the Airy pattern, and the baselines should be chosen accordingly.

It is important to note that since the data reduction scheme is analogous to a synchronous detection for the planetary signal, all these errors can be averaged out over long time integrations without being dominated by systematics.

3 Simulated Observations

Figure 2 shows the expected performance of the VLTI in the case of simulated observations with the following assumptions:

- A set of 11 baselines of VISA, the VLTI sub-array of 1.8 m auxiliary telescopes, is used;
- 5 nights are spent on each baseline, with 6 hours of observing time per night;
- The source is alternated with the reference every 5 mn;
- The planetary systems parameters are ($\alpha = 10^4$, $a = 3.6$ mas, $i = 75$ deg) for 51 Peg and ($\alpha = 3 \times 10^3$, $a = 2.4$ mas, $i = 75$ deg) for τ Boo. Two different orientations are tried: $\Omega = 0$ deg and $\Omega = 90$ deg.

The amount of observing time is considerable (55 nights in total), but they need not be consecutive and they involve only the more accessible sub-array. The relative error on the brightness ratio $\sigma(\alpha)/(\alpha)$ is plotted as a function of the interferometer accuracy (standard statistical error obtained in 5 mn of observing time). Even for the already demonstrated accuracy of 0.3 %, the photometry of τ Boo B can be measured within 20 % and the companion of 51 Peg can be marginally detected. The performance is expected to be substantially better when the VLTI is equipped with a fringe tracker.

Signal to noise estimates show that the bright star regime for which this simulation is valid, where the statistical noise is dominated by seeing noise,

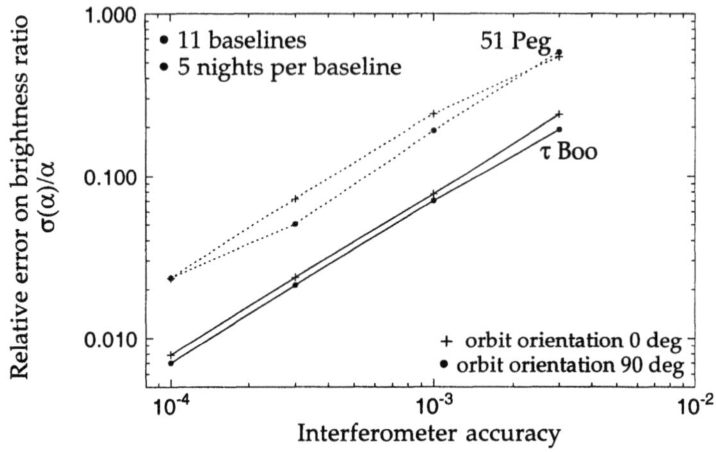

Fig. 2. Simulated performance of the VLTI on photometric measurements.

can be achieved with current detectors on 51 Peg and τ Boo for spectral resolutions of up to 250. This represents about 50 spectral channels in the photometric K band.

4 Conclusion

When the VLTI was conceived over 10 years ago, it was not expected to be able to directly observe exoplanets, because the observational difficulties seemed daunting. Interferometry was to be used only for indirect (astrometric) observations. The 1995 discovery of hot Jupiters around nearby stars, with their comparatively low contrast to their parent body, changed the situation. Indeed the simulations show that, provided a large number of nights can be dedicated to this program and the target companion has already been detected by radial velocity techniques, a VLTI sub-array equipped with a single-mode instrument and a fringe tracker has the sensitivity, the precision and the baselines required to directly observe selected exoplanets, and to perform low to medium resolution spectroscopy on them.

References

1. Perrin G., Coudé du Foresto V. et al. (1999) Interferometric observations of R Leonis in the K band. First direct detection of the photospheric pulsation and study of the atmospheric intensity distribution. Astron. Astrophys. **345**, 221–232
2. Von der Lühe O., Solanki S. et al. (1995) Observing stellar surface structures with the ESO-VLT interferometer, in IAU Symposium 176 "Stellar Surface Structure", Strassmeier K.G. ed., Vienna.

Direct Detection of Hot Extrasolar Planets Using Differential Interferometry

Bruno Lopez[1] and Romain G. Petrov[2]

[1] Observatoire de la Côte d'Azur, UMR 6528, BP 4229, F-06304 Nice cedex 4, France
[2] UMR 6525 "Astrophysique" du CNRS et de l'Université de Nice Sophia Antipolis, Parc Valrose, F-06108 Nice cedex 2, France

Abstract. We evaluate the potential of differential interferometry with the VLT or Keck interferometers for the direct detection and the study of atmospheric characteristics of hot giant extrasolar planets around nearby stars.

Differential interferometry has been shown to allow unbiassed measurements of phase and/or visibility variations with wavelength. For a star and planet system like for any binary, these measurements at several orbital phases yield the angular separation and the spectrum of the components.

We present an evaluation of the fundamental measurement uncertainty resulting from source and sky photon noise and detector noise. It shows that, with two 8 meters telescopes and a 40 meters baseline, we should detect Jupiter-like planets up to 0.1 or 0.2 astronomical units away from a solar type star at 10 pc.

1 Introduction

Since the first detection of a hot giant planet orbiting around the star 51 Peg, several very exciting results have been obtained in detecting planets outside our solar system. Numerous instruments are being proposed to extend and complement radial velocity measurements for the detection of extra solar planets and the study of their actual orbits, masses and other physical parameters. This includes coronographic instruments, photometers, such as Corot, and astrometric instruments. The most ambitious program with ground based instruments is very accurate interferometric astrometry with the VLT and Keck interferometers intended to measure the star reflex motion due to the planet ([5]).

Here, we focus on an alternative and quite promising method called Differential Interferometry (DI). Instead of measuring the absolute visibility ([4], [6]), DI is based on the variation with wavelength of the visibility and/or the phase of the interferometric fringes produced by a star and planet system. In practice one measures the ratio of complex coherence between one ore more channel(s) and a reference channel with a different central wavelength. This yields a sampling of the visibility ratio and phase difference. As we will see in section 2.1, these quantities are related to the spectrum and angular position of the planet, which can be separated if the measurement is repeated at several orbital phases. Thus, one of the main advantages of the method is that, in addition to the planet orbit, it gives a direct measurement of the planetary spectrum allowing to estimate the physical conditions in the planetary atmosphere. Section 2.2 discusses the fundamental sources of measurement uncertainty and shows that DI applied on a large apertures interferometer (see also [13] and [12]) has the potential to resolve planetary systems and to study Jupiter-like planets orbiting a Sun at 10 pc with semi major axes smaller than 0.2 AU. However, the method appears quite challenging because it needs an extremely high stability and/or a very accurate calibration of the instrument. Section 3 very briefly discusses this point and concludes that DI should yield exciting results in this field shortly after the AMBER and MIDI VLTI focal instrument start observing, given that some specific calibration tools are installed, such as for example a field inverter.

2 Potential Performances with the VLTI

2.1 The Measured Quantities

In its principle, it is a relative method which measures the visibility ratio and the phase difference of the fringes between two or several wavelengths. For insuring a relative calibration of the observation, the different wavelengths are observed simultaneously.

$$\frac{V(t/P,\vec{B}/\lambda_1)}{V(t/P,\vec{B}/\lambda_2)} = \left(\frac{I_*^2(\lambda_1)+{I_p}^2(t/P,\lambda_1)+2I_*(\lambda_1)I_p(t/P,\lambda_1)cos(2\pi\frac{\vec{B}}{\lambda_1}.\vec{\rho}(t/P))}{I_*^2(\lambda_2)+{I_p}^2(t/P,\lambda_2)+2I_*(\lambda_2)I_p(t/P,\lambda_2)cos(2\pi\frac{\vec{B}}{\lambda_2}.\vec{\rho}(t/P))}\right)^{0.5} \quad (1)$$

and,

$$\Delta\Phi(t/P,\lambda_2,\lambda_1) = \Phi(t/P,\vec{B}/\lambda_2) - \Phi(t/P,\vec{B}/\lambda_1) \quad (2)$$

which, due to the small value of the planet luminosity compared to its parent star, can be approximated by ,

$$\Delta\Phi(t/P,\lambda_2,\lambda_1) \simeq \frac{I_p(t/P,\lambda_2)}{I_*(\lambda_2)} \times sin(2\pi\frac{\vec{B}}{\lambda_2}.\vec{\rho}(t/P)) - \frac{I_p(t/P,\lambda_1)}{I_*(\lambda_1)} \times sin(2\pi\frac{\vec{B}}{\lambda_1}.\vec{\rho}(t/P)) \quad (3)$$

In the above relation, P, $I_p(t/P,\lambda)$ and $\vec{\rho}(t/P)$ represents the astrophysical parameters of interest that are constrained by the observation and which are respectively the period of the planet orbit, the intensity of the planet as a function of its phase and of its wavelength and the angular position of the planet. Several measurements at different angular positions of the planet are needed for separating actually the astrophysical parameters. They will be determined by fitting the measurements of $V(t/P, \vec{B}/\lambda_1)/V(t/P, \vec{B}/\lambda_2)$ and $\Delta\Phi(t/P, \lambda_2, \lambda_1)$ obtained at different epochs.

The first effect contributing to the variations of $V(t/P, \vec{B}/\lambda_1)/V(t/P, \vec{B}/\lambda_2)$ and $\Delta\Phi(t/P, \lambda_2, \lambda_1)$ is related to the wavelength dependence of the spatial frequency $\vec{B}/\lambda$. This provides an instantaneous covering of a part of the Fourier plane sometimes sufficient to detect a binary (star plus a planet) signature. An other contribution to the signal comes from the strong wavelength dependence of the flux ratio between the planet and its parent star : the planet and the star have two different effective temperatures and the planet has some specific spectral bands. For example strong and broad absorption bands are observed/predicted in J, H and K regions in the spectra of giant planets of various masses due to the absorption of water vapor and methane [1].

Figure 1 displays, as a function of the wavelengths and for different semi-major axis of the planet orbit, the variations of $V(t/P, \vec{B}/\lambda_1)/V(t/P, \vec{B}/\lambda_2)$ and $\Delta\Phi(t/P, \lambda_2, \lambda_1)$ assuming a channel of reference with $V(t/P, \vec{B}/\lambda_2) = 1$ and $\Phi(t/P, \lambda_2) = 0$.

The modelling assumes a sun like star at 10 pc of distance and several planet sizes (1 and 2 Jupiter radii for Figure 1). A slightly pessimistic value of 0.3 for the albedo has been taken. The study displayed in our curves are for passive planets (the conditions for a detections of a planet with an internal source of heat may be otherwise relaxed) radiating on all their surface. A synchronous planet radiating on half its surface has been considered and does not change much the estimates made here. The baseline has been set to 40 meters. The Figure 1 shows that when the system is resolved the visibility and the phase have basically the same shape while when the system becomes unresolved (i.e. when the wavelength increases) the visibility decreases faster than the phase.

2.2 The Fundamental Uncertainty of the Measure

The potential accuracy of the VLTI in J, H, K and N bands which will be covered by the AMBER (near-infrared) and MIDI (mid-infrared) instruments is shown in Figure 1.

The fundamental noises taken into account are the read out noise of the detector and the source and background photon noise. The uncertainty of a measurement, defined as its standard deviation of the measurement, is directly proportional to the inverse of the signal to noise ratio ([3], [7], [10]) estimated in the AMBER and MIDI reports ([8], [11]). Not considered here is the stability of the instrument that we expect to be strongly nulled by the relative differential method (see the section Discussion).

We assume that each of the J, H, K and N bands are split into 5 spectral channels. The estimation is made for two total observing times (0.5 and 5 hours) each divided in many short exposures of 100 ms. The VLTI baseline considered in the figures is 40 meters, other lengths have been considered and are discussed below. We assume that the 8 meter UT telescopes are used for this measurement.

2.3 Potential Performance

As a result, it appears that the K band offers the highest accuracy. For a Jupiter size planet, 0.5 hour of observations will permit a detection when the semi major axis between the planet and its star is less than about 0.1 AU. 5 hours of observations are needed for a detection of a planet if the semi major axis reaches about 0.2 AU. In the N band, about 20 hours of effective observations will be needed to detect a system with a separation ranging from 0.03 to 0.1 AU. For a planet twice the size of Jupiter, the situation is relaxed by a factor 4 in accuracy, allowing MIDI to reach the 0.2 AU limit.

The increase of baseline length from 40 meters (case considered here in Figure 1) to 80 and 120 meters (the maximum UTs VLTI baseline) brings a benefit mainly in the 10 microns regions of the spectrum, because at this wavelength the system changes from an unresolved or just barely resolved object to a better resolved one.

3 Discussion

The signal resulting from a Jupiter-like planet can be substantially higher than the minimum level of measurement uncertainty. Among the 18 planets already detected and suspected, 9 have separations smaller than the 0.2 AU typical limit for DI and 4 of them are supposed to be more massive than Jupiter. Half an hour of observations of a system like 51 Peg in the K band with AMBER will allow a 10 σ detection in each of the 0.1 μm spectral channel. However, the practical implementation of the experiment will be a challenge. Atmospheric chromatic effects, such as the dispersion will affect the result in a random way and will have to be calibrated. The dispersion of the air in the interferometric tunnels and in the optical fibers, the coupling between optical aberrations and the instrument stability, the variations of the detector gain table will introduce visibility and phase differences substantially higher than the astrophysical signal. We expect that the variations of the instrumental contribution to the signal is slow and that it can be eliminated by modulation techniques which modify the signal in a known way faster than the instrumental effects change.

A method which has been best proven in related techniques (Differential Speckle Interferometry) uses a device which inverses the field and therefore the signal without affecting the instrumental contribution. The field inversion device is an alternative to a method adapted to space interferometers/telescopes consisting in rotating all the interferometer about the axis to the star [2]. Such a device will be particularly useful to allow very accurate

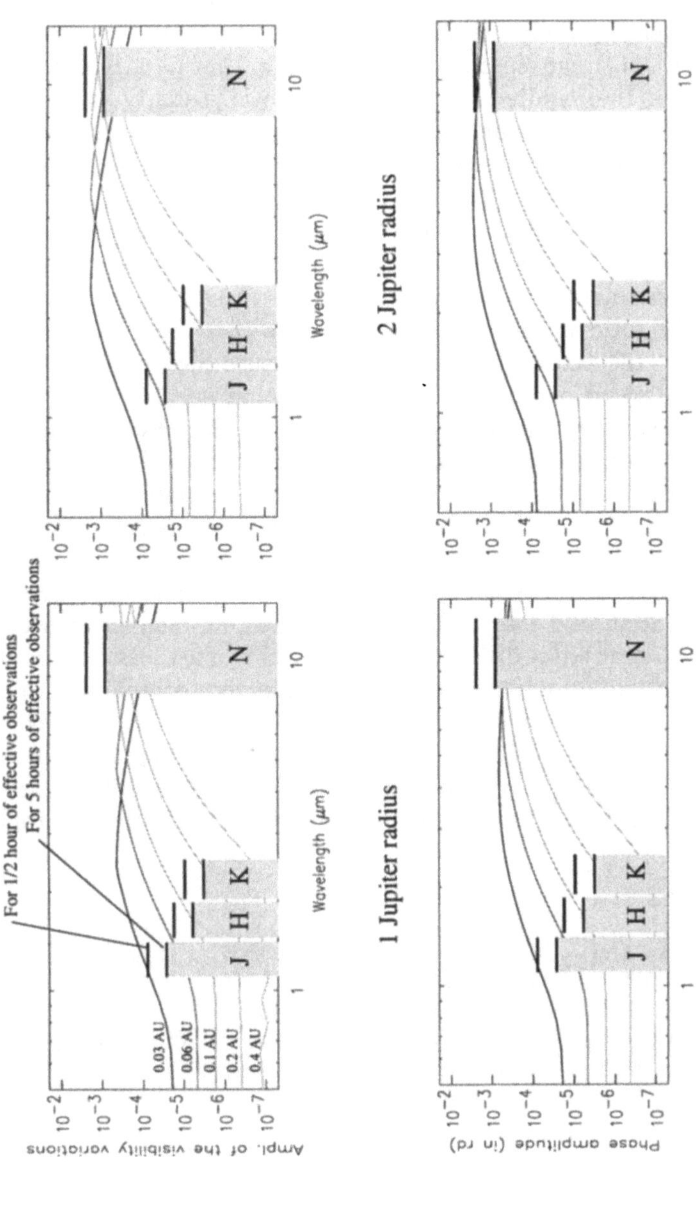

Fig. 1. The signal received from a Jupiter-like planet of two sizes and for different semi-major axis of its orbits compared to the estimated 1σ accuracy of the VLTI instruments AMBER and MIDI. The considered baseline is 40 meters.

calibrations of the phase differences between several instruments working at different wavelengths such as the VLTI instruments MIDI and/or AMBER and/or the VLTI fringe sensor. It is worth noting that for a 10 μm instrument such as MIDI, the requirements for instrumental stability are relaxed by a factor 100 with regard to the near infrared. Thus, even if the 10 μm situation appears less favorable with regard to the fundamental photon noises, it may result that it is easier to implement in practice.

Carefully modeling all the atmospheric and instrumental effects is the next step in the study of the feasibility of Differential Interferometric observation of hot giant extra solar planets. However, the present estimations of the limiting signal to noise ratio are encouraging. They show that there is a possibility not only to detect the extra solar planets and measure their orbits and therefore masses but also to measure their low resolution spectra and light curves which will yield very interesting constraints on the physical conditions of the planet atmosphere such as its temperature, composition, gravity. It is an interesting dream to think that we could by the proposed method estimate the radiating surface and see for example if it is consistent with the model predictions linking mass and planet size. Due to their proximity to their star, the atmospheres of these hot giant planets may be strongly perturbed by the star radiation and lose their mass in the same manner as a comet. A companion circumstellar disk possibly fed by this mass loss may be present.

The potential performances of the VLTI are very comparable to those of the Large Southern Array project supposed to detect hot Jupiter-like planets in a few 10 of hours ([9]). The new generation of infrared interferometers (VLTI, Keck) provides a set of adequate baselines and a sufficient sensitivity that must be used for this challenging and relevant astrophysical topic.

References

1. Burrows A., Marley M., Hubbard W. B. et al 1997, ApJ 491, 856
2. Bracewell R.N. 1978, Nature 274, 780
3. Chelli A. and Petrov R.G. 1995, A&AS 109, 401
4. Coudé du Foresto V. 1999, this symposium (Parallel Workshop 2)
5. Glindeman A., this symposium (Parallel Workshop 2)
6. Mariotti J.-M., Coudé du Foresto V., Perrin G. 1996, AAS DPS meeting 28, 1207
7. Lagarde, S., 1994, Ph. D. Thesis, University of Nice, France
8. Leinert C., Graser U. et al 1998, report of the MIDI VLTI instrument for the ESO concept review of October 1998
9. Menten K.M., this symposium (Plenary sessions)
10. Petrov, R.G., et all, 1991, ESO-NOAO conference on High Resolution Imaging By Interferometry II, Garching bei München, Germany
11. Petrov R.G. et al 1999, report of the AMBER VLTI instrument for the ESO concept review of January 1999
12. Quirrenbach A. 1999, this symposium (Parallel Workshop 2)
13. Van Belle G. 1999, this symposium (Parallel Workshop 2)

The CORALIE Planet-Search Sample

S. Udry[1], M. Mayor[1], D. Queloz[1,2], D. Naef[1], and N. Santos[1]

[1] Observatoire de Genève, CH-1290 Sauverny, Switzerland
[2] JPL, 4800 Oak Grove Drive, Pasadena, CA-91109, USA

This short contribution presents a description of the CORALIE sample selection of more than 1600 dwarf stars for the planet-search programme on the 1.2-m Euler Swiss telescope at La Silla. The instrumentation, aims and first discoveries of the CORALIE planet-search programme are discussed in Queloz et al. (this volume). The sample selection was performed according to the following criteria:

- Stars closer than 50 pc ($\pi \geq 20$ mas) with $\sigma_\pi \leq 5$ mas and F8$\leq$ Spectral type <M1 in the HIPPARCOS catalogue
- Photometric cut off of giant stars (2.5 mag off the main sequence)
- Removal of faint cool dwarfs by a colour-dependent distance cut off ($d_{\rm max} = f$(B-V), taking into account that, at a given precision level, with our cross-correlation technique less signal is needed for K dwarfs.

The final sample is presented in Fig. 1 displaying the stellar apparent magnitude and distance distributions. We note that ~80% of the selected stars are brighter than $m_V = 9$ and that faint objects are mostly K dwarfs for which CORALIE is the most efficient.

For the planet search itself, large amplitude binaries (CORAVEL) and potentially intrinsic variable stars (large activity index and rotational velocity) are gathered in low-priority subprogrammes. They however belong to the global volume-limited sample to be used for statistical purpose.

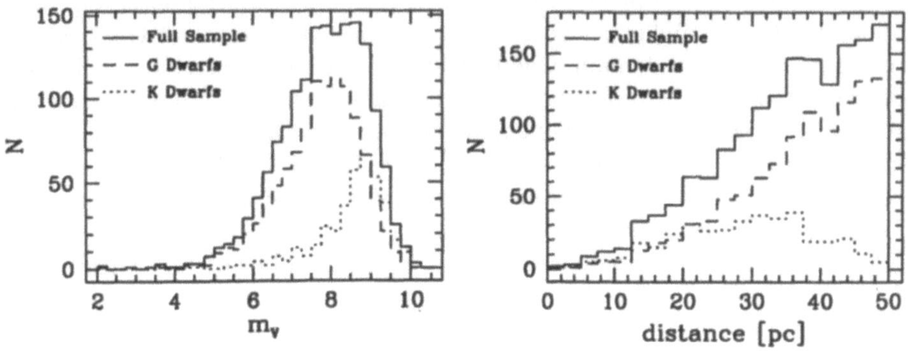

Fig. 1. Apparent magnitude and distance distributions of the CORALIE planet-search sample. 80% of the stars are brighter than $m_V = 9$

The HARPS Project

F. Pepe[1], M. Mayor[1], W. Benz[2], J.-L. Bertaux[3], J.-P. Sivan[4], D. Queloz[1], and S. Udry[1]

[1] Observatoire de Genève, CH-1290 Sauverny, Switzerland
[2] Physikalisches Institut of the University of Bern, CH-3012 Bern, Switzerland
[3] Service d'aéronomie, F-91371 Verrières le Buisson, France
[4] Observatoire de Haute-Provence, F-04870 St-Michel l'Observatoire, France

To pursue a competitive programme in the domain of extra-solar planet search, ESO proposed to build HARPS (High-Accuracy Radial-velocity Planetary Search), a new high-resolution spectrograph to be installed on the ESO 3.6-m telescope at La Silla and capable of achieving a $1\,\mathrm{ms}^{-1}$ precision for radial-velocity measurements.

Design and stability: HARPS is a two fiber-fed and cross-dispersed echelle spectrograph. The required $1\,\mathrm{ms}^{-1}$ long-term accuracy will only be attained if both *photon noise* and *instrumental* errors are below this limit. The spectrograph will therefore be operated *in vacuum* and will be *temperature controlled* in order to minimize drifts of the spectrum due to environmental effects. The precise wavelength calibration will be insured by the simultaneous ThAr method as for ELODIE or CORALIE.

Planet search at $1\,\mathrm{ms}^{-1}$: The amplitude of the periodic radial-velocity variations revealing the presence of an extra-solar planet scales with the mass of the planetary companion. The improvement in precision down to $1\,\mathrm{ms}^{-1}$ will thus largely contribute to remove the present detection biases for *light* and *long-period* planets. This is especially important for multi-planet systems.

Science with HARPS: The unequaled advantages of the HARPS facility (large telescope, high precision) combined with a substantial observing time allocation (~ 100 nights per year during 5 years) will permit a break through in the planet-search domain. One first aim is to derive a precise distribution function $f(m_2, a, e, [\chi_i/H])$ for solar-type stars. This is of prime importance to provide powerful constraints on stellar and planetary formation and evolution scenarios. In the planet-search domain, HARPS will also bring crucial contributions for the study of halo stars (influence of metallicity for planet formation), for the determination of "hot Jupiter" mean densities by combining transit and spectroscopic results, etc. HARPS will be useful to determine future targets for the interferometric programmes (VLTI).

In other domains as asteroseismology, pulsating stars, spectroscopic binaries, etc., the HARPS capabilities will also prove to be very efficient.

When available on the 3.6-m telescope, HARPS will provide ESO with an unequaled facility for the search of extra-solar planets

http://obswww.unige.ch/Instruments/harps

Author Index

Adelberger, K.L. 20
Allard, F. 399
Alonso, M.V. 298
Appenzeller, I. **3**, **164**
Arnaboldi, M. **126**, **204**
Artymowicz, P. **391**
Baraffe, I. 399
Barbuy, B. **357**
Barucci, M.A. **70**
Basri, G. **429**, 446
Beichman, C. 505
Béjar, V. 446
Bender, R. 3, **61**
Benz, W. 572
Bertaux, J.-L. 572
Böhnhardt, H. 3
Böhringer, H. **157**
Bonifacio, P. **338**
Borgani, S. **143**
Broadhurst, T. **138**
Burgasser, A. 505
Burnet, M. 548
Burrows, A. 505
Capaccioli, M. 126, 204
Cappellaro, E. **313**
Carlstrom, J.E. 150
Cavaliere, A. 176
Chabrier, G. **399**
Chincarini, G. 157, **198**
Chini, R. **303**
Chiosi, C. **323**
Clausen, J.V. **225**
Close, L. **518**
Cochran, W.D. 539
Collins, C.A. 157
Combes, F. 369
Confino, B. 548
Coudé du Foresto, V. **560**
Cristiani, S. 3, **184**
Cruddace, R. 157
Cutri, R. 505
da Costa, L. **192**
Dahn, C.C. 505
DeGrandi, S. 157
Dickinson, M. 20
Dietrich, M. 3
Drenkhahn, G. 259
Edge, A.C. 157
Ford, H. 126
Fort, B. **102**
Fouqué, P. 217
Freeman, K.C. 126
Fricke, K. 3
Fridlund, M. 535
Fuhrmann, K. **351**
Fürtig, W. 3
Gässler, W. 3
Gerhard, O. 126
Giacconi, R. 121
Giavalisco, M. 20
Gieren, W.P. **217**
Gilmozzi, R. 3
Girardi, L. 248, **294**
Gizis, J.E. 505
Glindemann, A. **468**
Gómez, M. 259
Gregg, M. 298
Groenewegen, M.A.T. 248
Gunn, J.E. 121
Guzzo, L. 157
Häfner, R. 3
Harke, R. 3
Hasinger, G. 121
Hatzes, A.P. **539**
Hauschildt, P.H. 399
Heidt, J. 3
Herrero, A. **331**
Hess, H.-J. 3
Hill, V. 345
Holder, G.P. 150
Holzapfel, W.L. 150
Hopp, U. 3

Hummel, W. 3
Jäger, K. 3
Joy, M.K. 150
Jürgens, P. 3
Jütte, M. 303
Käufl, H.U. **264**
Kümmel, M. 3
Kürster, M. **556**
Kalas, P. **415**
Kirkpatrick, J.D. 505
Kneer, R. 164
Krautter, J. 164
Kudritzki, R.-P. 3, 126, **236**, 331
Kurz, R. 204
Lagage, P.-O. **528**
Lagrange, A.M. **521**
Lattanzi, M.G. **479**
Laurance, R. 535
Leger, A. 535
Lehmann, I. **121**
Leibundgut, B. **43**
Leitch, E.M. 150
Lennon, D.J. 331
Leon, S. 369
Lévêque, S. 468
Liebert, J. **505**
Lin, D.N.C. 377
Lissauer, J.J. **377**
Lopez, B. **565**
MacGillivray, H.T. 157
Madau, P. **52**
Maeder, A. **317**
Mancini, D. 204
Mantel, K.-H. 3
Marigo, P. **248**
Martín, E.L. 446
Mayor, M. 548, 571, 572
McCarthy, J.K. 331
Mehlert, D. 3
Meisl, W. 3
Melnick, J. **252**
Menci, N. **176**
Méndez, R. 126
Méndez, R.H. **242**
Menten, K.M. **78**
Meylan, G. **369**
Miley, G. **32**
Minniti, D. **298**
Moellenhoff, C. 3
Mohr, J.J. **150**
Monet, D.G. 505
Monteverde, M.I. 331
Moorwood, A. **12**
Mouillet, D. 521
Mozurkewich, D. **474**
Muschielok, B. 3
Naef, D. 548, 571
Neumann, D.M. 157
Nicklas, H. 3
Nielbock, M. 303
Niemela, V. 232
Nissen, P.E. **363**
Norman, C. 180
Ortolani, S. **281**
Paresce, F. **422**
Pasquini, L. **211**
Patat, F. 313
Pelló, R. **131**
Penny, A. **535**
Pepe, F. **572**
Petrov, R.G. 565
Pettini, M. **20**
Pont, F. **269**
Potter, D. 518
Pritchard, J. **232**
Przybilla, N. 331
Queloz, D. **548**, 571, 572
Quirrenbach, A. **462**
Rebolo, R. 446
Reese, E.D. 150
Reid, I.N. 505
Reiprich, T.H. 157
Rejkuba, M. 298
Renzini, A. 3, **168**
Rich, R.M. **275**
Richichi, A. 264
Richter, S. 264
Richtler, T. **259**

Rosati, P. 3
Rupprecht, G. 3
Sackett, P.D. **436**
Saglia, R. 3
Santos, N. 548, 571
Scannapieco, E. 138
Scaramella, R. 204
Schindler, S. 157
Schmidt, M. 121
Schmutz, W. **307**
Schneider, D.P. 121
Schneider, J. **499**
Schuecker, P. 157
Sedmak, G. 204
Seggewiss, W. 259
Seifert, W. 3
Seitz, S. 3
Selman, F. 252
Shapley, A.E. 20
Shaver, P. 157
Sivan, J.-P. 572
Skrutskie, M. 505
Smartt, S.J. 331
Sozzetti, A. 479
Spagna, A. 479
Spite, M. **345**
Spyromilio, J. 3
Stahl, O. 3
Stanford, S.A. **115**
Stanke, T. 121
Stecklum, B. 264
Steidel, C.C. 20
Stockman, H.S. **94**
Storm, J. 217
Szeifert, T. 3
Tarantik, K. 3
Terlevich, E. 252
Terlevich, R. 252
Thiering, I. 164
Tokunaga, A. 518
Tolstoy, E. **287**
Tozzi, P. **180**
Trümper, J. 121
Udry, S. 548, **571**, 572
Unwin, S. **492**
Vacca, W.D. 307
van Belle, G.T. **450**
Venn, K. 331
Voges, W. 157
Wambsganss, J. 121
Weiss, A. 248
Woods, D. 121
Wuchterl, G. **408**
Zamorani, G. 121
Zapatero Osorio, M.R. **446**, **511**
Zickgraf, F.-J. 164
Zijlstra, A.A. 298

MIX
Papier aus verantwortungsvollen Quellen
Paper from responsible sources
FSC® C105338

If you have any concerns about our products, you can contact us on
ProductSafety@springernature.com

In case Publisher is established outside the EU, the EU authorized representative is:
Springer Nature Customer Service Center GmbH
Europaplatz 3, 69115 Heidelberg, Germany

Printed by Libri Plureos GmbH
in Hamburg, Germany